GENERAL LINEAR
METHODS FOR ORDINARY
DIFFERENTIAL EQUATIONS

GENERAL LINEAR METHODS FOR ORDINARY DIFFERENTIAL EQUATIONS

ZDZISŁAW JACKIEWICZ

WILEY

A JOHN WILEY & SONS, INC., PUBLICATION

Published by John Wiley & Sons, Inc., Hoboken, New Jersey.
Published simultaneously in Canada.

For general information on our other products and services or for technical support, please contact our Customer Care Department within the United States at (800) 762-2974, outside the United States at (317) 572-3993 or fax (317) 572-4002.

Wiley also publishes its books in a variety of electronic formats. Some content that appears in print may not be available in electronic format. For information about Wiley products, visit our web site at www.wiley.com.

Library of Congress Cataloging-in-Publication Data:

Jackiewicz, Zdzisław, 1950–
   General linear methods for ordinary differential equations / Zdzisław Jackiewicz.
      p. cm.
   Includes bibliographical references and index.
   ISBN 978-0-470-40855-1 (cloth)
   1. Differential equations, Linear. I. Title.
   QA372.J145 2009
   515'.352—dc22   2009007428

Printed in the United States of America.

10 9 8 7 6 5 4 3 2 1

*To my wife, Elżbieta
my son, Wojciech Tomasz
and my daughter, Hanna Katarzyna*

CONTENTS

Preface

This book is concerned with the theory, construction and implementation of general linear methods for ordinary differential equations. This is a very general class of methods which include the classical methods such as Runge-Kutta, linear multistep, and predictor-corrector methods as special cases. Some theoretical and practical aspects related to general linear methods are discussed in *Numerical Methods for Ordinary Differential Equations* by J.C. Butcher, in *Solving Ordinary Differential Equations I: Nonstiff Problems* by E. Hairer, S.P. Nørsett, and G. Wanner, and in *Solving Ordinary Differential Equations II: Stiff and Differential-Algebraic Problems* by E. Hairer and G. Wanner. However, these monographs cover the entire area of numerical solution of ordinary differential equations and devote only a limited amount of space to the discussion of general linear methods. This monograph is an attempt to present a complete analysis of some classes of general linear methods that have good potential for practical use. These classes include diagonally implicit multistage integration methods, two-step Runge-Kutta methods, and general linear methods with inherent Runge-Kutta stability.

In Chapter 1 we present a short introduction to ordinary differential equations, including existence and uniqueness theory, continuous dependence on the initial data and right-hand side, stability theory, and discussion of stiff differential equations and systems. Chapter 2 is an introduction to general linear methods. In particular, we discuss preconsistency, consistency, stage-consistency, zero-stability, convergence, order and stage order conditions, local discretization error, and linear stability theory, and present examples of methods that are appropriate for nonstiff or stiff differential systems in sequential or parallel computing environments. We also discuss briefly algebraic stability, the concept of an underlying one-step method, starting procedures, and codes based on general linear methods.

Chapters 3 to 8 constitute the main part of the book. In Chapters 3 and 4 we deal with the construction and implementation of diagonally implicit multistage integration methods. In Chapters 5 and 6 the theory and implementation of two-step Runge-Kutta methods is discussed. In Chapters 7 and 8 we describe the theory and implementation of general linear methods with inherent Runge-Kutta stability. The topics in these chapters related to the theory and construction of these methods include the derivation of order and stage order conditions, representation formulas for the coefficient matrices of these methods, construction of formulas with desirable accuracy and stability properties, and Nordsieck representation of these methods. The topics in these chapters related to implementation issues include the construction of appropriate starting procedures, local error estimation for small and large step sizes, step size and order changing strategies, construction of continuous interpolants of uniform high order, updating the vector of external approximations, and the solution of nonlinear systems of equations for stiff systems by the modified Newton method. We also present many examples of these methods of all types, mainly of order p and stage order $q = p$ or $q = p - 1$. Many implementation issues are illustrated by the results of numerical experiments with different classes of general linear methods.

I wish to acknowledge the support and assistance of many people during the last several years. I would like to thank Henryk Woźniakowski, who suggested, in somewhat unusual circumstances, that I should write this book. I would also like to express my gratitude to John Butcher, Zbigniew Ciesielski, Maksymilian Dryja, Ernst Hairer, Stanisław Kwapień, Marian Kwapisz, Wacław Marzantowicz, and Marian Mrożek for their help and understanding. I would like to express my gratitude to John Butcher, who has founded this area of research and from whom I have learned so much in the last several years. He was always very supportive of my work and provided invaluable insight into many topics. I would also like to thank him for his warm hospitality during my frequent visits to the University of Auckland and for creating stimulating research environment during these visits.

I would also like to thank my colleagues who offered comments and suggestions on earlier drafts of this manuscript or influenced my work in many ways. In particular, I would like to thank Peter Albrecht, Christopher Baker, Zbigniew Bartoszewski, Alfredo Bellen, Michał Braś, Kevin Burrage, Jeff Cash, Philippe Chartier, Joshua Chollom, Dajana Conte, Raffaele D'Ambrosio, Wayne Enright, Alan Feldstein, Luciano Galeone, Roberto Garrappa, Nicola Guglielmi, Laura Hewitt, Adrian Hill, Giuseppe Izzo, Marian Kwapisz, Edisanter

Lo, Stefano Maset, Hans Mittelmann, Brynjulf Owren, Beatrice Paternoster, Helmut Podhaisky, Rosemary Renaut, Elvira Russo, Larry Shampine, Stefania Tracogna, Jack VanWieren, Antonella Vecchio, Rossana Vermiglio, Jim Verner, Jan Verwer, Rüdiger Weiner, Bruno Welfert, William Wright, Marino Zennaro, and Barbara Zubik-Kowal. I would like to thank Irina Long for her help with editing of some figures, and the editorial and production group at John Wiley & Sons. Finally, I would like to acknowledge the support of National Science Foundation for many of my projects on numerical solution of ordinary differential equations.

Z. JACKIEWICZ

Arizona State University
April, 2009

CHAPTER 1

DIFFERENTIAL EQUATIONS AND SYSTEMS

1.1 THE INITIAL VALUE PROBLEM

Many problems in science and engineering can be modeled by the initial value problem for systems of ordinary differential equations (ODEs), which we write in autonomous form as follows:

$$
\begin{aligned}
u'(t) &= f\big(u(t)\big), \quad t \in [t_0, T], \\
u(t_0) &= u_0.
\end{aligned}
\tag{1.1.1}
$$

Here $f : \mathbb{R}^m \to \mathbb{R}^m$ is a given function that usually satisfies some regularity conditions, and $u_0 \in \mathbb{R}^m$ is a given initial vector. Introducing the notation

$$
u = \begin{bmatrix} u_1 \\ \vdots \\ u_m \end{bmatrix}, \ u' = \begin{bmatrix} u_1' \\ \vdots \\ u_m' \end{bmatrix}, \ f(u) = \begin{bmatrix} f_1(u_1, \ldots, u_m) \\ \vdots \\ f_m(u_1, \ldots, u_m) \end{bmatrix}, \ u_0 = \begin{bmatrix} u_{0,1} \\ \vdots \\ u_{0,m} \end{bmatrix},
$$

General Linear Methods for Ordinary Differential Equations. By Zdzisław Jackiewicz

(1.1.1) can be written in the following scalar form:

$$u_1' = f_1(u_1, \ldots, u_m), \qquad u_1(t_0) = u_{0,1},$$

$$\vdots \qquad\qquad \vdots$$

$$u_m' = f_m(u_1, \ldots, u_m), \qquad u_m(t_0) = u_{0,m},$$

$t \in [t_0, T]$, where we have suppressed dependence on the independent variable t in u_i and u_i', $i = 1, 2, \ldots, m$. Observe that the nonautonomous equation

$$y'(t) = g(t, y(t)), \quad t \in [t_0, T],$$

$$y(t_0) = y_0,$$

(1.1.2)

$g : \mathbb{R} \times \mathbb{R}^m \to \mathbb{R}^m$, $y_0 \in \mathbb{R}^m$, can always be reduced to a system of the form (1.1.1) of dimension $m + 1$ if we define

$$u = \begin{bmatrix} y \\ t \end{bmatrix}, \quad f(u) = \begin{bmatrix} g(t, y) \\ 1 \end{bmatrix}, \quad u_0 = \begin{bmatrix} y_0 \\ t_0 \end{bmatrix}.$$

Systems of the form (1.1.1) or (1.1.2) can also arise in practice from the conversion of initial value problem for differential equations of higher order. Consider, for example, an autonomous form of such a problem:

$$y^{(m)} = g(y, y^{(1)}, \ldots, y^{(m-1)}), \quad t \in [t_0, T],$$

$$y(t_0) = y_0, \; y^{(1)}(t_0) = y_0^{(1)}, \ldots, y^{(m-1)}(t_0) = y_0^{(m-1)},$$

where $y^{(i)}$ stands for the derivative of order i and we again suppressed dependence on t. Setting

$$u_1 = y, \; u_2 = y^{(1)}, \ldots, u_m = y^{(m-1)},$$

we obtain

$$
\begin{aligned}
u_1' &= u_2, & u_1(t_0) &= y_0, \\
u_2' &= u_3, & u_2(t_0) &= y_0^{(1)}, \\
&\;\vdots & &\;\vdots \\
u_{m-1}' &= u_m, & u_{m-1}(t_0) &= y_0^{(m-2)}, \\
u_m' &= g(u_1, u_2, \ldots, u_m), & u_m(t_0) &= y_0^{(m-1)},
\end{aligned}
$$

which is equivalent to (1.1.1) with

$$
f(u) = \begin{bmatrix} u_2 \\ u_3 \\ \vdots \\ u_m \\ g(u_1, u_2, \ldots, u_m) \end{bmatrix}, \quad u_0 = \begin{bmatrix} y_0 \\ y_0^{(1)} \\ \vdots \\ y_0^{(m-2)} \\ y_0^{(m-1)} \end{bmatrix}.
$$

In Section 1.3 we discuss the existence and uniqueness of solutions to (1.1.1) and (1.1.2) under various conditions on the functions f and g. In this discussion we often assume that these problems are defined not only for $t \in [t_0, T]$ but on the larger interval $t \in I$, where $I = \{t : |t - t_0| \le T\}$.

1.2   EXAMPLES OF DIFFERENTIAL EQUATIONS AND SYSTEMS

We list in this section several examples of differential equations and differential systems. These problems are used later in our numerical experiments with various algorithms for numerical solution of ODEs. These algorithms are based on some classes of general linear methods discussed herein. All equations in this section are examples of nonstiff equations and systems. Problem of stiffness and stiff differential equations and systems are discussed in Sections 1.7 and 1.8.

SCALAR — the scalar problem [143, p. 237]:

$$
\begin{aligned}
y'(t) &= -\mathrm{sign}(t)\big|1 - |t|\big|\, y^2, \quad t \in [-2, 2], \\
y(-2) &= 2/3.
\end{aligned}
\tag{1.2.1}
$$

The solution to this initial value problem has a discontinuity in the first derivative y' at the point $t = 0$ and discontinuities in the second derivative y'' at $t = -1$ and $t = 1$.

BUBBLE — a model of cavitating bubble [200, 257]:

$$
\begin{aligned}
\frac{dy_1}{ds} &= y_2, \\
\frac{dy_2}{ds} &= \frac{5\exp(-s/s^*) - 1 - 1.5y_2^2}{y_1} - \frac{a\,y_2 + D}{y_1^2} + \frac{1 + D}{y_1^{3\gamma+1}}, \\
y_1(0) &= 1, \quad y_2(0) = 0,
\end{aligned}
\tag{1.2.2}
$$

$t \in [0, T]$. Here s^*, a, D, and γ are real parameters. As observed by Shampine [257], this problem places great demands on the precision and step size control strategies of numerical algorithms.

AREN — Arenstorf orbit for the restricted three body problem [12, 13, 143]. This is an example from astronomy which describes the movement of two bodies of scaled masses $1 - \mu$ and μ in a circular rotation in a plane and the movement of a third body of negligible mass (e.g., satellite or spacecraft) in the same plane. The equations of motion are

$$y_1'' = y_1 + 2y_2' - (1 - \mu)\frac{y_1 + \mu}{D_1} - \mu\frac{y_1 - 1 + \mu}{D_2},$$

$$y_2'' = y_2 - 2y_1' - (1 + \mu)\frac{y_2}{D_1} - \mu\frac{y_2}{D_2}, \qquad (1.2.3)$$

$t \in [0, T]$, where

$$D_1 = \left((y_1 + \mu)^2 + y_2^2\right)^{3/2}, \quad D_2 = \left((y_1 - 1 + \mu)^2 + y_2^2\right)^{3/2}.$$

This problem with $\mu = 0.012277471$ corresponds to the earth-moon system. The periodic orbits of a satellite or spacecraft moving in such a system were discovered by Arenstorf [12, 13] by theoretical analysis of periodicity conditions and numerical calculations. Such periodic orbits may facilitate low-cost space exploration and are of interest to NASA. The initial conditions for which the solution to (1.2.3) is periodic are, for example,

$$y_1(0) = 0.994, \quad y_1'(0) = 0, \quad y_2(0) = 0, \quad y_2'(0) = -2.001585106379,$$

with the period of motion T_1 given by $T_1 = 17.06522$, or

$$y_1(0) = 0.994, \quad y_1'(0) = 0, \quad y_2(0) = 0, \quad y_2'(0) = -2.031732629557,$$

with the period of motion T_2 given by $T_2 = 11.1234$. Such orbits are plotted in, for example, [52, Fig. 102(i) and (ii)], and [143, Fig. 0.1].

LRNZ — the Lorenz model [209]. This is a system of three differential equations of the form

$$y_1' = -\sigma y_1 + \sigma y_2,$$

$$y_2' = -y_1 y_3 + r y_1 - y_2, \qquad (1.2.4)$$

$$y_3' = y_1 y_2 - b y_3,$$

$t \in [0, T]$. Here b, σ, and r are positive constants. For example, for $b = 8/3$, $\sigma = 10$, and $r = 28$, this system has aperiodic solutions.

EULR — Euler's equations of rotation of a rigid body [143]. This is a system of three differential equations given by

$$I_1 y_1' = (I_2 - I_3)y_2 y_3,$$

$$I_2 y_2' = (I_3 - I_1)y_3 y_1, \qquad (1.2.5)$$

$$I_3 y_3' = (I_1 - I_2)y_1 y_2 + f(t),$$

$t \in [0, T]$. Here y_1, y_2, and y_3 are the coordinates of the rotation vector; I_1, I_2, and I_3 are the principal moments of inertia; and the third coordinate has an additional exterior force $f(t)$.

PLEI — a celestial mechanics problem "the Pleiades" from [143, p. 245]. The equations of motion are

$$x_i'' = \sum_{j \neq i} m_j (x_j - x_i)/r_{ij},$$

$$y_i'' = \sum_{j \neq i} m_j (y_j - y_i)/r_{ij},$$

(1.2.6)

$t \in [0, 3]$, where

$$r_{ij} = \left((x_i - x_j)^2 + (y_i - y_j)^2 \right)^{3/2}, \quad i, j = 1, 2, \ldots, 7.$$

The initial conditions are

$$x_1(0) = 3, \quad x_2(0) = 3, \quad x_3(0) = -1, \quad x_4(0) = -3,$$
$$x_5(0) = 2, \quad x_6(0) = -2, \quad x_7(0) = 2,$$
$$y_1(0) = 3, \quad y_2(0) = -3, \quad y_3(0) = 2, \quad y_4(0) = 0,$$
$$y_5(0) = 0, \quad y_6(0) = -4, \quad y_7(0) = 4,$$

$x_i'(0) = y_i'(0) = 0$, for all i with the exception of

$$x_6'(0) = 1.75, \quad x_7'(0) = -1.5, \quad y_4'(0) = -1.25, \quad y_5'(0) = 1.$$

This problem describes the movement of seven stars in the plane with coordinates x_i, y_i and masses $m_i = i$, $i = 1, 2, \ldots, 7$. The trajectories of these stars are plotted by Hairer at al. [143, Fig. 10.2a], and speeds x_i' and y_i', $i = 1, 2, \ldots, 7$, [143, Fig. 10.2b].

ROPE — the movement of a hanging rope of length 1 under gravitation and the influence of horizontal $F_y(t)$ and vertical $F_x(t)$ forces [143]. As explained by Hairer at al. [143], the discretization of this problem leads to a system of differential equations of second order for the angles $\theta_l = \theta_l(t)$ between the tangents to the rope and the vertical axis at a discrete arc length s_l. This system takes the form

$$\sum_{k=1}^{n} a_{kl} \ddot{\theta}_k = -\sum_{k=1}^{n} b_{lk} \dot{\theta}_k^2 - n \left(n + \frac{1}{2} - l \right) \sin(\theta_l)$$

$$- n^2 \sin(\theta_l) F_x(t) + \begin{cases} n^2 \cos(\theta_l) F_y(t) & \text{if } l \leq 3n/4, \\ 0 & \text{if } l > 3n/4, \end{cases}$$

(1.2.7)

$t \in [0, T]$, $l = 1, 2, \ldots, n$, where

$$a_{lk} = g_{lk} \cos(\theta_l - \theta_k), \quad b_{lk} = g_{lk} \sin(\theta_l - \theta_k), \quad g_{lk} = n + \frac{1}{2} - \max\{l, k\}.$$

The horizontal force $F_y(t)$ acting at the point $s = 0.75$ is

$$F_y(t) = \left(\frac{1}{\cosh(4t - 2.5)} \right)^4,$$

and the vertical force $F_x(t)$ acting at the point $s = 1$ is

$$F_x(t) = 0.4.$$

This system will be solved for $n = 40$ with initial conditions

$$\theta_l(0) = \dot{\theta}_l(0) = 0, \quad l = 1, 2, \ldots, n,$$

on the interval $[0, 3.723]$.

Setting

$$A = \begin{bmatrix} a_{lk} \end{bmatrix}, \quad B = \begin{bmatrix} b_{lk} \end{bmatrix}, \quad \theta = \begin{bmatrix} \theta_1 & \theta_2 & \cdots & \theta_n \end{bmatrix}^T,$$

$$\dot{\theta} = \begin{bmatrix} \dot{\theta}_1 & \dot{\theta}_2 & \cdots & \dot{\theta}_n \end{bmatrix}^T, \quad \ddot{\theta} = \begin{bmatrix} \ddot{\theta}_1 & \ddot{\theta}_2 & \cdots & \ddot{\theta}_n \end{bmatrix}^T,$$

system (1.2.7) can be written in vector form as

$$A\ddot{\theta} = -B\dot{\theta}^2 + g(t, \theta), \quad t \in [0, T],$$
$$\theta(0) = \dot{\theta}(0) = 0,$$

$$(1.2.8)$$

where $\dot{\theta}^2$ denotes componentwise exponentiation and $g(t, \theta)$ is an appropriately defined vector function. The solution of (1.2.8) requires computation of the inverse matrix A^{-1}. As explained by Hairer at al. [143, 146] this can be done very efficiently in $O(n)$ operations, due to the special structure of the matrix A. It can be verified that

$$A + iB = \text{diag}\left(e^{i\theta_1}, e^{i\theta_2}, \ldots, e^{i\theta_n} \right) G \, \text{diag}\left(e^{-i\theta_1}, e^{-i\theta_2}, \ldots, e^{-i\theta_n} \right),$$

where $G = [g_{kl}]$. This matrix has the inverse

$$G^{-1} = \begin{bmatrix} 1 & -1 & & & \\ -1 & 2 & -1 & & \\ & \ddots & \ddots & \ddots & \\ & & -1 & 2 & -1 \\ & & & -1 & 3 \end{bmatrix}$$

and it follows that

$$(A + i\,B)^{-1} = C + i\,D$$

$$= \mathrm{diag}\!\left(e^{i\theta_1}, e^{i\theta_2}, \ldots, e^{i\theta_n}\right) G^{-1} \mathrm{diag}\!\left(e^{-i\theta_1}, e^{-i\theta_2}, \ldots, e^{-i\theta_n}\right),$$

where C and D are tridiagonal matrices of the form

$$C = \begin{bmatrix} 1 & c_{12} & & & & \\ c_{21} & 2 & c_{23} & & & \\ & \ddots & \ddots & \ddots & & \\ & & c_{n-1,n-2} & 2 & c_{n-1,n} \\ & & & c_{n,n-1} & 3 \end{bmatrix}$$

and

$$D = \begin{bmatrix} 0 & s_{12} & & & & \\ s_{21} & 0 & s_{23} & & & \\ & \ddots & \ddots & \ddots & & \\ & & s_{n-1,n-2} & 0 & s_{n-1,n} \\ & & & s_{n,n-1} & 0 \end{bmatrix}$$

with

$$c_{kl} = -\cos(\theta_k - \theta_l), \quad s_{kl} = -\sin(\theta_k - \theta_l).$$

Since $(A + i\,B)(C + i\,D) = I$, we have

$$AC - BD = I, \quad AD + BC = 0$$

and it follows that

$$C = A^{-1} + A^{-1}BD = A^{-1} + A^{-1}BCC^{-1}D = A^{-1} - DC^{-1}D$$

or $A^{-1} = C + DC^{-1}D$. We also have $A^{-1}B = -DC^{-1}$ and system (1.2.8) can be written as

$$\ddot{\theta} = DC^{-1}\!\left(\dot{\theta}^2 + Dg(t,\theta)\right) + Cg(t,\theta), \quad t \in [0, T].$$

As observed by Hairer at al. [143] this suggests the following efficient algorithm for computation of the acceleration vector $\ddot{\theta}$.

1. Compute $w = \dot{\theta}^2 + Dg(t,\theta)$.

2. Solve the tridiagonal system $Cu = w$.

3. Compute $\ddot{\theta} = Du + Cg(t,\theta)$.

BRUS — a reaction-diffusion equation (the Brusselator with diffusion) [143]. This is the system of partial differential equations of the form

$$\frac{\partial u}{\partial t} = 1 + u^2 v - 4.4u + \alpha\left(\frac{\partial^2 u}{\partial x^2} + \frac{\partial^2 u}{\partial y^2}\right),$$

$$\frac{\partial v}{\partial t} = 3.4u - u^2 v + \alpha\left(\frac{\partial^2 v}{\partial x^2} + \frac{\partial^2 v}{\partial y^2}\right),$$

$$(1.2.9)$$

$0 \le x \le 1$, $0 \le y \le 1$, $t \ge 0$, $\alpha = 2 \times 10^{-2}$, together with the Neumann boundary conditions

$$\frac{\partial u}{\partial \mathbf{n}} = 0, \quad \frac{\partial v}{\partial \mathbf{n}} = 0,$$

where $\mathbf{n}$ is the normal vector to the boundary of the region $[0, 1] \times [0, 1]$ and the initial conditions

$$u(x, y, 0) = 0.5 + y, \quad v(x, y, 0) = 1 + 5x.$$

Let $N > 1$ be an integer and define the grid in space variables x and y by

$$x_i = (i - 1)\Delta x, \quad y_j = (j - 1)\Delta y, \quad i, j = 1, 2, \ldots, N,$$

where $\Delta x = \Delta y = 1/(N - 1)$. Define also the functions

$$U_{ij}(t) = u(x_i, y_j, t), \quad V_{ij}(t) = v(x_i, y_j, t), \quad i, j = 1, 2, \ldots, N.$$

Discretizing (1.2.9) by the method of lines, where the space derivatives are approximated by finite differences of second order leads to the system of ordinary differential equations

$$
\begin{aligned}
U'_{ij} &= 1 + U_{ij}^2 V_{ij} - 4.4U_{ij} \\
&+ \frac{\alpha}{\Delta x^2}\left(U_{i+1,j} + U_{i-1,j} + U_{i,j+1} + U_{i,j-1} - 4U_{ij}\right), \\
V'_{ij} &= 3.4U_{ij} - U_{ij}^2 V_{ij} \\
&+ \frac{\alpha}{\Delta y^2}\left(V_{i+1,j} + V_{i-1,j} + V_{i,j+1} + V_{i,j-1} - 4V_{ij}\right),
\end{aligned}
$$

$$(1.2.10)$$

$t \in [0, T]$, $i, j = 1, 2, \ldots, N$, of dimension $2N^2$. The boundary conditions imply that

$$U_{0,j} = U_{2,j}, \quad U_{N+1,j} = U_{N-1,j}, \quad U_{i,0} = U_{i,2}, \quad U_{i,N+1} = U_{i,N-1},$$

$$V_{0,j} = V_{2,j}, \quad V_{N+1,j} = V_{N-1,j}, \quad V_{i,0} = V_{i,2}, \quad V_{i,N+1} = V_{i,N-1},$$

and the initial conditions are

$$U_{ij}(0) = 0.5 + y_j, \quad V_{ij}(0) = 1 + 5x_i, \quad i, j = 1, 2, \ldots, N.$$

1.3 EXISTENCE AND UNIQUENESS OF SOLUTIONS

In this section we formulate results regarding the existence and uniqueness of solutions to (1.1.2). We begin with the classical Peano existence theorem for the system (1.1.2), which assumes only that the function g is continuous in some domain. The proof of this result is based on the Arzela-Ascoli theorem about a family of vector-valued functions that is uniformly bounded and equicontinuous. Consider a family $\mathcal{F}$ of vector-valued functions $y = y(t)$ defined on an interval $I = \{t : |t - t_0| \leq T\}$. Define $\|y\| := \sup\{\|y(t)\| : t \in I\}$, where $\| \cdot \|$ is any norm on $\mathbb{R}^m$. We introduce the following definitions.

Definition 1.3.1 *A family $\mathcal{F}$ of vector-valued functions $y = y(t)$ is said to be uniformly bounded if there exists a constant M such that $\|y\| \leq M$ for every $y \in \mathcal{F}$.*

Definition 1.3.2 *A family $\mathcal{F}$ of vector-valued functions $y = y(t)$ is said to be equicontinuous if for every $\epsilon > 0$ there exists $\delta > 0$ such that the condition $|t - s| < \delta$, $t, s \in I$, implies that $\|y(t) - y(s)\| < \epsilon$ for all functions $y \in \mathcal{F}$.*

Theorem 1.3.3 (Arzela-Ascoli; see [212]) *Let $y_n(t)$, $n = 1, 2, \ldots$, be a uniformly bounded and equicontinuous sequence of vector functions defined on the interval I. Then there exists a subsequence $y_{n_j}(t)$, $j = 1, 2, \ldots$, which is uniformly convergent on I.*

We are now ready to formulate and prove the classical existence result for system (1.1.2).

Theorem 1.3.4 (Peano [235]) *Assume that the function $g(t, y)$ is continuous in the domain*

$$D = \Big\{(t, y) : \ |t - t_0| \leq T, \ \|y - y_0\| \leq K \Big\} \tag{1.3.1}$$

and that there exists a constant M such that $\|g(t, y)\| \leq M$ for $(t, y) \in D$. Then system (1.1.2) has at least one solution $y = y(t)$ defined for

$$|t - t_0| \leq T_1 := \min\{T, K/M\}$$

and passing through the point (t_0, y_0).

Proof: It is generally agreed that the original proof of Peano [235] was inadequate, and a satisfactory proof was found many years later (see e.g., Perron [236]). Here, we follow the presentation given by Birkhoff and Rota [21]. We will prove the theorem for the interval $[t_0, t_0 + T_1]$; the proof for the interval $[t_0 - T_1, t_0]$ is analogous. Consider the integral equation

$$y(t) = y_0 + \int_{t_0}^{t} g\big(s, y(s)\big) ds, \tag{1.3.2}$$

$t \in [t_0, t_0 + T_1]$, which is equivalent to (1.1.2). Define the sequence of functions $y_n = y_n(t)$, $n = 1, 2, \ldots$, by the formulas

$$y_n(t) = \begin{cases} y_0, & t \in \big[t_0, t_0 + T_1/n\big], \\ y_0 + \displaystyle\int_{t_0}^{t-T_1/n} g\big(s, y_n(s)\big)ds, & t \in \big(t_0 + T_1/n, t_0 + T_1\big]. \end{cases}$$

Observe that the right-hand side of the second formula above defines $y_n(t)$ for $t \in (t_0 + T_1/n, t_0 + T_1]$ in terms of $y_n(t)$ already defined for $t \in [t_0, t_0 + T_1/n]$. This sequence is well defined since

$$\big\|y_n(t) - y_0\big\| \leq \int_{t_0}^{t-T_1/n} \big\|g\big(s, y_n(s)\big)\big\|ds$$

$$\leq M\left(t - t_0 - \frac{T_1}{n}\right) \leq M(t - t_0) \leq MT_1 \leq K$$

and $y_n(t)$ are clearly continuous on $[t_0, t_0 + T_1]$. We have $\|y_n(t)\| \leq \|y_0\| + MT_1$, which shows that the sequence $y_n(t)$ is uniformly bounded. We also have

$$\big\|y_n(t_2) - y_n(t_1)\big\| \leq \int_{t_1 - T_1/n}^{t_2 - T_1/n} \big\|g\big(s, y_n(s)\big)\big\|ds \leq M|t_2 - t_1|$$

which shows that $y_n(t)$ is also equicontinuous. Hence, it follows from the Arzela-Ascoli theorem, Theorem 1.3.3, that there exists a subsequence $y_{n_j}(t)$, $j = 1, 2, \ldots$, which is uniformly convergent to a continuous function $\overline{y}(t)$; that is,

$$\lim_{j \to \infty} y_{n_j} = \overline{y}.$$

This subsequence satisfies the integral equation, which we write in the form

$$y_{n_j}(t) = \int_{t_0}^{t} g\big(s, y_{n_j}(s)\big)ds - \int_{t-T_1/n_j}^{t} g\big(s, y_{n_j}(s)\big)ds.$$

We have

$$\lim_{j \to \infty} \int_{t_0}^{t} g\big(s, y_{n_j}(s)\big)ds = \int_{t_0}^{t} g\big(s, \overline{y}(s)\big)ds$$

since the function $g(t, y)$ is uniformly continuous. We also have

$$\left\| \int_{t-T_1/n_j}^{t} g\big(s, y_{n_j}(s)\big)ds \right\| \leq M\frac{T_1}{n_j} \to 0 \quad \text{as} \quad j \to \infty.$$

Hence, passing to the limit as $j \to \infty$ in the integral equation for y_{n_j}, we obtain

$$\overline{y}(t) = y_0 + \int_{t_0}^{t} g\big(s, \overline{y}(s)\big)ds,$$

which proves that $\overline{y}(t)$ satisfies the integral equation (1.3.2); hence it also satisfies (1.1.2) for $t \in [t_0, t_0 + T_1]$. ∎

A solution whose existence is guaranteed by the Peano theorem, Theorem 1.3.4, is not necessarily unique. A simple example that illustrates this is given by a scalar initial value problem

$$y' = 3y^{2/3}, \quad y(0) = 0,$$

where $D = \{(t, y) : |t| \leq 1, |y| \leq 1\}$. Here the function $g(t, y) = y^{2/3}$ is continuous on D, but the problem has solutions $y_1(t) = 0$ and $y_2(t) = t^3$.

Assume that a function $g(t, y)$ is defined in some region $\mathcal{R} \subset \mathbb{R} \times \mathbb{R}^m$. To formulate uniqueness results for (1.1.2), we usually assume that the function $g(t, y)$ is not only continuous but satisfies some additional regularity properties. We introduce the following definitions.

Definition 1.3.5 *A function $g(t, y)$ satisfies a Lipschitz condition in $\mathcal{R}$ with a Lipschitz constant L if*

$$\big\| g(t, y_1) - g(t, y_2) \big\| \leq L \| y_1 - y_2 \| \tag{1.3.3}$$

for all $(t, y_1), (t, y_2) \in \mathcal{R}$, where $\| \cdot \|$ is any norm in $\mathbb{R}^m$.

Definition 1.3.6 *A function $g(t, y)$ satisfies a one-sided Lipschitz condition in $\mathcal{R}$ with a one-sided Lipschitz constant ν if*

$$\big(g(t, y_1) - g(t, y_2) \big)^T (y_1 - y_2) \leq \nu \| y_1 - y_2 \|^2 \tag{1.3.4}$$

for all $(t, y_1), (t, y_2) \in \mathcal{R}$. Here $\| \cdot \|$ is the Euclidean norm in $\mathbb{R}^m$; that is, $\| u \| := \sqrt{u^T u}$ for $u \in \mathbb{R}^m$.

One-sided Lipschitz condition (1.3.4) plays an important role in the analysis of numerical methods for stiff systems of ODEs (compare [109, 146]). Assume that the function $g(t, y)$ satisfies Lipschitz condition (1.3.3) in the Euclidean norm $\| \cdot \|$ with a constant L. Then using the Schwartz inequality and (1.3.3), we obtain

$$\big(g(t, y_1) - g(t, y_2) \big)^T (y_1 - y_2) \leq \big\| g(t, y_1) - g(t, y_2) \big\| \| y_1 - y_2 \| \leq L \| y_1 - y_2 \|^2$$

and it follows that $g(t, y)$ also satisfies one-sided Lipschitz condition (1.3.4) with the same constant L. However, as observed, for example, by Dekker and Verwer [109], the reverse is not true. A counterexample is provided by any monotonically nonincreasing function $g : \mathbb{R} \to \mathbb{R}$ which has, for some value of $\overline{y} \in \mathbb{R}$, an infinite slope. We then have

$$\big(g(y_1) - g(y_2) \big)(y_1 - y_2) \leq 0$$

for all $y_1, y_2 \in \mathbb{R}$ and it follows that $g(y)$ satisfies (1.3.4) with $\nu = 0$. However, this function does not satisfy (1.3.3) in any neighborhood of $\overline{y}$, where the slope is infinite.

Next we formulate a local existence and uniqueness theorem. We also show that the solution to (1.1.2) can be obtained as a limit of a uniformly convergent sequence of continuous functions starting with an arbitrary initial function that satisfies the appropriate initial condition.

Theorem 1.3.7 *Assume that the function $g(t, y)$ is continuous and satisfies a Lipschitz condition (1.3.3) in the domain D defined by (1.3.1). Set*

$$M = \max\Big\{\|g(t, y)\| : (t, y) \in D\Big\}.$$

Then (1.1.2) has a unique solution defined on the interval

$$|t - t_0| \le T_1 := \min\{T, K/M\}$$

passing through (t_0, y_0).

Proof: First consider the interval $[t_0, t_0 + T_1]$; the proof for the interval $[t_0 - T_1, t_0]$ is analogous. Define the integral operator

$$z(t) = \phi\big(y(t)\big) := y_0 + \int_{t_0}^{t} g\big(s, y(s)\big) ds, \tag{1.3.5}$$

$t \in [t_0, t_0 + T_1]$. Put $Y = \{y \in \mathbb{R}^m : \|y - y_0\| \le K\}$ and denote by $C([t_0, t_0 + T_1], Y)$ the space of continuous functions from $[t_0, t_0 + T_1]$ into Y with a uniform norm. Observe that if $y \in C([t_0, t_0 + T_1], Y)$, then

$$\big\|z(t) - y_0\big\| \le \int_{t_0}^{t} \big\|g\big(s, y(s)\big)\big\| ds \le MT_1 \le MK/M = K,$$

and it follows that the operator ϕ takes the functions from $C([t_0, t_0 + T_1], Y)$ into $C([t_0, t_0 + T_1], Y)$:

$$\phi : C\big([t_0, t_0 + T_1], Y\big) \to C\big([t_0, t_0 + T_1], Y\big).$$

Define the sequence of functions $y_n(t) \in C([t_0, t_0 + T_1], Y)$ by the formula

$$y_{n+1}(t) = \phi\big(y_n(t)\big) = y_0 + \int_{t_0}^{t} g\big(s, y_n(s)\big) ds, \tag{1.3.6}$$

$n = 0, 1, \ldots$, where $y_0(t) \equiv y_0$, $t \in [t_0, t_0 + T_1]$. Then we have the bound

$$\big\|y_n(t) - y_{n-1}(t)\big\| \le \frac{ML^{n-1}(t - t_0)^n}{n!}. \tag{1.3.7}$$

We prove (1.3.7) by induction with respect to n. Since

$$\big\|y_1(t) - y_0(t)\big\| \le \int_{t_0}^{t} \big\|g\big(s, y_0(s)\big)\big\| ds \le M(t - t_0),$$

this bound is true for $n = 1$. Assuming now that (1.3.7) is true for n we have

$$\left\|y_{n+1}(t) - y_n(t)\right\| \leq \int_{t_0}^t \left\|g\big(s, y_n(s)\big) - g\big(s, y_{n-1}(s)\big)\right\| ds$$

$$\leq L \int_{t_0}^t \left\|y_n(s) - y_{n-1}(s)\right\| ds$$

$$\leq L \int_{t_0}^t \frac{ML^{n-1}(s - t_0)^n}{n!} ds = \frac{ML^n(t - t_0)^{n+1}}{(n+1)!},$$

which is equivalent to (1.3.7) with n replaced by $n+1$. It follows from (1.3.7) that

$$\sum_{k=1}^{\infty} \left\|y_n(t) - y_{n-1}(t)\right\| \leq \frac{M}{L} \sum_{k=1}^{\infty} \frac{\big(L(t - t_0)\big)^n}{n!},$$

and since the series on the right-hand side is uniformly convergent on the interval $[t_0, t_0 + T_1]$ to the function $M(\exp(L(t - t_0)) - 1)/L$, we can also conclude that the series

$$y_0(t) + \sum_{k=1}^{\infty} \big(y_k(t) - y_{n-1}(t)\big)$$

whose nth partial sum is equal to $y_n(t)$ is also uniformly convergent on the interval $[t_0, t_0 + T_1]$. Denote by $\overline{y}(t)$ the limit of the sequence $y_n(t)$. Since

$$\left\|g\big(s, y_m(s)\big) - g\big(s, y_n(s)\big)\right\| \leq \left\|y_m(s) - y_n(s)\right\|,$$

the integral $\int_{t_0}^t g(s, y_n(s)) ds$ is uniformly convergent for $t \in [t_0, t_0 + T_1]$. Passing to the limit in (1.3.6) as $n \to \infty$, it follows that $\overline{y}(t)$ satisfies the integral equation

$$\overline{y}(t) = y_0 + \int_{t_0}^t g(s, \overline{y}(s)) ds,$$

$t \in [t_0, t_0 + T_1]$. Hence, $\overline{y}(t)$ also satisfies the equivalent initial value problem (1.1.2), which proves existence.

To prove uniqueness, assume that there are two solutions $x(t)$ and $y(t)$ to (1.1.2). Define a norm $\| \cdot \|_\alpha$ in the space $C([t_0, t_0 + T_1], Y)$ by the formula

$$\|y\|_\alpha := \sup \left\{ e^{-\alpha(t - t_0)} \|y(t)\| \, : \, t \in [t_0, t_0 + T_1] \right\},$$

where $\alpha > 0$ and $\| \cdot \|$ is any norm in $\mathbb{R}^m$. Subtracting the integral equations for $x(t)$ and $y(t)$ equivalent to (1.1.2), we obtain

$$x(t) - y(t) = \int_{t_0}^t \Big(g\big(s, x(s)\big) - g\big(s, y(s)\big)\Big) ds$$

and it follows that

$$\|x(t) - y(t)\| \leq L \int_{t_0}^{t} e^{\alpha(s-t_0)} e^{-\alpha(s-t_0)} \|x(s) - y(s)\| ds$$

$$\leq L \|x - y\|_\alpha \int_{t_0}^{t} e^{\alpha(s-t_0)} ds$$

$$= L \|x - y\|_\alpha \frac{e^{\alpha(t-t_0)} - 1}{\alpha} \leq \frac{L}{\alpha} \|x - y\|_\alpha e^{\alpha(t-t_0)}.$$

This leads to

$$e^{-\alpha(t-t_0)} \|x(t) - y(t)\| \leq \frac{L}{\alpha} \|x - y\|_\alpha,$$

and since the right-hand side of this inequality is independent of t, we obtain

$$\|x - y\|_\alpha \leq \frac{L}{\alpha} \|x - y\|_\alpha.$$

Choosing α such that $L/\alpha < 1$, we obtain $\|x - y\|_\alpha = 0$ or $x(t) \equiv y(t)$, $t \in [t_0, t_0 + T_1]$. This completes the proof. ∎

The proof of Theorem 1.3.7 is an illustration of the Banach contraction principle [212] for the metric space $C([t_0, t_0+T_1], Y)$ with the distance between x and y defined by $\|x - y\|_\alpha$. We have

$$\left\|\phi(x) - \phi(y)\right\|_\alpha \leq \frac{L}{\alpha} \|x - y\|_\alpha,$$

which shows that the mapping ϕ defined by (1.3.5) is contractive if α is chosen so that $L/\alpha < 1$. Hence, ϕ has a unique fixed point $\overline{y} = \phi(\overline{y})$, which is the limit of the sequence

$$y_{n+1}(t) = \phi(y_n(t)), \quad n = 1, 2, \ldots,$$

$$y_0(t) = y_0,$$

$t \in [t_0, t_0 + T_1]$.

The successive approximations defined by (1.3.6) are called Picard-Lindelöf iterations. These iterations and more general iteration schemes based on the appropriate splittings of the right-hand side of (1.1.2),

$$g(t, y) = g_1(t, y) + g_2(t, y),$$

form a basis for waveform relaxation iterations: numerical techniques for large differential systems, which can be implemented efficiently in a parallel computing environment. The continuous version of these iterations takes the form

$$y_{n+1}(t) = y_0 + \int_{t_0}^{t} g_1(s, y_{n+1}(s)) ds + \int_{t_0}^{t} g_2(s, y_n(s)) ds,$$

$n = 0, 1, \ldots,$ $y_0(t) \equiv y_0$, and they reduce to (1.3.6) if $g_1 \equiv 0$. The theoretical background for these techniques is given in a number of articles [213, 218, 219, 220] and monographs [31, 276, 290]. This approach was first proposed by Lelarasmee [205] and Lelarasmee et al. [206] in the engineering community as a practical algorithm for large differential systems modeling electrical networks.

The existence and uniqueness of solutions to (1.1.2) to the right of t_0 can also be established if the function $g(t, y)$ satisfies only a one-sided Lipschitz condition (1.3.4). To be more precise, we have the following theorem.

Theorem 1.3.8 *Assume that the function $g(t, y)$ is continuous and satisfies one-sided Lipschitz condition (1.3.4) in the domain D^+ defined by*

$$D^+ = \Big\{ (t, y) : \ t \in [t_0, t_0 + T], \ \|y - y_0\| \le K \Big\}.$$

Then initial value problem (1.1.2) has a unique solution defined on the interval $[t_0, t_0 + T_1]$, where

$$T_1 := \min\{T, K/M\} \quad and \quad M := \max \Big\{ \|g(t, y)\| : \ (t, y) \in D^+ \Big\}.$$

Proof: The existence of a solution follows from the Peano theorem, Theorem 1.3.4. To show uniqueness, assume that there are two solutions $x(t)$ and $y(t)$ to (1.1.2). Set

$$v(t) = \|x(t) - y(t)\|^2 e^{-2\nu(t-t_0)},$$

where ν is the one-sided Lipschitz constant and $\| \cdot \|$ stands for the Euclidean norm in $\mathbb{R}^m$. Then

$$\begin{aligned}
v'(t) &= 2\big(x'(t) - y'(t)\big)^T \big(x(t) - y(t)\big) e^{-2\nu(t-t_0)} \\
&\quad - 2\nu \|x(t) - y(t)\|^2 e^{-2\nu(t-t_0)} \\
&= 2\big(g(t, x(t)) - g(t, y(t))\big)^T \big(x(t) - y(t)\big) e^{-2\nu(t-t_0)} \\
&\quad - 2\nu \|x(t) - y(t)\|^2 e^{-2\nu(t-t_0)} \ \le \ 0.
\end{aligned}$$

Hence, the function $v(t)$ is nonincreasing and it follows that

$$\|x(t) - y(t)\|^2 e^{-2\nu(t-t_0)} \ \le \ \|x(t_0) - y(t_0)\|^2.$$

Since $x(t_0) = y(t_0) = y_0$, the inequality above implies that $x(t) \equiv y(t)$, $t \in [t_0, t_0 + T_1]$, which is our claim. ∎

If the function $g(t, y)$ satisfies a one-sided Lipschitz condition (1.3.4) in the domain D defined by (1.3.1), the solution to (1.1.2) is not necessarily unique to the left of t_0. The counterexample is provided by the initial value problem

$$y' = -2 \operatorname{sign}(y) \sqrt{|y|}, \quad y(0) = 0,$$

where the function on the right-hand side satisfies a one-sided Lipschitz condition with constant $\nu = 0$ for all $t, y \in \mathbb{R}$. This problem has two solutions, $y(t) \equiv 0$ and $y(t) = -t^2$, for $t \leq 0$. It follows from Theorem 1.3.8 that $y(t) \equiv 0$ is the unique solution to this problem for $t \geq 0$.

1.4   CONTINUOUS DEPENDENCE ON INITIAL VALUES AND THE RIGHT-HAND SIDE

In this section we show that the solution $y(t)$ to (1.1.2) depends continuously on the initial conditions and the right-hand side $g(t, y)$ of the differential equation. These investigations are aided by the following generalization of the Gronwall lemma [137].

Lemma 1.4.1 *Assume that* $\lambda, \mu \in C([t_0, \infty), \mathbb{R})$, $\mu(t) \geq 0$ *for* $t \geq t_0$, *and*

$$y(t) \leq \lambda(t) + \int_{t_0}^{t} \mu(s)y(s)ds, \quad t \geq t_0.$$

Then

$$y(t) \leq \lambda(t) + \int_{t_0}^{t} \lambda(s)\mu(s) \exp\left(\int_{s}^{t} \mu(\tau)d\tau\right)ds, \quad t \geq t_0. \tag{1.4.1}$$

Proof: Set

$$z(t) = \int_{t_0}^{t} \mu(s)y(s)ds, \quad t \geq t_0.$$

Then $y(t) \leq \lambda(t) + z(t)$ and

$$z'(t) = \mu(t)y(t) \leq \mu(t)\lambda(t) + \mu(t)z(t)$$

or

$$z'(t) - \mu(t)z(t) \leq \lambda(t)\mu(t), \quad t \geq t_0.$$

Multiplying both sides of this inequality by the integrating factor given by $\exp(-\int_{t_0}^{t} \mu(s)ds)$, we obtain

$$w'(t) \leq \lambda(t)\mu(t) \exp\left(-\int_{t_0}^{t} \mu(s)ds\right), \quad t \geq t_0,$$

where

$$w(t) = z(t) \exp\left(-\int_{t_0}^{t} \mu(s)ds\right).$$

Define the function $v(t)$ by

$$v'(t) = \lambda(t)\mu(t) \exp\left(-\int_{t_0}^{t} \mu(s)ds\right), \quad v(t_0) = 0.$$

Then $v'(t) - w'(t) \geq 0$, and since $v(t_0) - w(t_0) = 0$, it follows that

$$w(t) \leq v(t) = \int_{t_0}^{t} \lambda(s)\mu(s) \exp\left(-\int_{t_0}^{s} \mu(\tau)d\tau\right) ds.$$

Hence, using the definition of $w(t)$, we obtain

$$z(t) \leq \int_{t_0}^{t} \lambda(s)\mu(s) \exp\left(\int_{s}^{t} \mu(\tau)d\tau\right) ds.$$

The conclusion of the lemma follows from the inequality $y(t) \leq \lambda(t) + z(t)$. ∎

We have the following continuity theorem.

Theorem 1.4.2 *Assume that $x(t)$ and $y(t)$ satisfy the initial value problems*

$$x'(t) = f(t, x(t)), \quad x(t_0) = x_0, \tag{1.4.2}$$

$$y'(t) = g(t, y(t)), \quad y(t_0) = y_0, \tag{1.4.3}$$

for $t \in [t_0, t_0 + T]$, where the functions $f(t, x)$ and $g(t, y)$ are defined and continuous for $t \in [t_0, t_0 + T]$, $x, y \in Y \subset \mathbb{R}^m$. Let

$$\|f(t, z) - g(t, z)\| \leq \epsilon \tag{1.4.4}$$

for $t \in [t_0, t_0 + T]$ and $z \in Y$. Moreover, assume that $f(t, x)$ satisfies the Lipschitz condition

$$\|f(t, x) - f(t, y)\| \leq L\|x - y\|, \tag{1.4.5}$$

$t \in [t_0, t_0 + T]$, $x, y \in Y$. Then we have the inequality

$$\|x(t) - y(t)\| \leq \|x_0 - y_0\|e^{L(t-t_0)} + \frac{\epsilon}{L}\left(e^{L(t-t_0)} - 1\right). \tag{1.4.6}$$

Proof: Subtracting integral equations equivalent to (1.4.2) and (1.4.3) we obtain

$$x(t) - y(t) = x_0 - y_0 + \int_{t_0}^{t} \Big(f(s, x(s)) - g(s, y(s))\Big) ds$$

$$= x_0 - y_0 + \int_{t_0}^{t} \Big(f(s, x(s)) - f(s, y(s)) + f(s, y(s)) - g(s, y(s))\Big) ds.$$

Taking norms on both sides of the equation above and taking (1.4.4) and (1.4.5) into account leads to the inequality

$$\|x(t) - y(t)\| \leq \|x_0 - y_0\| + \epsilon(t - t_0) + L\int_{t_0}^{t} \|x(s) - y(s)\| ds.$$

Hence, using Lemma 1.4.1, we obtain

$$\|x(t) - y(t)\| \leq \|x_0 - y_0\| + \epsilon(t - t_0) + L \int_{t_0}^{t} \Big(\|x_0 - y_0\| + \epsilon(s - t_0)\Big) e^{L(t-s)} ds.$$

Conclusion (1.4.6) of the theorem follows after integrating by parts the inequality above.   ∎

Observe that the function $g(t, y)$ is not required to satisfy the Lipschitz condition.

Estimate (1.4.6) is still valid if the function $f(t, y)$ satisfies only one-sided Lipschitz condition (1.3.4), where the constant ν is not necessarily positive. To show this, we need the following lemmas.

Lemma 1.4.3 (Dahlquist [98]) *Assume that a vector-valued function $x(t)$ has the right-hand-side derivative $x'(t+0)$. Then $\|x(t)\|$ has a right-hand-side derivative, denoted by $\frac{d^+}{dt}\|x(t)\|$, which is equal to*

$$\frac{d^+}{dt}\|x(t)\| = \lim_{h \to 0+} \frac{\|x(t) + hx'(t+0)\| - \|x(t)\|}{h}.$$

Proof: Observe that the limit on the right-hand-side exists since the difference ratio is a bounded, monotonic function of h. We have

$$\left| \frac{\|x(t + h)\| - \|x(t)\|}{h} - \frac{\|x(t) + hx'(t + 0)\| - \|x(t)\|}{h} \right|$$

$$\leq \left\| \frac{x(t + h) - x(t) - hx'(t + 0)}{h} \right\| \to 0$$

as $h \to 0+$, which is our claim.   ∎

The next result is a fundamental lemma on differential inequalities.

Lemma 1.4.4 *Assume that the function $u(t)$ satisfies the inequality*

$$u'(t) \leq \nu u(t) + \epsilon, \quad t \geq t_0, \tag{1.4.7}$$

$u(t_0) = u_0$, *where $\nu \in \mathbb{R}$ is not necessarily positive. Then*

$$u(t) \leq u_0 e^{\nu(t-t_0)} + \frac{\epsilon}{\nu}\Big(e^{\nu(t-t_0)} - 1\Big) \tag{1.4.8}$$

if $\nu \neq 0$ or

$$u(t) \leq u_0 + \epsilon(t - t_0) \tag{1.4.9}$$

if $\nu = 0$.

Proof: Inequality (1.4.7) is equivalent to

$$w'(t) \leq \epsilon e^{-\nu(t-t_0)},$$

where $w(t) = e^{-\nu(t-t_0)}u(t)$. Consider the problem

$$v'(t) = \epsilon e^{-\nu(t-t_0)}, \quad v(t_0) = 0,$$

$t \geq t_0$, whose solution is

$$v(t) = \frac{\epsilon}{\nu}\left(1 - e^{-\nu(t-t_0)}\right)$$

if $\nu \neq 0$ or

$$v(t) = \epsilon(t - t_0)$$

if $\nu = 0$. Then $w'(t) - v'(t) \leq 0$ and the function $w(t) - v(t)$ is nonincreasing. Hence,

$$w(t) - v(t) \ \leq \ w(t_0) - v(t_0) = u_0$$

and it follows that

$$w(t) \ \leq \ u_0 + \frac{\epsilon}{\nu}\left(1 - e^{-\nu(t-t_0)}\right)$$

if $\nu \neq 0$ or

$$w(t) \ \leq \ u_0 + \epsilon(t - t_0)$$

if $\nu = 0$. Since $w(t) = e^{-\nu(t-t_0)}u(t)$, these inequalities are equivalent to (1.4.8) and (1.4.9), respectively. This completes the proof. ∎

We introduce the following definition.

Definition 1.4.5 *For any square matrix $A \in \mathbb{R}^{m \times m}$ and matrix norm $\| \cdot \|$, the limit*

$$\mu(A) = \lim_{h \to 0+} \frac{\|I + hA\| - 1}{h}$$

is called the logarithmic norm of A.

This limit always exists since the difference quotient is a monotonic, bounded function of h (compare [94, 98], and [109, Lemma 1.5.1]).

The logarithmic norm can take negative values, so it is not a norm. However, it has many properties similar to the properties of a norm: for example,

$$\mu(\alpha A) = \alpha\mu(A) \quad \text{if} \quad \alpha \geq 0, \qquad |\mu(A)| \leq \|A\|,$$

$$\mu(A + B) \leq \mu(A) + \mu(B), \qquad |\mu(A) - \mu(B)| \leq \|A - B\|,$$

(compare [94, 98, 109]). Another interesting property of $\mu(A)$ is given in the following result.

Lemma 1.4.6 *Assume that $\| \cdot \|$ is an Euclidean norm in the space $\mathbb{R}^m$ (that is, $\|x\| = \sqrt{x^T x}$), and denote by the same symbol the associated matrix norm. Then*

$$\mu(A) = \max_{x \neq 0} \frac{x^T A x}{\|x\|^2}.$$

Proof: We have

$$
\begin{aligned}
\mu(A) &= \lim_{h\to 0+} \frac{\|I + hA\| - 1}{h} = \lim_{h\to 0+} \max_{x\neq 0} \frac{\|x + hAx\| - \|x\|}{h\|x\|} \\[2mm]
&= \lim_{h\to 0+} \max_{x\neq 0} \frac{\|x + hAx\|^2 - \|x\|^2}{h\|x\|\big(\|x + hAx\| + \|x\|\big)} \\[2mm]
&= \lim_{h\to 0+} \max_{x\neq 0} \frac{(x + hAx)^T(x + hAx) - x^T x}{h\|x\|\big(\|x + hAx\| + \|x\|\big)} \\[2mm]
&= \lim_{h\to 0+} \max_{x\neq 0} \frac{2x^T Ax + hx^T A^T Ax}{\|x\|\big(\|x + hAx\| + \|x\|\big)} = \max_{x\neq 0} \frac{x^T Ax}{\|x\|^2},
\end{aligned}
$$

which is our claim. ∎

For the three most commonly used norms, $\|\cdot\|_1$, $\|\cdot\|_2$, and $\|\cdot\|_\infty$, the explicit expressions for $\mu_1(A)$, $\mu_2(A)$, and $\mu_\infty(A)$, are given by

$$
\mu_1(A) = \max_{1\leq i\leq m} \left(a_{ii} + \sum_{j=1, j\neq i}^{m} |a_{ij}| \right),
$$

$$
\mu_2(A) = \lambda_{\max}\left(\frac{A + A^T}{2} \right),
$$

$$
\mu_\infty(A) = \max_{1\leq j\leq m} \left(a_{jj} + \sum_{i=1, i\neq j}^{m} |a_{ij}| \right),
$$

again compare [94, 98, 109]. Here, $\lambda_{\max}(M)$ stands for the largest eigenvalue of the matrix M.

The next result is the relationship between the one-sided Lipschitz condition for the function $f(t, y)$ and the logarithmic norm of the integral of $\frac{\partial f}{\partial y}(t, \eta)$.

Lemma 1.4.7 *Assume that the function $f(t, y)$ is continuously differentiable with respect to y. Then the condition*

$$
(x - y)^T \big(f(t, x) - f(t, y) \big) \leq \nu\|x - y\|^2 \tag{1.4.10}
$$

implies that

$$
\mu\left(\int_0^1 \frac{\partial f}{\partial y}\big(t, \theta x + (1 - \theta)y\big)d\theta \right) \leq \nu. \tag{1.4.11}
$$

Proof: It follows from the mean value theorem for vector functions that

$$
f(t, x) - f(t, y) = \int_0^1 \frac{\partial f}{\partial y}\big(t, \theta x + (1 - \theta)y\big)d\theta\,(x - y)
$$

(compare [229, page 71]). Hence, inequality (1.4.10) implies that

$$\frac{(x-y)^T \int_0^1 \frac{\partial f}{\partial y}\Big(t, \theta x + (1-\theta)y\Big) d\theta\, (x-y)}{\|x-y\|^2} \leq \nu,$$

and it follows that

$$\max_{z \neq 0} \frac{z^T \int_0^1 \frac{\partial f}{\partial y}\Big(t, \theta x + (1-\theta)y\Big) d\theta\, z}{\|z\|^2} \leq \nu.$$

By Lemma 1.4.6 this inequality is equivalent to (1.4.11). This completes the proof. ∎

We are now ready to formulate and prove the main result of this section, which is an analog of Theorem 1.4.2.

Theorem 1.4.8 *Assume that $x(t)$ and $y(t)$ satisfy the initial value problems*

$$x'(t) = f\big(t, x(t)\big), \quad x(t_0) = x_0,$$

$$y'(t) = g\big(t, y(t)\big), \quad y(t_0) = y_0,$$

for $t \in [t_0, t_0 + T]$, where $x'(t)$ and $y'(t)$ stands for the right-hand-side derivative at $t = t_0$. Assume that the functions $f(t, x)$ and $g(t, y)$ are defined and continuous and that

$$\|f(t, z) - g(t, z)\| \leq \epsilon$$

for $t \in [t_0, t_0 + T]$, $x, y \in Y \subset \mathbb{R}^m$. Assume also that the function $f(t, y)$ is continuously differentiable with respect to the second argument and satisfies a one-sided Lipschitz condition (1.4.10) for $t \in [t_0, T]$ and $x, y \in Y$. Then we have the following inequality, which is an analog of (1.4.6):

$$\|x(t) - y(t)\| \leq \|x_0 - y_0\| e^{\nu(t-t_0)} + \frac{\epsilon}{\nu}\Big(e^{\nu(t-t_0)} - 1\Big), \qquad (1.4.12)$$

$t \in [t_0, T]$, if $\nu \neq 0$ and

$$\|x(t) - y(t)\| \leq \|x_0 - y_0\| + \epsilon(t - t_0), \qquad (1.4.13)$$

$t \in [t_0, T]$, if $\nu = 0$.

Proof: It follows from Lemma 1.4.3 that

$$
\frac{d^+}{dt}\left\| x(t) - y(t) \right\| = \lim_{h\to 0+} \frac{\left\| x(t) - y(t) + h\frac{d^+}{dt}\left(x(t) - y(t)\right) \right\| - \left\| x(t) - y(t) \right\|}{h}
$$

$$
= \lim_{h\to 0+} \frac{\left\| x(t) - y(t) + h\left(f\left(t, x(t)\right) - g\left(t, y(t)\right)\right) \right\| - \left\| x(t) - y(t) \right\|}{h}
$$

$$
\leq \lim_{h\to 0+} \frac{\left\| x(t) - y(t) + h\left(f\left(t, x(t)\right) - f\left(t, y(t)\right)\right) \right\| - \left\| x(t) - y(t) \right\|}{h} + \epsilon.
$$

Set $z(t) = x(t) - y(t)$. Then

$$
\frac{d^+}{dt}\left\| z(t) \right\| \leq \lim_{h\to 0+} \frac{\left\| z(t) + h\int_0^1 \frac{\partial f}{\partial y}\left(t, \theta x(t) + (1-\theta)y(t)\right) d\theta\, z(t) \right\| - \left\| z(t) \right\|}{h} + \epsilon
$$

$$
= \lim_{h\to 0+} \frac{\left\| \left(I + h\int_0^1 \frac{\partial f}{\partial y}\left(t, \theta x(t) + (1-\theta)y(t)\right) d\theta\right) z(t) \right\| - \left\| z(t) \right\|}{h} + \epsilon
$$

$$
= \mu\left(\int_0^1 \frac{\partial f}{\partial y}\left(t, \theta x(t) + (1-\theta)y(t)\right) d\theta\right) \leq \nu\left\| z(t) \right\| + \epsilon.
$$

Inequalities (1.4.12) and (1.4.13) now follow from Lemma 1.4.4. ∎

1.5 DERIVATIVES WITH RESPECT TO PARAMETERS AND INITIAL VALUES

In this section we investigate the differentiability of solutions to differential systems with respect to parameters and initial values. Assume that the function $g(t, y, \lambda)$ is defined in some open set D and consider the initial value problem

$$
y' = g(t, y, \lambda_0), \quad y(t_0) = y_0, \tag{1.5.1}
$$

where $(t_0, y_0, \lambda_0) \in D \subset \mathbb{R} \times \mathbb{R}^m \times \mathbb{R}^s$. Denote by $y(t) = y(t; t_0, y_0, \lambda_0)$ the solution to (1.5.1), which is defined on a compact interval J containing t_0. We have the following theorem.

Theorem 1.5.1 *Assume that the function $g(t, y, \lambda)$ has continuous partial derivatives*

$$
\frac{\partial g}{\partial y} = \left[\frac{\partial g_i}{\partial y_j}\right]_{i=1,\dots,m, j=1,\dots,m} \quad and \quad \frac{\partial g}{\partial \lambda} = \left[\frac{\partial g_i}{\partial \lambda_j}\right]_{i=1,\dots,m, j=1,\dots,s}
$$

at all points $(t, y(t; t_0, y_0, \lambda_0), \lambda_0)$ *with* $t \in J$. *Then for all* λ *sufficiently close to* λ_0, *the initial value problem*

$$y' = g(t, y, \lambda), \quad y(t_0) = y_0, \tag{1.5.2}$$

has a unique solution $y(t) = y(t; t_0, y_0, \lambda)$ *which is defined on the interval* J. *Moreover, for all* $t \in J$, *the partial derivative* $\frac{\partial y}{\partial \lambda}(t; t_0, y_0, \lambda_0)$ *exists and satisfies the initial value problem for the linear differential system*

$$\eta' = \frac{\partial g}{\partial y}\Big(t, y(t; t_0, y_0, \lambda_0), \lambda_0\Big)\eta + \frac{\partial g}{\partial \lambda}\Big(t, y(t; t_0, y_0, \lambda_0), \lambda_0\Big), \quad \eta(t_0) = 0, \tag{1.5.3}$$

$\eta : J \to R^m \times \mathbb{R}^s$.

Proof: Since $\partial g/\partial y$ and $\partial g/\partial \lambda$ are continuous in (t, y, λ) and $y(t; t_0, y_0, \lambda_0)$ is continuous in t, there exists for each $\epsilon > 0$ and each $s \in J$ a corresponding $\delta = \delta(s, \epsilon) > 0$ such that

$$\left\|\frac{\partial g}{\partial y}(t, y, \lambda) - \frac{\partial g}{\partial y}\Big(t, y(t; t_0, y_0, \lambda_0), \lambda_0\Big)\right\| \le \epsilon,$$

$$\left\|\frac{\partial g}{\partial \lambda}(t, y, \lambda) - \frac{\partial g}{\partial \lambda}\Big(t, y(t; t_0, y_0, \lambda_0), \lambda_0\Big)\right\| \le \epsilon \tag{1.5.4}$$

if

$$|t - s| \le \delta, \quad \|y - y(t; t_0, y_0, \lambda_0)\| \le \delta, \quad \|\lambda - \lambda_0\| \le \delta.$$

Since interval J is compact, it can be covered by a finite number of intervals I_s of the form $I_s = \{t : |t - s| \le \delta(s, \epsilon)\}$. Hence, there exists $\delta' = \delta'(\epsilon)$ (independent of s) such that inequalities (1.5.4) are satisfied for all $(t, y, \lambda) \in Q$, where set Q is defined by

$$Q = \Big\{(t, y, \lambda) : t \in J, \ \|y - y(t; t_0, y_0, \lambda_0)\| \le \delta', \ \|\lambda - \lambda_0\| \le \delta'\Big\}.$$

Set Q is closed and it follows that $\partial g/\partial y$ and $\partial g/\partial \lambda$ are bounded on it; that is, there exist constants A and B such that

$$\left\|\frac{\partial g}{\partial y}\right\| \le A, \quad \left\|\frac{\partial g}{\partial \lambda}\right\| \le B.$$

In particular, $g(t, y, \lambda)$ satisfies a Lipschitz condition with respect to y, and, if $\|\lambda - \lambda_0\| \le \delta'$, it follows from Theorem 1.3.7 that initial value problem (1.5.2) has a unique solution $y(t; t_0, y_0, \lambda_0)$. We can assume without loss of generality that this solution stays in the region

$$Q_\lambda = \{(t, y) : (t, y, \lambda) \in J\}$$

since if this is not the case, we can always restrict the size of the interval J. We have

$$y'(t; t_0, y_0, \lambda) = g\big(t, y(t; t_0, y_0, \lambda), \lambda\big),$$

$$y'(t; t_0, y_0, \lambda_0) = g\big(t, y(t; t_0, y_0, \lambda_0), \lambda_0\big),$$

and since

$$\|g(t, y, \lambda) - g(t, y, \lambda_0)\| \le B\|\lambda - \lambda_0\|$$

for all y such that $\|y - y(t; t_0, y_0, \lambda_0)\| \le \delta'$, it follows from Theorem 1.4.2 that

$$\|y(t; t_0, y_0, \lambda) - y(t; t_0, y_0, \lambda_0)\| \le C\|\lambda - \lambda_0\|, \tag{1.5.5}$$

where $C = B(e^{Ah} - 1)/A$ and h is the length of the interval J. Set

$$\chi(t; \lambda) = y(t; t_0, y_0, \lambda) - y(t; t_0, y_0, \lambda_0) - \eta(t)(\lambda - \lambda_0), \tag{1.5.6}$$

where $\eta(t)$ is the solution to (1.5.3). Then $\chi(t_0; \lambda) = 0$ and

$$
\begin{aligned}
\chi'(t; \lambda) &= g\big(t, y(t; t_0, y_0, \lambda), \lambda\big) - g\big(t, y(t; t_0, y_0, \lambda_0), \lambda_0\big) \\
&\quad - \frac{\partial g}{\partial y}\Big(t, y(t; t_0, y_0, \lambda_0), \lambda_0\Big)\eta(t)(\lambda - \lambda_0) \\
&\quad - \frac{\partial g}{\partial \lambda}\Big(t, y(t; t_0, y_0, \lambda_0), \lambda_0\Big)(\lambda - \lambda_0).
\end{aligned}
$$

Computing $\eta(t)(\lambda - \lambda_0)$ from (1.5.6) and substituting it into the equation above, we obtain

$$\chi'(t; \lambda) = \frac{\partial g}{\partial y}\Big(t, y(t; t_0, y_0, \lambda_0), \lambda_0\Big)\chi(t; \lambda) + \xi,$$

where

$$
\begin{aligned}
\xi &= g\big(t, y(t; t_0, y_0, \lambda), \lambda\big) - g\big(t, y(t; t_0, y_0, \lambda_0), \lambda\big) \\
&\quad + g\big(t, y(t; t_0, y_0, \lambda_0), \lambda\big) - g\big(t, y(t; t_0, y_0, \lambda_0), \lambda_0\big) \\
&\quad - \frac{\partial g}{\partial y}\Big(t, y(t; t_0, y_0, \lambda_0), \lambda_0\Big)\big(y(t; t_0, y_0, \lambda) - y(t; t_0, y_0, \lambda_0)\big) \\
&\quad - \frac{\partial g}{\partial \lambda}\Big(t, y(t; t_0, y_0, \lambda_0), \lambda_0\Big)(\lambda - \lambda_0).
\end{aligned}
$$

It follows from the integral form of the mean value theorem [229] that

$$\xi = \int_0^1 \left(\frac{\partial g}{\partial y}\Big(t, y(t; t_0, y_0, \lambda_0) + \theta\big(y(t; t_0, y_0, \lambda) - y(t; t_0, y_0, \lambda_0)\big), \lambda\Big) \right.$$

$$\left. - \frac{\partial g}{\partial y}\Big(t, y(t; t_0, y_0, \lambda_0), \lambda_0\Big) \right) \big(y(t; t_0, y_0, \lambda) - y(t; t_0, y_0, \lambda_0)\big) d\theta$$

$$+ \int_0^1 \left(\frac{\partial g}{\partial \lambda}\Big(t, y(t; t_0, y_0, \lambda_0), \lambda_0 + \theta(\lambda - \lambda_0)\Big) \right.$$

$$\left. - \frac{\partial g}{\partial \lambda}\Big(t, y(t; t_0, y_0, \lambda_0), \lambda_0\Big) \right) (\lambda - \lambda_0) d\theta.$$

Hence, using (1.5.4) and (1.5.5), we obtain

$$\|\xi\| \le \epsilon\,(C+1)\|\lambda - \lambda_0\|.$$

Consider also the differential system

$$\overline{\chi}' = \frac{\partial g}{\partial y}\Big(t, y(t; t_0, y_0, \lambda_0), \lambda_0\Big)\overline{\chi}, \quad \overline{\chi}(t_0) = 0,$$

which has a zero solution $\overline{\chi}(t, \lambda) \equiv 0$. Applying Theorem 1.4.2 again, it follows that

$$\|\chi(t; \lambda)\| \le \epsilon\,\widetilde{C}\|\lambda - \lambda_0\|,$$

or using definition (1.5.6) of $\chi(t; \lambda)$, we have

$$\frac{\left\| y(t; t_0, y_0, \lambda) - y(t; t_0, y_0, \lambda_0) - \eta(t)(\lambda - \lambda_0) \right\|}{\|\lambda - \lambda_0\|} \le \epsilon\widetilde{C},$$

where $\widetilde{C} = (C + 1)(e^{Ah} - 1)/A$. Hence, since ϵ can be arbitrarily small, it follows from the definition of differentiability that $\frac{\partial y}{\partial \lambda}(t; t_0, y_0, \lambda_0)$ exists and is equal to $\eta(t)$ for all $t \in J$. This completes the proof.   ∎

Consider next the system

$$y' = g(t, y), \tag{1.5.7}$$

where the function $g(t, y)$ is defined in a domain D and denote by $y(t; \tau, \xi)$ the solution to (1.5.7) passing through the point $(\tau, \xi) \in D$. We have the following theorem.

Theorem 1.5.2 *Assume that the function $g(t, y)$ is continuous in D. Assume also that $y(t; t_0, y_0)$, $(t_0, y_0) \in D$, exists on a compact interval J and that $g(t, y)$ has continuous partial derivative $\frac{\partial g}{\partial y}(t, y(t; t_0, y_0))$ for all $t \in J$. Then*

$y(t; \tau, \xi)$ *exists and is unique on* J *for all points* (τ, ξ) *that are near* (t_0, y_0). *Moreover, for all* $t \in J$ *the partial derivative* $\frac{\partial y}{\partial \xi}(t; t_0, y_0)$ *exists and is equal to the solution of the homogeneous linear matrix differential system*

$$\eta' = \frac{\partial g}{\partial y}(t, y(t; t_0, y_0))\eta, \quad \eta(t_0) = I, \tag{1.5.8}$$

while $\frac{\partial y}{\partial \tau}(t; t_0, y_0)$ *exists and*

$$\frac{\partial y}{\partial \tau}(t; t_0, y_0) = -\frac{\partial y}{\partial \xi}(t; t_0, y_0)g(t_0, y_0). \tag{1.5.9}$$

Proof: Set $y = x + \xi$. Then (1.5.7) with initial condition $y(\tau) = \xi$ is transformed into the initial value problem

$$x' = g(t, x + \xi) = G(t, x, \xi), \quad x(\tau) = 0, \tag{1.5.10}$$

where ξ is some parameter. Let us denote the solution to this problem by $x(t) = x(t; \tau, 0, \xi)$. Then

$$y(t; \tau, \xi) = x(t; \tau, 0, \xi) + \xi.$$

It follows from Theorem 1.5.1 that $\partial x/\partial \xi$ satisfies the initial value problem

$$\frac{d}{dt}\frac{\partial x}{\partial \xi}(t; t_0, 0, y_0) = \frac{\partial G}{\partial x}\left(t, x(t; t_0, 0, y_0), y_0\right)x(t; t_0, 0, y_0)$$

$$+ \frac{\partial G}{\partial \xi}\left(t, x(t; t_0, 0, y_0), y_0\right),$$

$$\frac{\partial x}{\partial \xi}(t_0; t_0, 0, y_0) = 0.$$

Since

$$\frac{\partial G}{\partial x}\left(t, x(t; t_0, 0, y_0), y_0\right) = \frac{\partial G}{\partial \xi}\left(t, x(t; t_0, 0, y_0), y_0\right)$$

$$= \frac{\partial g}{\partial y}(t, x(t; t_0, 0, y_0) + y_0) = \frac{\partial g}{\partial y}(t, y(t; t_0, y_0))$$

and

$$\frac{\partial y}{\partial \xi}(t_0; t_0, y_0) = \frac{\partial x}{\partial \xi}(t_0; t_0, 0, y_0) + I,$$

it follows that

$$\frac{\partial y}{\partial \xi}(t; t_0, y_0) = \frac{\partial g}{\partial y}(t, y(t; t_0, y_0))\frac{\partial y}{\partial \xi}(t; t_0, y_0), \quad \frac{\partial y}{\partial \xi}(t_0; t_0, y_0) = I.$$

This proves that $\frac{\partial y}{\partial \xi}(t; t_0, y_0)$ satisfies (1.5.8).

Similarly as in the proof of Theorem 1.5.1, we can argue that $\partial g/\partial y$ is bounded in the neighborhood of (t_0, y_0). Hence, g satisfies a Lipschitz condition in this neighborhood, and the solution $y(t; \tau, \xi)$ to (1.5.7) exists and is uniquely determined on the interval $[t_0, \tau]$ if (τ, ξ) is sufficiently close to (t_0, y_0). It also follows from Theorem 1.5.1 that this solution exists and is uniquely determined on J.

Set $\tau = t_0 + h$ and define y_1 by $y_1 = y(t_0; t_0 + h, y_0)$. Then $y_1 \to y_0$ as $h \to 0$. Moreover,

$$y_1 - y_0 = y(t_0; t_0+h, y_0) - y(t_0+h; t_0+h, y_0) = -\int_{t_0}^{t_0+h} g\big(s, y(s; t_0+h, y_0)\big) ds.$$

It follows from the mean value theorem that there exists $0 < \theta < 1$ such that

$$\int_{t_0}^{t_0+h} g\big(s, y(s; t_0 + h, y_0)\big) ds = g\big(t_0 + \theta h, y(t_0 + \theta h; t_0 + h, y_0)\big) h,$$

and taking into account that g and y are continuous, we obtain

$$y_1 - y_0 = -g(t_0, y_0)h + o(h) \quad \text{as} \quad h \to 0.$$

Since the points $(t_0 + h, y_0)$ and (t_0, y_1) lie on the same solution, we obtain

$$y(t; t_0 + h, y_0) - y(t; t_0, y_0) = y(t; t_0, y_1) - y(t; t_0, y_0)$$

$$= \left(\frac{\partial y}{\partial \xi}(t; t_0, y_0) + o(1) \right)(y_1 - y_0)$$

$$= -\left(\frac{\partial y}{\partial \xi}(t; t_0, y_0) + o(1) \right) g(t_0, y_0)h + o(h).$$

Hence,

$$\frac{y(t; t_0 + h, y_0) - y(t; t_0, y_0)}{h} = -\left(\frac{\partial y}{\partial \xi}(t; t_0, y_0) + o(1) \right) g(t_0, y_0),$$

and taking the limit as $h \to 0$, we obtain (1.5.9). This completes the proof. $\blacksquare$

1.6   STABILITY THEORY

In this section we follow mainly the account of stability theory for differential equations as presented by Coppel [94, 95]. Let $y(t)$ be a solution to the differential system (1.1.2), which is defined for all $t \geq t_0$. We have the following definitions.

Definition 1.6.1 *Solution $y(t)$ to (1.1.2) is said to be stable over the interval $[t_0, \infty)$ if for each $\epsilon > 0$ there exists a $\delta = \delta(\epsilon) > 0$ such that any solution $\widetilde{y}(t)$*

to (1.1.2) that satisfies the inequality $\|\widetilde{y}(t_0) - y(t_0)\| < \delta$ exists and satisfies the inequality $\|\widetilde{y}(t) - y(t)\| < \epsilon$ for all $t \geq t_0$.

Definition 1.6.2 *Solution $y(t)$ to (1.1.2) is said to be asymptotically stable if it is stable and, in addition, $\|\widetilde{y}(t) - y(t)\| \to 0$ as $t \to \infty$ whenever $\|\widetilde{y}(t_0) - y(t_0)\|$ is sufficiently small.*

Definition 1.6.3 *Solution $y(t)$ to (1.1.2) is said to be uniformly stable if for each $\epsilon > 0$ there exists a $\delta = \delta(\epsilon) > 0$ such that any solution $\widetilde{y}(t)$ to (1.1.2) that satisfies the inequality $\|\widetilde{y}(t_1) - y(t_1)\| < \delta$ for some $t_1 \geq t_0$ exists and satisfies the inequality $\|\widetilde{y}(t) - y(t)\| < \epsilon$ for all $t \geq t_1$.*

Definition 1.6.4 *Solution $y(t)$ to (1.1.2) is said to be uniformly asymptotically stable if it is uniformly stable and, in addition, there is a $\delta_0 > 0$, and for each $\epsilon > 0$ a corresponding $T = T(\epsilon) > 0$, such that if $\|\widetilde{y}(t_1) - y(t_1)\| < \delta_0$ for some $t_1 \geq t_0$, then $\|\widetilde{y}(t) - y(t)\| < \epsilon$ for all $t \geq t_1 + T$.*

Assume that $y(t)$ is a particular solution to (1.1.2) whose solution we are studying. Then by the change of variables

$$x = y - y(t) \tag{1.6.1}$$

this problem can always be reduced to the problem of studying the stability of the zero solution $x(t) \equiv 0$ of the corresponding new system

$$x' = G(t, x) := g\big(t, x + y(t)\big) - g\big(t, y(t)\big), \quad t \geq t_0. \tag{1.6.2}$$

The condition $\|\widetilde{y}(t) - y(t)\| < \delta$ now takes the form $\|\widetilde{x}(t)\| < \delta$, and the condition $\|\widetilde{y}(t) - y(t)\| < \epsilon$ takes the form $\|\widetilde{x}(t)\| < \epsilon$. Similarly, the condition $\|\widetilde{y}(t) - y(t)\| \to 0$ as $t \to \infty$ is now $\|\widetilde{x}(t)\| \to 0$ as $t \to \infty$. This implies that the solution $y(t)$ to (1.1.2) is stable, asymptotically stable, uniformly stable, or uniformly asymptotically stable if and only if the same holds for the solution $x(t) \equiv 0$ to (1.6.2).

We discuss first the stability of solutions to the linear system

$$y' = A(t)y + b(t), \quad t \geq t_0, \tag{1.6.3}$$

where the matrix $A(t)$ and the vector $b(t)$ are continuous functions of t for $t \geq t_0$. Then every solution to this system is defined for all $t \geq t_0$. Let $y(t)$ be a particular solution to (1.6.3). Then by making the substitution (1.6.1), this system is transformed into the homogeneous system

$$x' = A(t)x, \quad t \geq t_0, \tag{1.6.4}$$

which has a solution $x(t) \equiv 0$ corresponding to the solution $y(t)$. It follows from the form of (1.6.4) that it is possible to express the conditions for various stability properties in a way independent of any particular solution $y(t)$ to (1.6.3) and the inhomogeneous term $b(t)$. We will then say that (1.6.4) possesses some stability property if the same is true for the zero solution of this system.

Denote by $X(t)$ a fundamental matrix of (1.6.4), i.e., the matrix with linearly independent columns that satisfies the matrix differential system

$$X'(t) = A(t)X(t), \quad t \geq t_0.$$

We have the following theorem.

Theorem 1.6.5 *System (1.6.4) is*
(i) stable if and only if there exists a positive constant K such that

$$\|X(t)\| \leq K \quad \text{for} \quad t \geq t_0, \tag{1.6.5}$$

(ii) uniformly stable if and only if there exists a positive constant K such that

$$\|X(t)X^{-1}(s)\| \leq K \quad \text{for} \quad t_0 \leq s \leq t < \infty, \tag{1.6.6}$$

(iii) asymptotically stable if and only if

$$\|X(t)\| \to 0 \quad \text{as} \quad t \to \infty, \tag{1.6.7}$$

(iv) uniformly asymptotically stable if and only if there exist positive constants K and α such that

$$\|X(t)X^{-1}(s)\| \leq Ke^{-\alpha(t-s)} \quad \text{for} \quad t_0 \leq s \leq t < \infty. \tag{1.6.8}$$

Proof: We can assume without loss of generality that $X(t_0) = I$. Then the solution to (1.6.4) with initial condition $x(t_0) = x_0$ is given by

$$x(t) = X(t)x_0, \quad t \geq t_0.$$

Assume first that (1.6.5) holds. Then

$$\|x(t)\| \leq K\|x_0\| < \epsilon \quad \text{if} \quad \|x_0\| < \frac{\epsilon}{K},$$

which shows that the zero solution to (1.6.4) is stable. Conversely, if $\|X(t)x_0\| < \epsilon$ for all x_0 such that $\|x_0\| < \delta$, then

$$\|X(t)\| = \sup_{x \neq 0} \frac{\|X(t)x\|}{\|x\|} = \frac{1}{\delta}\sup_{x \neq 0}\left\|X(t)\frac{\delta x}{\|x\|}\right\| = \frac{1}{\delta}\sup_{\|\eta\|=\delta}\|X(t)\eta\| \leq \frac{\epsilon}{\delta},$$

which is equivalent to (1.6.5) with $K = \epsilon/\delta$.

Assume now that (1.6.6) holds. The solution to (1.6.4) such that $x(s) = \xi$ is

$$x(t) = X(t)X^{-1}(s)\xi, \quad t_0 \leq s \leq t < \infty,$$

and it follows that

$$\|x(t)\| \leq K\|\xi\| < \epsilon \quad \text{if} \quad \|x(s)\| = \|\xi\| < \frac{\epsilon}{K},$$

$t_0 \leq s \leq t < \infty$. This proves that the zero solution to (1.6.4) is uniformly stable. Conversely, if $\|x(t)\| = \|X(t)X^{-1}(s)\xi\| < \epsilon$ for $t_0 \leq s \leq t < \infty$ and all ξ such that $\|\xi\| = \|x(s)\| < \delta$ then, similarly as before,

$$\|X(t)X^{-1}(s)\| = \frac{1}{\delta} \sup_{x \neq 0} \left\| X(t)X^{-1}(s)\frac{\delta x}{\|x\|} \right\| = \frac{1}{\delta} \sup_{\|\eta\|=\delta} \|X(t)X^{-1}(s)\eta\| < \frac{\epsilon}{\delta},$$

which is equivalent to (1.6.6) with $K = \epsilon/\delta$.

Assume next that (1.6.7) holds. Then any solution to (1.6.4) such that $x(t_0) = x_0$ satisfies

$$\|x(t)\| = \|X(t)x_0\| \leq \|X(t)\|\|x_0\| \to 0 \quad \text{as} \quad t \to \infty,$$

which proves that the zero solution to (1.6.4) is asymptotically stable. Conversely, assume that the zero solution to (1.6.4) is asymptotically stable. Then for any solution $x(t)$ such that $x(t_0) = x_0$, $\|x_0\| \leq \delta$, we have $\|x(t)\| \to 0$ as $t \to \infty$. Hence, there exists $\overline{x}_0$ such that $\|\overline{x}_0\| = \delta$ and

$$\|X(t)\| = \frac{1}{\delta} \sup_{\|x_0\|=\delta} \|X(t)x_0\| = \frac{1}{\delta}\|X(t)\overline{x}_0\| \to 0 \quad \text{as} \quad t \to \infty.$$

This proves (1.6.7).

Assume finally that (1.6.8) holds. Then (1.6.4) is also uniformly stable. Consider a solution $x(t)$ to (1.6.4) such that $x(s) = \xi$, where $\|\xi\| \leq \delta = 1$, $t_0 \leq s \leq t < \infty$. Then for each $0 < \epsilon < K$ we have

$$\|x(t)\| = \|X(t)X^{-1}(s)\xi\| \leq Ke^{-\alpha(t-s)} < \epsilon$$

for $t \geq s + T$, where $T = -\ln(\epsilon/K)/\alpha$. This proves that (1.6.4) is uniformly asymptotically stable. Conversely, assume now that (1.6.4) is uniformly asymptotically stable and consider the solution to (1.6.4) such that $x(s) = \xi$. Then there exists a $\delta > 0$ and for each $0 < \epsilon < \delta$ a corresponding $T > \epsilon$ such that if $\|x(s)\| = \|\xi\| \leq \delta$ for $s \geq t_0$, then

$$\|x(t)\| = \|X(t)X^{-1}(s)\xi\| < \epsilon \quad \text{for} \quad t \geq s + T.$$

Hence,

$$\|X(t)X^{-1}(s)\| = \frac{1}{\delta} \sup_{\|\xi\|=1} \|X(s+T)X^{-1}(s)\xi\| \leq \frac{\epsilon}{\delta} < 1 \qquad (1.6.9)$$

for $s \geq t_0$. Moreover, it follows from (1.6.6) that there exists a positive constant $\overline{K}$ such that

$$\|X(s+h)X^{-1}(s)\| \leq \overline{K} \quad \text{for} \quad s \geq t_0, \quad 0 \leq h \leq T. \qquad (1.6.10)$$

We will show that (1.6.9) and (1.6.10) imply (1.6.8). If $t \geq s$, we can write

$$s + nT \leq t < s + (n+1)T$$

for some nonnegative integer n. Then it follows that

$$\|X(t)X^{-1}(s)\| \leq \|X(t)X^{-1}(s+nT)\| \, \|X(s+nT)X^{-1}(s+(n-1)T)\|$$

$$\cdots \|X(s+2T)X^{-1}(s+T)\| \, \|X(s+T)X^{-1}(s)\|$$

$$\leq \overline{K}\left(\frac{\epsilon}{\delta}\right)^n.$$

Define a constant α by the relation

$$\frac{\epsilon}{\delta} = e^{-\alpha T} \quad \text{or} \quad \alpha = -\frac{\ln(\epsilon/\delta)}{T}.$$

Then $\alpha > 0$,

$$\left(\frac{\epsilon}{\delta}\right)^{n+1} = e^{-\alpha(n+1)T} \leq e^{-\alpha(t-s)},$$

and it follows that

$$\|X(t)X^{-1}(s)\| \leq \frac{\overline{K}\delta}{\epsilon} e^{-\alpha(t-s)}, \quad t_0 \leq s \leq t < \infty.$$

This is equivalent to (1.6.8) with $K = \overline{K}\delta/\epsilon$. ∎

Consider next a linear homogeneous and autonomous system of the form

$$x' = Ax, \quad t \geq t_0. \tag{1.6.11}$$

It is easy to verify that for this system, stability is equivalent to uniform stability and asymptotic stability is equivalent to uniform asymptotic stability. We have the following theorem.

Theorem 1.6.6 *System (1.6.11) is stable if and only if every eigenvalue of the constant matrix A has real part less than or equal to zero, and those eigenvalues with zero real parts are simple. This system is asymptotically stable if and only if every eigenvalue of A has a negative real part.*

Proof: Let T be a nonsingular matrix that transforms A into Jordan canonical form

$$T^{-1}AT = J,$$

where

$$J = \begin{bmatrix} J_1 & & & \\ & J_2 & & \\ & & \ddots & \\ & & & J_\nu \end{bmatrix}, \quad J_i = \begin{bmatrix} \lambda_i & 1 & & \\ & \lambda_i & \ddots & \\ & & \ddots & 1 \\ & & & \lambda_i \end{bmatrix},$$

$i = 1, 2, \ldots, \nu$, and λ_i is an eigenvalue of the matrix A corresponding to the Jordan block J_i of dimension μ_i. The fundamental matrix of (1.6.11) is

$$X(t) = e^{A(t-t_0)} = T e^{J(t-t_0)} T^{-1}.$$

It can be verified that $e^{J(t-t_0)}$ is a block diagonal matrix with blocks of the form

$$D_i = \begin{bmatrix} e^{\lambda_i(t-t_0)} & (t-t_0)e^{\lambda_i(t-t_0)} & \cdots & \frac{(t-t_0)^{\mu_i}}{\mu_i!}e^{\lambda_i(t-t_0)} \\ & e^{\lambda_i(t-t_0)} & \cdots & \frac{(t-t_0)^{\mu_i-1}}{(\mu_i-1)!}e^{\lambda_i(t-t_0)} \\ & & \ddots & \vdots \\ & & & e^{\lambda_i(t-t_0)} \end{bmatrix},$$

$i = 1, 2, \ldots, \nu$. Hence, $\|D_i\| < \infty$ and, as a result, $\|e^{A(t-t_0)}\| < \infty$ if $\mathrm{Re}(\lambda_i) \le 0$ and λ_i is simple (i.e., $\mu_i = 1$), which is equivalent to stability. Similarly, $\|D_i\| \to 0$ as $t \to \infty$, and consequently, $\|e^{A(t-t_0)}\| \to 0$ as $t \to \infty$ if $\mathrm{Re}(\lambda_i) < 0$, which is equivalent to asymptotic stability. ∎

The characterization of stability given in Theorem 1.6.6 is no longer possible if matrix A in (1.6.11) depends on t. A counterexample given by Vinograd [282] and reproduced by Dekker and Verwer [109] is provided by the homogeneous system (1.6.4) with matrix $A(t)$ of the form

$$A(t) = \begin{bmatrix} -1 - 9\cos^2(6t) + 6\sin(12t) & 12\cos^2(6t) + \frac{9}{2}\sin(12t) \\ -12\sin^2(6t) + \frac{9}{2}\sin(12t) & -1 - 9\sin^2(6t) - 6\sin(12t) \end{bmatrix}.$$

The eigenvalues of this matrix are $\lambda_1 = -1$ and $\lambda_2 = -10$ for any t. But the fundamental matrix corresponding to $A(t)$ is

$$X(t) = \begin{bmatrix} e^{2t}\big(\cos(6t) + 2\sin(6t)\big) & e^{-13t}\big(\sin(6t) - 2\cos(6t)\big) \\ e^{2t}\big(2\cos(6t) - \sin(6t)\big) & e^{-13t}\big(2\sin(6t) + \cos(6t)\big) \end{bmatrix},$$

and as a result the corresponding system (1.6.4) is unstable. Other examples that illustrate this point are given by Coppel [95] and Hairer at al. [143].

We can also establish stability criteria for (1.6.4) which are based on the logarithmic norm $\mu(A(t))$ introduced in Definition 1.4.5. Some of these criteria are a consequence of the following result given by Coppel [94].

Theorem 1.6.7 *Assume that $A(t)$ in (1.6.4) is a continuous function of t for $t \geq t_0$. Then for any solution $x(t)$ to (1.6.4), we have the bounds*

$$\|x(t_0)\| \exp\left(-\int_{t_0}^{t} \mu(-A(s))ds\right)$$

$$\leq \|x(t)\| \leq \|x(t_0)\| \exp\left(\int_{t_0}^{t} \mu(A(s))ds\right). \tag{1.6.12}$$

Proof: It follows from the property $|\mu(A) - \mu(B)| \leq \|A - B\|$ in Section 1.4 that $\mu(A(t))$ is a continuous function of t for $t \geq t_0$. It follows from Lemma 1.4.3 that $\|x(t)\|$ has a right-hand-side derivative that is given by

$$\frac{d^+}{dt}\|x(t)\| = \lim_{h \to 0+} \frac{\|x(t) + hx'(t+0)\| - \|x(t)\|}{h}.$$

Since

$$\|x(t) + hx'(t+0)\| = \|(I + hA(t))x(t)\| \leq \|I + hA(t)\| \, \|x(t)\|,$$

it follows that

$$\frac{d^+}{dt}\|x(t)\| \leq \lim_{h \to 0+} \frac{\|I + hA(t)\| - 1}{h}\|x(t)\| = \mu(A(t)) \, \|x(t)\|$$

or

$$\frac{d^+}{dt}\|x(t)\| - \mu(A(t)) \, \|x(t)\| \leq 0.$$

Multiplying both sides of this inequality by the integrating factor equal to $\exp(-\int_{t_0}^{t} \mu(A(s))ds)$, we obtain

$$\frac{d^+}{dt}w(t) \leq 0,$$

where

$$w(t) = \exp\left(-\int_{t_0}^{t} \mu(A(s))ds\right)\|x(t)\|.$$

Hence, $w(t)$ is a nonincreasing function of t, and as a result, $w(t) \leq w(t_0)$, $t \geq t_0$, which is equivalent to the upper bound in (1.6.12). The lower bound can be proved similarly by the change of variables $t \to -t$.   ∎

This theorem implies the following stability criteria [94].

Corollary 1.6.8 *Linear system (1.6.4) is*
(i) stable if

$$\limsup_{t \to \infty} \int_{t_0}^{t} \mu(A(s))ds < +\infty,$$

(ii) asymptotically stable if

$$\lim_{t \to \infty} \int_{t_0}^{t} \mu\big(A(s)\big)ds = -\infty,$$

(iii) uniformly stable if

$$\mu\big(A(t)\big) \leq 0 \quad for \quad t \geq t_0,$$

(iv) uniformly asymptotically stable if

$$\mu\big(A(t)\big) \leq -\alpha < 0 \quad for \quad t \geq t_0.$$

We investigate next the stability of solutions to a nonlinear system of the form

$$x' = A(t)x + f(t,x), \quad t \geq t_0, \tag{1.6.13}$$

where $A(t)$ is a continuous function of t for $t \geq t_0$ and $f(t,x)$ is continuous in the region

$$\tilde{D} = \Big\{(t,x) : \ t \geq t_0, \ \|x\| \leq M\Big\} \subset \mathbb{R} \times \mathbb{R}^m.$$

Then it follows from the Peano existence theorem, Theorem 1.3.4, that for any point $(s, \xi) \in \tilde{D}$, system (1.6.13) has a local solution $x(t)$ such that $x(s) = \xi$.

It is of interest to investigate under what conditions various stability properties of homogeneous linear system (1.6.4) are inherited by the solutions to the "perturbed" system (1.6.13). For uniform stability and asymptotic stability, an answer is given by the following theorem [94].

Theorem 1.6.9 *Assume that (1.6.4) is uniformly stable and let $f(t,x)$ satisfy the inequality*

$$\big\|f(t,x)\big\| \leq \gamma(t)\|x\|, \quad t \geq t_0, \tag{1.6.14}$$

where $\gamma(t)$ is a continuous nonnegative function such that

$$\int_{t_0}^{t} \gamma(s)ds < \infty. \tag{1.6.15}$$

(In particular, $f(t,0) = 0$ and $x \equiv 0$ is a solution to (1.6.13)). Then for each $0 < \epsilon < M$ there exists a positive constant Q such that if $t_1 \geq t_0$, any solution $x(t)$ to (1.6.13) for which $\|x(t_1)\| < \delta = Q^{-1}\epsilon$ is defined and satisfies

$$\big\|x(t)\big\| \leq Q\|x(t_1)\| < \epsilon \quad for \ all \quad t \geq t_1. \tag{1.6.16}$$

In particular, the zero solution to (1.6.13) is uniformly stable. If, in addition, the zero solution to (1.6.4) is asymptotically stable, i.e., $X(t) \to 0$ as $t \to \infty$, then $x(t) \to 0$ as $t \to \infty$, i.e., the zero solution to (1.6.13) is asymptotically stable.

Proof: Assume that $x(t)$ is a solution to (1.6.13). Then $x(t)$ is also a solution to the nonhomogeneous linear system

$$x' = A(t)x + b(t), \quad t \geq t_0,$$

where the vector function $b(t)$ is defined by $b(t) = f(t, x(t))$. By the variation of constants formula, the solution $x(t)$ satisfies the relation

$$x(t) = X(t)X^{-1}(t_1)x(t_1) + \int_{t_1}^{t} X(t)X^{-1}(s)f(s, x(s))ds, \quad t \geq t_1, \quad (1.6.17)$$

where $X(t)$ is the fundamental matrix of (1.6.4) such that $X(t_0) = I$. Since (1.6.4) is uniformly stable, it follows from Theorem 1.6.5 that there exists a positive constant K such that

$$\|X(t)X^{-1}(s)\| \leq K \quad \text{for} \quad t_0 \leq s \leq t < \infty.$$

Taking norms on both sides of inequality (1.6.17) and using the inequality above and (1.6.14), we obtain

$$\|x(t)\| \leq K\|x(t_1)\| + K \int_{t_1}^{t} \gamma(s)\|x(s)\|ds, \quad t \geq t_1.$$

Application of the Gronwall lemma (1.4.1) now leads to

$$\|x(t)\| \leq K\|x(t_1)\| \exp\left(K \int_{t_1}^{t} \gamma(s)ds\right) \leq Q\|x(t_1)\| < \epsilon,$$

where

$$Q = K \exp\left(K \int_{t_1}^{\infty} \gamma(s)ds\right).$$

This proves that the zero solution to (1.6.13) is uniformly stable.

Assume next that $X(t) \to 0$ as $t \to \infty$. It follows from the first part of the theorem that if the solution $x(t)$ to (1.6.13) exists for $t \geq t_0$, it is bounded, i.e., $\|x(t)\| \leq C$, $t \geq t_0$ for some constant C. It follows from the variation of constants formula that this solution satisfies the relation

$$x(t) = X(t)x(t_0) + \int_{t_0}^{t} X(t)X^{-1}(s)f(s, x(s))ds, \quad t \geq t_0.$$

Hence, for any $t_1 \geq t_0$, we obtain

$$\|x(t)\| \leq \|X(t)\| \, \|x(t_0)\|$$

$$+ \|X(t)\| \int_{t_0}^{t_1} \|X^{-1}(s)\| \|\gamma(s)\| \|x(s)\|ds + K \int_{t_1}^{t} \gamma(s)\|x(s)\|ds$$

$$\leq \|X(t)\| \, \|x(t_0)\| + C\|X(t)\| \int_{t_0}^{t_1} \|X^{-1}(s)\| \gamma(s)ds + CK \int_{t_1}^{\infty} \gamma(s)ds.$$

It follows from (1.6.15) that for any $\epsilon > 0$, we can choose sufficiently large $t_1 \geq t_0$ such that

$$CK \int_{t_1}^{\infty} \gamma(s)ds < \frac{\epsilon}{2}.$$

Since $X(t) \to 0$ as $t \to \infty$, we can then choose $t_2 \geq t_1$ such that

$$\left(\|x(t_0)\| + C \int_{t_0}^{t_1} \|X^{-1}(s)\| \gamma(s)ds \right) \|X(t)\| \leq \frac{\epsilon}{2}$$

for $t \geq t_2$. Hence, $\|x(t)\| < \epsilon$ for $t \geq t_2$, which proves that $x(t) \to 0$ as $t \to \infty$. ∎

For uniform asymptotic stability, an answer is given by the following result.

Theorem 1.6.10 *Assume that (1.6.4) is uniformly asymptotically stable, that is,*

$$\|X(t)X^{-1}(s)\| \leq Ke^{-\alpha(t-s)}, \quad t_0 \leq s \leq t < \infty,$$

for some constants K and α. Let $f(t,x)$ satisfy the inequality

$$\|f(t,x)\| \leq \gamma\|x\|,$$

where the constant $\gamma < \alpha/K$. Then for each $0 < \epsilon < M$, every solution to (1.6.13) for which

$$\|x(t_0)\| < \delta = \frac{\epsilon}{K}$$

is defined for all $t \geq t_0$ and

$$\|x(t)\| \leq Ke^{-\beta(t-s)}\|x(s)\|, \quad t_0 \leq s \leq t < \infty, \tag{1.6.18}$$

where $\beta = \alpha - \gamma K > 0$. In particular, the zero solution to (1.6.13) is uniformly asymptotically stable.

Proof: Similarly as in the proof of Theorem 1.6.9, the solution $x(t)$ to (1.6.13) also satisfies the integral equation

$$x(t) = X(t)X^{-1}(t_1)x(t_1) + \int_{t_1}^{t} X(t)X^{-1}(s)f(s,x(s))ds, \quad t \geq t_1.$$

Hence,

$$\|x(t)\| \leq Ke^{-\alpha(t-t_1)}\|x(t_1)\| + \gamma K \int_{t_1}^{t} e^{-\alpha(t-s)}\|x(s)\|ds$$

or

$$w(t) \leq Kw(t_1) + \gamma K \int_{t_1}^{t} w(s)ds,$$

where the scalar function $w(t)$ is defined by

$$w(t) = e^{\alpha(t-t_1)}\|x(t)\|.$$

Application of the Gronwall inequality implies that

$$w(t) \le Kw(t_1)e^{\gamma K(t-t_1)}, \quad t \ge t_1,$$

which is equivalent to (1.6.18). Assume that $\|x(s)\| < \delta$ and for any $0 < \epsilon < K\delta$ define $T > 0$ by $T = -\ln(\epsilon/(K\delta))/\beta$. Then it follows from (1.6.18) that

$$\|x(t)\| < Ke^{-\beta T}\delta = \epsilon \quad \text{for} \quad t \ge s + T.$$

This proves that the zero solution to (1.6.13) is uniformly asymptotically stable. ∎

Important tools in the stability theory for differential systems (1.6.2) are based on the Lyapunov functions $V(t, x)$, which satisfy the conditions:
(i) $V(t, 0) \equiv 0$,
(ii) $V(t, x) \ge a(\|x\|)$, where $a(r)$ is a continuous, monotonically increasing function, and $a(0) = 0$,
(iii) $V(t, x(t))$ is monotonically decreasing for all solutions $x(t)$ to (1.6.2) for which $x(t_0)$ is sufficiently small.

It can be proved, for example, that the existence of such a function is a sufficient condition for stability of the zero solution to (1.6.2). This theory is not reviewed here. For a good account of stability theory based on this approach, we refer the reader to the classical book by Halanay [149].

1.7 STIFF DIFFERENTIAL EQUATIONS AND SYSTEMS

In this section we discuss stiff differential equations and the phenomenon of stiffness. It is difficult to define stiffness in a mathematically rigorous manner; various more or less successful attempts at this may be found in the literature on the subject. Miranker [215] observes that stiff differential equations are equations that are ill-conditioned in a computational sense, and that they seriously defy traditional numerical methods. Shampine in Aiken [2] and Burrage [31] observe that stiff equations are problems with large $L(T - t_0)$, where L is the Lipschitz constant of the differential equation and $[t_0, T]$ is the interval of integration. Shampine and Gordon [262] state that roughly speaking, a stiff problem is one in which the solution components of interest are slowly varying but solutions with very rapidly changing components are possible. They also point out the difficulties of codes based on explicit methods to deal with such problems in an efficient manner. Similar comments are made by Dekker and Verwer [109], who observe that the essence of stiffness is that the solution to be computed is slowly varying but that perturbations exist which are rapidly damped and that the presence of such perturbations complicates the numerical computation of the slowly varying solution. This is reiterated by Butcher

[41], who points out that systems whose solutions contain rapidly decaying components are referred to as stiff differential equations. He adds that such problems are important in numerical analysis because they frequently arise in practical problems and because they are difficult to solve by traditional numerical methods. Burrage [30] observes that stiffness is a difficult concept to define since it manifests itself in so many different ways — but the crucial point is that while the solution to be computed is slowly changing, there exist perturbations that are rapidly damped but which complicate computation of the slowly changing solution. Lambert [195] points out that stiffness occurs when stability requirements rather than those of accuracy constrain the step length, and that stiffness occurs when some components of the solution decay much more rapidly than others. Then he proposes a definition that relates to what we observe in practice: If a numerical method with a finite region of absolute stability, applied to a system with any initial conditions, is forced in a certain interval of integration to use a step length that is excessively small in relation to the smoothness of the exact solution in that interval the system is said to be stiff in that interval. This definition was also adopted by Quarteroni et al. [241]. Similar observations are made by Iserles [172], who states that an ODE system is stiff if its numerical solution by some methods requires (perhaps in a portion of the solution interval) a significant depression of the step size to avoid instability, and by Moler [216], who says that a problem is stiff if the solution being sought varies slowly, but there are nearby solutions that vary rapidly, so that the numerical method must take small steps to obtain satisfactory results. Hairer and Wanner [146] point out that stiff differential equations are problems for which explicit methods don't work. Ascher and Petzold [14] observe that, loosely speaking, the initial value problem is said to be stiff if the absolute stability requirement dictates a much smaller step size than is needed to satisfy approximation requirements alone. They also remark that scientists often describe stiffness in terms of multiple time scales: If the problem has widely varying time scales, and the phenomena (or solution modes) that change on fast scales are stable, the problem is stiff. They also add that the concept of stiffness is best understood in qualitative rather than quantitative terms and that, in general, stiffness is defined in terms of the behavior of an explicit difference method. LeVeque [208] observes that the difficulty in integrating stiff systems arises from the fact that many numerical methods, including all explicit methods, are unstable in the sense of absolute stability unless the time step is small relative to the time scale of the rapid transient, which in a stiff problem is much smaller than the time scale of the solution we are trying to compute. He also adds that a stiff ODE can be characterized by the property that $f'(u)$ is much larger, in absolute value or norm, than $u'(t)$, and that the latter quantity measures the smoothness of the solution $u(t)$ being computed, while $f'(u)$ measures how rapidly f varies as we move away from this particular solution. Dormand [116] observes that the problem is said to be stiff if the eigenvalues of the Jacobian $J = f'(u)$ differ greatly and that this can present severe problems for an integrator.

As observed by Shampine and Gear [259], Dekker and Verwer [109], and Burrage [31], considerable insight into the problem of stiffness can be obtained by considering the Prothero-Robinson problem [240] as solved by the (explicit) forward Euler and the (implicit) backward Euler methods. This is a scalar problem of the form

$$y' = \lambda\big(y - F(t)\big) + F'(t), \quad t \geq t_0,$$
$$y(t_0) = y_0,$$

(1.7.1)

where $\operatorname{Re}(\lambda)$ is large and negative and $F(t)$ is a slowly varying function on the interval $[t_0, \infty)$. The exact solution to this problem is

$$y(t) = \big(y_0 - F(t_0)\big)e^{\lambda(t-t_0)} + F(t), \quad t \geq t_0.$$

(1.7.2)

Analysis of numerical approximations to (1.7.1) by forward and backward Euler methods will be aided by a linear difference equation of the first-order,

$$y_{n+1} = a y_n + \varphi_n, \quad n = 0, 1, \ldots,$$

(1.7.3)

whose exact solution is given by

$$y_n = a^n y_0 + \sum_{j=0}^{n-1} a^{n-j-1} \varphi_j, \quad n = 0, 1, \ldots.$$

(1.7.4)

Consider first the forward Euler method,

$$y_{n+1} = y_n + h g(t_n, y_n), \quad n = 0, 1, \ldots,$$

for the numerical solution of (1.1.2). The local discretization error of this method has the form

$$\frac{1}{2} h^2 y''(t) + O(h^3),$$

where $y(t)$ is the exact solution to (1.1.2). This method applied to (1.7.1) takes the form

$$y_{n+1} = (1 + h\lambda) y_n - h\big(\lambda F(t_n) - F'(t_n)\big), \quad n = 0, 1, \ldots,$$

(1.7.5)

which is of the form (1.7.3) with

$$a = 1 + h\lambda, \quad \varphi_n = -h\big(\lambda F(t_n) - F'(t_n)\big).$$

It follows from (1.7.4) that the solution to (1.7.5) is given by

$$y_n = (1 + h\lambda)^n y_0 - h \sum_{j=0}^{n-1} (1 + h\lambda)^{n-j-1}\big(\lambda F(t_j) - F'(t_j)\big),$$

(1.7.6)

$n = 0, 1, \ldots$. We analyze first the behavior of y_n as $n \to \infty$, assuming that $nh = t - t_0$ is fixed. Then

$$n = \frac{t - t_0}{h}, \quad j = \frac{t_j - t_0}{h},$$

and (1.7.6) can be written in the form

$$y_n = \left(1 + \frac{\lambda(t - t_0)}{n}\right)^n y_0 - h \sum_{j=0}^{n-1} \left((1 + h\lambda)^{1/h}\right)^{t - t_j - h} \left(\lambda F(t_j) - F'(t_j)\right),$$

$n = 0, 1, \ldots$. Routine calculations yield

$$\lim_{n \to \infty, nh = t - t_0} y_n = e^{\lambda(t - t_0)} y_0 - \int_{t_0}^{t} e^{\lambda(t-s)} \left(\lambda F(s) - F'(s)\right) ds$$

$$= e^{\lambda(t - t_0)} \left(y_0 - F(t_0)\right) + F(t),$$

which shows that y_n is convergent to the solution (1.7.2) of (1.7.1) at the point t as $n \to \infty$. A more careful analysis reveals that

$$y_n = e^{\lambda(t - t_0)} \left(y_0 - F(t_0)\right) + F(t) + O(h)$$

as $h \to 0$, which demonstrates convergence of order 1.

We analyze next the behavior of y_n given by (1.7.6) as $n \to \infty$ but the step size h is fixed. Since the function $F(t)$ is slowly varying, we can assume that

$$\lim_{t \to \infty} F(t) = \varphi, \quad \lim_{t \to \infty} F'(t) = 0. \tag{1.7.7}$$

These assumptions imply that for every $\epsilon > 0$ there exists N such that for $j > N$ we have

$$\left|F(t_j) - \varphi\right| < \frac{\epsilon}{2}, \quad \left|F'(t_j)\right| < \frac{\lambda \epsilon}{2},$$

or, equivalently,

$$F(t_j) = \varphi + \theta_{1,j} \frac{\epsilon}{2}, \quad F'(t_j) = -\theta_{2,j} \frac{\lambda \epsilon}{2}, \tag{1.7.8}$$

where $|\theta_{1,j}| \le 1$ and $|\theta_{2,j}| \le 1$ for $j > N$. For $n > N$, splitting the sum in (1.7.6) into two parts and substituting the above relations for $F(t_j)$ and $F'(t_j)$ y_n, can be written in the form

$$y_n = (1 + h\lambda)^n y_0 - h \sum_{j=0}^{N} (1 + h\lambda)^{n-j-1} \left(\lambda F(t_j) - F'(t_j)\right)$$

$$- h\lambda\varphi \sum_{j=N+1}^{n-1} (1 + h\lambda)^{n-j-1} - \frac{h\lambda\epsilon}{2} \sum_{j=N+1}^{n-1} (1 + h\lambda)^{n-j-1} (\theta_{1,j} + \theta_{2,j}).$$

Since

$$\sum_{j=N+1}^{n-1} (1+h\lambda)^{n-j-1} = \frac{(1+h\lambda)^{n-N-1}-1}{h\lambda},$$

y_n can be written in the form

$$y_n = (1+h\lambda)^n y_0 - h\sum_{j=0}^{N}(1+h\lambda)^{n-j-1}\big(\lambda F(t_j) - F'(t_j)\big)$$

$$+ \varphi\big(1 - (1+h\lambda)^{n-N-1}\big) + r_n,$$

where r_n is given by

$$r_n = -\frac{h\lambda\epsilon}{2}\sum_{j=N+1}^{n-1}(1+h\lambda)^{n-j-1}(\theta_{1,j}+\theta_{2,j}).$$

We have

$$|r_n| \le C_n\epsilon,$$

where

$$C_n = |h\lambda|\sum_{j=N+1}^{n-1}|1+h\lambda|^{n-j-1} = |h\lambda|\frac{1-|1+h\lambda|^{n-N-1}}{1-|1+h\lambda|},$$

and it follows that

$$C_n \le C := \frac{|h\lambda|}{1-|1+h\lambda|}$$

for $n > N$ and

$$\lim_{n\to\infty} y_n = \lim_{t\to\infty} y(t) = \varphi$$

only if we impose a restriction on the step size of integration h of the form

$$|1+h\lambda| < 1. \tag{1.7.9}$$

Observe that $C = 1$ if $\lambda < 0$ and $1 + h\lambda > 0$. Hence, the forward Euler method can resolve the proper behavior of the solution $y(t)$ given by (1.7.2) to problem (1.7.1) for large t only if condition (1.7.9) is satisfied. This condition is a severe restriction on the step size h if $\text{Re}(\lambda)$ is large and negative (i.e., if problem (1.7.1) is stiff). In contrast, as argued by Shampine and Gear [259], the accuracy requirement, i.e., the requirement that the principal part of the local discretization error

$$\frac{1}{2}h^2 y''(t) = \frac{1}{2}h^2\left((y_0 - F(t_0))\lambda^2 e^{\lambda(t-t_0)} + F''(t)\right)$$

is approximately equal to the given accuracy tolerance Tol, is easy to satisfy except in a small initial transient for which the term $(y_0 - F(t_0))\lambda^2 e^{\lambda(t-t_0)}$ dominates. For small $t - t_0$ the requirement $\frac{1}{2}h^2|y''(t)| = $ Tol leads to

$$h \approx \left(\frac{2\,\text{Tol}}{|(y_0 - F(t_0))\lambda^2|}\right)^{1/2} \tag{1.7.10}$$

and the step size needed for accuracy must be quite small to resolve the initial rapid change in the solution. For large $t-t_0$ the exponential term is negligible, i.e.,

$$\frac{1}{2}h^2 y''(t) \approx \frac{1}{2}h^2 F''(t),$$

and the requirement $\frac{1}{2}h^2|F''(t)| = $ Tol leads to

$$h = \left(\frac{2\,\text{Tol}}{|F''(t)|}\right)^{1/2}, \tag{1.7.11}$$

so after the small initial transient the step size becomes quite large and independent of λ. For stiff problems this step size does not satisfy condition (1.7.9).

Consider next the backward Euler method for the numerical solution of (1.7.1). This method takes the form

$$y_{n+1} = y_n + hg(t_{n+1}, y_{n+1}), \quad n = 0, 1, \dots,$$

and its local discretization error is given by

$$-\frac{1}{2}h^2 y''(t) + O(h^3).$$

Application of this method to (1.7.1) leads to the recurrence relation

$$y_{n+1} = y_n + h\left(\lambda(y_{n+1} + F(t_{n+1})) + F'(t_{n+1})\right),$$

$n = 0, 1, \dots,$ or

$$y_{n+1} = \frac{1}{1 - h\lambda}y_n - h\frac{\lambda F(t_{n+1}) - F'(t_{n+1})}{1 - h\lambda}, \quad h\lambda \neq 1. \tag{1.7.12}$$

This equation is of the form (1.7.3) with

$$a = \frac{1}{1 - h\lambda}, \quad \varphi_n = -h\frac{\lambda F(t_{n+1}) - F'(t_{n+1})}{1 - h\lambda},$$

and it follows from (1.7.4) that the solution to (1.7.12) is given by

$$y_n = \left(\frac{1}{1 - h\lambda}\right)^n y_0 - h\sum_{j=0}^{n-1}\left(\frac{1}{1 - h\lambda}\right)^{n-j}(\lambda F(t_{j+1}) - F'(t_{j+1})), \tag{1.7.13}$$

$n = 0, 1, \ldots$. Consider first the behavior of y_n given by (1.7.13) as $n \to \infty$ and $nh = t - t_0$ is held fixed. It is then easy to verify that the first term on the right-hand side of (1.7.13) tends to

$$e^{\lambda(t-t_0)}$$

and the second term on the right-hand side of (1.7.13) tends to the expression

$$-\int_{t_0}^{t} e^{\lambda(t-s)}\left(\lambda F(s) - F'(s)\right)ds.$$

Hence,

$$\lim_{n\to\infty, nh=t-t_0} y_n = e^{\lambda(t-t_0)}\left(y_0 - F(t_0)\right) + F(t),$$

and it follows that y_n is convergent to the solution (1.7.2) of (1.7.1). Moreover, similar to the case of the forward Euler method, a more careful analysis reveals that

$$y_n = e^{\lambda(t-t_0)}\left(y_0 - F(t_0)\right) + F(t) + O(h)$$

as $h \to 0$, which again demonstrates convergence of order 1.

As in the case of the forward Euler method, consider next the behavior of y_n given by (1.7.13) as $n \to \infty$ but with the step size h kept constant. We again assume (1.7.7) so that the relation (1.7.8) holds with $\theta_{1,j}$ and $\theta_{2,j}$ bounded in modulus by 1 for large j. Splitting the sum in (1.7.13) into two parts and substituting (1.7.8) into the second part, we obtain

$$y_n = \left(\frac{1}{1-h\lambda}\right)^n y_0 - h\sum_{j=0}^{N}\left(\frac{1}{1-h\lambda}\right)^{n-j}\left(\lambda F(t_{j+1}) - F'(t_{j+1})\right)$$

$$- h\lambda\varphi\sum_{j=N+1}^{n-1}\left(\frac{1}{1-h\lambda}\right)^{n-j} - \frac{h\lambda\epsilon}{2}\sum_{j=N+1}^{n-1}\left(\frac{1}{1-h\lambda}\right)^{n-j}(\theta_{1,j} + \theta_{2,j}).$$

Since

$$\sum_{j=N+1}^{n-1}\left(\frac{1}{1-h\lambda}\right)^{n-j} = \frac{1}{h\lambda}\left(\left(\frac{1}{1-h\lambda}\right)^{n-N} - 1\right),$$

y_n can be written in the form

$$y_n = \left(\frac{1}{1-h\lambda}\right)^n y_0 - h\sum_{j=0}^{N}\left(\frac{1}{1-h\lambda}\right)^{n-j}\left(\lambda F(t_{j+1}) - F'(t_{j+1})\right)$$

$$+ \varphi\left(1 - \left(\frac{1}{1-h\lambda}\right)^{n-N}\right) + r_n,$$

$n = 0, 1, \ldots,$ where

$$r_n = -\frac{h\lambda\epsilon}{2}\sum_{j=N+1}^{n-1}\left(\frac{1}{1-h\lambda}\right)^{n-j}(\theta_{1,j} + \theta_{2,j}).$$

We have
$$|r_n| \le C_n \epsilon,$$
where
$$C_n = |h\lambda| \sum_{j=N+1}^{n-1} \left(\frac{1}{1-h\lambda}\right)^{n-j} = 1 - \left(\frac{1}{1-h\lambda}\right)^{n-N} \le 1$$

for $\lambda < 0$ and $n > N$. As a consequence,
$$\lim_{n \to \infty} y_n = \lim_{t \to \infty} y(t) = \varphi;$$

that is, the backward Euler method can resolve the proper behavior of the solution (1.7.2) to (1.7.1) for large $t - t_0$ without any restrictions on the step size of integration h. Hence, the step size h can be chosen to satisfy only the accuracy requirement
$$\frac{1}{2}h^2|y''(t)| = \text{Tol},$$

which, similarly as in the case of the forward Euler method, leads to formula (1.7.10) in an initial transient when $t - t_0$ is small and to formula (1.7.11) when $t - t_0$ is large.

As explained in Section 2.6 condition (1.7.9) defines the region of absolute stability of the forward Euler method, while the condition
$$\left|\frac{1}{1-h\lambda}\right| < 1 \tag{1.7.14}$$

defines the corresponding region of absolute stability of the backward Euler method. This illustrates again that for stiff equations (i.e., equations for which $\text{Re}(\lambda)$ is large and negative) the step size h is restricted by stability (condition (1.7.9)) rather than accuracy in the case of the forward Euler method. In contrast, the step size is restricted only by accuracy in the case of the backward Euler method since stability condition (1.7.14) is satisfied automatically if $\text{Re}(h\lambda) < 0$.

We can also obtain stability conditions on the step size of numerical algorithms considering the error propagation of the forward and backward Euler methods (compare [259]). Define the global discretization error e_n at the point t_n by
$$e_n = y(t_n) - y_n,$$

where $y(t)$ is the solution to (1.7.1) and y_n is an approximation to $y(t_n)$. For the forward Euler method we have
$$y_{n+1} = y_n + h\left(\lambda(y_n - F(t_n)) + F'(t_n)\right),$$
$$y(t_{n+1}) = y(t_n) + h\left(\lambda(y(t_n) - F(t_n)) + F'(t_n)\right) + \frac{1}{2}y''(t_n) + O(h^3),$$

$n = 0, 1, \ldots$, and subtracting these equations, we obtain the error equation,

$$e_{n+1} = (1 + h\lambda)e_n + \frac{1}{2}h^2 y''(t_n) + O(h^3). \tag{1.7.15}$$

It follows from this relation that the error e_{n+1} at the next grid point $t_{n+1} = t_n + h$ consists of the error $(1 + h\lambda)e_n$ propagated from the previous step t_n and the local discretization error $\frac{1}{2}h^2 y''(t_n) + O(h^3)$ at the point t_{n+1}. The propagated error is damped whenever $|1+h\lambda| < 1$, which leads to the stability condition (1.7.9) for the forward Euler method. In contrast, for the backward Euler method, we have

$$y_{n+1} = y_n + h\Big(\lambda\big(y_{n+1} - F(t_{n+1})\big) + F'(t_{n+1})\Big),$$

$$y(t_{n+1}) = y(t_n) + h\Big(\lambda\big(y(t_{n+1}) - F(t_{n+1})\big) + F'(t_{n+1})\Big) - \frac{1}{2}h^2 y''(t_n) + O(h^3),$$

and the error equation takes the form

$$e_{n+1} = \frac{1}{1 - h\lambda}e_n - \frac{1}{2(1 - h\lambda)}h^2 y''(t_n) + O(h^3), \tag{1.7.16}$$

$n = 0, 1, \ldots$. Hence, it follows that the propagated error is damped whenever $1/|1-h\lambda| < 1$, which leads to the stability condition (1.7.14) for the backward Euler method. As already observed, this condition is satisfied automatically if $\mathrm{Re}(\lambda) < 0$, so there are no stability limitations on the method and the step size is restricted only by accuracy requirements.

Additional insight into the phenomenon of stiffness can be obtained by considering a family of local problems

$$\begin{aligned} y' &= \lambda\big(y - F(t)\big) + F'(t), \quad t \in [t_n, t_{n+1}], \\ y(t_n) &= y_n = F(t_n) + \delta_n, \end{aligned} \tag{1.7.17}$$

$n = 0, 1, \ldots$, defined on the subintervals $[t_n, t_{n+1}]$, whose solutions $y_{\mathrm{loc}}(t)$ are given by

$$y_{\mathrm{loc}}(t) = \delta_n e^{\lambda(t-t_n)} + F(t), \quad t \in [t_n, t_{n+1}]. \tag{1.7.18}$$

These solutions are shown by thin lines in Fig. 1.7.1. This figure corresponds to $t_n = t_0 + nh$, $h = 1/2$, $\delta_n = (-1)^n$, $n = 1, 2, \ldots, 19$, $\lambda = -10$, and the slowly varying function $F(t)$ defined by

$$F(t) = 2e^{-t/4} \cos(t).$$

We also show by a thick line the global solution to (1.7.1) corresponding to the initial condition $y(t_0) = y_0 = -1$.

We can observe that the perturbation δ_n in (1.7.18) is rapidly damped by the negative exponential $e^{\lambda(t-t_n)}$, and the local solutions $y_{\mathrm{loc}}(t)$ to (1.7.17) are rapidly convergent to the global solution $y(t)$ of (1.7.1), which is approximately

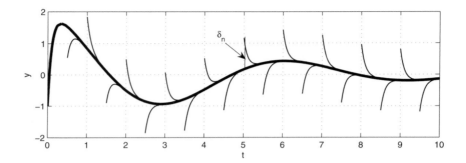

Figure 1.7.1   Global and local solutions to the Prothero-Robinson problem (1.7.1) for $t \in [0, 10]$, $y(0) = -1$, $F(t) = 2e^{-t/4} \cos(t)$, and $\lambda = -10$

equal to $F(t)$ for stiff problems, except for a short initial transient. Since the forward Euler method follows the slope $y'_{\mathrm{loc}}(t_n) = \lambda \delta_n + F'(t_n)$ of the local solution at the point t_n, any small perturbation δ_n is greatly amplified by λ if the problem is stiff. As a result, $y'_{\mathrm{loc}}(t_n)$ is a poor approximation to $y'(t_n) \approx F'(t_n)$, which leads to instability of the forward Euler method if the step size is not drastically reduced. The perturbation $\delta_{n+1} = y_{n+1} - F(t_{n+1})$ propagates according to the formula

$$\delta_{n+1} = (1 + h\lambda)\delta_n + F(t_n) + hF'(t_n) - F(t_{n+1}),$$

which resembles the error equation (1.7.15), and we again obtain the stability restriction (1.7.9). In contrast, the backward Euler method follows the slope $y'_{\mathrm{loc}}(t_{n+1}) = \lambda \delta_n e^{\lambda h} + F'(t_{n+1})$ of the local solution at t_{n+1}, which is a much better approximation to $y'(t_{n+1}) \approx F'(t_{n+1})$ because $\lambda \delta_n$ is rapidly damped by the exponential $e^{\lambda h}$, especially if the step size is large. This is the case once the initial transient is resolved and the step size is selected according to (1.7.11). The perturbations propagate according to the formula

$$\delta_{n+1} = \frac{1}{1 - h\lambda}\delta_n + \frac{1}{1 - h\lambda}\left(F(t_n) + hF'(t_{n+1}) - F(t_{n+1})\right),$$

which resembles error equation (1.7.16), and we again obtain the condition (1.7.14), which does not lead to step size restrictions.

Following Dekker and Verwer [109], we consider next the linear system of differential equations

$$y' = Ay, \qquad t \geq t_0,$$
$$y(t_0) = y_0,$$

(1.7.19)

with matrix A given by

$$
A = \begin{bmatrix} d & \dfrac{1}{\epsilon} \\ 0 & -\dfrac{1}{\epsilon} \end{bmatrix},
$$

where $|d| < 1$ is a moderate constant, $\epsilon > 0$ is close to zero, and $d\epsilon + 1 \neq 0$. The solution to this problem is

$$
y(t) = \begin{bmatrix} e^{d(t-t_0)} & \dfrac{e^{d(t-t_0)} - e^{-(t-t_0)/\epsilon}}{d\epsilon + 1} \\ 0 & e^{-(t-t_0)/\epsilon} \end{bmatrix} y_0, \quad t \geq t_0.
$$

This solution has a rapidly varying component $e^{-(t-t_0)/\epsilon}$, which dies out after a short initial transient, and as a result the smooth component $e^{d(t-t_0)}$ dominates the solution for large $t - t_0$. This leads to stiffness. However, the problem is not called stiff in the transient phase, where $e^{-(t-t_0)/\epsilon}$ is still active. In this phase it is then natural to use small step sizes to resolve rapidly changing components.

We also have the following relationship between the solution $y(t)$ to (1.7.19) at the points t_n and $t_{n+1} = t_n + h$:

$$
y(t_{n+1}) = X y(t_n),
$$

where

$$
X = \begin{bmatrix} e^{dh} & \dfrac{e^{dh} - e^{-h/\epsilon}}{d\epsilon + 1} \\ 0 & e^{-h/\epsilon} \end{bmatrix}.
$$

It can be verified that matrix X can be diagonalized by a similarity transformation PXP^{-1}, where the matrices P and P^{-1} are

$$
P = \begin{bmatrix} 1 & -\dfrac{1}{d\epsilon + 1} \\ 0 & 1 \end{bmatrix}, \quad P^{-1} = \begin{bmatrix} 1 & \dfrac{1}{d\epsilon + 1} \\ 0 & 1 \end{bmatrix}. \tag{1.7.20}
$$

Hence, it follows from [228, Theorem 1.3.8] that there exists a matrix norm $\|\cdot\|$ such that $\rho(X) = \|X\|$, where $\rho(X)$ is the spectral radius of X. Denote by the same symbol, $\|\cdot\|$, a consistent vector norm (i.e., a norm such that $\|Xv\| \leq \|X\|\|v\|$). The existence of such norm was demonstrated by, for example, Householder [163, p. 42]. Hence,

$$
\|y(t_{n+1})\| \leq \max\{e^{dh}, e^{-h/\epsilon}\}\|y(t_n)\| = e^{dh}\|y(t_n)\|.
$$

It follows from this relation that the norm $\|y(t)\|$ of the solution $y(t)$ to (1.7.19) can be amplified over the interval $[t_n, t_{n+1}]$ by a factor e^{dh} if $d > 0$, and it is damped if $d < 0$.

We now consider approximations to (1.7.9) computed by the forward and the backward Euler methods. The forward Euler method takes the form

$$y_{n+1} = X_{FE}y_n, \quad n = 0, 1, \ldots,$$

where

$$X_{FE} = \begin{bmatrix} 1 + hd & \dfrac{h}{\epsilon} \\ 0 & 1 - \dfrac{h}{\epsilon} \end{bmatrix}.$$

We also have

$$y(t_{n+1}) = X_{FE}y(t_n) + \frac{1}{2}h^2 y''(t_n) + O(h^3),$$

and subtracting these relations we obtain the error equation,

$$e_{n+1} = X_{FE}e_n + \frac{1}{2}h^2 y''(t_n) + O(h^3).$$

It can be verified that the matrix X_{FE} can be diagonalized by the same similarity transformation $PX_{FE}P^{-1}$ as the matrix X with P and P^{-1} given by (1.7.20). Hence, there exists a matrix norm $\|\cdot\|$ such that $\rho(X_{FE}) = \|X_{FE}\|$ and the corresponding consistent vector norm $\|\cdot\|$, and we have

$$\|e_{n+1}\| \leq \max\left\{|1 + hd|, \left|1 - \frac{h}{\epsilon}\right|\right\}\|e_n\| + \frac{1}{2}h^2\|y''(t_n)\| + O(h^3).$$

If $d > 0$, the norm $\|e_n\|$ of the error should grow no faster than the norm of the solution $\|y(t_n)\|$, which grows at most by a factor $e^{dh} = 1 + hd + O(h^2)$. This leads to the condition

$$\left|1 - \frac{h}{\epsilon}\right| < 1 + hd,$$

and we obtain the step size restriction

$$h < \frac{2\epsilon}{1 - \epsilon d} < \frac{2}{1 - \epsilon_0 d}\epsilon,$$

for $\epsilon < \epsilon_0$, where $\epsilon_0 > 0$ is a constant. If $d < 0$, the norm $\|e_n\|$ of the error should be damped. This leads to the conditions

$$|1 + hd| < 1 \quad \text{and} \quad \left|1 - \frac{1}{\epsilon}\right| < 1$$

and we obtain the step size restrictions

$$h < -\frac{2}{d} \quad \text{and} \quad h < 2\epsilon.$$

In either case, the step size must be restricted by some constant times ϵ and the forward Euler method becomes very inefficient once the initial transient is resolved.

Consider next the approximation to (1.7.19) by the the backward Euler method. This leads to the recurrence relation

$$y_{n+1} = X_{BE}y_n, \quad n = 0, 1, \ldots,$$

with the matrix X_{BE} given by

$$X_{BE} = \begin{bmatrix} \dfrac{1}{1-dh} & \dfrac{h}{(1-dh)(\epsilon+h)} \\ 0 & \dfrac{\epsilon}{\epsilon+h} \end{bmatrix}.$$

We also have

$$y(t_{n+1}) = X_{BE}y(t_n) - \frac{1}{2}h^2 y''(t_n) + O(h^3),$$

and the error equation takes the form

$$e_{n+1} = X_{BE}e_n - \frac{1}{2}h^2 y''(t_n) + O(h^3).$$

As before, it can be verified that X_{BE} can be diagonalized by a similarity transformation $PX_{BE}P^{-1}$, where P and P^{-1} are defined by (1.7.20). Hence,

$$\|e_{n+1}\| \leq \max\left\{\left|\frac{1}{1-hd}\right|, \left|\frac{\epsilon}{\epsilon+h}\right|\right\}\|e_n\| + \frac{1}{2}h^2\|y''(t_n)\| + O(h^3),$$

where we have again chosen a matrix norm $\|\cdot\|$ such that $\rho(X_{BE}) = \|X_{BE}\|$ and the corresponding consistent vector norm $\|\cdot\|$. If $d > 0$ and $h < 1/d$, then for any $\epsilon > 0$ the norm of the error $\|e_n\|$ can grow at most by a factor $1/(1-hd)$, which is consistent with the growth factor $e^{dh} = 1/(1-hd) + O(h^2)$ of the solution. If $d < 0$, the norm of the error $\|e_n\|$ is always damped for any ϵ. Hence, in either case, there are no restrictions on the step size, which depend on the stiffness parameter ϵ even if we admit infinite stiffness as $\epsilon \to \infty$, and the step size of the backward Euler method can be chosen only according to accuracy requirements.

We consider next the initial value problem for linear systems of differential equations

$$\begin{aligned} y' &= Ay + \varphi(t), \quad t \geq t_0, \\ y(t_0) &= y_0 \in \mathbb{R}^m, \end{aligned} \tag{1.7.21}$$

where $A \in \mathbb{R}^{m \times m}$ is a constant matrix and $\varphi(t)$ is a time-dependent forcing term. Assuming that A has m distinct eigenvalues $\lambda_1, \lambda_2, \ldots, \lambda_m$ with corresponding eigenvectors $v_1, v_2, \ldots, v_m$, the general solution to (1.7.21) has the form

$$y(t) = \sum_{i=1}^{m} c_i v_i e^{\lambda_i (t-t_0)} + \psi(t), \quad t \geq t_0, \tag{1.7.22}$$

where the c_i are arbitrary constants and $\psi(t)$ is a particular solution. If $\mathrm{Re}(\lambda_i)$ is large in absolute value and negative, the corresponding exponential term $c_i v_i e^{\lambda_i (t-t_0)}$ in (1.7.22) is rapidly damped which indicates stiffness. This motivates to characterize the stiffness of (1.7.21) in terms of the conditions imposed on the spectrum of the matrix A. For example, Dekker and Verwer [109] call the problem (1.7.21) stiff if the following conditions are satisfied:

1. There exist eigenvalues λ_i of A for which $\mathrm{Re}(\lambda_i)$ is negative and large in absolute value.

2. There exist λ_i of moderate size (i.e., $|\lambda_i|$ is small compared with the modulus of eigenvalues satisfying 1).

3. There are no eigenvalues for which $\mathrm{Re}(\lambda_i)$ is large and positive.

4. There are no eigenvalues for which $\mathrm{Im}(\lambda_i)$ is large unless $\mathrm{Re}(\lambda_i)$ is also large in absolute value and negative.

It is also customary to define the stiffness ratio r_s as

$$r_s := \frac{\max\limits_{i=1,2,\ldots,m} |\lambda_i|}{\min\limits_{i=1,2,\ldots,m} |\lambda_i|}. \tag{1.7.23}$$

Somewhat different definitions of stiffness which appear in the literature may require that $\mathrm{Re}(\lambda_i) < 0$ for all i (compare [116, 195, 241]).

Condition 4 in the Dekker and Verwer [109] definition was added to exclude highly oscillatory problems. Such problems require special treatment, and their numerical solution has been discussed in, e.g., [142, 173, 174, 203, 215].

1.8   EXAMPLES OF STIFF DIFFERENTIAL EQUATIONS AND SYSTEMS

In this section we have listed some examples of stiff differential systems that are used in a numerical analysis community to test and compare various codes for such equations. The scalar stiff differential equation (1.7.1) due to Prothero and Robinson [240] was already discussed in Section 1.7 to provide a simple introduction to the phenomenon of stiffness. What follows are examples of stiff differential systems of small and moderate dimension taken from the monograph by Hairer and Wanner [146].

VDPOL — the van der Pohl equation [146]. This corresponds to the differential equation of the second order which describes oscillations in an electrical circuit. This equation, written as a first-order system, takes the form

$$y_1' = y_2, \qquad\qquad\qquad y_1(0) = 2,$$
$$y_2' = \left((1 - y_1^2)y_2 - y_1\right)/\epsilon, \quad y_2(0) = 0,$$

$$(1.8.1)$$

$t \in [0, t_{\text{out}}]$, $t_{\text{out}} = 1, 2, 3, \ldots, 11$. Here ϵ corresponds to the stiffness parameter and the typical value is $\epsilon = 10^{-6}$.

ROBER — a system describing the chemical reactions of Robertson [247]. This is a system of three differential equations of the form

$$y_1' = -0.04y_1 + 10^4 y_2 y_3, \qquad\qquad y_1(0) = 1,$$
$$y_2' = 0.04y_1 - 10^4 y_2 y_3 - 3 \cdot 10^7 y_2^2, \quad y_2(0) = 0, \qquad (1.8.2)$$
$$y_3' = 3 \cdot 10^7 y_2^2, \qquad\qquad\qquad y_3(0) = 0,$$

where the interval of integration is usually chosen as $t \in [0, t_{\text{out}}]$, $t_{\text{out}} = 40$. However, to test the reliability of codes for stiff equations, this system is also solved for $t_{\text{out}} = 1, 10, 10^2, \ldots, 10^{11}$, and less reliable codes are unable to reach the end of the interval of integration, due to the overflow to $-\infty$ of the component y_2.

OREGO — the Oregonator. This is a system of three ordinary differential equations describing the Belusov-Zhabotinskii reaction [125, 128]. This system takes the form

$$y_1' = 77.27\left(y_2 + y_1(1 - 8.375 \cdot 10^{-6} y_1 - y_2)\right), \quad y_1(0) = 1,$$
$$y_2' = \frac{1}{77.27}\left(y_3 - (1 + y_1)y_2\right), \qquad\qquad\qquad y_2(0) = 2, \qquad (1.8.3)$$
$$y_3' = 0.161(y_1 - y_2), \qquad\qquad\qquad\qquad y_3(0) = 3,$$

$t \in [0, t_{\text{out}}]$, where $t_{\text{out}} = 30, 60, \ldots, 360$.

HIRES — a chemical reaction that describes the growth and differentiation of plant tissue independent of photosynthesis at high levels of irradiance by light. This system of eight equations proposed by Schäfer [248] takes the form

$$
\begin{aligned}
y_1' &= -1.71\,y_1 + 0.43\,y_2 + 8.32\,y_3 + 0.0007, & y_1(0) &= 1, \\
y_2' &= 1.71\,y_1 - 8.75\,y_2, & y_2(0) &= 0, \\
y_3' &= -10.03\,y_3 + 0.43\,y_4 + 0.035\,y_5, & y_3(0) &= 0, \\
y_4' &= 8.32\,y_2 + 1.71\,y_3 - 1.12\,y_4, & y_4(0) &= 0, \\
y_5' &= -1.745\,y_5 + 0.43\,y_6 + 0.43\,y_7, & y_5(0) &= 0, & (1.8.4) \\
y_6' &= -280\,y_6\,y_8 + 0.69\,y_4 + 1.71\,y_5 \\
&\quad - 0.43\,y_6 + 0.69\,y_7, & y_6(0) &= 0, \\
y_7' &= 280\,y_6\,y_8 - 1.81\,y_7, & y_7(0) &= 0, \\
y_8' &= -280\,y_6\,y_8 + 1.81\,y_7, & y_8(0) &= 0.0057,
\end{aligned}
$$

$t \in [0, t_{\text{out}}]$, where $t_{\text{out}} = 321.8122$ and $t_{\text{out}} = 421.8122$.

PLATE — a boundary value problem for the partial differential equation that describes the movement of a rectangular plate under the load of a car passing across it [146]. This problem is defined by

$$
\begin{aligned}
\frac{\partial^2 u}{\partial t^2} + \omega \frac{\partial u}{\partial t} + \sigma \Delta \Delta u &= f(x, y, t), & (x, y) &\in \Omega, & t &\in [0, t_{\text{out}}], \\
u\Big|_{\partial \Omega} = 0, \quad \Delta u\Big|_{\partial \Omega} &= 0, & t &\in [0, t_{\text{out}}], & & (1.8.5) \\
u(x, y, 0) = 0, \quad \frac{\partial u}{\partial t}(x, y, 0) &= 0, & (x, y) &\in \Omega,
\end{aligned}
$$

$t_{\text{out}} = 7$, where

$$
\Omega = \left\{ (x, y) : \ 0 \leq x \leq 2, \ 0 \leq y \leq \frac{4}{3} \right\}.
$$

Following Hairer and Wanner [146] we consider the load $f(x, y, t)$, which is idealized by the sum of two Gaussian curves that move in the x-direction and reside on four wheels. This load takes the form

$$
f(x, y, t) =
\begin{cases}
200\left(e^{-5(t-x-2)^2} + e^{-5(t-x-5)^2}\right) & \text{if } y = y_2 \text{ or } y_4, \\
0 & \text{for all other } y.
\end{cases}
$$

This problem is discretized on a grid of 8×5 interior points $x_i = ih$, $y_j = jh$, $h = 2/9$, where $\partial^2 u / \partial t^2$ is approximated by finite differences of the second order, $\partial u / \partial t$ by forward difference, and the plate operator $\Delta \Delta$ is discretized by the standard computational molecule, defined by

$$
\begin{array}{ccccc}
 & & 1 & & \\
 & 2 & -8 & 2 & \\
1 & -8 & 20 & -8 & 1 \\
 & 2 & -8 & 2 & \\
 & & 1 & &
\end{array} \ .
$$

The friction and the stiffness parameters are $\omega = 1000$ and $\sigma = 100$. This leads to a linear and nonautonomous example of medium stiffness and medium size. The resulting system is then of dimension 80 with negative real as well as complex eigenvalues λ that satisfy $-500 \le \mathrm{Re}(\lambda) < 0$.

BEAM — a problem that describes the discretization of a stiff elastic beam clamped at arc length $s = 0$ with external force acting at the free end $s = 1$. As explained by Hairer and Wanner [146], this results in a system of a second-order differential equations for the angles $\theta_l = \theta_l(t)$ between the tangents to the beam and vertical axis at discrete arcs lengths s_l, similarly to the ROPE problem described in Section 1.2. This system takes the form

$$
\sum_{k=1}^{n} a_{kl}\ddot{\theta}_k \;=\; -\sum_{k=1}^{n} b_{lk}\dot{\theta}_k^2 \;+\; n^4\left(\theta_{l-1} - 2\theta_l + \theta_{l+1}\right)
$$

$$
+ \; n^2\left(\cos(\theta_l)F_y - \sin(\theta_l)F_x\right),
\tag{1.8.6}
$$

$$
\theta_0 = -\theta_1, \qquad \theta_{n+1} = \theta_n,
$$

$t \in [0, T]$, $l = 1, 2, \ldots, n$, where as in the ROPE problem

$$
a_{lk} = g_{lk}\cos(\theta_l - \theta_k), \quad b_{lk} = g_{lk}\sin(\theta_l - \theta_k), \quad g_{lk} = n + \frac{1}{2} - \max\{l, k\},
$$

and the external forces $F = (F_x, F_y)$ acting at the free end $s = 1$ take the form

$$
F_x = -\varphi(t), \quad F_y = \varphi(t), \quad \varphi(t) = \begin{cases} 1.5\sin^2(t), & 0 \le t \le \pi, \\ 0, & t \ge \pi. \end{cases}
$$

This system will be solved for $n = 40$ with initial conditions

$$
\theta_l(0) = \dot{\theta}_l(0) = 0, \quad l = 1, 2, \ldots, n,
$$

on the interval $[0, 5]$. As for the ROPE problem, setting

$$
A = \begin{bmatrix} a_{lk} \end{bmatrix}, \quad B = \begin{bmatrix} b_{lk} \end{bmatrix}, \quad \theta = \begin{bmatrix} \theta_1 & \theta_2 & \cdots & \theta_n \end{bmatrix}^T,
$$

$$\dot{\theta} = \begin{bmatrix} \dot{\theta}_1 & \dot{\theta}_2 & \cdots & \dot{\theta}_n \end{bmatrix}^T, \quad \ddot{\theta} = \begin{bmatrix} \ddot{\theta}_1 & \ddot{\theta}_2 & \cdots & \ddot{\theta}_n, \end{bmatrix}^T,$$

system (1.8.6) can be written in vector form as

$$A\ddot{\theta} = -B\dot{\theta}^2 + v(t,\theta), \quad t \in [0, T],$$

$$\theta(0) = \dot{\theta}(0) = 0,$$

(1.8.7)

where $\dot{\theta}^2$ denotes componentwise exponentiation and

$$v(t,\theta) = v = \begin{bmatrix} v_1 & v_2 & \cdots & v_n \end{bmatrix}^T$$

is defined by

$$v_l = n^4\left(\theta_{l-1} - 2\theta_l + \theta_{l+1}\right) + n^2\left(\cos(\theta_l)F_y - \sin(\theta_l)F_x\right),$$

$l = 1, 2, \ldots, n$, with the relations $\theta_0 = -\theta_1$, $\theta_{n+1} = \theta_n$ (compare (1.8.6)). The eigenvalues of the Jacobian matrix for this problem vary between $-6400i$ and $6400i$ (compare [146]).

The acceleration vector $\ddot{\theta}$ can be computed by the algorithm already described in Section 1.2 for the PLATE problem. This algorithm takes the following form:

1. Compute $w = Dv + \dot{\theta}^2$.

2. Solve the tridiagonal system $Cu = w$.

3. Compute $\ddot{\theta} = Cv + Du$.

Here matrices C and D are defined in the description of the PLATE problem in Section 1.2.

HEAT — a system of differential equations resulting from discretization of the boundary-value problem for the heat equation in one dimension. This problem takes the form

$$\frac{\partial u}{\partial t}(x,t) = \sigma\frac{\partial^2 u}{\partial t^2}(x,t), \quad 0 \le x \le 1, \quad t \ge 0,$$

$$u(0,t) = u(1,t) = 0, \quad t \ge 0,$$

(1.8.8)

$$u(0,x) = g(x), \quad 0 \le x \le 1,$$

where $\sigma > 0$ is a given constant and $g(x)$ is a given initial function.

Setting $x_i = i\Delta x$, $i = 0, 1, \ldots, N+1$, where $(N+1)\Delta x = 1$ and approximating $\partial^2 u/\partial t^2$ at the point x_i by the finite difference of second order,

$$\frac{u(x_{i+1},t) - 2u(x_i,t) + u(x_{i-1},t)}{(\Delta x)^2},$$

leads to the initial value problem for the system of differential equations of the form

$$y_i'(t) = \frac{\sigma}{(\Delta x)^2}\Big(y_{i+1}(t) - 2y_i(t) + y_{i-1}(t)\Big),$$

$$(1.8.9)$$

$$y_i(0) = g(x_i),$$

$i = 1, 2, \ldots, N$, where $y_0(t) = 0$, $y_{N+1}(t) = 0$, $y_i(t) \approx u(x_i, t)$, $i = 1, 2, \ldots, N$. Introducing the notation

$$y = \begin{bmatrix} y_1 & y_2 & \cdots & y_N \end{bmatrix}^T, \quad y_0 = \begin{bmatrix} g(x_1) & g(x_2) & \cdots & g(x_N) \end{bmatrix}^T,$$

and

$$Q_N = \frac{\sigma}{(\Delta x)^2}\widetilde{Q}_N,$$

where

$$\widetilde{Q}_m = \begin{bmatrix} 2 & -1 & 0 & \cdots & 0 & 0 \\ -1 & 2 & -1 & \cdots & 0 & 0 \\ 0 & -1 & 2 & \cdots & 0 & 0 \\ \vdots & \vdots & \vdots & \ddots & \vdots & \vdots \\ 0 & 0 & 0 & \cdots & 2 & -1 \\ 0 & 0 & 0 & \cdots & -1 & 2 \end{bmatrix} \in \mathbb{R}^{m \times m}, \qquad (1.8.10)$$

problem (1.8.9) can be written in vector form as

$$y'(t) + Q_N y(t) = 0, \quad t \geq 0,$$

$$(1.8.11)$$

$$y(0) = y_0.$$

Problems of this type also describe the interconnected resistor-capacitor line (see [202, 204]).

To investigate the stiffness of (1.8.11), we compute first the eigenvalues of the matrix $\widetilde{Q}_N$. Set

$$P_m(\xi) = \det\left(\xi I_m - \widetilde{Q}_m\right),$$

where I_m is identity matrix of dimension m and $\widetilde{Q}_m$ is defined by (1.8.10), and define $P_0(\xi) = 1$. Then $P_1(\xi) = \xi - 2$, and expanding $\det(\xi I_m - \widetilde{Q}_m)$ with respect to the first row it can be verified that the polynomials $P_m(\xi)$ satisfy the triple recurrence relation

$$P_m(\xi) - (\xi - 2)P_{m-1}(\xi) + P_{m-2}(\xi) = 0, \quad m \geq 2.$$

Set $\bar{\xi} = \xi - 2$ and $\overline{P}_m(\bar{\xi}) = \overline{P}_m(\xi - 2) = P_m(\xi)$. Then this recurrence relation can be written as

$$\overline{P}_m(\bar{\xi}) - \bar{\xi}\,\overline{P}_{m-1}(\bar{\xi}) + \overline{P}_{m-2}(\bar{\xi}) = 0, \quad m \geq 2,$$

$$\overline{P}_0(\bar{\xi}) = 1, \quad \overline{P}_1(\bar{\xi}) = \bar{\xi}. \tag{1.8.12}$$

The characteristic polynomial of (1.8.12) takes the form

$$r^2 - \bar{\xi}\,r + 1 = 0$$

and has real roots

$$r_1 = \frac{\bar{\xi} - \sqrt{\bar{\xi}^2 - 4}}{2}, \quad r_2 = \frac{\bar{\xi} + \sqrt{\bar{\xi}^2 - 4}}{2}$$

for $-2 < \bar{\xi} < 2$. It follows that

$$\overline{P}_m(\bar{\xi}) = A\,r_1^m + B\,r_2^m,$$

where the constants A and B satisfy the system of equations corresponding to the initial conditions in (1.8.12). This system takes the form

$$\overline{P}_0(\bar{\xi}) = A + B = 1,$$

$$\overline{P}_1(\bar{\xi}) = A\,r_1 + B\,r_2 = \bar{\xi},$$

and its solution is

$$A = \frac{-\bar{\xi} + \sqrt{\bar{\xi}^2 - 4}}{2\sqrt{\bar{\xi}^2 - 4}}, \quad B = \frac{\bar{\xi} + \sqrt{\bar{\xi}^2 - 4}}{2\sqrt{\bar{\xi}^2 - 4}}.$$

Substituting A, B, r_1, and r_2 into $\overline{P}_m(\bar{\xi})$, after some computations we obtain

$$\overline{P}_m(\bar{\xi}) = \frac{1}{\sqrt{\bar{\xi}^2 - 4}} \left(\frac{\bar{\xi} - \sqrt{\bar{\xi}^2 - 4}}{2}\right)^{m+1} \left(\left(\frac{\bar{\xi} + \sqrt{\bar{\xi}^2 - 4}}{\bar{\xi} - \sqrt{\bar{\xi}^2 - 4}}\right)^{m+1} - 1\right).$$

Since $\bar{\xi} - \sqrt{\bar{\xi}^2 - 4} \neq 0$ for $-2 < \bar{\xi} < 2$, it follows that the roots of $\overline{P}_m(\bar{\xi})$ satisfy the relation

$$\frac{\bar{\xi} + \sqrt{\bar{\xi}^2 - 4}}{\bar{\xi} - \sqrt{\bar{\xi}^2 - 4}} = \exp\left(\frac{2\pi k i}{m + 1}\right),$$

$k = 0, 1, \ldots, m$. But $k = 0$ implies that $\sqrt{\bar{\xi}^2 - 4} = 0$, which must be ruled out. Put

$$\theta_k = \exp\left(\frac{2\pi k i}{m+1}\right), \quad k = 1, 2, \ldots, m.$$

Then

$$\sqrt{\bar{\xi}^2 - 4}\,(1 + \theta_k) = \bar{\xi}\,(\theta_k - 1)$$

or

$$\left(\bar{\xi}^2 - 4\right)\left(1 + \theta_k\right)^2 = \bar{\xi}^2\,(\theta_k - 1)^2.$$

Solving this equation with respect to $\bar{\xi}$, we obtain

$$\bar{\xi}^2 = \frac{(1 + \theta_k)^2}{\theta_k}$$

or

$$\bar{\xi}^2 = \frac{\left(1 + \cos(\varphi_k) + i\sin(\varphi_k)\right)^2}{\cos(\varphi_k) + i\sin(\varphi_k)},$$

where $\varphi_k = 2k\pi/(m+1)$. This relation can be simplified to

$$\bar{\xi}^2 = 4\cos^2\left(\frac{\varphi_k}{2}\right),$$

and it follows that the roots of $\overline{P}_m(\bar{\xi})$ are given by

$$\bar{\xi}_k = 2\cos\left(\frac{\pi k}{m+1}\right), \quad k = 1, 2, \ldots, m.$$

Hence, the roots of $P_m(\xi)$ are

$$\xi_k = 2\left(1 - \cos\left(\frac{\pi k}{m+1}\right)\right), \quad k = 1, 2, \ldots, m,$$

and since $Q_N = (\sigma/(\Delta x)^2)\widetilde{Q}_N$, the eigenvalues λ_k of the matrix $-Q_N$ take the form

$$\lambda_k = \frac{2\sigma}{(\Delta x)^2}\left(\cos\left(\frac{\pi k}{N+1}\right) - 1\right), \quad k = 1, 2, \ldots, N.$$

Since $\Delta x = 1/(N+1)$, we have

$$\lambda_1 = 2\sigma(N+1)^2\left(\cos\left(\frac{\pi}{N+1}\right) - 1\right) = -\sigma\pi^2 \frac{\sin^2\left(\frac{\pi}{2(N+1)}\right)}{\left(\frac{\pi}{2(N+1)}\right)^2} \approx -\sigma\pi^2,$$

and

$$\lambda_N = 2\sigma(N+1)^2\left(\cos\left(\frac{\pi N}{N+1}\right) - 1\right) \approx -4\sigma(N+1)^2$$

for large N. Hence, the stiffness ratio r_s defined by $(1.7.23)$ is

$$r_s = \left| \frac{\lambda_N}{\lambda_1} \right| \approx \frac{4}{\pi^2}(N+1)^2,$$

and problem $(1.8.11)$ is becoming increasingly stiff for large values of N, which determines the accuracy in space variable x.

CHAPTER 2

INTRODUCTION TO GENERAL LINEAR METHODS

2.1 REPRESENTATION OF GENERAL LINEAR METHODS

To motivate the class of general linear methods (GLMs), we quote after Butcher [40]: "Following the advice of Aristotle, we look for the greatest good as a mean between extremes. Of the various methods devised as generalizations of the classical method of Euler, two extreme approaches are usually followed. One is to generalize the Euler method through the use of multistep methods; the other is to increase the complexity of one-step methods as in the Runge-Kutta methods. General linear methods are introduced as a middle ground between these types of generalization."

Consider the initial value problem for an autonomous system of differential equations, which we write in this chapter in the form

$$y'(t) = f\big(y(t)\big), \quad t \in [t_0, T],$$
$$y(t_0) = y_0,$$

(2.1.1)

$f : \mathbb{R}^m \to \mathbb{R}^m$, $y_0 \in \mathbb{R}^m$. GLMs for (2.1.1) can be represented by the abscissa vector $\mathbf{c} = [c_1, \ldots, c_s]^T$, and four coefficient matrices $\mathbf{A} = [a_{ij}]$, $\mathbf{U} = [u_{ij}]$,

General Linear Methods for Ordinary Differential Equations. By Zdzisław Jackiewicz **59**
Copyright © 2009 John Wiley & Sons, Inc.

$\mathbf{B} = [b_{ij}]$, and $\mathbf{V} = [v_{ij}]$, where

$$\mathbf{A} \in \mathbb{R}^{s \times s}, \quad \mathbf{U} \in \mathbb{R}^{s \times r}, \quad \mathbf{B} \in \mathbb{R}^{r \times s}, \quad \mathbf{V} \in \mathbb{R}^{r \times r}.$$

On the uniform grid $t_n = t_0 + nh$, $n = 0, 1, \ldots, N$, $Nh = T - t_0$, these methods take the form

$$
\begin{aligned}
Y_i^{[n]} &= h \sum_{j=1}^{s} a_{ij} f(Y_j^{[n]}) + \sum_{j=1}^{r} u_{ij} y_j^{[n-1]}, \quad i = 1, 2, \ldots, s, \\
y_i^{[n]} &= h \sum_{j=1}^{s} b_{ij} f(Y_j^{[n]}) + \sum_{j=1}^{r} v_{ij} y_j^{[n-1]}, \quad i = 1, 2, \ldots, r,
\end{aligned}
\tag{2.1.2}
$$

$n = 0, 1, \ldots, N$, where s is the number of internal stages and r is the number of external stages, which propagate from step to step. Here, h is a step size, $Y_i^{[n]}$ is an approximation (possibly of low order) to $y(t_{n-1} + c_i h)$, and $y_i^{[n]}$ is an approximation to the linear combination of the derivatives of y at the point t_n. This will be made more precise later. As discussed by Butcher and Burrage [33, 41, 52], method (2.1.2) can be represented conveniently by the abscissa vector $\mathbf{c}$ and a partitioned $(s + r) \times (s + r)$ matrix

$$\left[\begin{array}{c|c} \mathbf{A} & \mathbf{U} \\ \hline \mathbf{B} & \mathbf{V} \end{array} \right].$$

Introducing the notation

$$
Y^{[n]} = \begin{bmatrix} Y_1^{[n]} \\ \vdots \\ Y_s^{[n]} \end{bmatrix}, \quad
F(Y^{[n]}) = \begin{bmatrix} f(Y_1^{[n]}) \\ \vdots \\ f(Y_s^{[n]}) \end{bmatrix}, \quad
y^{[n]} = \begin{bmatrix} y_1^{[n]} \\ \vdots \\ y_r^{[n]} \end{bmatrix},
$$

(2.1.2) can be written in the vector form

$$
\left[\begin{array}{c} Y^{[n]} \\ \hline y^{[n]} \end{array} \right] =
\left[\begin{array}{c|c} \mathbf{A} \otimes \mathbf{I} & \mathbf{U} \otimes \mathbf{I} \\ \hline \mathbf{B} \otimes \mathbf{I} & \mathbf{V} \otimes \mathbf{I} \end{array} \right]
\left[\begin{array}{c} hF(Y^{[n]}) \\ \hline y^{[n-1]} \end{array} \right].
$$

Here $\mathbf{I}$ is the identity matrix of dimension m (i.e., the dimension of the ODE system (2.1.1)), and the Kronecker or tensor product of two matrices $\mathbf{A} \in \mathbb{R}^{m_1 \times n_1}$ and $\mathbf{B} \in \mathbb{R}^{m_2 \times n_2}$ is defined as a block matrix of the form

$$
\mathbf{A} \otimes \mathbf{B} = \begin{bmatrix}
a_{11}\mathbf{B} & a_{12}\mathbf{B} & \cdots & a_{1,n_1}\mathbf{B} \\
a_{21}\mathbf{B} & a_{22}\mathbf{B} & \cdots & a_{2,n_1}\mathbf{B} \\
\vdots & \vdots & \ddots & \vdots \\
a_{m_1,1}\mathbf{B} & a_{m_1,2}\mathbf{B} & \cdots & a_{m_1,n_1}\mathbf{B}
\end{bmatrix} \in \mathbb{R}^{m_1 m_2 \times n_1 n_2}
$$

(compare [197]). These methods include as special cases many known methods for ODEs. For example, Runge-Kutta (RK) methods given by

$$Y_i^{[n]} = y_{n-1} + h \sum_{j=1}^{s} a_{ij} f(Y_j^{[n]}), \quad i = 1, 2, \ldots, s,$$

$$y_n = y_{n-1} + h \sum_{j=1}^{s} b_j f(Y_j^{[n]}),$$

(2.1.3)

$n = 0, 1, \ldots, N$, or by the Butcher tableaux

$$\frac{\mathbf{c} \mid \mathbf{A}}{\mid \mathbf{b}^T} = \begin{array}{c|ccc} c_1 & a_{11} & \cdots & a_{1s} \\ \vdots & \vdots & \ddots & \vdots \\ c_s & a_{s1} & \cdots & a_{ss} \\ \hline & b_1 & \cdots & b_s \end{array}$$

can be represented as GLM (2.1.2) with $r = 1$ in the form

$$\left[\begin{array}{c|c} \mathbf{A} & \mathbf{e} \\ \hline \mathbf{b}^T & 1 \end{array} \right] = \left[\begin{array}{ccc|c} a_{11} & \cdots & a_{1s} & 1 \\ \vdots & \ddots & \vdots & \vdots \\ a_{s1} & \cdots & a_{ss} & 1 \\ \hline b_1 & \cdots & b_s & 1 \end{array} \right].$$

For a thorough discussion of RK methods we refer readers to the monographs [41, 52, 143, 146, 199].

Consider next the class of linear multistep methods defined by

$$y_n = \sum_{j=1}^{k} \alpha_j y_{n-j} + h \sum_{j=0}^{k} \beta_j f(y_{n-j}),$$

(2.1.4)

$n = k, k + 1, \ldots, N$. Putting $Y^{[n]} = y_n$ and

$$y^{[n]} = \left[\begin{array}{ccccccc} y_n & y_{n-1} & \cdots & y_{n-k+1} & hf(y_n) & hf(y_{n-1}) & \cdots & hf(y_{n-k+1}) \end{array} \right]^T,$$

$$y^{[n-1]} = \left[\begin{array}{ccccccc} y_{n-1} & y_{n-2} & \cdots & y_{n-k} & hf(y_{n-1}) & hf(y_{n-2}) & \cdots & hf(y_{n-k}) \end{array} \right]^T,$$

method (2.1.4) for scalar ODEs has a representation with $r = 2k$ and $s = 1$ of the form

$$
\left[\begin{array}{c|c} \mathbf{A} & \mathbf{U} \\ \hline \mathbf{B} & \mathbf{V} \end{array} \right] =
\left[\begin{array}{c|ccccccccc}
\beta_0 & \alpha_1 & \cdots & \alpha_{k-1} & \alpha_k & \beta_1 & \cdots & \beta_{k-1} & \beta_k \\
\hline
\beta_0 & \alpha_1 & \cdots & \alpha_{k-1} & \alpha_k & \beta_1 & \cdots & \beta_{k-1} & \beta_k \\
0 & 1 & \cdots & 0 & 0 & 0 & \cdots & 0 & 0 \\
\vdots & \vdots & \ddots & \vdots & \vdots & \vdots & \ddots & \vdots & \vdots \\
0 & 0 & \cdots & 1 & 0 & 0 & \cdots & 0 & 0 \\
1 & 0 & \cdots & 0 & 0 & 0 & \cdots & 0 & 0 \\
0 & 0 & \cdots & 0 & 0 & 1 & \cdots & 0 & 0 \\
\vdots & \vdots & \ddots & \vdots & \vdots & \vdots & \ddots & \vdots & \vdots \\
0 & 0 & \cdots & 0 & 0 & 0 & \cdots & 1 & 0
\end{array} \right].
$$

This representation was first proposed by Burrage and Butcher [33] (compare also [54]). More compact representation is possible for special cases of (2.1.4). For example, Adams methods can be represented with $r = k+1$ and backward differentiation formulas (BDFs) with $r = k$ (compare [52, 293]). A more compact representation of general linear multistep formulas (2.1.4) with $r = k$ and $s = 1$ was discovered recently by Butcher and Hill [63]. Following [63], we define

$$
y_i^{[n-1]} = \sum_{j=k-i+1}^{k} \left(\alpha_j y_{n+k-i-j} + h\beta_j f(y_{n+k-i-j}) \right),
$$

$i = 1, 2, \ldots, k$. This formula was previously proposed by Skeel [264], although not in the context of GLMs. Then method (2.1.4) can be written in the form

$$
y_n = h\beta_0 f(y_n) + \sum_{j=1}^{k} \left(\alpha_j y_{n-j} + h\beta_j f(y_{n-j}) \right) = h\beta_0 f(y_n) + y_k^{[n-1]},
$$

and we have

$$
\begin{aligned}
y_i^{[n]} &= \sum_{j=k-i+1}^{k} \left(\alpha_j y_{n+1+k-i-j} + h\beta_j f(y_{n+1+k-i-j}) \right) \\
&= \alpha_{k-i+1} y_n + h\beta_{k-i+1} f(y_n) \\
&\quad + \sum_{j=k-i+2}^{k} \left(\alpha_j y_{n+1+k-i-j} + h\beta_j f(y_{n+1+k-i-j}) \right) \\
&= \left(\alpha_{k-i+1} \beta_0 + \beta_{k-i+1} \right) h f(y_n) + \alpha_{k-i+1} y_k^{[n-1]} + y_{i-1}^{[n-1]},
\end{aligned}
$$

$i = 1, 2, \ldots, k$. This leads to the representation

$$
\begin{bmatrix} y_n \\ \hline y_1^{[n]} \\ y_2^{[n]} \\ y_3^{[n]} \\ \vdots \\ y_{k-1}^{[n]} \\ y_k^{[n]} \end{bmatrix}
=
\left[\begin{array}{c|cccccc}
\beta_0 & 0 & 0 & 0 & \cdots & 0 & 1 \\
\hline
\alpha_k\beta_0 + \beta_k & 0 & 0 & 0 & \cdots & 0 & \alpha_k \\
\alpha_{k-1}\beta_0 + \beta_{k-1} & 1 & 0 & 0 & \cdots & 0 & \alpha_{k-1} \\
\alpha_{k-2}\beta_0 + \beta_{k-2} & 0 & 1 & 0 & \cdots & 0 & \alpha_{k-2} \\
\vdots & \vdots & \vdots & \vdots & \ddots & \ddots & \vdots & \vdots \\
\alpha_2\beta_0 + \beta_2 & 0 & 0 & 0 & \ddots & 0 & \alpha_2 \\
\alpha_1\beta_0 + \beta_1 & 0 & 0 & 0 & \cdots & 1 & \alpha_1
\end{array} \right]
\begin{bmatrix} hf(y_n) \\ \hline y_1^{[n-1]} \\ y_2^{[n-1]} \\ y_3^{[n-1]} \\ \vdots \\ y_{k-1}^{[n-1]} \\ y_k^{[n-1]} \end{bmatrix} .
$$

To simplify the error analysis for linear multistep methods (2.1.4) for stiff differential systems Dahlquist [100, 102] introduced the corresponding class of one-leg methods defined by

$$
\alpha_0 y_n = \sum_{j=1}^{k} \alpha_j y_{n-j} + h\beta f\left(\frac{1}{\beta} \sum_{j=0}^{k} \beta_j y_{n-j} \right), \tag{2.1.5}
$$

$n = k, k+1, \ldots, N$, $\alpha_0 = 1$, $\beta = \sum_{j=0}^{k} \beta_j$, which require only one evaluation of the function f per step. As pointed out by Hundsdorfer and Steininger [169], compared with the corresponding linear multistep method, the one-leg method (2.1.5) may have stronger nonlinear stability properties, such as G-stability, and more robust behavior on nonuniform grids (compare [101, 221]). Set

$$
Y^{[n]} = \frac{1}{\beta} \sum_{j=0}^{k} \beta_j y_{n-j}.
$$

Then substituting (2.1.5) into this relation for $Y^{[n]}$, we obtain

$$
\begin{aligned}
Y^{[n]} &= \frac{1}{\beta}\left(\beta_0 y_n + \sum_{j=1}^{k} \beta_j y_{n-j} \right) \\
&= \frac{1}{\beta}\left(\beta_0 \Big(\sum_{j=1}^{k} \alpha_j y_{n-j} + h\beta f(Y^{[n]}) \Big) + \sum_{j=1}^{k} \beta_j y_{n-j} \right) \\
&= \frac{1}{\beta} \sum_{j=1}^{k} (\beta_0 \alpha_j + \beta_j) y_{n-j} + h\beta_0 f(Y^{[n]}).
\end{aligned}
$$

Setting

$$
y^{[n]} = \begin{bmatrix} y_n & y_{n-1} & \cdots & y_{n-k+1} \end{bmatrix}^T,
$$

$$
y^{[n-1]} = \begin{bmatrix} y_{n-1} & y_{n-2} & \cdots & y_{n-k} \end{bmatrix}^T,
$$

method (2.1.5) for scalar ODEs has a representation as GLM (2.1.2) with $r = k$ and $s = 1$ of the form

$$
\left[\begin{array}{c|c} \mathbf{A} & \mathbf{U} \\ \hline \mathbf{B} & \mathbf{V} \end{array} \right] = \left[\begin{array}{c|ccccc} \beta_0 & \frac{\beta_0\alpha_1+\beta_1}{\beta} & \frac{\beta_0\alpha_2+\beta_2}{\beta} & \cdots & \frac{\beta_0\alpha_{k-1}+\beta_{k-1}}{\beta} & \frac{\beta_0\alpha_k+\beta_k}{\beta} \\ \hline \beta & \alpha_1 & \alpha_2 & \cdots & \alpha_{k-1} & \alpha_k \\ 0 & 1 & 0 & \cdots & 0 & 0 \\ 0 & 0 & 1 & \cdots & 0 & 0 \\ \vdots & \vdots & \vdots & \ddots & \vdots & \vdots \\ 0 & 0 & 0 & \cdots & 1 & 0 \end{array} \right]
$$

(compare [41]). A different representation of one-leg method (2.1.5) as GLM (2.1.2), where the vector of external approximations was specified in reverse order and a different normalization $\alpha_0 \neq 0$ and $\beta = 1$ was used, was considered by Butcher and Hill [63]. In this representation the coefficient matrix $(\mathbf{A}, \mathbf{U}, \mathbf{B}, \mathbf{V})$ for the one-leg method written as a GLM is the transpose of the coefficient matrix of the linear multistep method written as a GLM with k inputs, subject to this different normalization.

An interesting class of extended backward differentiation formulas (EBDFs) suitable for the numerical solution of stiff differential systems was introduced by Cash [85] (see also [145, 146]). These methods involve approximations at future point t_{n+k+1} and take the form

$$
\sum_{j=0}^{k} \alpha_j y_{n+j} = h\beta_k f_{n+k} + h\beta_{k+1} f_{n+k+1}, \tag{2.1.6}
$$

$f_{n+k} = f(t_{n+k}, y_{n+k})$, $f_{n+k+1} = f(t_{n+k+1}, y_{n+k+1})$, with coefficients α_j, $j = 0, 1, \ldots, k$, β_k, β_{k+1}, computed by solving the appropriate order conditions for the order $p = k + 1$ and with the normalization $\alpha_k = 1$. These coefficients are listed in [85]. The resulting methods are A- and L-stable for $k = 1$, 2, and 3, and $A(\alpha)$-stable for $k = 4$, 5, 6, 7, and 8, and the regions of absolute stability are plotted Cash [85], and by Hairer and Wanner [145].

Assume that the approximate solutions $y_n, y_{n+1}, \ldots, y_{n+k-1}$ are already available. Then algorithm based on EBDF methods takes the following form.
(i) Compute $\overline{y}_{n+k}$ as the solution of the conventional BDF method

$$
\overline{y}_{n+k} + \sum_{j=0}^{k-1} \widehat{\alpha}_j y_{n+j} = h\widehat{\beta}_k \overline{f}_{n+k}, \tag{2.1.7}
$$

$\overline{f}_{n+k} = f(t_{n+k}, \overline{y}_{n+k})$.
(ii) Compute $\overline{y}_{n+k+1}$ as the solution of the same BDF formula advanced one step, that is,

$$
\overline{y}_{n+k+1} + \widehat{\alpha}_{k-1}\overline{y}_{n+k} + \sum_{j=0}^{k-2} \widehat{\alpha}_j y_{n+j+1} = h\widehat{\beta}_k \overline{f}_{n+k+1}, \tag{2.1.8}
$$

$\overline{f}_{n+k+1} = f(t_{n+k+1}, \overline{y}_{n+k+1}).$

(iii) Discard $\overline{y}_{n+k}$, insert $\overline{f}_{n+k+1}$ into EBDF method (2.1.6), and solve for y_{n+k}:

$$y_{n+k} + \sum_{j=0}^{k-1} \alpha_j y_{n+j} = h\beta_k f_{n+k} + h\beta_{k+1}\overline{f}_{n+k+1}. \tag{2.1.9}$$

As observed by Cash [86] and Hairer and Wanner[146], the disadvantage of the algorithm given above is that stages (i) and (ii) represent nonlinear systems with the same Jacobian $I - h\widehat{\beta}_k J$, $J = \partial f/\partial y$, but stage (iii) has a different Jacobian, $I - h\beta_k J$, which requires extra LU decomposition. To remedy this situation, Cash [86] proposed an algorithm where the last stage (iii) was replaced by a modified EBDF (MEBDF) method of the form

$$\sum_{j=0}^{k} \alpha_j y_{n+j} = h\widehat{\beta}_k f_{n+k} + h(\beta_k - \widehat{\beta}_k)\overline{f}_{n+k} + h\beta_{k+1}\overline{f}_{n+k+1}. \tag{2.1.10}$$

These methods are also A- and L-stable for $k = 1$, 2, and 3, and $A(\alpha)$-stable for $k = 4$, 5, 6, 7, and 8, with a larger angles α than that of the corresponding EBDF methods. These angles for BDF, EBDF, and MEBDF methods are listed in [85, 86, 145, 146], and reproduced in Table 2.1.1, where an asterisk indicates that the method is not $A(\alpha)$-stable. The stability regions of MEBDF methods have been plotted by Hairer and Wanner [146].

k	1	2	3	4	5	6	7	8
α for BDF	90°	90°	88°	73°	51°	18°	*	*
α for EBDF	90°	90°	90°	87.61°	80.21°	67.73°	48.82°	19.98°
α for MEBDF	90°	90°	90°	88.36°	83.07°	74.48°	61.98°	42.87°

Table 2.1.1 Angles α of $A(\alpha)$-stability for BDF, EBDF, and MEBDF formulas

Substituting (2.1.7) into (2.1.8), we obtain

$$\overline{y}_{n+k+1} = \widehat{\alpha}_{k-1}\widehat{\alpha}_0 y_n + \sum_{j=1}^{k-1}\left(\widehat{\alpha}_{k-1}\widehat{\alpha}_j - \widehat{\alpha}_{j-1}\right)y_{n+j}$$

$$- h\widehat{\alpha}_{k-1}\widehat{\beta}_k\overline{f}_{n+k} + h\widehat{\beta}_k\overline{f}_{n+k+1}. \tag{2.1.11}$$

An algorithm based on formulas (2.1.7), (2.1.11), and (2.1.10) can be written as a GLM of the form (2.1.2) with

$$
Y^{[n]} = \begin{bmatrix} \overline{y}_{n+k} \\ \overline{y}_{n+k+1} \\ y_{n+k} \end{bmatrix}, \quad
F(Y^{[n]}) = \begin{bmatrix} \overline{f}_{n+k} \\ \overline{f}_{n+k+1} \\ f_{n+k} \end{bmatrix}, \quad
y^{[n]} = \begin{bmatrix} y_{n+k} \\ y_{n+k-1} \\ \vdots \\ y_{n+1} \end{bmatrix},
$$

and with the coefficient matrices **A**, **U**, **B**, and **V** given by

$$
\mathbf{A} = \begin{bmatrix}
\widehat{\beta}_k & 0 & 0 \\
-\widehat{\alpha}_{k-1}\widehat{\beta}_k & \widehat{\beta}_k & 0 \\
\beta_k - \widehat{\beta}_k & \beta_{k+1} & \widehat{\beta}_k
\end{bmatrix},
$$

$$
\mathbf{U} = \begin{bmatrix}
-\widehat{\alpha}_{k-1} & -\widehat{\alpha}_{k-2} & \cdots & -\widehat{\alpha}_1 & -\widehat{\alpha}_0 \\
\widehat{\alpha}_{k-1}\widehat{\alpha}_{k-1}-\widehat{\alpha}_{k-2} & \widehat{\alpha}_{k-1}\widehat{\alpha}_{k-2}-\widehat{\alpha}_{k-3} & \cdots & \widehat{\alpha}_{k-1}\widehat{\alpha}_1-\widehat{\alpha}_0 & \widehat{\alpha}_{k-1}\widehat{\alpha}_0 \\
-\alpha_{k-1} & -\alpha_{k-2} & \cdots & -\alpha_1 & -\alpha_0
\end{bmatrix},
$$

$$
\mathbf{B} = \begin{bmatrix}
\beta_k - \widehat{\beta}_k & \beta_{k+1} & \widehat{\beta}_k \\
0 & 0 & 0 \\
\vdots & \vdots & \vdots \\
0 & 0 & 0 \\
0 & 0 & 0
\end{bmatrix}, \quad
\mathbf{V} = \begin{bmatrix}
-\alpha_{k-1} & -\alpha_{k-2} & \cdots & -\alpha_1 & -\alpha_0 \\
1 & 0 & \cdots & 0 & 0 \\
\vdots & \vdots & \ddots & \vdots & \vdots \\
0 & 0 & \cdots & 0 & 0 \\
0 & 0 & \cdots & 1 & 0
\end{bmatrix}.
$$

Consider next the general class of two-step Runge-Kutta (TSRK) methods introduced by Jackiewicz and Tracogna [183] (see also [270]). These methods depend on stage values at two consecutive steps and have the form

$$
\begin{aligned}
Y_i^{[n]} &= (1 - u_i)y_{n-1} + u_i y_{n-2} \\
&\quad + h\sum_{j=1}^{s}\left(a_{ij}f(Y_j^{[n]}) + b_{ij}f(Y_j^{[n-1]})\right), \\
y_n &= (1 - \vartheta)y_{n-1} + \vartheta y_{n-2} \\
&\quad + h\sum_{j=1}^{s}\left(v_j f(Y_j^{[n]}) + w_j f(Y_j^{[n-1]})\right),
\end{aligned}
\tag{2.1.12}
$$

$i = 1, 2, \ldots, s$, $n = 2, 3, \ldots, N$, where $Y_i^{[n]}$ are approximations to $y(t_{n-1} + c_i h)$. Method (2.1.12) can be represented by the abscissa vector $\mathbf{c} = [c_1, \ldots, c_s]^T$

and the tableaux

$$
\begin{array}{c|c|c}
\mathbf{u} & \mathbf{A} & \mathbf{B} \\
\hline
\vartheta & \mathbf{v}^T & \mathbf{w}^T
\end{array}.
$$

For scalar ODEs these methods can be represented as GLMs (2.1.2) with $r = s + 2$ in the form

$$
\begin{bmatrix}
Y^{[n]} \\
\hline
y_n \\
y_{n-1} \\
hF(Y^{[n]})
\end{bmatrix}
=
\begin{bmatrix}
\mathbf{A} & \mathbf{e} - \mathbf{u} & \mathbf{u} & \mathbf{B} \\
\hline
\mathbf{v}^T & 1 - \vartheta & \vartheta & \mathbf{w}^T \\
0 & 1 & 0 & 0 \\
\mathbf{I} & 0 & 0 & 0
\end{bmatrix}
\begin{bmatrix}
hF(Y^{[n]}) \\
\hline
y_{n-1} \\
y_{n-2} \\
hF(Y^{[n-1]})
\end{bmatrix}
$$

(compare [293]). Here $\mathbf{I}$ is the identity matrix of dimension s and $\mathbf{0}$ is a zero matrix or vector of appropriate dimensions. Different representation of (2.1.12) as GLMs are considered in [15, 183, 189, 270].

Burrage [28, 29] and Burrage and Sharp [36] studied multistep Runge-Kutta (MRK) methods defined by

$$
\begin{aligned}
Y_i^{[n]} &= h \sum_{j=1}^{s} a_{ij} f(Y_j^{[n]}) + \sum_{j=1}^{k} u_{ij} y_{n+1-j}, \\
y_{n+1} &= h \sum_{j=1}^{s} b_j f(Y_j^{[n]}) + \sum_{j=1}^{k} v_j y_{n+1-j},
\end{aligned}
\tag{2.1.13}
$$

$i = 1, 2, \ldots, s$, $n = k - 1, k, \ldots, N - 1$. Here $Y_i^{[n]}$ is an approximation to $y(t_n + c_i h)$. For $k = 1$ these methods reduce to the RK formulas (2.1.3). However, for $k = 2$ they do not reduce to the class of TSRK methods (2.1.12) since MRK methods (2.1.13) depend only on stage values $Y_i^{[n]}$ on the current step, whereas TSRK methods (2.1.12) depend on stage values $Y_i^{[n]}$ and $Y_i^{[n-1]}$ on two consecutive steps. Setting

$$
y^{[n]} = \begin{bmatrix} y_{n+1} & y_n & \cdots & y_{n-k+2} \end{bmatrix}^T,
$$

$$
y^{[n-1]} = \begin{bmatrix} y_n & y_{n-1} & \cdots & y_{n-k+1} \end{bmatrix}^T,
$$

MRK methods (2.1.13) for scalar ODEs have the representation as GLMs (2.1.2) of the form

$$
\left[\begin{array}{c|c} \mathbf{A} & \mathbf{U} \\ \hline \mathbf{B} & \mathbf{V} \end{array}\right] =
\left[\begin{array}{cccc|cccc}
a_{11} & a_{12} & \cdots & a_{1s} & u_{11} & u_{12} & \cdots & u_{1,k-1} & u_{1k} \\
a_{21} & a_{22} & \cdots & a_{2s} & u_{21} & u_{22} & \cdots & u_{2,k-1} & u_{2k} \\
\vdots & \vdots & \ddots & \vdots & \vdots & \vdots & \ddots & \vdots & \vdots \\
a_{s1} & a_{s2} & \cdots & a_{ss} & u_{s1} & u_{s2} & \cdots & u_{s,k-1} & u_{sk} \\
\hline
b_1 & b_2 & \cdots & b_s & v_1 & v_2 & \cdots & v_{k-1} & v_k \\
0 & 0 & \cdots & 0 & 1 & 0 & \cdots & 0 & 0 \\
\vdots & \vdots & \ddots & \vdots & \vdots & \vdots & \ddots & \vdots & \vdots \\
0 & 0 & \cdots & 0 & 0 & 0 & \cdots & 0 & 0 \\
0 & 0 & \cdots & 0 & 0 & 0 & \cdots & 1 & 0
\end{array}\right].
$$

Weiner and coworkers investigated various classes of peer methods first introduced in [250], i.e., methods in which all stages have the same properties and no extraordinary solution variable is used. For example, the two-step peer methods investigated by Weiner at al. [286, 289], on the uniform grid take the form

$$
Y_i^{[n]} = \sum_{j=1}^s b_{ij} Y_j^{[n]} + h \sum_{j=1}^s a_{ij} f(Y_j^{[n-1]}) + h \sum_{j=1}^s r_{ij} f(Y_j^{[n]}), \tag{2.1.14}
$$

$i = 1, 2, \ldots, s$, $n = 1, 2, \ldots$, or in vector form,

$$
Y^{[n]} = (\mathbf{B} \otimes \mathbf{I}) Y^{[n-1]} + h(\mathbf{A} \otimes \mathbf{I}) F(Y^{[n-1]}) + h(\mathbf{R} \otimes \mathbf{I}) F(Y^{[n]}), \tag{2.1.15}
$$

$n = 1, 2, \ldots$, where

$$
Y^{[n]} \approx y(t_n + c_i h), \quad i = 1, 2, \ldots, s.
$$

The representation of (2.1.14) or (2.1.15) as a GLM for the scalar differential problem takes the form

$$
\left[\begin{array}{c} Y^{[n]} \\ \hline Y^{[n]} \\ hF(Y^{[n]}) \end{array}\right] =
\left[\begin{array}{c|c|c} \mathbf{R} & \mathbf{B} & \mathbf{A} \\ \hline \mathbf{R} & \mathbf{B} & \mathbf{A} \\ \hline \mathbf{I} & 0 & 0 \end{array}\right] =
\left[\begin{array}{c} hF(Y^{[n]}) \\ \hline Y^{[n-1]} \\ hF(Y^{[n-1]}) \end{array}\right].
$$

Variable step size variant of these methods have also been considered in which the matrices $\mathbf{A} = \mathbf{A}(\delta)$, $\mathbf{B} = \mathbf{B}(\delta)$, and $\mathbf{R} = \mathbf{R}(\delta)$ depend on the ratio of step sizes $\delta = \delta_n = h_n/h_{n-1}$ [286, 289]. Different classes of peer methods have also been introduced: for example, parallel peer two-step W-methods [288], Rosenbrock-type peer two-step methods [238], multi-implicit peer two-step W-methods [252], and implicit parallel peer methods for stiff systems [251]. The numerical experiments presented in these papers indicate that various classes of peer methods have a potential as building blocks of software which can be competitive with that currently in use based on RK or predictor-corrector methods.

Other examples of methods that can be represented as GLMs include predictor-corrector methods in various implementation modes (e.g., $P(EC)^m$ or $P(EC)^m E$) [194, 195]; the generalized multistep or hybrid methods of Butcher [37], Gear [130], and Gragg and Stetter [135]; the split linear multistep methods discussed by Cash [87] and Voss and Casper [283]; the cyclic composite methods of Donelson and Hansen [115]; the pseudo Runge-Kutta methods of Byrne and Lambert [84] and Byrne [82]; special cases of two-step Runge-Kutta methods [181, 182, 190, 244]; the diagonally implicit single-eigenvalue methods (DIMSEMs) introduced by Enenkel and Jackson [126, 127]; and the Almost Runge-Kutta methods [47, 49, 76]. Representation of some of these methods as GLMs (2.1.2) have been discussed in [41, 52, 53, 143, 146, 293].

2.2 PRECONSISTENCY, CONSISTENCY, STAGE-CONSISTENCY, AND ZERO-STABILITY

To identify useful GLMs (2.1.2), we have to impose some accuracy and stability conditions. To find some minimal accuracy conditions, we assume that there exist vectors $\mathbf{q}_0$ and $\mathbf{q}_1$,

$$
\mathbf{q}_0 = \left[\begin{array}{cccc} q_{1,0} & q_{2,0} & \cdots & q_{r,0} \end{array} \right]^T, \quad \mathbf{q}_1 = \left[\begin{array}{cccc} q_{1,1} & q_{2,1} & \cdots & q_{r,1} \end{array} \right]^T,
$$

such that the components of the input vector $y^{[n-1]}$ satisfy

$$
y_i^{[n-1]} = q_{i,0}y(t_{n-1}) + q_{i,1}hy'(t_{n-1}) + O(h^2), \quad i = 1, 2, \ldots, r.
$$

We then request that the components of the stage vector $Y^{[n]}$ and the output vector $y^{[n]}$ satisfy

$$
Y_i^{[n]} = y(t_{n-1} + c_i h) + O(h^2), \quad i = 1, 2, \ldots, s,
$$

and

$$
y_i^{[n]} = q_{i,0}y(t_n) + q_{i,1}hy'(t_n) + O(h^2), \quad i = 1, 2, \ldots, r.
$$

Observe that the condition for $Y_i^{[n]}$ is more general than the condition

$$
Y_i^{[n]} = y(t_n) + O(h), \quad i = 1, 2, \ldots, s
$$

considered in [31, 52]. Substituting these relations into (2.1.2), we obtain

$$
y(t_{n-1}) + hc_i y'(t_{n-1}) = h \sum_{j=1}^{s} a_{ij} y'(t_{n-1})
$$
$$
+ \sum_{j=1}^{r} u_{ij} \Big(q_{j,0} y(t_{n-1}) + h q_{j,1} y'(t_{n-1}) \Big) + O(h^2),
$$

$i = 1, 2, \ldots, s$, and

$$
q_{i,0} y(t_n) + q_{i,1} h y'(t_n) = h \sum_{j=1}^{s} b_{ij} y'(t_n)
$$
$$
+ \sum_{j=1}^{r} v_{ij} \Big(q_{j,0} y(t_{n-1}) + q_{j,1} h y'(t_{n-1}) \Big) + O(h^2),
$$

$i = 1, 2, \ldots, r$. Comparing $O(1)$ and $O(h)$ terms in these relations, we obtain

$$
\sum_{j=1}^{r} u_{ij} q_{j,0} = 1, \quad i = 1, 2, \ldots, s, \qquad \sum_{j=1}^{r} v_{ij} q_{j,0} = q_{i,0}, \quad i = 1, 2, \ldots, r,
$$

and

$$
\sum_{j=1}^{s} a_{ij} + \sum_{j=1}^{r} u_{ij} q_{j,1} = c_i, \quad i = 1, 2, \ldots, s,
$$

$$
\sum_{j=1}^{s} b_{ij} + \sum_{j=1}^{r} v_{ij} q_{j,1} = q_{i,0} + q_{i,1}, \quad i = 1, 2, \ldots, r.
$$

Considerations above motivate the following definitions.

Definition 2.2.1 *GLM* $(\mathbf{c}, \mathbf{A}, \mathbf{U}, \mathbf{B}, \mathbf{V})$ *is preconsistent if there exists a vector* $\mathbf{q}_0$ *such that*

$$
\mathbf{U} \mathbf{q}_0 = \mathbf{e}, \quad \mathbf{V} \mathbf{q}_0 = \mathbf{q}_0, \tag{2.2.1}
$$

where $\mathbf{e} = [1, \ldots, 1]^T \in \mathbb{R}^s$. *The vector* $\mathbf{q}_0$ *is called the preconsistency vector.*

Definition 2.2.2 *GLM* $(\mathbf{c}, \mathbf{A}, \mathbf{U}, \mathbf{B}, \mathbf{V})$ *is consistent if it is preconsistent with preconsistency vector* $\mathbf{q}_0$ *and there exists a vector* $\mathbf{q}_1$ *such that*

$$
\mathbf{B} \mathbf{e} + \mathbf{V} \mathbf{q}_1 = \mathbf{q}_0 + \mathbf{q}_1, \tag{2.2.2}
$$

where $\mathbf{e} = [1, \ldots, 1]^T \in \mathbb{R}^r$. *The vector* $\mathbf{q}_1$ *is called the consistency vector.*

Definition 2.2.3 *GLM* $(\mathbf{c}, \mathbf{A}, \mathbf{U}, \mathbf{B}, \mathbf{V})$ *is stage-consistent if*

$$
\mathbf{A} \mathbf{e} + \mathbf{U} \mathbf{q}_1 = \mathbf{c}. \tag{2.2.3}
$$

It can be verified that for an s-stage RK method (2.1.3) $\mathbf{q}_0 = 1$, $\mathbf{q}_1 = 0$ and the stage-consistency and consistency conditions are equivalent to

$$\mathbf{A}\mathbf{e} = \mathbf{c}, \quad \mathbf{b}^T\mathbf{e} = 1.$$

The condition $\mathbf{A}\mathbf{e} = \mathbf{c}$ is not necessary and RK methods that do not satisfy it have been investigated in [224, 297] (see also [109]).

In the case of linear multistep methods (2.1.4),

$$\mathbf{q}_0 = \begin{bmatrix} 1 & \cdots & 1 & 0 & \cdots & 0 \end{bmatrix}^T \in \mathbb{R}^{2k},$$

$$\mathbf{q}_1 = \begin{bmatrix} 0 & -1 & \cdots & -(k-1) & 1 \cdots & 1 \end{bmatrix}^T \in \mathbb{R}^{2k},$$

and the preconsistency, consistency, and stage-consistency conditions take the form

$$\sum_{j=1}^{k} \alpha_j = 1, \quad \sum_{j=1}^{k} j\alpha_j = \sum_{j=0}^{k} \beta_j, \quad \mathbf{c} = 1 \tag{2.2.4}$$

(compare [97, 153, 194, 195]). For more compact representation of linear multistep methods discovered by Butcher and Hill [63] and discussed in Section 2.1, the components of the vector of external approximations have the expansions

$$y_i^{[n]} = \sum_{j=k-i+1}^{k} \alpha_j y(t_n) + h \sum_{j=k-i+1}^{k} (\beta_j - (j-k+i-1)\alpha_j)y'(t_n) + O(h^2),$$

$i = 1, 2, \ldots, k$, and it follows that the preconsistency and consistency vectors take the form

$$\mathbf{q}_0 = \begin{bmatrix} \alpha_k \\ \alpha_{k-1} + \alpha_k \\ \vdots \\ \sum_{j=1}^{k} \alpha_j \end{bmatrix} \in \mathbb{R}^k, \quad \mathbf{q}_1 = \begin{bmatrix} \beta_k \\ \beta_{k-1} + \beta_k - \alpha_k \\ \vdots \\ \sum_{j=1}^{k} \beta_j - \sum_{j=1}^{k}(j-1)\alpha_j \end{bmatrix} \in \mathbb{R}^k.$$

Similarly to a representation with $2k$ inputs, the preconsistency, consistency, and stage-consistency conditions take the form (2.2.4).

For the class of one-leg methods (2.1.5), we have

$$\mathbf{q}_0 = \mathbf{e} \in \mathbb{R}^k, \quad \mathbf{q}_1 = \begin{bmatrix} 0 & -1 & \cdots & -(k-1) \end{bmatrix}^T \in \mathbb{R}^k,$$

and the preconsistency, consistency, and stage-consistency conditions take the form

$$\sum_{j=1}^{k} \alpha_j = 1, \quad \sum_{j=1}^{k} j\alpha_j = \sum_{j=0}^{k} \beta_j, \quad \mathbf{c} = 1 - \frac{1}{\beta}\sum_{j=1}^{k} j\beta_j.$$

For the MEBDF methods of Cash [86] discussed in Section 2.1, we have

$$\mathbf{q}_0 = \mathbf{e} \in \mathbb{R}^k, \quad \mathbf{q}_1 = \begin{bmatrix} 0 & -1 & \cdots & -(k-1) \end{bmatrix}^T \in \mathbb{R}^k,$$

which have the same form as for one-leg methods (2.1.5). These expressions follow by expanding the vector of external approximations $y^{[n]}$ corresponding to these methods around the point t_{n+k}. Then the preconsistency conditions take the form

$$1 + \sum_{j=0}^{k-1} \widehat{\alpha}_j = 0, \quad 1 + \sum_{j=0}^{k-1} \alpha_j = 0,$$

the consistency condition is

$$\beta_k + \beta_{k+1} + \sum_{j=0}^{k-1} (k-j)\alpha_j = 0,$$

and the stage-consistency conditions are

$$c_1 = \widehat{\beta}_k + \sum_{j=0}^{k-1} (k-j)\widehat{\alpha}_j + 1,$$

$$c_2 = (\widehat{\alpha}_{k-1} - 1)\left(2 - \widehat{\beta}_k + \sum_{j=0}^{k-1} (k-j)\widehat{\alpha}_j\right),$$

and

$$c_3 = \beta_k + \beta_{k+1} + \sum_{j=0}^{k-1} (k-j)\alpha_j + 1.$$

For the class of TSRK methods (2.1.12), we have

$$\mathbf{q}_0 = \begin{bmatrix} 1 & 1 & | & 0 & \cdots & 0 \end{bmatrix}^T \in \mathbb{R}^{s+2}, \quad \mathbf{q}_1 = \begin{bmatrix} 0 & -1 & | & 1 & \cdots & 1 \end{bmatrix}^T \in \mathbb{R}^{s+2},$$

the preconsistency conditions are satisfied automatically and the consistency and stage-consistency and conditions take the form

$$(\mathbf{v}^T + \mathbf{w}^T)\mathbf{e} = 1 + \vartheta, \quad (\mathbf{A} + \mathbf{B})\mathbf{e} - \mathbf{u} = \mathbf{c}$$

(compare [183, 270]).

For MRK methods investigated by Burrage et al. [28, 29, 36] and written as GLMs in Section 2.1, the preconsistency and consistency vectors have the same form as for one-leg methods (2.1.5) and MEBDF methods, that is,

$$\mathbf{q}_0 = \mathbf{e} \in \mathbb{R}^k, \quad \mathbf{q}_1 = \begin{bmatrix} 0 & -1 & \cdots & -(k-1) \end{bmatrix}^T \in \mathbb{R}^k.$$

For these methods the preconsistency conditions are

$$\sum_{j=1}^{k} u_{ij} = 1, \quad i = 1, 2, \ldots, s, \quad \sum_{j=1}^{s} v_j = 1,$$

the consistency condition is

$$\sum_{j=1}^{s} b_j - \sum_{j=1}^{k} j v_j = 0,$$

and the stage-consistency conditions are

$$c_i = 1 + \sum_{j=1}^{s} - \sum_{j=1}^{k} j u_{ij}, \quad i = 1, 2, \ldots, s.$$

For the two-step peer methods discussed by Weiner et al. [250], [286], [289] and in Section 2.1, the preconsistency and consistency vectors are

$$\mathbf{q}_0 = \begin{bmatrix} \mathbf{e} \\ \mathbf{0} \end{bmatrix} \in \mathbb{R}^{2s}, \quad \mathbf{q}_1 = \begin{bmatrix} \mathbf{c} - \mathbf{e} \\ \mathbf{e} \end{bmatrix} \in \mathbb{R}^{2s}.$$

The preconsistency condition for this class of methods takes the form

$$\mathbf{B}\mathbf{e} = \mathbf{e}$$

and the consistency and stage-consistency conditions reduce to the same relation of the form

$$(\mathbf{R} + \mathbf{A})\mathbf{e} + \mathbf{B}(\mathbf{c} - \mathbf{e}) = \mathbf{c}.$$

Next we investigate the zero-stability of GLMs. To find minimal stability conditions, we apply GLM (2.1.2) to the equation $y' = 0$, $t \geq t_0$. This leads to

$$y^{[n]} = \mathbf{V} y^{[n-1]} = \mathbf{V}^n y^{[0]}, \quad n = 0, 1, \ldots,$$

and motivates the following definition.

Definition 2.2.4 *GLM* $(\mathbf{c}, \mathbf{A}, \mathbf{U}, \mathbf{B}, \mathbf{V})$ *is zero-stable if there exists a constant C such that*

$$\|\mathbf{V}^n\| \leq C \quad \quad (2.2.5)$$

for all $n = 0, 1, \ldots$.

It is well known that condition (2.2.5) is equivalent to the following criterion.

Theorem 2.2.5 (compare [31, 41]) *GLM* $(\mathbf{c}, \mathbf{A}, \mathbf{U}, \mathbf{B}, \mathbf{V})$ *is zero-stable if the minimal polynomial of the coefficient matrix $\mathbf{V}$ has no zeros with magnitude greater than 1 and all zeros with magnitude equal to 1 are simple.*

We also introduce the following definition.

Definition 2.2.6 *GLM* $(\mathbf{c}, \mathbf{A}, \mathbf{U}, \mathbf{B}, \mathbf{V})$ *is strictly zero-stable if it is zero-stable and the coefficient matrix* $\mathbf{V}$ *has exactly one eigenvalue on the unit circle that is equal to 1.*

For RK methods (2.2.2) the minimal polynomial is $p(w) = w - 1$, and these methods are always zero-stable. It can be verified that for the linear multistep method (2.2.3) written as GLM with $2k$ or k inputs, the minimal polynomials are

$$p(w) = w^k \rho(w) \quad \text{or} \quad p(w) = \rho(w),$$

where

$$\rho(w) = w^k - \sum_{j=1}^{k} \alpha_j w^{k-j}$$

is the first characteristic polynomial of (2.2.3). For one-leg methods the minimal polynomial is $p(w) = \rho(w)$. Hence, linear multistep methods and one-leg methods are zero-stable if $\rho(w)$ satisfies the root condition. This means that $\rho(w)$ has no root with modulus greater than 1, and that every root with modulus 1 is simple (compare [194, 195]). For the TSRK methods (2.1.12), the minimal polynomial is

$$p(w) = w\big(w^2 - (1 - \vartheta)w - \vartheta\big).$$

The roots of this polynomial are $w = 0$, $w = 1$, and $w = -\vartheta$ and it follows that (2.1.12) is zero-stable if and only if $-1 < \vartheta \leq 1$ (compare [183, 270]).

2.3 CONVERGENCE

To investigate the convergence of GLM (2.1.2), following Butcher [52] we assume only that there exists a starting procedure

$$S_h : \mathbb{R}^m \to \mathbb{R}^{mr}$$

which associates with every step size $h > 0$ a starting vector $y^{[0]} = y^{[0]}(h) \in \mathbb{R}^{mr}$ such that

$$\lim_{h \to 0} S_h(y_0) = \lim_{h \to 0} y^{[0]} = (\mathbf{q}_0 \otimes \mathbf{I})y(t_0) \tag{2.3.1}$$

for some nonzero vector $\mathbf{q}_0$ (preconsistency vector). Here $y_0 \in \mathbb{R}^m$ is the initial value and y is a solution to (2.1.1). We then investigate the conditions under which the sequence of vectors $y^{[n]}$ computed using n steps of GLM (2.1.2) with step size h such that $nh = \bar{t} - t_0$ converges to $\mathbf{q}_0 y(\bar{t})$ for any fixed $\bar{t} \in [t_0, T]$. We introduce the following definition.

Definition 2.3.1 *GLM* $(\mathbf{c}, \mathbf{A}, \mathbf{U}, \mathbf{B}, \mathbf{V})$ *is convergent if for any initial value problem (2.1.1) satisfying the Lipschitz condition, there exists a nonzero vector*

$\mathbf{q}_0 \in \mathbb{R}^r$ *and a starting procedure* S_h *satisfying (2.3.1), such that the sequence of vectors* $y^{[n]}$ *computed using* n *steps of GLM (2.1.2) with* $y^{[0]} = S_h(y_0)$ *and* $h = (\bar{t} - t_0)/n$ *converges to* $\mathbf{q}_0 y(\bar{t})$ *for any* $\bar{t} \in [t_0, T]$. *Here* y *is a solution to* *(2.1.1).*

In the next two theorems we establish that zero-stability and consistency are necessary and sufficient conditions for convergence. The proofs of these theorems follow closely Butcher's presentation [52].

Theorem 2.3.2 *Assume that GLM* $(\mathbf{c}, \mathbf{A}, \mathbf{U}, \mathbf{B}, \mathbf{V})$ *is convergent. Then it is zero-stable and consistent.*

Proof: We show first that convergence implies zero-stability. Suppose, on the contrary, that the sequence $\|\mathbf{V}^n\|$, $n = 1, 2, \ldots$, is unbounded. Then, since $\|\mathbf{V}^n\| = \max_{\|w\|=1} \|\mathbf{V}^n w\|$, there exists a sequence of vectors w_n, $\|w_n\| = 1$, such that the sequence $\|\mathbf{V}^n w_n\|$, $n = 1, 2, \ldots$, is unbounded. Consider the solution of the initial value problem

$$y'(t) = 0, \quad y(0) = 0,$$

$t \geq 0$, at the point $\bar{t} = 1$, where n steps are taken by GLM (2.1.2) with a step size $h = 1/n$ and with the initial vector $y^{[0]}$ given by

$$y^{[0]} = S_h(0) = \frac{w_n}{\max\limits_{1 \leq i \leq n} \|\mathbf{V}^i w_i\|}.$$

Observe that $\lim_{h \to 0} S_h(0) = 0$ since the sequence $\|\mathbf{V}^n w_n\|$ is unbounded. The approximation $y^{[n]}$ after n steps is then given by

$$y^{[n]} = \mathbf{V}^n S_h(0) = \frac{\mathbf{V}^n w_n}{\max\limits_{1 \leq i \leq n} \|\mathbf{V}^i w_i\|}$$

with norm

$$\|y^{[n]}\| = \|\mathbf{V}^n S_h(0)\| = \frac{\|\mathbf{V}^n w_n\|}{\max\limits_{1 \leq i \leq n} \|\mathbf{V}^i w_i\|}.$$

Since $\|\mathbf{V}^n w_n\|$ is unbounded, there exists a monotonically increasing subsequence $\|\mathbf{V}^{n_j} w_{n_j}\|$ such that $\lim_{j \to \infty} \|\mathbf{V}^{n_j} w_{n_j}\| = \infty$ and $\max_{1 \leq i \leq n_j} \|\mathbf{V}^i w_i\| = \|\mathbf{V}^{n_j} w_{n_j}\|$. Hence,

$$\|y^{[n_j]}\| = \|\mathbf{V}_{n_j} S(1/n_j)\| = 1,$$

arbitrarily often, which contradicts convergence.

We show next that convergence implies consistency. Consider the solution of the initial value problem

$$y'(t) = 1, \quad y(0) = 0,$$

$t \geq 0$, at the point $\bar{t} = 1$, where n steps are taken by GLM (2.1.2) with step size $h = 1/n$ and with the initial vector $y^{[0]}$ given by

$$y^{[0]} = S_h(0) = \mathbf{q}_0 y(0) = \mathbf{0}.$$

The output vectors $y^{[i]}$ are generated according to the formula

$$y^{[i]} = h\mathbf{Be} + \mathbf{V}y^{[i-1]}, \quad i = 1, 2, \ldots, n,$$

which leads to

$$y^{[n]} = h(\mathbf{I} + \mathbf{V} + \cdots + \mathbf{V}^{n-1})\mathbf{Be}.$$

Since $\mathbf{V}^i \mathbf{q}_0 = \mathbf{q}_0$, $i = 0, 1, \ldots$, we have

$$h(\mathbf{I} + \mathbf{V} + \cdots + \mathbf{V}^{n-1})\mathbf{q}_0 = hn\mathbf{q}_0 = \mathbf{q}_0,$$

and it follows that

$$y^{[n]} - \mathbf{q}_0 = h(\mathbf{I} + \mathbf{V} + \cdots + \mathbf{V}^{n-1})(\mathbf{Be} - \mathbf{q}_0).$$

We have already proved in the first part of this theorem that the coefficient matrix $\mathbf{V}$ is power bounded. This implies that there exists a nonsingular matrix $\mathbf{P}$ such that

$$\mathbf{V} = \mathbf{P}^{-1} \begin{bmatrix} \mathbf{I} & \mathbf{0} \\ \mathbf{0} & \widetilde{\mathbf{V}} \end{bmatrix} \mathbf{P},$$

where $\mathbf{I}$ is the identity matrix of dimension $\tilde{r} \leq r$ and $\widetilde{\mathbf{V}}$ is power bounded and such that $1 \notin \sigma(\widetilde{\mathbf{V}})$. Here $\sigma(\widetilde{\mathbf{V}})$ stands for the spectrum of the matrix $\widetilde{\mathbf{V}}$. Hence,

$$\begin{aligned}
y^{[n]} - \mathbf{q}_0 &= \mathbf{P}^{-1} \begin{bmatrix} \mathbf{I} & \mathbf{0} \\ \mathbf{0} & h(\mathbf{I} + \widetilde{\mathbf{V}} + \cdots + \widetilde{\mathbf{V}}^{n-1}) \end{bmatrix} \mathbf{P}(\mathbf{Be} - \mathbf{q}_0) \\
&= \mathbf{P}^{-1} \begin{bmatrix} \mathbf{I} & \mathbf{0} \\ \mathbf{0} & h(\mathbf{I} - \widetilde{\mathbf{V}})^{-1}(\mathbf{I} - \widetilde{\mathbf{V}}^n) \end{bmatrix} \mathbf{P}(\mathbf{Be} - \mathbf{q}_0).
\end{aligned}$$

Passing to the limit as $n \to \infty$, $nh = 1$, in the relation above and taking into account that

$$\lim_{n \to \infty, nh=1} y^{[n]} = \mathbf{q}_0 y(1) = \mathbf{q}_0,$$

we obtain

$$\begin{bmatrix} \mathbf{I} & \mathbf{0} \\ \mathbf{0} & \mathbf{0} \end{bmatrix} \mathbf{P}(\mathbf{Be} - \mathbf{q}_0) = \begin{bmatrix} \mathbf{0} \\ \mathbf{0} \end{bmatrix}.$$

This relation implies that the vector $\mathbf{P}(\mathbf{Be} - \mathbf{q}_0)$ has only zeros in its first $\tilde{r}$ components. Since the matrix $\mathbf{I} - \tilde{\mathbf{V}}$ is nonsingular, we can write this vector in the form

$$\mathbf{P}(\mathbf{Be} - \mathbf{q}_0) = \begin{bmatrix} \mathbf{0} \\ (\mathbf{I} - \tilde{\mathbf{V}})\tilde{\mathbf{q}}_1 \end{bmatrix}$$

for some vector $\tilde{\mathbf{q}}_1 \in \mathbb{R}^{r-\tilde{r}}$. Define the vector $\mathbf{q}_1 \in \mathbb{R}^r$ by

$$\mathbf{q}_1 = \mathbf{P}^{-1} \begin{bmatrix} \mathbf{0} \\ \tilde{\mathbf{q}}_1 \end{bmatrix}.$$

Then

$$\mathbf{P}(\mathbf{Be} - \mathbf{q}_0) = \begin{bmatrix} \mathbf{0} & \mathbf{0} \\ \mathbf{0} & \mathbf{I} - \tilde{\mathbf{V}} \end{bmatrix} \begin{bmatrix} \mathbf{0} \\ \tilde{\mathbf{q}}_1 \end{bmatrix} = \left(\mathbf{I} - \begin{bmatrix} \mathbf{0} & \mathbf{0} \\ \mathbf{0} & \tilde{\mathbf{V}} \end{bmatrix} \right) \mathbf{P}\mathbf{q}_1$$

$$= \mathbf{P}\left(\mathbf{I} - \mathbf{P}^{-1} \begin{bmatrix} \mathbf{0} & \mathbf{0} \\ \mathbf{0} & \tilde{\mathbf{V}} \end{bmatrix} \mathbf{P} \right) \mathbf{q}_1 = \mathbf{P}(\mathbf{I} - \mathbf{V})\mathbf{q}_1.$$

Hence, $\mathbf{Be} + \mathbf{Vq}_1 = \mathbf{q}_0 + \mathbf{q}_1$, which is the consistency condition with consistency vector $\mathbf{q}_1$. This completes the proof. ∎

To prove that zero-stability and consistency are also sufficient conditions for convergence, we need the following variant of a lemma on stable sequences.

Lemma 2.3.3 *Assume that* $Q \in \mathbb{R}^{\nu \times \nu}$ *is a stable matrix with a bound on* $\|Q^n\|$ *given by*

$$\|Q^n\| \leq C_1, \quad n = 0, 1, \ldots,$$

where C_1 *is a constant such that* $C_1 \geq 1$. *Let sequences* $u_n, w_n \in \mathbb{R}^\nu$ *be such that*

$$u_n = Qu_{n-1} + w_n, \quad \|w_n\| \leq C_2\|u_{n-1}\| + \delta,$$

$n = 0, 1, \ldots,$ *for constants* $C_2 \geq 0$ *and* $\delta \geq 0$. *Then*

$$\|u_n\| \leq C_1(\|u_0\| + n\delta) \tag{2.3.2}$$

if $C_2 = 0$ *and*

$$\|u_n\| \leq C_1(1 + C_1 C_2)^n \|u_0\| + \frac{\delta}{C_2}\left((1 + C_1 C_2)^n - 1\right) \tag{2.3.3}$$

if $C_2 > 0$, $n = 0, 1, \ldots$.

Proof: Applying the formula for u_n n times, we obtain

$$u_n = Q^n u_0 + \sum_{j=1}^{n} Q^{n-j} w_j,$$

$n = 0, 1, \ldots$. Taking norms on both sides of this equation and then bounding $\|Q^{n-j}\|$ by C_1 and using the inequality for $\|w_n\|$ leads to

$$\|u_n\| \leq C_1 \left(\|u_0\| + C_2 \sum_{j=1}^{n} \|u_{n-j}\| + n\delta \right).$$

Bound (2.3.2) follows immediately for $C_2 = 0$. For $C_2 > 0$ we prove bound (2.3.3) by induction with respect to n. Substituting the bounds for $\|u_{n-j}\|$ given by (2.3.3) for $j = 1, 2, \ldots, n$ into the last inequality, we obtain

$$
\begin{aligned}
\|u_n\| \;\leq\; & C_1 \Bigg(\|u_0\| + C_2 \sum_{j=1}^{n} \Big(C_1 (1 + C_1 C_2)^{j-1} \|u_0\| \\
& + \frac{\delta}{C_2} \big((1 + C_1 C_2)^{j-1} - 1 \big) \Big) + n\delta \Bigg) \\
=\; & C_1 \Big(1 + C_1 C_2 \sum_{j=1}^{n} (1 + C_1 C_2)^{j-1} \Big) \|u_0\| + C_1 \delta \sum_{j=1}^{n} (1 + C_1 C_2)^{j-1} \\
=\; & C_1 (1 + C_1 C_2)^n \|u_0\| + \frac{\delta}{C_2} \big((1 + C_1 C_2)^n - 1 \big),
\end{aligned}
$$

which is bound (2.3.3). This completes the proof.   ∎

Theorem 2.3.4 *A zero-stable and consistent GLM (2.1.2) is convergent.*

Proof: Define the vectors $\widehat{y}^{[n-1]}$, $\widehat{y}^{[n]} \in \mathbb{R}^{mr}$,

$$\widehat{y}^{[n-1]} = \begin{bmatrix} \widehat{y}_1^{[n-1]} \\ \vdots \\ \widehat{y}_r^{[n-1]} \end{bmatrix}, \quad \widehat{y}^{[n]} = \begin{bmatrix} \widehat{y}_1^{[n]} \\ \vdots \\ \widehat{y}_r^{[n]} \end{bmatrix},$$

by

$$\widehat{y}_i^{[n-1]} = q_{i,0} y(t_{n-1}) + q_{i,1} h y'(t_{n-1}), \quad \widehat{y}_i^{[n]} = q_{i,0} y(t_n) + q_{i,1} h y'(t_n),$$

where $q_{i,0}$ and $q_{i,1}$ are components of preconsistency and consistency vectors $\mathbf{q}_0$ and $\mathbf{q}_1$, respectively. We define vectors $\xi_i(h)$ and $h\eta_i(h)$ as residua obtained by replacing $y_i^{[n-1]}$, $y_i^{[n]}$ in (2.1.2) by $\widehat{y}_i^{[n-1]}$, $\widehat{y}_i^{[n]}$ and $Y_i^{[n]}$ by $y(t_{n-1} + c_i h)$; that is,

$$
\begin{aligned}
y(t_{n-1} + c_i h) \;=\; & h \sum_{j=1}^{s} a_{ij} y'(t_{n-1} + c_j h) \\
& + \sum_{j=1}^{r} u_{ij} \Big(q_{i,0} y(t_{n-1}) + q_{i,1} h y'(t_{n-1}) \Big) + \xi_i(h),
\end{aligned}
$$

$i = 1, 2, \ldots, s$, and

$$\widehat{y}_i^{[n]} = h \sum_{j=1}^{s} b_{ij} y'(t_{n-1} + c_j h) + \sum_{j=1}^{r} v_{ij} \widehat{y}_j^{[n-1]} + h\eta_i(h)$$

or

$$q_{i,0} y(t_n) + q_{i,1} hy'(t_n) \;=\; h \sum_{j=1}^{s} b_{ij} y'(t_{n-1} + c_i h)$$
$$+ \; \sum_{j=1}^{r} v_{ij} \Big(q_{i,0} y(t_{n-1}) + q_{i,1} hy'(t_{n-1}) \Big) + h\eta_i(h),$$

$i = 1, 2, \ldots, r$. Expanding $y(t_{n-1} + c_i h)$ and $y'(t_{n-1} + c_i h)$ around t_{n-1} in the equations that define $\xi_i(h)$, $i = 1, 2, \ldots, s$, and using the preconsistency condition $\mathbf{U}\mathbf{q}_0 = \mathbf{e}$, it follows that

$$\xi_i(h) = O(h), \quad i = 1, 2, \ldots, s.$$

Expanding $y(t_n)$ and $y'(t_n)$ around the point t_{n-1} in the equations that define $\eta_i(h)$, $i = 1, 2, \ldots, r$, and using the preconsistency condition $\mathbf{V}\mathbf{q}_0 = \mathbf{q}_0$ and consistency condition $\mathbf{B}\mathbf{e} + \mathbf{V}\mathbf{q}_1 = \mathbf{q}_0 + \mathbf{q}_1$, we can conclude that

$$\eta_i(h) = O(h), \quad i = 1, 2, \ldots, r.$$

Subtracting the equations for $y_i^{[n]}$ and $\widehat{y}_i^{[n]}$, we obtain

$$y_i^{[n]} - \widehat{y}_i^{[n]} = h \sum_{j=1}^{s} b_{ij} \Big(f(Y_j^{[n]}) - f\big(y(t_{n-1} + c_i h)\big) \Big) + \sum_{j=1}^{r} v_{ij}(y_j^{[n]} - \widehat{y}_j^{[n]}) - h\eta_i(h),$$

$i = 1, 2, \ldots, r$. Using the notation introduced in Section 2.1, this can be written in vector form as

$$y^{[n]} - \widehat{y}^{[n]} = (\mathbf{V} \otimes \mathbf{I})(y^{[n-1]} - \widehat{y}^{[n-1]})$$
$$+ \; h(\mathbf{B} \otimes \mathbf{I})\Big(F(Y^{[n]}) - F\big(y(t_{n-1} + \mathbf{c}h)\big) \Big) - h\eta(h),$$

where

$$y(t + \mathbf{c}h) = \begin{bmatrix} y(t + c_1 h) \\ \vdots \\ y(t + c_s h) \end{bmatrix}, \quad F\big(y(t + \mathbf{c}h)\big) = \begin{bmatrix} f\big(y(t + c_1 h)\big) \\ \vdots \\ f\big(y(t + c_s h)\big) \end{bmatrix},$$

$\mathbf{I}$ is an identity matrix of dimension m, and $\eta(h) \in \mathbb{R}^{mr}$ is composed from the vectors $\eta_i(h)$, $i = 1, 2, \ldots, r$. To analyze the behavior of the sequence

$y^{[n]} - \widehat{y}^{[n]}$ we will use, for convenience, the spectral norm $\|\cdot\| = \|\cdot\|_2$ for which $\|\mathbf{Q} \otimes \mathbf{I}\| = \|\mathbf{Q}\|$ for any matrix Q, and

$$\left\|F(Y) - F(\widehat{Y})\right\| \leq L\|Y - \widehat{Y}\|,$$

where L is the Lipschitz constant of the function f. Introducing the notation

$$u_n = y^{[n]} - \widehat{y}^{[n]}, \quad w_n = h(\mathbf{B} \otimes \mathbf{I})\Big(F(Y^{[n]}) - F\big(y(t_{n-1} + \mathbf{c}h)\big)\Big) - h\eta(h),$$

we obtain

$$u_n = (\mathbf{V} \otimes \mathbf{I})u_{n-1} + w_n$$

and

$$\left\|w_n\right\| \leq hL\|\mathbf{B}\| \left\|Y^{[n]} - y(t_{n-1} + \mathbf{c}h)\right\| + h\|\eta(h)\|.$$

Subtracting the relations for $Y^{[n]}$ and $y(t_{n-1} + \mathbf{c}h)$, we get

$$\begin{aligned}
Y^{[n]} - y(t_{n-1} + \mathbf{c}h) &= h(\mathbf{A} \otimes \mathbf{I})\Big(F(Y^{[n]}) - F\big(y(t_{n-1} + \mathbf{c}h)\big)\Big) \\
&+ (\mathbf{U} \otimes \mathbf{I})(y^{[n-1]} - \widehat{y}^{[n-1]}) - \xi(h),
\end{aligned}$$

where $\xi(h) \in \mathbb{R}^{mr}$ is composed from $\xi_i(h)$. Hence,

$$\begin{aligned}
\left\|Y^{[n]} - y(t_{n-1} + \mathbf{c}h)\right\| &\leq hL\|\mathbf{A}\| \left\|Y^{[n]} - y(t_{n-1} + \mathbf{c}h)\right\| \\
&+ \|\mathbf{U}\| \left\|y^{[n-1]} - \widehat{y}^{[n-1]}\right\| + \|\xi(h)\|.
\end{aligned}$$

Assume that h_0 is a step size such that $h_0 L\|\mathbf{A}\| < 1$. Then for $h < h_0$ we have

$$\left\|Y^{[n]} - y(t_{n-1} + \mathbf{c}h)\right\| \leq \frac{\|\mathbf{U}\|}{1 - h_0 L\|\mathbf{A}\|}\left\|y^{[n-1]} - \widehat{y}^{[n-1]}\right\| + \frac{\|\xi(h)\|}{1 - h_0 L\|\mathbf{A}\|},$$

and substituting this inequality into the inequality for $\|w_n\|$ it follows that

$$\|w_n\| \leq hD\|u_n\| + h\delta(h),$$

where D and $\delta(h)$ are defined by

$$D = \frac{L\|\mathbf{B}\|\|\mathbf{U}\|}{1 - h_0 L\|\mathbf{A}\|}, \quad \delta(h) = \frac{L\|\mathbf{B}\|\|\xi(h)\|}{1 - h_0 L\|\mathbf{A}\|} + \|\eta(h)\|.$$

Taking into account that the norms $\|\mathbf{V}^n\|$ are bounded by a constant C (compare Definition 2.2.4), application of Lemma 2.3.3 leads to the following bounds on $\|u_n\|$:

$$\|u_n\| \leq C\Big(\|u_0\| + nh\delta(h)\Big)$$

if $D = 0$ or

$$\|u_n\| \le C(1 + hCD)^n \|u_0\| + \frac{(1 + hCD)^n - 1}{D} \delta(h)$$

if $D > 0$. Since $hn = \bar{t} - t_0$, we have

$$(1 + hCD)^n \le e^{CD(\bar{t} - t_0)}$$

and the bounds on $\|u_n\| = \|y^{[n]} - \widehat{y}^{[n]}\|$ can be written in the form

$$\|u_n\| \le C\Big(\|u_0\| + (\bar{t} - t_0)\delta(h)\Big)$$

if $D = 0$ or

$$\big\|y^{[n]} - \widehat{y}^{[n]}\big\| \le Ce^{CD(\bar{t} - t_0)}\big\|y^{[0]} - \widehat{y}^{[0]}\big\| + \frac{e^{CD(\bar{t} - t_0)} - 1}{D}\delta(h)$$

if $D > 0$. Using triangle and inverse triangle inequalities

$$\big|\|x\| - \|y\|\big| \le \|x + y\| \le \|x\| + \|y\|,$$

we obtain

$$\big\|y_i^{[n]} - q_{i,0}y(\bar{t})\big\| \le C\Big(\big\|y_i^{[0]} - q_{i,0}y(t_0)\big\| + (\bar{t} - t_0)\delta(h)\Big) + O(h)$$

if $D = 0$ or

$$\big\|y_i^{[n]} - q_{i,0}y(\bar{t})\big\| \le Ce^{CD(\bar{t} - t_0)}\big\|y_i^{[0]} - q_{i,0}y(t_0)\big\| + \frac{e^{CD(\bar{t} - t_0)} - 1}{D}\delta(h) + O(h)$$

if $D > 0$, $i = 1, 2, \ldots, r$. The convergence now follows since

$$\lim_{h \to 0} \big\|y_i^{[0]} - q_{i,0}y(t_0)\big\| = 0$$

and $\delta(h) = O(h)$. ∎

2.4 ORDER AND STAGE ORDER CONDITIONS

In assessing the accuracy of GLMs (2.1.2), we identify two integers: p, the order of the method, and q, the stage order. To formulate order and stage order conditions, we assume that the components of the input vector $y_i^{[n-1]}$ for the next step satisfy

$$y_i^{[n-1]} = \sum_{k=0}^{p} q_{ik} h^k y^{(k)}(t_{n-1}) + O(h^{p+1}), \quad i = 1, 2, \ldots, r, \tag{2.4.1}$$

for some real parameters q_{ik}, $i = 1, 2, \ldots, r$, $k = 0, 1, \ldots, p$. We then request that the components of the internal stages $Y_i^{[n]}$ be approximations of order q to the solution y of (2.1.1) at the points $t_{n-1} + c_i h$; that is,

$$Y_i^{[n]} = y(t_{n-1} + c_i h) + O(h^{q+1}), \quad i = 1, 2, \ldots, s. \tag{2.4.2}$$

We also request that the components of the output vector $y_i^{[n]}$ satisfy

$$y_i^{[n]} = \sum_{k=0}^{p} q_{ik} h^k y^{(k)}(t_n) + O(h^{p+1}), \quad i = 1, 2, \ldots, r, \tag{2.4.3}$$

for the same parameters q_{ik}.

The formula for $y_i^{[n-1]}$ given by (2.4.1) was chosen here because this representation is convenient for the purpose of investigating the structure of order conditions. However, this representation also suggests that to start the integration we need to compute the initial vector $y^{[0]}$ by a starting procedure $S_h : \mathbb{R}^m \to \mathbb{R}^{mr}$ (compare Section 2.3) satisfying (2.4.1) for $n = 1$; that is, in vector notation,

$$y^{[0]} = S_h(y_0) = \sum_{k=0}^{p} (q_k \otimes \mathbf{I}) h^k y^{(k)}(t_0) + O(h^{p+1}), \quad i = 1, 2, \ldots, r. \tag{2.4.4}$$

Observe that this is a stronger requirement on S_h than condition (2.3.1) which was required for convergence. In practice, such $y^{[0]}$ will usually be found from approximations to the solution values at equally spaced arguments without direct computation of approximations to the scaled derivatives $h^k y^{(k)}(t_0)$. Moreover, we describe some families of GLMs of various orders and stage orders for which it will be easy to increase the order one unit at a time. In such a case, in the first step of the integration a method of order $p = 1$ will be used for which computation of the required starting vector $y^{[0]}$ will easily be effected from knowledge of the initial value y_0.

The derivation of order and stage order conditions for general p and q is quite complicated and requires sophisticated algebraic tools developed by Butcher [41, 52] (see also [143, 293]). However, this analysis is quite simple for methods of high stage order(i.e., methods with $q = p$ or $q = p - 1$). In these cases order and stage order conditions can be expressed conveniently using the theory of functions of a complex variable.

It is convenient to collect the parameters q_{ik} in the matrix $\mathbf{W}$ defined by

$$\mathbf{W} = \begin{bmatrix} \mathbf{q}_0 & \mathbf{q}_1 & \cdots & \mathbf{q}_p \end{bmatrix} = \begin{bmatrix} q_{10} & q_{11} & \cdots & q_{1p} \\ q_{20} & q_{21} & \cdots & q_{2p} \\ \vdots & \vdots & \ddots & \vdots \\ q_{r0} & q_{r1} & \cdots & q_{rp} \end{bmatrix}.$$

We recognize $\mathbf{q}_0$ and $\mathbf{q}_1$ as preconsistency and consistency vectors, respectively. We also introduce the notation

$$e^{\mathbf{c}z} = \begin{bmatrix} e^{c_1 z} & e^{c_2 z} & \cdots & e^{c_s z} \end{bmatrix}^T$$

and define the vector $\mathbf{w} = \mathbf{w}(z)$ by

$$\mathbf{w} = \mathbf{w}(z) = \sum_{k=0}^{p} \mathbf{q}_k z^k.$$

We have the following theorems.

Theorem 2.4.1 (Butcher [44]) *Assume that $y^{[n-1]}$ satisfies (2.4.1). Then the GLM (2.1.2) of order p and stage order $q = p$ satisfies (2.4.2) and (2.4.3) if and only if*

$$e^{\mathbf{c}z} = z\mathbf{A}e^{\mathbf{c}z} + \mathbf{U}\mathbf{w}(z) + O(z^{p+1}), \qquad (2.4.5)$$

$$e^z \mathbf{w}(z) = z\mathbf{B}e^{\mathbf{c}z} + \mathbf{V}\mathbf{w}(z) + O(z^{p+1}). \qquad (2.4.6)$$

Proof: Since $Y_i^{[n]} = y(t_{n-1} + c_i h) + O(h^{p+1})$, $i = 1, 2, \ldots, s$, it follows that

$$\begin{aligned} hf(Y_i^{[n]}) &= hy'(t_{n-1} + c_i h) + O(h^{p+2}) \\ &= \sum_{k=1}^{p+1} \frac{c_i^{k-1}}{(k-1)!} h^k y^{(k)}(t_{n-1}) + O(h^{p+2}) \\ &= \sum_{k=1}^{p} \frac{c_i^{k-1}}{(k-1)!} h^k y^{(k)}(t_{n-1}) + O(h^{p+1}). \end{aligned}$$

Furthermore, using a Taylor series expansion, (2.4.3) can be written in the form

$$y_i^{[n]} = \sum_{k=0}^{p} \left(\sum_{l=0}^{k} \frac{1}{l!} q_{i,k-l} \right) h^k y^{(k)}(t_{n-1}) + O(h^{p+1}).$$

Substituting $y_i^{[n-1]}$ defined by (2.4.1) and the relations above for $f(Y_i^{[n]})$ and $y_i^{[n]}$, into the equation (2.1.2), which define computations performed in the nth step, we obtain

$$\sum_{k=0}^{p} \left(c_i^k - \sum_{j=1}^{s} k a_{ij} c_j^{k-1} - \sum_{j=1}^{r} k! u_{ij} q_{jk} \right) \frac{h^k}{k!} y^{(k)}(t_{n-1}) = O(h^{p+1})$$

and

$$\sum_{k=0}^{p} \left(\sum_{l=0}^{k} \frac{k!}{l!} q_{i,k-l} - \sum_{j=1}^{s} k b_{ij} c_j^{k-1} - \sum_{j=1}^{r} k! v_{ij} q_{jk} \right) \frac{h^k}{k!} y^{(k)}(t_{n-1}) = O(h^{p+1}).$$

Equating coefficients of $h^k y^{(k)}(t_{n-1})/k!$, $k = 0, 1, \ldots, p$, to zero and then multiplying these coefficients by $z^k/k!$ and adding them from $k = 0$ to $k = p$, we obtain

$$e^{c_i z} - \sum_{j=1}^{s} z a_{ij} e^{c_j z} - \sum_{j=1}^{r} u_{ij} w_j = O(z^{p+1}) \quad i = 1, 2, \ldots, s,$$

and

$$e^z w_i - \sum_{j=1}^{s} z b_{ij} e^{c_j z} - \sum_{j=1}^{r} v_{ij} w_j = O(z^{p+1}), \quad i = 1, 2, \ldots, r.$$

These relations are equivalent to (2.4.5) and (2.4.6). This completes the proof. ∎

Theorem 2.4.2 (Butcher and Jackiewicz [65]) *Assume that $y^{[n-1]}$ satisfies (2.4.1). Then GLM (2.1.2) of order p and stage order $q = p-1$ satisfies (2.4.2) and (2.4.3) if and only if*

$$e^{\mathbf{c}z} = z\mathbf{A}e^{\mathbf{c}z} + \mathbf{U}\mathbf{w}(z) + \left(\frac{\mathbf{c}^p}{p!} - \frac{\mathbf{A}\mathbf{c}^{p-1}}{(p-1)!} - \mathbf{U}\mathbf{q}_p \right) z^p + O(z^{p+1}) \qquad (2.4.7)$$

and (2.4.6).

Proof: Proceeding similarly as in the proof of Theorem 2.4.1, we obtain

$$\sum_{k=0}^{p} \left(c_i^k - \sum_{j=1}^{s} k a_{ij} c_j^{k-1} - \sum_{j=1}^{r} k! u_{ij} q_{jk} \right) \frac{h^k}{k!} y^{(k)}(t_{n-1}) = O(h^p)$$

and

$$\sum_{k=0}^{p} \left(\sum_{l=0}^{k} \frac{k!}{l!} q_{i,k-l} - \sum_{j=1}^{s} k b_{ij} c_j^{k-1} - \sum_{j=1}^{r} k! v_{ij} q_{jk} \right) \frac{h^k}{k!} y^{(k)}(t_{n-1}) = O(h^{p+1}).$$

The right-hand side of the first of the equations above is of order $O(h^p)$ only because the stage order q is one less than order p. We obtain the relations

$$\sum_{k=0}^{p} \left(c_i^k - \sum_{j=1}^{s} k a_{ij} c_j^{k-1} - \sum_{j=1}^{r} k! u_{ij} q_{jk} \right) \frac{z^k}{k!} = \left(c_i^p - \sum_{j=1}^{s} p a_{ij} c_j^{p-1} - \sum_{j=1}^{r} p! u_{ij} q_{jp} \right) \frac{z^p}{p!}$$

and

$$\sum_{k=0}^{p} \left(\sum_{l=0}^{k} \frac{k!}{l!} q_{i,k-l} - \sum_{j=1}^{s} k b_{ij} c_j^{k-1} - \sum_{j=1}^{r} k! v_{ij} q_{jk} \right) \frac{z^k}{k!} = 0,$$

which are equivalent to (2.4.7) and (2.4.6), respectively. ∎

It follows from the proofs of Theorems 2.4.1 and 2.4.2 that conditions (2.4.5), (2.4.6), and (2.4.7) are equivalent to

$$\mathbf{c}^k - k\mathbf{A}\mathbf{c}^{k-1} - k!\mathbf{U}\mathbf{q}_k = 0, \quad k = 0, 1, \ldots, p, \qquad (2.4.8)$$

$$\sum_{l=0}^{k} \frac{k!}{l!} \mathbf{q}_{k-l} - k\mathbf{B}\mathbf{c}^{k-1} - k!\mathbf{V}\mathbf{q}_k = 0, \quad k = 0, 1, \ldots, p, \tag{2.4.9}$$

and

$$\mathbf{c}^k - k\mathbf{A}\mathbf{c}^{k-1} - k!\mathbf{U}\mathbf{q}_k = 0, \quad k = 0, 1, \ldots, p-1. \tag{2.4.10}$$

Define vectors $\widehat{y}^{[n-1]}$ and $\widehat{y}^{[n]}$ composed of

$$\widehat{y}_i^{[n-1]} = q_{i,0}y(t_{n-1}) + q_{i,1}hy'(t_{n-1}) + \cdots + q_{i,p}h^p y^{(p)}(t_{n-1}),$$

$$\widehat{y}_i^{[n]} = q_{i,0}y(t_n) + q_{i,1}hy'(t_n) + \cdots + q_{i,p}h^p y^{(p)}(t_n).$$

We have the following theorem.

Theorem 2.4.3 *Assume that GLM (2.1.2) has order p and stage order $q = p$ or $q = p - 1$. Then*

$$\left\| y_i^{[n]} - \widehat{y}_i^{[n]} \right\| = O(h^p), \quad i = 1, 2, \ldots, r$$

as $h \to 0$, $h_n = \bar{t} - t_0$, provided that

$$\left\| y_i^{[0]} - \widehat{y}_i^{[0]} \right\| = O(h^p), \quad i = 1, 2, \ldots, r$$

as $h \to 0$. Moreover,

$$\left\| Y_i^{[n]} - y(t_{n-1} + c_i h) \right\| = O(h^p), \quad i = 1, 2, \ldots, s.$$

Proof: Similarly to the proof of Theorem 2.3.4, define $\xi_i(h)$ and $h\eta_i(h)$ by

$$y(t_{n-1} + c_i h) = h \sum_{j=1}^{s} a_{ij} y'(t_{n-1} + c_j h) + \sum_{j=1}^{r} u_{ij} \widehat{y}_j^{[n-1]} + \xi_i(h),$$

$$\widehat{y}_i^{[n]} = h \sum_{j=1}^{s} b_{ij} y'(t_{n-1} + c_j h) + \sum_{j=1}^{r} v_{ij} \widehat{y}_j^{[n-1]} + h\eta_i(h).$$

Expanding $y(t_{n-1}+c_i h)$, $y'(t_{n-1}+c_i h)$, and $y^{(k)}(t_n)$, $k = 0, 1, \ldots, p$, appearing in $\widehat{y}^{[n]}$ into a Taylor series around t_{n-1} and using (2.4.8) and (2.4.9) if $q = p$ or (2.4.10) and (2.4.9) if $q = p-1$, it follows that $\xi_i(h) = O(h^{p+1})$ if $q = p$ and $\xi_i(h) = O(h^p)$ if $q = p - 1$, $i = 1, 2, \ldots, s$, and $\eta_i(h) = O(h^p)$, $i = 1, 2, \ldots, r$. Proceeding in the same way as in the proof of Theorem 2.3.4, we obtain

$$u_n = (\mathbf{V} \otimes \mathbf{I})u_{n-1} + w_n, \quad \|w_n\| \le D\|u_n\| + \delta(h),$$

where now

$$u_n = y^{[n]} - \widehat{y}^{[n]}, \quad w_n = h(\mathbf{B} \otimes \mathbf{I})\Big(F(Y^{[n]}) - F\big(y(t_{n-1} + \mathbf{c}h)\big)\Big) - h\eta(h)$$

and D and $\delta(h)$ are defined as before. We have $\delta(h) = O(h^p)$, and the conclusion of the theorem follows from the application of Lemma 2.3.3 on stable sequences and the inequality

$$\left\| Y^{[n]} - y(t_{n-1} + \mathbf{c}h) \right\| \leq \frac{\|\mathbf{U}\|}{1 - h_0\|\mathbf{A}\|}\|u_n\| + \frac{\|\xi(h)\|}{1 - h_0\|\mathbf{A}\|},$$

which holds for $h_0 L\|\mathbf{A}\| < 1$ and $h < h_0$. ■

Next we examine conditions (2.4.8), (2.4.9), and (2.4.10) in the context of RK methods (2.1.3). The analysis of these conditions reveals that except for some special cases of low-order methods, RK methods (2.1.3) cannot have stage order equal to p or $p - 1$, where p is the order. For example, it follows from (2.4.9) that for RK methods with $p > 2$, we must have

$$\mathbf{q}_0 = 1, \ \mathbf{q}_1 = 0, \ \ldots, \ \mathbf{q}_p = 0$$

to satisfy the quadrature order conditions

$$\mathbf{b}^T \mathbf{c}^{k-1} = \frac{1}{k}, \quad k = 1, 2, \ldots, p.$$

For explicit RK methods, this is in conflict, however, with (2.4.8) or (2.4.10) for $k = 2$ since

$$\mathbf{c}^2 - 2\mathbf{A}\mathbf{c} \neq 0$$

if we assume that $c_2 \neq 0$. The situation is somewhat better for implicit RK formulas, but also in this case there are no methods with $q = p$ or $q = p - 1$ for $p > 3$ (compare [41, 52]).

Next we apply Theorem 2.4.1 to derive order conditions for linear multistep methods (2.1.4). It follows from the form of the vector of external approximations, or from the analysis of relations (2.4.8) and (2.4.9), that the vectors $\mathbf{q}_l$ take the form

$$\mathbf{q}_l = \frac{(-1)^{l-1}}{(l-1)!} \left[0 \ \ -\frac{1}{l} \ \ -\frac{2^l}{l} \ \ \cdots \ \ -\frac{(k-1)^l}{l} \ \Big| \ 0 \ \ 1 \ \ 2^{l-1} \ \ \cdots \ \ (k-1)^{l-1} \right]^T,$$

$l = 0, 1, \ldots, p$. The preconsistency and consistency vectors $\mathbf{q}_0$ and $\mathbf{q}_1$ were already determined in Section 2.2. Substituting these vectors into (2.4.8) with $\mathbf{c} = 1$ and $\mathbf{A} = \beta_0$ leads to the following order conditions for method (2.1.4):

$$(-1)^l \sum_{j=1}^{k}(j-1)^l \alpha_j + (-1)^{l-1}l \sum_{j=1}^{k}(j-1)^{l-1}\beta_j = 1 - l\beta_0, \qquad (2.4.11)$$

$l = 0, 1, \ldots, p$. Using more compact representation of linear multistep methods with k inputs discovered by Butcher and Hill [63] and discussed in Section 2.1, the vectors $\mathbf{q}_l$ take the form

$$
\mathbf{q}_l = \begin{bmatrix}
0 \\
\dfrac{(-1)^l}{l!}\alpha_k + \dfrac{(-1)^{l-1}}{(l-1)!}\beta_k \\
\vdots \\
\dfrac{(-1)^l}{l!}\sum_{j=1}^{k}(j-1)^l\alpha_j + \dfrac{(-1)^{l-1}}{(l-1)!}\sum_{j=k-1}^{k}(j-1)^{l-1}\beta_j
\end{bmatrix},
$$

$l = 2, 3, \ldots, p$, with the preconsistency and consistency vectors determined in Section 2.2. Substituting these vectors into (2.4.8) with $\mathbf{c} = \mathbf{1}$ and appropriate coefficients matrices $\mathbf{A}$ and $\mathbf{U}$ corresponding to this representation we again obtain order conditions (2.4.11). These conditions could be obtained directly by expanding local discretization error of (2.1.4), defined by

$$
\mathrm{lte}(t_n) = y(t_n) - \sum_{j=1}^{k}\alpha_j y(t_{n-j}) - h\sum_{j=0}^{k}\beta_j y'(t_{n-j})
$$

into a Taylor series around the point t_{n-1} and equating to zero the terms of order $O(h^l)$, $l = 0, 1, \ldots, p$. Expanding $\mathrm{lte}(t_n)$ around t_n leads to a simpler form of order conditions for (2.1.4), given by

$$
\sum_{j=1}^{k}\alpha_j = 1, \quad \sum_{j=1}^{k}j^l\alpha_j = l\sum_{j=0}^{k}j^{l-1}\beta_j, \quad l = 1, 2, \ldots, p,
$$

which are equivalent to (2.4.11) (compare the discussion in [194, 195]).

Theorems 2.4.1 and 2.4.2 can also be used to obtain stage order and order conditions for other classes of GLMs: for example, one-leg methods, EBDFs, MEBDFs, TSRK methods, MRK methods, and peer methods, considered in Section 2.1. However, this is not developed further here. Order conditions for the general class of TSRK methods (2.1.12) are discussed in Chapter 5. These order conditions reduce to those obtained from Theorems 2.4.1 and 2.4.2 for TSRK methods with $q = p$ or $q = p - 1$. Order conditions for two-step peer methods using the GLM point of view have been discussed by Izzo and Jackiewicz [175].

2.5 LOCAL DISCRETIZATION ERROR OF METHODS OF HIGH STAGE ORDER

Assume that the solution y to (2.1.1) is sufficiently smooth and that GLM (2.1.2) has order p and stage order $q = p$ or $q = p-1$. The local discretization or truncation error $\text{le}_i(t_n)$ of the ith external stage $y_i^{[n]}$ of GLM (2.1.2) at the point t_n is defined by

$$\text{le}_i(t_n) = \widehat{y}_i^{[n]} - h\sum_{j=1}^{s} b_{ij} f(\widehat{Y}_j^{[n]}) - \sum_{j=1}^{r} v_{ij}\widehat{y}_j^{[n-1]}, \tag{2.5.1}$$

$i = 1, 2, \ldots, r$, where $\widehat{y}_i^{[n-1]}$ and $\widehat{y}_i^{[n]}$ are defined as in Section 2.4 (before Theorem 2.4.3) and the $\widehat{Y}_i^{[n]}$ are given by

$$\widehat{Y}_i^{[n]} = h\sum_{j=1}^{s} a_{ij} f(\widehat{Y}_j^{[n]}) + \sum_{j=1}^{r} u_{ij}\widehat{y}_j^{[n-1]}, \tag{2.5.2}$$

$i = 1, 2, \ldots, s$. Denote by $\text{le}(t_n) \in \mathbb{R}^{mr}$ the vector composed of $\text{le}_i(t_n)$. We have the following theorem.

Theorem 2.5.1 *The local discretization error $\text{le}(t_n)$ of GLM (2.1.2) at the point t_n is given by*

$$\text{le}(t_n) = (\varphi_p \otimes \mathbf{I})h^{p+1}y^{(p+1)}(t_{n-1}) + O(h^{p+2}) \tag{2.5.3}$$

if $q = p$ and by

$$\begin{aligned} \text{le}(t_n) &= (\varphi_p \otimes \mathbf{I})h^{p+1}y^{(p+1)}(t_{n-1}) \\ &+ (\psi_p \otimes \mathbf{I})h^{p+1}\frac{\partial f}{\partial y}(y(t_{n-1}))y^{(p)}(t_{n-1}) + O(h^{p+2}) \end{aligned} \tag{2.5.4}$$

if $q = p - 1$, where

$$\varphi_p = \sum_{k=1}^{p+1} \frac{\mathbf{q}_{p+1-k}}{k!} - \frac{\mathbf{B}c^p}{p!} \tag{2.5.5}$$

and

$$\psi_p = \mathbf{B}\left(\frac{c^p}{p!} - \frac{\mathbf{A}c^{p-1}}{(p-1)!} - \mathbf{U}\mathbf{q}_p\right). \tag{2.5.6}$$

Proof: First we establish the relationship between $\widehat{Y}_i^{[n]}$ and $y(t_{n-1} + c_ih)$. We have

$$y(t_{n-1} + c_ih) = h\sum_{j=1}^{s} a_{ij} f(y(t_{n-1} + c_jh)) + \sum_{j=1}^{r} u_{ij}\widehat{y}_j^{[n-1]} + \xi_i(h),$$

where $\xi_i(h) = O(h^{p+1})$ if $q = p$ and $\xi_i(h) = O(h^p)$ if $q = p - 1$ (compare the proof of Theorem 2.4.3). Subtracting the equation above from (2.5.2), we obtain

$$\widehat{Y}_i^{[n]} - y(t_{n-1} + c_ih) = h\sum_{j=1}^{s} a_{ij} \left(f(\widehat{Y}_j^{[n]}) - f(y(t_{n-1} + c_ih))\right) - \xi_i(h),$$

$i = 1, 2, \ldots, s$. Hence, assuming that $h_0 L \|\mathbf{A}\| < 1$, where L is a Lipschitz constant of f, we have for $h \le h_0$

$$\|\widehat{Y}^{[n]} - y(t_{n-1} + \mathbf{c}h)\| \le \frac{\|\xi(h)\|}{1 - h_0 L \|\mathbf{A}\|},$$

and it follows that $\|\widehat{Y}^{[n]} - y(t_{n-1} + c_i h)\| = O(h^{p+1})$ provided that $q = p$ and that $\|\widehat{Y}^{[n]} - y(t_{n-1} + c_i h)\| = O(h^p)$ provided that $q = p - 1$.

Consider first the case $q = p$. Substituting $\widehat{Y}_j^{[n]} = y(t_{n-1} + c_j h) + O(h^{p+1})$ into (2.5.1), we obtain

$$\sum_{k=0}^{p} q_{ik} h^k y^{(k)}(t_n) = h \sum_{j=1}^{s} b_{ij} y'(t_{n-1} + c_j h) + \sum_{j=1}^{r} \sum_{k=0}^{p} v_{ij} q_{jk} h^k y^{(k)}(t_{n-1}) - \mathrm{le}_i(t_n).$$

Expanding $y^{(k)}(t_n)$ and $y'(t_{n-1} + c_j h)$ into a Taylor series around t_{n-1} and collecting the corresponding terms, we get

$$\sum_{k=0}^{p} \left(\sum_{l=0}^{k} \frac{k!}{l!} q_{i,k-l} - \sum_{j=1}^{s} k b_{ij} c_j^{k-1} - \sum_{j=1}^{r} k! v_{ij} q_{jk} \right) \frac{h^k}{k!} y^{(k)}(t_{n-1})$$

$$+ \left(\sum_{l=1}^{p+1} \frac{q_{i,p+1-l}}{l!} - \sum_{j=1}^{s} \frac{b_{ij} c_j^p}{p!} \right) h^{p+1} y^{(p+1)}(t_{n-1}) = \mathrm{le}_i(t_n) + O(h^{p+2}).$$

It follows from (2.4.8) that all the terms up to order $O(h^p)$ vanish, and the local discretization error takes the form

$$\mathrm{le}_i(t_n) = \left(\sum_{l=1}^{p+1} \frac{q_{i,p+1-l}}{l!} - \sum_{j=1}^{s} \frac{b_{ij} c_j^p}{p!} \right) h^{p+1} y^{(p+1)}(t_{n-1}) + O(h^{p+2}),$$

which is equivalent to (2.5.3) with φ_p given by (2.5.5).

Consider next the case $q = p - 1$. We have

$$\widehat{Y}_i^{[n]} = y(t_{n-1} + c_i h) + \gamma_i(t_{n-1}) h^p + O(h^{p+1}),$$

where $\gamma_i(t_{n-1})$ stands for the principal part of the error in stage values $\widehat{Y}_i^{[n]}$. Substituting this relation into (2.5.2), we get

$$y(t_{n-1} + c_i h) + \gamma_i(t_{n-1}) h^p$$

$$= h \sum_{j=1}^{s} a_{ij} y'(t_{n-1} + c_j h) + \sum_{k=0}^{p} \sum_{j=1}^{r} u_{ij} q_{jk} h^k y^{(k)}(t_{n-1}) + O(h^{p+2}).$$

Expanding $y(t_{n-1} + c_i h)$ and $y'(t_{n-1} + c_i h)$ around t_{n-1} and comparing terms of order $O(h^p)$, it follows that

$$\gamma_i(t_{n-1}) = -\left(\frac{c_i^p}{p!} - \sum_{j=1}^{s} \frac{a_{ij} c_j^{p-1}}{(p-1)!} - \sum_{j=1}^{r} u_{ij} q_{jp} \right) y^{(p)}(t_{n-1}).$$

Substituting $\widehat{Y}_i^{[n]}$ given above in terms of $y(t_{n-1} + c_i h)$ and $\gamma_i(t_{n-1})$ and

$$hf(\widehat{Y}_i^{[n]}) = hy'(t_{n-1} + c_i h) + \frac{\partial f}{\partial y}\big(y(t_{n-1})\big)\gamma_i(t_{n-1})h^{p+1} + O(h^{p+2})$$

into (2.5.1) leads to the relation

$$\sum_{k=0}^{p} q_{ik} h^k y^{(k)}(t_n) = h \sum_{j=1}^{s} b_{ij} y'(t_{n-1} + c_j h)$$

$$+ \sum_{j=1}^{s} b_{ij} \frac{\partial f}{\partial y}\big(y(t_{n-1})\big)\gamma_j(t_{n-1})h^{p+1} \sum_{j=1}^{r}\sum_{k=0}^{p} v_{ij} q_{jk} h^k y^{(k)}(t_{n-1}) - \mathrm{le}_i(t_n).$$

Proceeding similarly as before, we obtain

$$\mathrm{le}_i(t_n) = \left(\sum_{l=1}^{p+1} \frac{q_{i,p+1-l}}{l!} - \sum_{j=1}^{s} \frac{b_{ij} c_j^p}{p!}\right) h^{p+1} y^{(p+1)}(t_{n-1})$$

$$+ \sum_{j=1}^{s} b_{ij}\left(\frac{c_j^p}{p!} - \sum_{k=1}^{s} \frac{a_{jk} c_k^{p-1}}{(p-1)!} - \sum_{k=1}^{r} u_{jk} q_{kp}\right) \frac{\partial f}{\partial y}\big(y(t_{n-1})\big) h^{p+1} y^{(p)}(t_{n-1})$$

$$+ O(h^{p+2}),$$

which is equivalent to (2.5.4) with φ_p and ψ_p given by (2.5.5) and (2.5.6), respectively. This completes the proof. $\blacksquare$

2.6   LINEAR STABILITY THEORY OF GENERAL LINEAR METHODS

In this section we investigate stability properties of GLMs (2.1.2) with respect to the standard linear test equation

$$y' = \xi y, \quad t \geq 0, \tag{2.6.1}$$

where ξ is a complex parameter. Applying (2.1.2) to (2.6.1), we obtain in vector notation

$$Y^{[n]} = h\xi \mathbf{A} Y^{[n]} + \mathbf{U} y^{[n-1]}, \tag{2.6.2}$$

$$y^{[n]} = h\xi \mathbf{B} Y^{[n]} + \mathbf{V} y^{[n-1]}, \tag{2.6.3}$$

$n = 1, 2, \ldots$. Put $z = h\xi$ and assume that the matrix $\mathbf{I} - z\mathbf{A}$ is nonsingular. Then it follows from (2.6.2) that

$$Y^{[n]} = (\mathbf{I} - z\mathbf{A})^{-1}\mathbf{U} y^{[n-1]},$$

and substituting this relation into (2.6.3), we get

$$y^{[n]} = \mathbf{M}(z) y^{[n-1]},$$

$n = 1, 2, \ldots$, where the matrix $\mathbf{M}(z)$ is defined by

$$\mathbf{M}(z) = \mathbf{V} + z\mathbf{B}(\mathbf{I} - z\mathbf{A})^{-1}\mathbf{U}. \qquad (2.6.4)$$

The $\mathbf{M}(z)$ is called the stability matrix of GLM (2.1.2). We also define the stability function $p(w, z)$ by the formula

$$p(w, z) = \det\left(w\mathbf{I} - \mathbf{M}(z)\right), \qquad (2.6.5)$$

where $w \in \mathbb{C}$. We introduce the following standard definitions.

Definition 2.6.1 *GLM (2.1.2) is said to be absolutely stable for given $z \in \mathbb{C}$ if for that z, all the roots $w_i = w_i(z)$, $i = 1, 2, \ldots, r$, of stability function $p(w, z)$ are inside the unit circle.*

Definition 2.6.2 *Region $\mathcal{A}$ of absolute stability of (2.1.2) is the set of all $z \in \mathbb{C}$ such that the method is absolutely stable, i.e.,*

$$\mathcal{A} = \left\{z \in \mathbb{C} : \ |w_i(z)| < 1, \ i = 1, 2, \ldots, r\right\}.$$

An interval $\mathcal{I}$ of absolute stability is the intersection of $\mathcal{A}$ with a real axis (i.e., $\mathcal{I} = \mathcal{A} \cap \mathbb{R}$).

Definition 2.6.3 *GLM (2.1.2) is said to be A-stable if its region of absolute stability $\mathcal{A}$ contains a negative half-plane:*

$$\left\{z \in \mathbb{C} : \ \mathrm{Re}(z) < 0\right\} \subset \mathcal{A}.$$

Definition 2.6.4 *GLM (2.1.2) is said to be L-stable if it is A-stable and, in addition,*

$$\lim_{z \to \infty} \rho\left(\mathbf{M}(z)\right) = 0,$$

where $\rho(\mathbf{M}(z))$ stands for the spectral radius of stability matrix $\mathbf{M}(z)$.

For RK methods (2.1.3) the stability matrix $\mathbf{M}(z)$ reduces to the stability function

$$R(z) = 1 + z\mathbf{b}^T(\mathbf{I} - z\mathbf{A})^{-1}\mathbf{e}. \qquad (2.6.6)$$

For linear multistep methods (2.1.4) the stability matrix $\mathbf{M}(z)$ takes the form

$$
\left[
\begin{array}{cccc|cccc}
\dfrac{\alpha_1}{1-z\beta_0} & \cdots & \dfrac{\alpha_{k-1}}{1-z\beta_0} & \dfrac{\alpha_k}{1-z\beta_0} & \dfrac{\beta_1}{1-z\beta_0} & \cdots & \dfrac{\beta_{k-1}}{1-z\beta_0} & \dfrac{\beta_k}{1-z\beta_0} \\
1 & \cdots & 0 & 0 & 0 & \cdots & 0 & 0 \\
\vdots & \ddots & \vdots & \vdots & \vdots & \ddots & \vdots & \vdots \\
0 & \cdots & 1 & 0 & 0 & \cdots & 0 & 0 \\
\hline
\dfrac{\alpha_1 z}{1-z\beta_0} & \cdots & \dfrac{\alpha_{k-1} z}{1-z\beta_0} & \dfrac{\alpha_k z}{1-z\beta_0} & \dfrac{\beta_1 z}{1-z\beta_0} & \cdots & \dfrac{\beta_{k-1} z}{1-z\beta_0} & \dfrac{\beta_k z}{1-z\beta_0} \\
0 & \cdots & 0 & 0 & 1 & \cdots & 0 & 0 \\
\vdots & \ddots & \vdots & \vdots & \vdots & \ddots & \vdots & \vdots \\
0 & \cdots & 0 & 0 & 0 & \cdots & 1 & 0
\end{array}
\right],
$$

where we partitioned this matrix into blocks of dimension $k \times k$. Let $\rho(w)$ and $\sigma(w)$ be the first and second characteristic polynomials of (2.1.4):

$$
\rho(w) = w^k - \sum_{j=1}^{k} \alpha_j w^{k-j}, \quad \sigma(w) = \sum_{j=0}^{k} \beta_j w^{k-j}.
$$

We already encountered the polynomial $\rho(w)$ when we discussed zero-stability of linear multistep methods in Section 2.2. We have the following theorem.

Theorem 2.6.5 *The stability function $p(w, z) = \det(w\mathbf{I} - \mathbf{M}(z))$ of the linear multistep method (2.1.4) written as GLM with $2k$ inputs is given by*

$$
p(w, z) = \frac{1}{1 - \beta_0 z} w^k \big(\rho(w) - z\sigma(w)\big). \tag{2.6.7}
$$

Proof: First consider the case $\beta_0 = 0$, which corresponds to explicit methods. Then the stability function $p(w, z)$, given by

$$
\det
\left[
\begin{array}{cccc|cccc}
w - \alpha_1 & \cdots & -\alpha_{k-1} & -\alpha_k & -\beta_1 & \cdots & -\beta_{k-1} & -\beta_k \\
-1 & \cdots & 0 & 0 & 0 & \cdots & 0 & 0 \\
\vdots & \ddots & \vdots & \vdots & \vdots & \ddots & \vdots & \vdots \\
0 & \cdots & -1 & w & 0 & \cdots & 0 & 0 \\
\hline
-\alpha_1 z & \cdots & -\alpha_{k-1} z & -\alpha_k z & w - \beta_1 z & \cdots & -\beta_{k-1} z & -\beta_k z \\
0 & \cdots & 0 & 0 & -1 & \cdots & 0 & 0 \\
\vdots & \ddots & \vdots & \vdots & \vdots & \ddots & \vdots & \vdots \\
0 & \cdots & 0 & 0 & 0 & \cdots & -1 & w
\end{array}
\right]
$$

is linear in z. Hence,

$$p(w, z) = p(w, 0) + \frac{\partial p}{\partial z}(w, 0)z.$$

It can be verified that $p(w, 0) = w^k \rho(w)$ (compare the discussion in Section 2.2). Subtracting the $(k+1)$st row from the first row of $\frac{\partial p}{\partial z}(w, 0)$, we also have

$$\frac{\partial p}{\partial z}(w, 0) = \det \left[\begin{array}{cccc|cccc} w & \cdots & 0 & 0 & 0 & \cdots & 0 & 0 \\ -1 & \cdots & 0 & 0 & 0 & \cdots & 0 & 0 \\ \vdots & \ddots & \vdots & \vdots & \vdots & \ddots & \vdots & \vdots \\ 0 & \cdots & -1 & w & 0 & \cdots & 0 & 0 \\ \hline -\alpha_1 & \cdots & -\alpha_{k-1} & -\alpha_k & -\beta_1 & \cdots & -\beta_{k-1} & -\beta_k \\ 0 & \cdots & 0 & 0 & -1 & \cdots & 0 & 0 \\ \vdots & \ddots & \vdots & \vdots & \vdots & \ddots & \vdots & \vdots \\ 0 & \cdots & 0 & 0 & 0 & \cdots & -1 & w \end{array} \right]$$

and the routine computations yield

$$\frac{\partial p}{\partial z}(w, 0) = w^k \left(-\beta_1 w^{k-1} - \beta_2 w^{k-2} - \cdots - \beta_k \right) = -w^k \sigma(w).$$

Hence, $p(w, z) = w^k(\rho(w) - z\sigma(w))$, which corresponds to (2.6.7) with $\beta_0 = 0$.

Next consider the case $\beta_0 \neq 0$, which corresponds to implicit methods. Putting

$$\widetilde{\alpha}_i = \frac{\alpha_i}{1 - \beta_0 z}, \quad \widetilde{\beta}_i = \frac{\beta_i}{1 - \beta_0 z}, \quad i = 1, 2, \ldots, k,$$

$p(w, z)$ takes the form

$$\det \left[\begin{array}{cccc|cccc} w - \widetilde{\alpha}_1 & \cdots & -\widetilde{\alpha}_{k-1} & -\widetilde{\alpha}_k & -\widetilde{\beta}_1 & \cdots & -\widetilde{\beta}_{k-1} & -\widetilde{\beta}_k \\ -1 & \cdots & 0 & 0 & 0 & \cdots & 0 & 0 \\ \vdots & \ddots & \vdots & \vdots & \vdots & \ddots & \vdots & \vdots \\ 0 & \cdots & -1 & w & 0 & \cdots & 0 & 0 \\ \hline -\widetilde{\alpha}_1 z & \cdots & -\widetilde{\alpha}_{k-1} z & -\widetilde{\alpha}_k z & w - \widetilde{\beta}_1 z & \cdots & -\widetilde{\beta}_{k-1} z & -\widetilde{\beta}_k z \\ 0 & \cdots & 0 & 0 & -1 & \cdots & 0 & 0 \\ \vdots & \ddots & \vdots & \vdots & \vdots & \ddots & \vdots & \vdots \\ 0 & \cdots & 0 & 0 & 0 & \cdots & -1 & w \end{array} \right].$$

Proceeding similarly as before, we obtain

$$
\begin{aligned}
p(w,z) &= w^k \left(w^k - \sum_{j=1}^{k} \widetilde{\alpha}_j w^{k-j} - z \sum_{j=1}^{k} \widetilde{\beta}_j w^{k-j} \right) \\
&= \frac{1}{1 - \beta_0 z} w^k \left(w^k (1 - \beta_0 z) - \sum_{j=1}^{k} \alpha_j w^{k-j} - z \sum_{j=1}^{k} \beta_j w^{k-j} \right) \\
&= \frac{1}{1 - \beta_0 z} w^k \big(\rho(w) - z\sigma(w) \big),
\end{aligned}
$$

which is again (2.6.7) and proves our claim. ∎

Using the representation of linear multistep methods with k inputs proposed by Butcher and Hill [63] discussed in Section 2.1, the stability matrix $\mathbf{M}(z)$ takes the form

$$
\mathbf{M}(z) =
\begin{bmatrix}
0 & \cdots & 0 & \alpha_k + \dfrac{z}{1 - \beta_0 z}(\alpha_k \beta_0 + \beta_k) \\
1 & \cdots & 0 & \alpha_{k-1} + \dfrac{z}{1 - \beta_0 z}(\alpha_{k-1} \beta_0 + \beta_{k-1}) \\
\vdots & \ddots & \vdots & \vdots \\
0 & \cdots & 1 & \alpha_1 + \dfrac{z}{1 - \beta_0 z}(\alpha_1 \beta_0 + \beta_1)
\end{bmatrix}.
$$

The characteristic polynomial of this matrix is described in the following result.

Theorem 2.6.6 *The stability function $p(w,z) = \det(w\mathbf{I} - \mathbf{M}(z))$ of the linear multistep method (2.1.4) written as GLM with k inputs is given by*

$$
p(w,z) = \frac{1}{1 - \beta_0 z} \big(\rho(w) - z\sigma(w) \big). \tag{2.6.8}
$$

Proof: Expanding $p(w,z) = \det(w\mathbf{I} - \mathbf{M}(z))$ with respect to the last row, we obtain

$$
\begin{aligned}
p(w,z) &= w^k - \sum_{j=1}^{k} \alpha_j w^{k-j} - \frac{z}{1 - \beta_0 z} \sum_{j=1}^{k} (\alpha_j \beta_0 + \beta_j) w^{k-j} \\
&= \rho(w) - \frac{\beta_0 z}{1 - \beta_0 z} (w^k - \rho(w)) - \frac{z}{1 - \beta_0 z} (\sigma(w) - \beta_0 w^k) \\
&= \left(1 + \frac{\beta_0 z}{1 - \beta_0 z} \right) \rho(w) - \frac{z}{1 - \beta_0 z} \sigma(w) \\
&= \frac{1}{1 - \beta_0 z} \big(\rho(w) - z\sigma(w) \big),
\end{aligned}
$$

which is our claim. ∎

The polynomial
$$\pi(w, z) := \rho(w) - z\,\sigma(w)$$
is called the characteristic polynomial of linear multistep method (2.1.4) (compare [194, 195]). Denote by $w_i(z)$, $i = 1, 2, \ldots, k$, the roots of this polynomial. Then it follows from Theorem 2.6.5 or from Theorem 2.6.6 that the region of absolute stability of (2.1.4) is given by

$$\mathcal{A} = \Big\{ z \in \mathbb{C} : \ \big|w_i(z)\big| < 1, \ i = 1, 2, \ldots, k \Big\},$$

which is in agreement with the classical theory of linear multistep methods described by Lambert [194, 195].

For one-leg methods (2.1.5), the stability matrix $\mathbf{M}(z)$ takes the form

$$\mathbf{M}(z) = \begin{bmatrix} m_{11}(z) & \cdots & m_{1,k-1}(z) & m_{1,k}(z) \\ 1 & \cdots & 0 & 0 \\ \vdots & \ddots & \vdots & \vdots \\ 0 & \cdots & 1 & 0 \end{bmatrix}$$

with
$$m_{1j}(z) = \alpha_j + \frac{z}{1 - \beta_0 z}(\beta_0 \alpha_j + \beta_j),$$

$j = 1, 2, \ldots, k$, and it follows that the stability function $p(w, z)$ has the form (2.6.8) as described in Theorem 2.6.6 for linear multistep methods written as GLMs with k inputs.

For TSRK methods (2.1.12), the stability matrix $\mathbf{M}(z)$ takes the form

$$\mathbf{M}(z) = \begin{bmatrix} 1 - \vartheta + z\mathbf{v}^T\mathbf{S}(z)(\mathbf{e} - \mathbf{u}) & \vartheta + z\mathbf{v}^T\mathbf{S}(z)\mathbf{u} & \mathbf{w}^T + z\mathbf{v}^T\mathbf{S}(z)\mathbf{B} \\ 1 & 0 & \mathbf{0} \\ z\mathbf{S}(z)(\mathbf{e} - \mathbf{u}) & z\mathbf{S}(z)\mathbf{u} & z\mathbf{S}(z)\mathbf{B} \end{bmatrix},$$

where $\mathbf{S}(z) = (\mathbf{I} - z\mathbf{A})^{-1}$. The construction of TSRK methods with appropriate stability properties is discussed in Chapter 5.

We can also analyze linear stability properties of other classes of GLMs, but this is not discussed further here. Stability properties of two-step peer methods from the GLM point of view are discussed by Izzo and Jackiewicz [175].

2.7   TYPES OF GENERAL LINEAR METHODS

The coefficient matrix $\mathbf{A}$ plays a role similar to the coefficient matrix $\mathbf{A}$ of RK formulas and determines the implementation costs of GLMs. Depending on its structure, we divide methods (2.1.2) into four types, depending on the nature of the differential system to be solved and the computer architecture that is used to implement these methods. For GLMs of type 1 or 2, matrix $\mathbf{A}$ has the form

$$\mathbf{A} = \begin{bmatrix} \lambda & & & \\ a_{21} & \lambda & & \\ \vdots & \vdots & \ddots & \\ a_{s1} & a_{s2} & \cdots & \lambda \end{bmatrix}, \tag{2.7.1}$$

where $\lambda = 0$ or $\lambda > 0$, respectively. Such methods are appropriate for nonstiff or stiff differential systems in a sequential computing environment. For type 3 or 4 methods, matrix $\mathbf{A}$ takes the form

$$\mathbf{A} = \mathrm{diag}(\lambda, \lambda, \ldots, \lambda) = \lambda \mathbf{I}, \tag{2.7.2}$$

where $\lambda = 0$ or $\lambda > 0$, respectively. Such methods are appropriate for nonstiff or stiff differential systems in a parallel computing environment.

Type 1 RK methods correspond to explicit formulas and type 2 methods to singly diagonally implicit Runge-Kutta (SDIRK) formulas [9]. The concepts of type 3 and type 4 methods are of little relevance in the context of RK formulas. In the context of linear multistep methods (2.1.4), type 1 or type 3 methods correspond to explicit formulas, and type 2 or type 4 methods correspond to implicit formulas.

It will often be our goal to construct GLMs that will have the same stability properties as explicit RK formulas or SDIRK methods. For this reason we review below some relevant facts about the stability of these RK methods.

Applying any RK method with a step size h to the test equation

$$y' = \xi y, \quad t \geq 0, \quad y(0) = 1,$$

whose exact solution is $y(t) = \exp(t)$, we obtain after one step $y_1 = R(z)$, where $z = h\xi$ and $R(z)$ is the stability function defined by (2.6.6). Hence, assuming that the RK method has order p, we can conclude that

$$R(z) = e^z + O(z^{p+1}) \quad \text{as} \quad z \to 0.$$

For explicit s-stage RK methods of order p the function $R(z)$ is a polynomial of degree s of the form

$$R(z) = 1 + z + \frac{z^2}{2!} + \cdots + \frac{z^p}{p!} + c_{p+1} z^{p+1} + \cdots + c_s z^s,$$

where $c_{p+1}, c_{p+2}, \ldots, c_s$ are some constants. In particular, for explicit RK formulas with $p = s \leq 4$, this function has the unique form

$$R(z) = 1 + z + \frac{z^2}{2!} + \cdots + \frac{z^s}{s!},$$

which is equal to the polynomial approximation of degree s and order $p = s$ to the exponential function $\exp(z)$. Stability regions of the corresponding RK methods are plotted in Fig. 2.7.1.

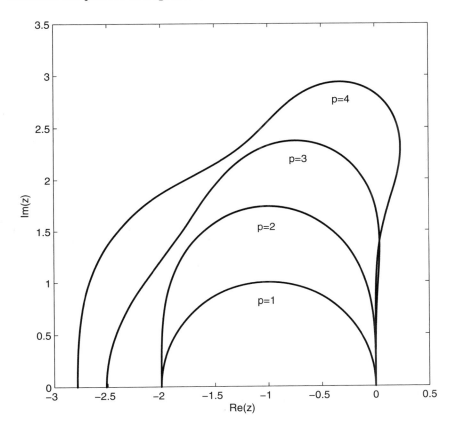

Figure 2.7.1   Stability regions of explicit RK methods with $p = s \leq 4$

For implicit RK methods $R(z)$ is the rational function

$$R(z) = \frac{P(z)}{Q(z)},$$

where $P(z)$ and $Q(z)$ are polynomials of degree $\leq s$. It follows from the maximum principle applied to the negative half-plane $\mathbb{C}^-$ that the RK method is A-stable if and only if it is stable on the imaginary axis (I-stable):

$$|R(iy)| \leq 1 \quad \text{for all} \quad y \in \mathbb{R},$$

and $R(z)$ is analytic for $\text{Re}(z) < 0$ (i.e., $Q(z)$ does not have roots with negative or zero real parts). I-stability is equivalent to the fact that the Nørsett polynomial defined by

$$E(y) := |Q(iy)|^2 - |P(iy)|^2 = Q(iy)Q(-iy) - P(iy)P(-iy)$$

satisfies

$$E(y) \geq 0 \quad \text{for all} \quad y \in \mathbb{R}. \tag{2.7.3}$$

Consider next SDIRK methods, which can be represented by the Butcher tableaux

$$\begin{array}{c|ccccc}
c_1 & \lambda \\
c_2 & a_{21} & \lambda \\
\vdots & \vdots & \vdots & \ddots \\
c_{s-1} & a_{s-1,1} & a_{s-1,2} & \cdots & \lambda \\
c_s & a_{s1} & a_{s2} & \cdots & a_{s,s-1} & \lambda \\
\hline
& b_1 & b_2 & \cdots & b_{s-1} & b_s
\end{array} \tag{2.7.4}$$

Assume that SDIRK method (2.7.4) has order $p = s$. Then it can be demonstrated (compare, e.g., [146]) that its stability function takes the form

$$R(z) = \frac{P(z)}{(1 - \lambda z)^s}, \quad P(z) = (-1)^s \sum_{j=0}^{s} L_s^{(s-j)}\left(\frac{1}{\lambda}\right)(\lambda z)^j, \tag{2.7.5}$$

where

$$L_s(x) = \sum_{j=0}^{s} (-1)^j \binom{s}{j} \frac{x^j}{j!} \tag{2.7.6}$$

is the Laguerre polynomial of degree s and $L_s^{(k)}(x)$ stands for the kth derivative. The Nørsett polynomials, denoted by $E_s(y)$, for SDIRK methods with $p = s \leq 4$ are given below (compare [146]), where the explicit formula for $E_s(y)$ is also given.

$$E_1(y) = y^2(2\lambda - 1),$$

$$E_2(y) = y^4\left(-\frac{1}{4} + 2\lambda - 5\lambda^2 + 4\lambda^3\right) = y^4(2\lambda - 1)^2\left(\lambda - \frac{1}{4}\right),$$

$$E_3(y) = y^4\left(\frac{1}{12} - \lambda + 3\lambda^2 - 2\lambda^3\right)$$
$$+ y^6\left(-\frac{1}{36} + \frac{1}{2}\lambda - \frac{13}{4}\lambda^2 + \frac{28}{3}\lambda^3 - 12\lambda^4 + 6\lambda^5\right),$$

$$E_4(y) = y^6\left(\frac{1}{72} - \frac{1}{3}\lambda + \frac{17}{6}\lambda^2 - \frac{32}{3}\lambda^3 + 17\lambda^4 - 8\lambda^5\right)$$
$$+ y^8\left(-\frac{1}{576} + \frac{1}{8}\lambda - \frac{25}{36}\lambda^2 + \frac{13}{3}\lambda^3 - \frac{173}{12}\lambda^4 + \frac{76}{3}\lambda^5 - 22\lambda^6 + 8\lambda^7\right).$$

Since function (2.7.5) is analytic in $\mathbb{C}^-$ for $\lambda > 0$, it follows that A-stability is equivalent to $E_s(y) > 0$ for all $y \in \mathbb{R}$ (compare (2.7.3)); that is, all coefficients of $E_s(y)$ are nonnegative. We reproduce in Table 2.7.1 the results given by Hairer and Wanner [146] (compare Table 6.3, p. 97) about A-stability of SDIRK methods of order $p \geq s$ for $s = 1, 2, \ldots, 8$.

s	A-stability and $p = s$	A-stability and $p = s + 1$
1	$1/2 \leq \lambda < \infty$	$1/2$
2	$1/4 \leq \lambda < \infty$	$(3 + \sqrt{3})/6$
3	$1/3 \leq \lambda \leq 1.06857902$	1.06857902
4	$0.39433757 \leq \lambda \leq 1.28057976$	——
5	$\begin{cases} 0.24650519 \leq \lambda \leq 0.36180340 \\ 0.42078251 \leq \lambda \leq 0.47326839 \end{cases}$	0.47326839
6	$0.28406464 \leq \lambda \leq 0.54090688$	——
7	——	——
8	$0.21704974 \leq \lambda \leq 0.26471425$	——

Table 2.7.1 A-stability of the functions $R(z)$ defined by (2.7.5) for $p \geq s$

s	L-stability and $p = s - 1$	L-stability and $p = s$
2	$(2 - \sqrt{2})/2 \leq \lambda \leq (2 + \sqrt{2})/2$	$\lambda = (2 \pm \sqrt{2})/2$
3	$0.18042531 \leq \lambda \leq 2.18560010$	$\lambda = 0.43586652$
4	$0.22364780 \leq \lambda \leq 0.57281606$	$\lambda = 0.57281606$
5	$0.24799464 \leq \lambda \leq 0.67604239$	$\lambda = 0.27805384$
6	$0.18391465 \leq \lambda \leq 0.33414237$	$\lambda = 0.33414237$
7	$0.20408345 \leq \lambda \leq 0.37886489$	——
8	$0.15665860 \leq \lambda \leq 0.23437316$	$\lambda = 0.23437316$

Table 2.7.2 L-stability of $R(z)$ with $P(z)$ defined by (2.7.7) for $p \geq s - 1$

We conclude this section with a discussion of stiffly accurate SDIRK methods, i.e., methods for which

$$a_{sj} = b_j, \quad j = 1, 2, \ldots, s.$$

This property implies that the numerical approximation y_{n+1} is equal to the last stage, $Y_s^{[n]}$. For these methods the numerator $P(z)$ of the stability function $R(z)$ takes the form

$$P(z) = (-1)^s \sum_{j=0}^{s-1} L_s^{(s-j)}\left(\frac{1}{\lambda}\right)(\lambda z)^j \qquad (2.7.7)$$

(compare [146]). This is a polynomial of degree $s - 1$ and it follows that

$$\lim_{z \to \infty} R(z) = 0.$$

Hence, if these methods are A-stable, they are also L-stable. We reproduce in Table 2.7.2 the results from [146] (compare Table 6.4, p. 98) about L-stability of stiffly accurate SDIRK methods of order $p \geq s - 1$ for $s = 2, 3, \ldots, 8$.

2.8   ILLUSTRATIVE EXAMPLES OF GENERAL LINEAR METHODS

In this section we describe the construction of GLMs of all four types of order p and stage order $q = p - 1$ and $q = p$ with some desired stability properties. To illustrate various design criteria in the construction of such methods we restrict our attention in this section to methods with $p = r = s = 2$ and $q = 1$ or 2. Moreover, we always assume that abscissa vector $\mathbf{c} = [0, 1]^T$ and that the coefficient matrix $\mathbf{U} = \mathbf{I}$, where $\mathbf{I}$ is an identity matrix of dimension 2.

2.8.1   Type 1: $p = r = s = 2$ and $q = 1$ or 2

These methods take the form

$$\left[\begin{array}{c|c} \mathbf{A} & \mathbf{U} \\ \hline \mathbf{B} & \mathbf{V} \end{array}\right] = \left[\begin{array}{cc|cc} 0 & 0 & 1 & 0 \\ a_{21} & 0 & 0 & 1 \\ \hline b_{11} & b_{12} & v_{11} & v_{12} \\ b_{21} & b_{22} & v_{21} & v_{22} \end{array}\right],$$

where $a_{21} \neq 0$. Consider first the case $p = q + 1 = r = s = 2$. Solving stage order and order conditions with $q = 1$ and $p = 2$ given by (2.4.10) and (2.4.9) leads to a five-parameter family of methods depending on a_{21}, v_{11}, v_{21}, q_{12}, and q_{22}. The stability polynomial of these methods takes the form

$$p(w, z) = w^2 - (p_{10} + p_{11}z + p_{12}z^2)w + p_{00} + p_{01}z + p_{02}z^2,$$

where the p_{ij} depend on the free parameters of the methods. As a design criterion we look first for methods that have the same stability properties as explicit RK methods of order $p = 2$ with two stages. This leads to the system of equations

$$p_{00} = 0, \quad p_{01} = 0, \quad p_{02} = 0,$$

where p_{ij} are polynomials with respect to a_{21}, v_{21}, q_{12}, and q_{22}. The solution of this system leads to two families of methods, depending on v_{11}. These families are given by

$$
\left[\begin{array}{c|c} \mathbf{A} & \mathbf{U} \\ \hline \mathbf{B} & \mathbf{V} \end{array}\right] =
\left[\begin{array}{cc|cc}
0 & 0 & 1 & 0 \\
1 & 0 & 0 & 1 \\
\hline
\frac{1}{2} & \frac{1}{2} & v_{11} & 1 - v_{11} \\
\frac{1}{2} & \frac{1}{2} & v_{11} & 1 - v_{11}
\end{array}\right]
$$

and

$$
\left[\begin{array}{c|c} \mathbf{A} & \mathbf{U} \\ \hline \mathbf{B} & \mathbf{V} \end{array}\right] =
\left[\begin{array}{cc|cc}
0 & 0 & 1 & 0 \\
\frac{1}{2v_{11}(1-v_{11})} & 0 & 0 & 1 \\
\hline
\frac{2v_{11}^3 + 1}{2v_{11}} & v_{11}(1 - v_{11}) & v_{11} & 1 - v_{11} \\
\frac{2v_{11}^4 - 2v_{11}^2 + 2v_{11} - 1}{2v_{11}(v_{11} - 1)} & \frac{-2v_{11}^3 + 2v_{11} - 1}{2v_{11}} & v_{11} & 1 - v_{11}
\end{array}\right].
$$

Observe that the spectrum of the coefficient matrix $\mathbf{V}$ is $\sigma(\mathbf{V}) = \{1, 0\}$, and as a consequence, these methods are zero-stable. It follows from Theorem 2.5.1 that the local discretization error $\mathrm{le}(t_n)$ of these methods takes the form

$$
\mathrm{le}(t_n) = \left((\varphi_2 \otimes \mathbf{I}) y'''(t_{n-1}) + (\psi_2 \otimes \mathbf{I}) \frac{\partial f}{\partial y}(y(t_{n-1})) y''(t_{n-1}) \right) h^3 + O(h^4).
$$

Here, for the first family of methods,

$$
\varphi_2 = \left[\begin{array}{cc} q_{22} - \frac{1}{12} & q_{22} - \frac{1}{12} \end{array}\right]^T, \quad
\psi_2 = \left[\begin{array}{cc} \frac{1}{4} - q_{22} & \frac{1}{4} - q_{22} \end{array}\right]^T,
$$

and choosing $q_{22} = \frac{1}{4}$, the local discretization error takes a simple form,

$$
\mathrm{le}(t_n) = (\varphi_2 \otimes \mathbf{I}) h^3 y'''(t_{n-1}) + O(h^4)
$$

with $\varphi_2 = [\frac{1}{6}, \frac{1}{6}]$. The expressions for φ_2 and ψ_2 are more complicated for the second family of methods and are not listed here. It can be verified that for this family $\psi_2 = 0$ if and only if $q_{22} = \frac{1}{2}$ and $v_{11} = \frac{1}{2}$. This corresponds to the method with stage order $q = 2$ discussed below.

We would like to reiterate that by design all these methods have Runge-Kutta stability (RK stability) properties: that is, theirs stability polynomials take the form

$$
p(w, z) = w(w - R(z)),
$$

where $R(z) = 1 + z + \frac{1}{2}z^2$ is a stability function of explicit RK method of order $p = 2$. All these methods are examples of DIMSIMs, which are the subject of Chapter 3.

Consider next the case $p = q = r = s = 2$. Solving stage order and order conditions (2.4.8) and (2.4.9) with $q = 2$ and $p = 2$ leads to a three-parameter family of methods, depending on a_{21}, v_{11}, and v_{21}. This family takes the form

$$
\left[\begin{array}{c|c} \mathbf{A} & \mathbf{U} \\ \hline \mathbf{B} & \mathbf{V} \end{array} \right] = \left[\begin{array}{cc|cc} 0 & 0 & 1 & 0 \\ a_{21} & 0 & 0 & 1 \\ \hline a_{21}(1 - v_{11}) + \frac{1}{2}v_{11} & \frac{1}{2}v_{11} & v_{11} & 1 - v_{11} \\ \frac{1}{2}(2a_{21} - 1)(1 - v_{21}) & \frac{1}{2}(v_{21} + 3) - a_{21} & v_{21} & 1 - v_{21} \end{array} \right].
$$

The spectrum of the coefficient matrix $\mathbf{V}$ is $\sigma(\mathbf{V}) = \{1, v_{11} - v_{21}\}$ and we have to impose the condition

$$-1 \leq v_{11} - v_{21} < 1 \tag{2.8.1}$$

to guarantee zero-stability. The stability polynomial of these methods takes the form

$$p(w, z) = w^2 - p_1(z)w + p_0(z),$$

where $p_1(z)$ and $p_0(z)$ are polynomials in z given by

$$p_1(z) = 1 + v_{11} - v_{21} + \left(\frac{1}{2}v_{11}(1 - 2a_{21}) + \frac{1}{2}(v_{21} + 3) \right)z + \frac{1}{2}v_{11}a_{21}z^2,$$

$$p_0(z) = v_{11} - v_{21} + \left(\frac{1}{2}(1 + 3v_{11}) - a21v_{11} - \frac{1}{2}v_{21} \right)z + \frac{1}{2}v_{11}(2 - a_{21})z^2.$$

These methods have RK stability if $p_0(z) \equiv 0$, which leads to the system

$$v_{11} - v_{21} = 0, \quad 1 + 3v_{11} - 2a_{21}v_{11} - v_{21} = 0, \quad v_{11}(2 - a_{21}) = 0.$$

This system has a unique solution given by $a_{21} = 2$, $v_{11} = v_{21} = \frac{1}{2}$ and the resulting GLM is

$$
\left[\begin{array}{c|c} \mathbf{A} & \mathbf{U} \\ \hline \mathbf{B} & \mathbf{V} \end{array} \right] = \left[\begin{array}{cc|cc} 0 & 0 & 1 & 0 \\ 2 & 0 & 0 & 1 \\ \hline \frac{5}{4} & \frac{1}{4} & \frac{1}{2} & \frac{1}{2} \\ \frac{3}{4} & -\frac{1}{4} & \frac{1}{2} & \frac{1}{2} \end{array} \right].
$$

The local discretization error of this method is

$$\mathrm{le}(t_n) = (\varphi_2 \otimes \mathbf{I})y'''(t_n)h^3 + O(h^4) \quad \text{with} \quad \varphi_2 = \left[\begin{array}{cc} \frac{1}{24} & \frac{7}{24} \end{array} \right]^T.$$

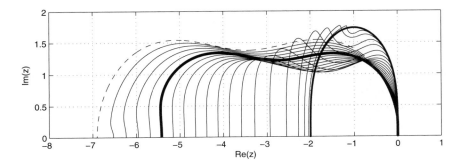

Figure 2.8.1   Stability regions of GLMs of type 1 with $p = q = r = s = 2$

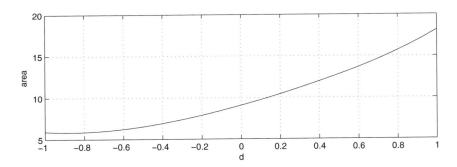

Figure 2.8.2   Area of stability region versus d for GLMs of type 1 with $p = q = r = s = 2$

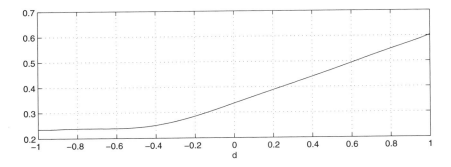

Figure 2.8.3   Norm of φ_2 versus d for GLMs of type 1 with $p = q = r = s = 2$

We can also consider different design criteria in the construction of GLMs. For example, instead of searching for methods with RK stability, we may try to construct methods with maximal regions or intervals of absolute stability. In the context of methods discussed above, this can be done by a computer search in the parameter space a_{21}, v_{11}, and v_{21} subject to the constraint (2.8.1) to guarantee zero-stability. Accordingly, we choose $d = v_{11} - v_{21}$ in advance from the interval $[-1, 1)$ and then search in the parameter space (a_{21}, v_{11}) of dimension 2, for methods with maximal regions of absolute stability. Stability regions of the resulting methods are plotted in Fig. 2.8.1 for $d = -1 + i\Delta d$, $i = 0, 1, \ldots, N$, $\Delta d = 2/N$, $N = 20$, where we show by a thick solid line the stability region corresponding to GLM with $d = 1/2$. The coefficients of this method expressed in rational format are

$$
\left[\begin{array}{c|c} \mathbf{A} & \mathbf{U} \\ \hline \mathbf{B} & \mathbf{V} \end{array}\right] = \left[\begin{array}{cc|cc} 0 & 0 & 1 & 0 \\ \frac{2490}{1943} & 0 & 0 & 1 \\ \hline \frac{2723542}{2656081} & \frac{224}{1367} & \frac{448}{1367} & \frac{919}{1367} \\ \frac{9733585}{10624324} & \frac{1406013}{10624324} & -\frac{471}{2734} & \frac{3205}{2734} \end{array}\right].
$$

For comparison, we also show by a thick solid line the stability region of two-stage RK methods of order $p = 2$. The method corresponding to $d = 1$ is not zero-stable, and its stability region is plotted by a dashed line. In Fig. 2.8.2 we have plotted the area of stability regions versus d, and in Fig. 2.8.3 the Euclidean norm of the vector φ_2 appearing in the local discretization error versus d.

2.8.2   Type 2: $p = r = s = 2$ and $q = 1$ or 2

These methods take the form

$$
\left[\begin{array}{c|c} \mathbf{A} & \mathbf{U} \\ \hline \mathbf{B} & \mathbf{V} \end{array}\right] = \left[\begin{array}{cc|cc} \lambda & 0 & 1 & 0 \\ a_{21} & \lambda & 0 & 1 \\ \hline b_{11} & b_{12} & v_{11} & v_{12} \\ b_{21} & b_{22} & v_{21} & v_{22} \end{array}\right],
$$

where $\lambda \neq 0$ and $a_{21} \neq 0$. As before, consider first the case $p = q + 1 = r = s = 2$. Solving stage order and order conditions with $q = 1$ and $p = 2$ given by (2.4.10) and (2.4.9) leads to a six-parameter family of methods depending on λ, a_{21}, v_{11}, v_{21}, q_{12}, and q_{22}. To analyze stability properties of these methods it is more convenient to work with a polynomial $(1 - \lambda z)^2 p(w, z)$, which we denote by the same symbol $p(w, z)$, than with the rational stability function, and we adopt this convention for methods of type 2 and 4.

This stability polynomial takes the form

$$p(w, z) = (1 - \lambda z)^2 w^2 - (p_{10} + p_{11}z + p_{12}z^2)w + p_{00} + p_{01}z + p_{02}z^2,$$

where the p_{ij} depend on the free parameters of the methods. Similarly as for type 1 methods, as a design criterion we look first for methods that have the same stability properties as SDIRK methods of order $p = 2$ with two stages. This leads to the system of equations

$$p_{00} = 0, \quad p_{01} = 0, \quad p_{02} = 0,$$

where the p_{ij} are polynomials with respect to λ, a_{21}, v_{21}, q_{12}, and q_{22}. The solution of this system leads to two families of methods, depending on λ and v_{11}. These families are given by

$$\left[\begin{array}{c|c} \mathbf{A} & \mathbf{U} \\ \hline \mathbf{B} & \mathbf{V} \end{array}\right] = \left[\begin{array}{cc|cc} \lambda & 0 & 1 & 0 \\ 1 & \lambda & 0 & 1 \\ \hline \frac{1}{2} + \lambda & \frac{1}{2} - \lambda & v_{11} & 1 - v_{11} \\ \frac{1}{2} + \lambda & \frac{1}{2} - \lambda & v_{11} & 1 - v_{11} \end{array}\right]$$

and

$$\left[\begin{array}{c|c} \mathbf{A} & \mathbf{U} \\ \hline \mathbf{B} & \mathbf{V} \end{array}\right] = \left[\begin{array}{cc|cc} \lambda & 0 & 1 & 0 \\ \frac{1 - 2\lambda}{2v_{11}(1 - v_{11})} & \lambda & 0 & 1 \\ \hline \frac{2v_{11}^3 - 2\lambda + 1}{2v_{11}} & v_{11}(1 - v_{11}) & v_{11} & 1 - v_{11} \\ b_{21} & b_{22} & v_{11} & 1 - v_{11} \end{array}\right],$$

with

$$b_{21} = \frac{2v_{11}^4 - 2v_{11}^2 + 2v_{11} - 1 + 2\lambda - 4v_{11}\lambda}{2v_{11}(v_{11} - 1)},$$

$$b_{22} = \frac{-2v_{11}^3 + 2v_{11} - 1 + 2\lambda}{2v_{11}}.$$

For the first family of methods, vectors φ_2 and ψ_2 in the local discretization error take the form

$$\varphi_2 = \left[\begin{array}{cc} q_{22} - \frac{1}{12} & q_{22} - \frac{1}{12} \end{array}\right]^T,$$

$$\psi_2 = \left[\begin{array}{cc} \lambda^2 - \lambda + \frac{1}{4} - q_{22} & \lambda^2 - \lambda + \frac{1}{4} - q_{22} \end{array}\right]^T,$$

and choosing $q_{22} = \lambda^2 - \lambda + \frac{1}{4}$ leads to methods with $\psi_2 = 0$. The expressions for φ_2 and ψ_2 are more complicated for the second family of methods and are not listed.

The stability polynomial for both families of methods is independent of v_{11} and takes the form

$$p(w, z) = w\left((1 - \lambda z)^2 w - 1 - (1 - 2\lambda)z - \left(\frac{1}{2} - 2\lambda + \lambda^2\right)z^2\right).$$

We would like to stress again that, by design, the nonzero solution to $p(w, z) = 0$ is equal to the stability function of SDIRK method of order $p = 2$ with two stages. Hence, it follows from the theory of SDIRK methods that both families of GLMs are A-stable if $\lambda \geq \frac{1}{4}$ (compare Table 2.7.1). These methods are also L-stable if $\lambda = (2 \pm \sqrt{2})/2$ (compare Table 2.7.2).

Next consider the case $p = q = r = s = 2$. Solving stage order and order conditions (2.4.8) and (2.4.9) with $q = 2$ and $p = 2$ and imposing RK stability, we obtain a one-parameter family of methods given by

$$\left[\begin{array}{c|c} \mathbf{A} & \mathbf{U} \\ \hline \mathbf{B} & \mathbf{V} \end{array}\right] = \left[\begin{array}{cc|cc} \lambda & 0 & 1 & 0 \\ \frac{2}{1+2\lambda} & \lambda & 0 & 1 \\ \hline \frac{8\lambda^3+12\lambda^2-2\lambda+5}{4(2\lambda+1)} & \frac{1-4\lambda^2}{4} & \frac{1}{2}+\lambda & \frac{1}{2}-\lambda \\ \frac{8\lambda^3+20\lambda^2-2\lambda+3}{4(2\lambda+1)} & \frac{-8\lambda^3-12\lambda^2+10\lambda-1}{4(2\lambda+1)} & \frac{1}{2}+\lambda & \frac{1}{2}-\lambda \end{array}\right].$$

The vector φ_2 takes the form

$$\varphi_2 = \left[\begin{array}{cc} \frac{1}{2}\lambda^2 - \frac{1}{2}\lambda + \frac{1}{24} & \frac{1}{2}\lambda^2 - \lambda + \frac{7}{24} \end{array}\right]^T.$$

These methods are A-stable if $\lambda \geq \frac{1}{4}$ and attain order $p = 3$ if $\lambda = (3+\sqrt{3})/6$ (compare Table 2.7.1). These methods are also L-stable for $\lambda = (2 \pm \sqrt{2})/2$ (compare Table 2.7.2).

2.8.3 Type 3: $p = r = s = 2$ and $q = 1$ or 2

These methods take the form

$$\left[\begin{array}{c|c} \mathbf{A} & \mathbf{U} \\ \hline \mathbf{B} & \mathbf{V} \end{array}\right] = \left[\begin{array}{cc|cc} 0 & 0 & 1 & 0 \\ 0 & 0 & 0 & 1 \\ \hline b_{11} & b_{12} & v_{11} & v_{12} \\ b_{21} & b_{22} & v_{21} & v_{22} \end{array}\right].$$

Consider first the case $p = q + 1 = r = s = 2$. Solving stage order and order conditions (2.4.10) and (2.4.9) with $q = 1$ and $p = 2$ leads to a four-parameter family of methods depending on v_{11}, v_{21}, q_{12}, and q_{22}. Similarly as for type 1 formulas, the stability polynomial takes the form

$$p(w, z) = w^2 - (p_{10} + p_{11}z + p_{12}z^2)w + p_{00} + p_{01}z + p_{02}z^2,$$

where the p_{ij} depend on the free parameters of the methods. It can be verified that $p_{00} = 0$ if $v_{21} = v_{11}$ and then that $p_{01} - p_{02} = \frac{1}{2}$. This implies that type 3 methods with RK stability do not exist. There exists, however, a two-parameter family of methods depending on q_{12} and q_{22} such that $\psi_2 = 0$. The coefficients of these methods are not listed here.

Consider next the case $p = q = r = s = 2$. Solving stage order and order conditions (2.4.8) and (2.4.9) with $q = 2$ and $p = 2$, we obtain a two-parameter family of methods of the form

$$
\left[\begin{array}{c|c} \mathbf{A} & \mathbf{U} \\ \hline \mathbf{B} & \mathbf{V} \end{array}\right] = \left[\begin{array}{cc|cc} 0 & 0 & 1 & 0 \\ 0 & 0 & 0 & 1 \\ \hline \frac{v_{11}}{2} & \frac{v_{11}}{2} & v_{11} & 1 - v_{11} \\ \frac{v_{21}-1}{2} & \frac{v_{21}+3}{2} & v_{21} & 1 - v_{21} \end{array}\right].
$$

For these methods the vector φ_2 appearing in the local discretization error is

$$
\varphi_2 = \left[\begin{array}{cc} \frac{1}{6} - \frac{1}{4}v_{11} & \frac{5}{12} - \frac{1}{4}v_{21} \end{array}\right]^T,
$$

and the stability polynomial $p(w, z)$ takes the form

$$
\begin{aligned}
p(w, z) = {}& w^2 - \left(1 + v_{11} - v_{21} + \left(\frac{1}{2}v_{11} + \frac{1}{2}v_{21} + \frac{3}{2}\right)z\right)w \\
& + v_{11} - v_{21} + \left(\frac{3}{2}v_{11} - \frac{1}{2}v_{21} + \frac{1}{2}\right)z + v_{11}z^2.
\end{aligned}
$$

As in Section 2.8.1, we have searched for type 3 methods with maximal

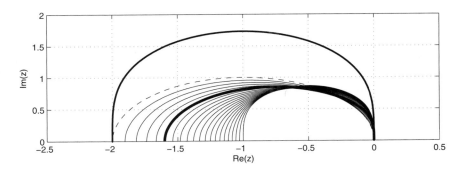

Figure 2.8.4   Stability regions of GLMs of type 3 with $p = q = r = s = 2$

regions of absolute stability. Choosing in advance the parameter $d = v_{11} - v_{21}$, which is equal to the eigenvalue of the coefficient matrix $\mathbf{V}$, this search

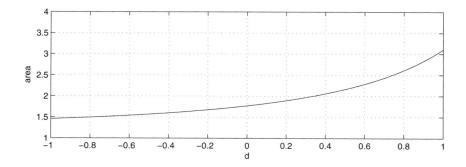

Figure 2.8.5   Area of stability region versus d for GLMs of type 3 with $p = q = r = s = 2$

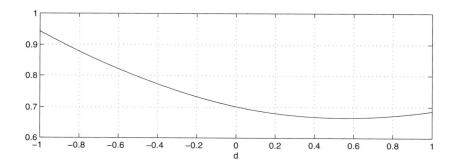

Figure 2.8.6   Norm of φ_2 versus d for GLMs of type 3 with $p = q = r = s = 2$

depends on only one parameter, v_{11}. The results of this search are presented in Figs. 2.8.4, 2.8.5, and 2.8.6. In Fig. 2.8.4 we have plotted stability regions of the resulting methods for $d = -1 + i\Delta$, $i = 1, 2, \ldots, N$, $\Delta d = 2/N$, $N = 20$. The stability region corresponding to $d = 1/2$ is shown by a thick solid line. The coefficients of this method in rational format are

$$
\left[\begin{array}{c|c} \mathbf{A} & \mathbf{U} \\ \hline \mathbf{B} & \mathbf{V} \end{array} \right] = \left[\begin{array}{cc|cc} 0 & 0 & 1 & 0 \\ 0 & 0 & 0 & 1 \\ \hline -\dfrac{2371}{15160} & -\dfrac{2371}{15160} & -\dfrac{2371}{7580} & \dfrac{9951}{7580} \\ -\dfrac{13741}{15160} & \dfrac{16579}{15160} & -\dfrac{6161}{7580} & \dfrac{13741}{7580} \end{array} \right].
$$

As before, we have also shown by a thick line the stability region of a two-stage RK method of order $p = 2$. In Fig 2.8.5 we have plotted the area of the region of absolute stability versus d, and in Fig. 2.8.6 we have plotted the norm of the vector φ_2 versus d.

2.8.4   Type 4: $p = r = s = 2$ and $q = 1$ or 2

These methods take the form

$$
\left[\begin{array}{c|c} \mathbf{A} & \mathbf{U} \\ \hline \mathbf{B} & \mathbf{V} \end{array} \right] = \left[\begin{array}{cc|cc} \lambda & 0 & 1 & 0 \\ 0 & \lambda & 0 & 1 \\ \hline b_{11} & b_{12} & v_{11} & v_{12} \\ b_{21} & b_{22} & v_{21} & v_{22} \end{array} \right],
$$

where $\lambda \neq 0$. Consider first the case $p = q + 1 = r = s = 2$. Solving stage order and order conditions with $q = 1$ and $p = 2$ given by (2.4.10) and (2.4.9) leads to a five-parameter family of methods depending on λ, v_{11}, v_{21}, q_{12}, and q_{22}. Next solving the system

$$
p_{00} = 0, \quad p_{01} = 0, \quad p_{02} = 0
$$

with respect to v_{21}, q_{12}, and λ we obtain a family of type 4 methods with RK stability. Here p_{00}, p_{01} and p_{02} are defined as in Section 2.8.2. The coefficients of these methods are given by

$$
\left[\begin{array}{c|c} \mathbf{A} & \mathbf{U} \\ \hline \mathbf{B} & \mathbf{V} \end{array} \right] = \left[\begin{array}{cc|cc} \frac{1}{2} & 0 & 1 & 0 \\ 0 & \frac{1}{2} & 0 & 1 \\ \hline v_{11}^2 & v_{11}(1 - v_{11}) & v_{11} & 1 - v_{11} \\ v_{11}(1 + v_{11}) & 1 - v_{11}^2 & v_{21} & 1 - v_{21} \end{array} \right].
$$

We have to impose the condition (2.8.1) to guarantee zero-stability. Choosing $q_{22} = -v_{11}^2$, we have $\psi_2 = 0$ and

$$
\varphi_2 = \left[\begin{array}{cccc} \frac{1}{2}v_{11} - \frac{1}{2}v_{11}^2 - \frac{1}{12} & -\frac{1}{2}v_{11}^2 - \frac{1}{12} \end{array} \right]^T.
$$

The stability function of these methods is

$$
p(w, z) = w\left(\left(1 - \frac{1}{2}z\right)^2 w - \left(1 - \frac{1}{4}z^2\right) \right),
$$

which has the roots $w = 0$ and

$$
w = \frac{1 - \frac{1}{4}z^2}{\left(1 - \frac{1}{2}z\right)^2} = \frac{1 + \frac{1}{2}z}{1 - \frac{1}{2}z}.
$$

This corresponds to the stability function of the trapezoidal rule,

$$y_{n+1} = y_n + \frac{1}{2}\big(f(y_n) + f(y_{n+1})\big),\tag{2.8.2}$$

and it follows that these methods are A-stable but not L-stable for any v_{11}. For $v_{11} = 0$ the GLM has stage order $q = 2$ and it is equivalent to the trapezoidal rule. It can be verified that for $\lambda \neq \frac{1}{2}$, methods of type 4 cannot have RK stability. This case is not discussed further here.

Consider next the case $p = q = r = s = 2$. Solving stage order and order conditions (2.4.8) and (2.4.9) with $q = 2$ and $p = 2$ leads to a three-parameter family of methods depending on v_{11}, v_{21}, and λ. Trying to construct methods that will be A- as well as L-stable, we determine the parameters v_{11} and v_{21} from the system of equations

$$p_{02} = 0, \quad p_{12} = 0,$$

where p_{02} and p_{12} are defined as in Section 2.8.2. This leads to a family of methods depending on λ. The coefficients of these methods are given by

$$\left[\begin{array}{c|c} \mathbf{A} & \mathbf{U} \\ \hline \mathbf{B} & \mathbf{V} \end{array}\right] = \left[\begin{array}{cc|cc} \lambda & 0 & 1 & 0 \\ 0 & \lambda & 0 & 1 \\ \hline \frac{\lambda(2\lambda+1)(3\lambda-2)}{2(2\lambda-1)} & \frac{\lambda(2-3\lambda)}{2} & \frac{\lambda(3\lambda-2)}{2\lambda-1} & \frac{(3\lambda-1)(\lambda-1)}{2\lambda-1} \\ \frac{10\lambda^3-7\lambda^2-6\lambda+4}{2(2\lambda-1)} & \frac{\lambda(6-5\lambda)}{2} & \frac{(5\lambda-3)(\lambda-1)}{2\lambda-1} & \frac{-5\lambda^2+10\lambda-4}{2\lambda-1} \end{array}\right].$$

$\lambda \neq \frac{1}{2}$ and the vector φ_2 takes the form

$$\varphi_2 = \left[\begin{array}{cc} \frac{3}{4}\lambda^2 - \lambda + \frac{1}{6} & \frac{5}{4}\lambda^2 - 3\lambda + \frac{7}{6} \end{array}\right]^T.$$

For $\lambda = \frac{1}{2}$ we obtain the GLM discussed above, which is equivalent to the trapezoidal rule (2.8.2). The stability polynomial of the methods corresponding to $\lambda \neq \frac{1}{2}$ is

$$\begin{aligned} p(w, z) &= (1 - \lambda z)^2 w^2 + \frac{2(\lambda^2 - 4\lambda + 2)(\lambda z - 1)}{2\lambda - 1} w \\ &+ \frac{2\lambda^3 - 6\lambda^2 + 6\lambda - 2}{2\lambda - 1} z - \frac{2\lambda^2 - 6\lambda + 3}{2\lambda - 1}. \end{aligned}$$

The eigenvalues of the coefficient matrix $\mathbf{V}$ are

$$\sigma_1 = 1 \quad \text{and} \quad \sigma_2 = \frac{2\lambda^2 - 6\lambda + 3}{1 - 2\lambda}$$

and it is easy to verify that $\sigma_2 \in [-1, 1)$ if

$$2 - \sqrt{2} \leq \lambda \leq 2 + \sqrt{2} \quad \text{and} \quad \lambda \neq 1.\tag{2.8.3}$$

We will use the Schur criterion [253] (compare also [118, 194, 214]) to investigate for which values of λ this polynomial is A-acceptable (i.e., it has roots w_1 and w_2 inside the unit circle for all $z \in \mathbb{C}$ such that $\text{Re}(z) \leq 0$). If this is the case, the corresponding GLM is A-stable. This criterion for a polynomial of any degree k can be formulated as follows. Consider the polynomial

$$\phi(w) = c_k w^k + c_{k-1} w^{k-1} + \cdots + c_1 w + c_0,$$

where c_i are complex coefficients, $c_k \neq 0$ and $c_0 \neq 0$. $\phi(w)$ is said to be a Schur polynomial if all its roots w_i, $i = 1, 2, \ldots, k$, are inside the unit circle. Define

$$\widehat{\phi}(w) = \overline{c}_0 w^k + \overline{c}_1 w^{k-1} + \cdots + \overline{c}_{k-1} w + \overline{c}_k,$$

where $\overline{c}_i$ is the complex conjugate of c_i. Define also the polynomial

$$\phi_1(w) = \frac{1}{w}\left(\widehat{\phi}(0)\phi(w) - \phi(0)\widehat{\phi}(w)\right)$$

of degree at most $k - 1$. We have the following theorem.

Theorem 2.8.1 (Schur [253]) *Polynomial $\phi(w)$ is a Schur polynomial if and only if*

$$\left|\widehat{\phi}(0)\right| > |\phi(0)|$$

and $\phi_1(w)$ is a Schur polynomial.

To apply this criterion to the polynomial $p(w, z)$ corresponding to a one-parameter family of GLMs of type 4 observe first that for $\lambda > 0$ the roots $w_1 = w_1(z)$ and $w_2 = w_2(z)$ are analytic functions of z for $\text{Re}(z) \leq 0$. Hence, it follows from the maximum principle for analytic functions (compare, e.g., [119]) that these roots are inside the unit circle for $\text{Re}(z) \leq 0$ if and only if they are inside the unit circle for z on the imaginary axis (i.e., for $z = iy$, where $y \in \mathbb{R}$). Hence, we can reduce the problem to an easier one, and we only have to investigate for which values of λ the polynomial $q(w, y)$ defined by

$$q(w, y) = p(w, iy)$$

is a Schur polynomial for all $y \in \mathbb{R}$. Define

$$\widehat{q}(w, y) = w^2 q\left(\frac{1}{w}, -y\right),$$

$$q_1(w, y) = \frac{1}{w}\left(\widehat{q}(0, y)q(w, y) - q(0, y)\widehat{q}(w, y)\right),$$

$$\widehat{q}_1(w, y) = wq_1\left(\frac{1}{w}, -y\right),$$

$$q_0(y) = \frac{1}{w}\left(\widehat{q}_1(0, y)q_1(w, y) - q_1(0, y)\widehat{q}_1(w, y)\right).$$

Then it follows from Theorem 2.8.1 that all roots of $q(w, y)$ are in the unit circle for all $y \in \mathbb{R}$ if and only if

$$q_0(y) > 0 \quad \text{and} \quad |\hat{q}(0, y)| > |q(0, y)|.$$

A careful examination of these conditions reveals that this is again the case if λ satisfies (2.8.3) and the corresponding methods are zero- and A-stable as well as L-stable.

2.9   ALGEBRAIC STABILITY OF GENERAL LINEAR METHODS

We start this section with a brief review of the AN-, B-, BN-, and algebraic stability of RK methods (2.1.3). The concept of AN-stability is a generalization of A-stability and is related to the scalar, linear, nonautonomous test equation

$$y' = \xi(t)y, \quad t \geq 0,$$
$$y(0) = y_0,$$
(2.9.1)

$\text{Re}(\xi(t)) \leq 0$, where $\xi(t)$ is an arbitrarily varying complex-valued function. Application of the RK method (2.1.3) to (2.9.1) yields

$$y_{n+1} = (1 + \mathbf{b}^T \xi (\mathbf{I} - \mathbf{A}\xi)^{-1} \mathbf{e}) y_n,$$
(2.9.2)

$n = 0, 1, \ldots$, where the diagonal matrix $\xi \in \mathbb{C}^{s \times s}$ is given by

$$\xi = \text{diag}(\xi_1, \ldots, \xi_s) = \text{diag}(h\xi(t_n + c_1 h), \ldots, h\xi(t_n + c_s h)).$$
(2.9.3)

Set

$$K(\xi) = 1 + \mathbf{b}^T \xi (\mathbf{I} - \mathbf{A}\xi)^{-1} \mathbf{e}.$$

Relation (2.9.2) motivates the following definition.

Definition 2.9.1 ([32],[249]) *RK method (2.1.3) is said to be AN-stable if the function $K(\xi)$ satisfies*

$$|K(\xi)| \leq 1$$

for all $\xi = diag(\xi_1, \ldots, \xi_s)$ such that $\xi_i = \xi_j$ whenever $c_i = c_j$ and such that $\text{Re}(\xi_i) \leq 0$ for $i = 1, 2, \ldots, s$.

Observe that for $\xi = z\mathbf{I}$, $K(\xi)$ reduces to the stability function $R(z)$ defined by (2.6.6) of RK method (2.1.3). Hence, it follows that AN-stability implies A-stability.

Consider next the initial value problem (1.1.2) defined for $t \geq 0$

$$y'(t) = g(t, y(t)), \quad t \geq 0,$$
$$y(0) = y_0,$$
(2.9.4)

$g : \mathbb{R} \times \mathbb{R}^m \to \mathbb{R}^m$, where the function g satisfies the one-sided Lipschitz condition (1.3.4) with a one-sided Lipschitz constant $\nu = 0$:

$$\big(g(t, y_1) - g(t, y_2)\big)^T (y_1 - y_2) \le 0 \tag{2.9.5}$$

for all $t \ge 0$ and $y_1, y_2 \in \mathbb{R}^m$. Equation (2.9.4) with the function g satisfying (2.9.5) was first proposed by Dahlquist [100] for the analysis of stability properties of one-leg methods (2.1.5) for ODEs. Denote by $y(t)$ and $\widetilde{y}(t)$ solutions to (2.9.4) with initial conditions y_0 and $\widetilde{y}_0$, respectively. Then the condition (2.9.5) implies that

$$\big\|y(t_2) - \widetilde{y}(t_2)\big\| \le \big\|y(t_1) - \widetilde{y}(t_1)\big\| \tag{2.9.6}$$

for $0 \le t_1 \le t_2$ (compare [54, 109]). Differential equations (2.9.4) with this property are called dissipative.

Let $\{y_n\}_{n=0}^{\infty}$, $\{Y^{[n]}\}_{n=0}^{\infty}$ be the numerical solution obtained by applying to (2.9.4) the RK method (2.1.3), and let $\{\widetilde{y}_n\}_{n=0}^{\infty}$, $\{\widetilde{Y}^{[n]}\}_{n=0}^{\infty}$ be a solution obtained by perturbing (2.1.3) or by using a different initial value $\widetilde{y}_0$. Following Butcher [38] and Burrage and Butcher [32], we introduce the concepts of B- and BN-stability.

Definition 2.9.2 *RK method (2.1.3) is said to be B-stable if for all autonomous problems (2.9.4) satisfying (2.9.5); that is, problems for which the function g is independent of t, we have*

$$\|y_{n+1} - \widetilde{y}_{n+1}\| \le \|y_n - \widetilde{y}_n\|, \tag{2.9.7}$$

$n = 0, 1, \ldots,$ *for all $h > 0$.*

Definition 2.9.3 *RK method (2.1.3) is said to be BN-stable if for all problems (2.9.4) with g satisfying (2.9.5) the inequality (2.9.7) holds for all $t \ge 0$.*

Clearly, BN-stability implies B-stability. The algebraic characterization of B-stable methods was discovered by Butcher [38] and Crouzeix [96] and the characterization of BN-stability by Burrage and Butcher [32]. Consider the matrix $\mathbf{M}$ defined by

$$\mathbf{M} = \mathbf{B}\mathbf{A} + \mathbf{A}^T\mathbf{B} - \mathbf{b}^T\mathbf{b}, \tag{2.9.8}$$

where $\mathbf{A}$ and $\mathbf{b}$ are coefficients of the RK method (2.1.3) and the matrix $\mathbf{B}$ is defined by

$$\mathbf{B} = \operatorname{diag}(b_1, b_2, \ldots, b_s).$$

Definition 2.9.4 *RK method (2.1.3) is said to be algebraically stable if $\mathbf{B}$ and $\mathbf{M}$ are nonnegative definite.*

We recall that a matrix $\mathbf{A} \in \mathbb{R}^{s \times s}$ is positive definite if $\mathbf{x}^T \mathbf{A} \mathbf{x} > 0$ for all nonzero vectors $\mathbf{x} \in \mathbb{R}^s$. A matrix $\mathbf{A} \in \mathbb{R}^{s \times s}$ is nonnegative definite if

$\mathbf{x}^T\mathbf{A}\mathbf{x} \geq 0$ for all $\mathbf{x} \in \mathbb{R}^s$ (compare [136, 227]). We write $\mathbf{A} > 0$ for a positive definite matrix and $\mathbf{A} \geq 0$ for a nonnegative definite matrix.

The significance of the definition of algebraic stability follows from the following theorem.

Theorem 2.9.5 (compare [32, 38, 96]) *If RK method (2.1.3) is algebraically stable, it is BN-stable.*

The two-stage Gauss-Legendre formula [41, 52, 109]

$$
\begin{array}{c|cc}
\mathbf{c} & \mathbf{A} \\
\hline
& \mathbf{b}^T
\end{array}
=
\begin{array}{c|cc}
\frac{3-\sqrt{3}}{6} & \frac{1}{4} & \frac{3-2\sqrt{3}}{12} \\
\frac{3+\sqrt{3}}{6} & \frac{3+2\sqrt{3}}{12} & \frac{1}{4} \\
\hline
& \frac{1}{2} & \frac{1}{2}
\end{array}
$$

is algebraically stable since $b_i > 0$, $i = 1, 2$, and

$$
\mathbf{M} = \frac{1}{24}\begin{bmatrix} 3 & 3-2\sqrt{3} \\ 3+2\sqrt{3} & 3 \end{bmatrix} + \frac{1}{24}\begin{bmatrix} 3 & 3+2\sqrt{3} \\ 3-2\sqrt{3} & 3 \end{bmatrix} - \frac{1}{4}\begin{bmatrix} 1 & 1 \\ 1 & 1 \end{bmatrix} = \mathbf{0}.
$$

Other examples of algebraically stable RK methods are given in the papers [32, 41, 52, 109].

An excellent discussion of various contractivity and stability properties of RK methods for stiff systems and the relationships between various stability concepts is given by Dekker and Verwer [109].

We now turn our attention to the generalization of concepts of *AN-*, *B-*, *BN-*, and algebraic stability which are relevant in the context of GLMs (2.1.2). This is discussed by Hairer and Wanner [146, Chap. V.9], and in recent papers [53, 54, 158].

Applying GLM (2.1.2) to test equation (2.9.1), we obtain

$$
y^{[n+1]} = \mathbf{S}(\xi)y^{[n]},
$$

$n = 0, 1, \ldots$, where ξ is defined by (2.9.3) and

$$
\mathbf{S}(\xi) = \mathbf{V} + \mathbf{B}\xi(\mathbf{I} - \mathbf{A}\xi)^{-1}\mathbf{U}.
$$

To define *AN-*, *B-*, and *BN-*stability, let $\mathbf{G} = [g_{ij}]_{i,j=1}^r$ be a real, symmetric, and positive definite matrix, and for a vector $y \in \mathbb{R}^{mr}$,

$$
y = \begin{bmatrix} y_1 \\ \vdots \\ y_r \end{bmatrix}, \quad y_i \in \mathbb{R}^m, \quad i = 1, 2, \ldots, r,
$$

consider the inner product norm

$$\|y\|_G^2 = \sum_{i=1}^{r} \sum_{j=1}^{r} g_{ij} y_i^T y_j. \tag{2.9.9}$$

Definition 2.9.6 ([146]) *GLM (2.1.2) is said to be AN-stable if there exists a real, symmetric, and positive definite matrix* **G** *such that*

$$\big\| \mathbf{S}(\xi) y \big\|_G \leq \|y\|_G$$

for all $\xi = diag(\xi_1, \ldots, \xi_s)$ *such that* $\xi_i = \xi_j$ *whenever* $c_i = c_j$ *and such that* $\mathrm{Re}(\xi_i) \leq 0$ *for* $i = 1, 2, \ldots, s$.

Different variants of this definition have been considered by Butcher [41, 42, 43]. Observe that for $\xi = z\mathbf{I}$, $\mathbf{S}(\xi)$ reduces to the stability matrix $\mathbf{M}(z)$ defined by (2.6.4) of GLM (2.1.2). Hence, if a GLM is *AN*-stable it is also *A*-stable.

Denote by $\{y^{[n]}\}_{n=0}^{N}$ the solution to (2.1.2) and by $\{\widetilde{y}^{[n]}\}_{n=0}^{N}$ the solution obtained by perturbing (2.1.2) or by using a different initial value. The behavior of a numerical method that inherits property (2.9.6) of the solution $y(t)$ to (2.9.4) in the norm $\| \cdot \|_G$ given by (2.9.9) is defined as *G*-stability. As mentioned earlier this definition was introduced by Dahlquist [100] in the context of one-leg methods (2.1.5). For GLMs (2.1.2) this definition takes the following form.

Definition 2.9.7 *GLM (2.1.2) is said to be G-stable if there exists a real, symmetric, and positive definite matrix* $\mathbf{G} \in \mathbb{R}^{r \times r}$ *such that for two numerical solutions,* $\{y^{[n]}\}_{n=0}^{N}$ *and* $\{\widetilde{y}^{[n]}\}_{n=0}^{N}$, *we have*

$$\big\| y^{[n+1]} - \widetilde{y}^{[n+1]} \big\|_G \leq \big\| y^{[n]} - \widetilde{y}^{[n]} \big\|_G,$$

where $\| \cdot \|_G$ *is the norm defined by (2.9.9) for all step sizes* $h > 0$ *and for all differential equations (2.9.4) with the function g satisfying (2.9.5).*

For given $\mathbf{G} \in \mathbb{R}^{r \times r}$ and $\mathbf{D} \in \mathbb{R}^{s \times s}$, define the matrix $\mathbf{M}$ by the formula

$$\mathbf{M} := \left[\begin{array}{c|c} \mathbf{DA} + \mathbf{A}^T\mathbf{D} - \mathbf{B}^T\mathbf{GB} & \mathbf{DU} - \mathbf{B}^T\mathbf{GV} \\ \hline \mathbf{U}^T\mathbf{D} - \mathbf{V}^T\mathbf{GB} & \mathbf{G} - \mathbf{V}^T\mathbf{GV} \end{array} \right]. \tag{2.9.10}$$

We now introduce the definition of algebraic stability of GLMs.

Definition 2.9.8 ([33, 43]) *GLM (2.1.2) is said to be algebraically stable if there exist a real, symmetric, and positive definite matrix* **G** *and a real, diagonal, and positive definite matrix* **D** *such that the matrix* **M** *of (2.9.10) is nonnegative definite.*

It has been proved by Burrage and Butcher [33] that for a preconsistent and algebraically stable GLM (2.1.2), matrices **G** and **D** are not independent but necessarily related by the equation

$$\mathbf{De} = \mathbf{B}^T\mathbf{Gq}_0,$$

where $\mathbf{q}_0$ is the preconsistency vector (compare Definition 2.2.1). Moreover, $\mathbf{Gq}_0$ is a left eigenvector of the coefficient matrix $\mathbf{V}$ corresponding to the eigenvalue 1:

$$(\mathbf{I} - \mathbf{V}^T)\mathbf{Gq}_0 = \mathbf{0}$$

(compare [146], part ii) of Lemma 9.5).

It follows from the results of Burrage and Butcher [33] that the algebraic stability of GLMs (2.1.2) implies G-stability. Moreover, for a large class of GLMs the concepts of algebraic, G-, and AN-stability are equivalent. We have the following theorem.

Theorem 2.9.9 (Butcher [43]) *For a preconsistent and nonconfluent GLM (2.1.2) (i.e., method with distinct abscissas c_i, $i = 1, 2, \ldots, s$), algebraic, G-, and AN-stability are equivalent.*

In general, it is quite difficult to verify if a given GLM method (2.1.2) is algebraically stable; that is, it is difficult to find a real, symmetric, and positive definite matrix $\mathbf{G}$ and a real, diagonal, and positive definite matrix $\mathbf{D}$ such that the matrix $\mathbf{M}$ defined by (2.9.10) is nonnegative definite. An interesting example of algebraically stable GLM was constructed by Dekker [108]. This is the method of order $p = 4$ and stage order $q = 3$ whose coefficients are given by

$$\left[\begin{array}{c|c} \mathbf{A} & \mathbf{U} \\ \hline \mathbf{B} & \mathbf{V} \end{array}\right] = \left[\begin{array}{cc|cc} \frac{2}{3} & 0 & 1 & -\frac{7}{6} \\ \frac{2}{3} & \frac{2}{3} & 1 & \frac{1}{6} \\ \hline \frac{1}{2} & \frac{1}{2} & 1 & 0 \\ -\frac{1}{11} & \frac{7}{11} & 0 & \frac{5}{11} \end{array}\right]. \tag{2.9.11}$$

Then for the matrices

$$\mathbf{G} = \left[\begin{array}{cc} 1 & 0 \\ 0 & \frac{11}{12} \end{array}\right], \quad \mathbf{D} = \left[\begin{array}{cc} \frac{1}{2} & 0 \\ 0 & \frac{1}{2} \end{array}\right]$$

the matrix $\mathbf{M}$ of (2.9.10) takes the form

$$\mathbf{M} = \left[\begin{array}{cc|cc} 0 & 0 & 0 & 0 \\ 0 & \frac{8}{11} & -\frac{6}{11} & -\frac{2}{11} \\ \hline 0 & -\frac{6}{11} & \frac{9}{22} & \frac{3}{22} \\ 0 & -\frac{2}{11} & \frac{3}{22} & \frac{1}{22} \end{array}\right].$$

This matrix is real, symmetric, and nonnegative definite which proves the algebraic stability of GLM (2.9.11).

An interesting family of algebraically stable GLMs of arbitrarily high order was constructed by Burrage [28], where such methods were found in the class of MRK methods (2.1.13) of order $p = 2s$ using an elegant extension of the collocation approach for RK methods. In what follows we describe these results following the presentation by Hairer and Wanner [146], which simplifies the original exposition of Burrage [28].

Consider the bilinear form defined by

$$\langle f, g \rangle = \sum_{j=1}^{k} v_j \int_{1-j}^{1} f(x)g(x)dx = \int_{1-k}^{1} \omega(x)f(x)g(x)dx, \qquad (2.9.12)$$

where $\omega(x)$ is the step function defined on the interval $[1-k, 1]$ by the formulas

$$\omega(x) = \begin{cases} v_k, & 1-k \leq x \leq 1-k+1, \\ v_k + v_{k-1}, & 1-k+1 < x \leq 1-k+2, \\ \vdots \\ v_k + \cdots + v_2, & -1 < x \leq 0, \\ v_k + \cdots + v_1, & 0 < x \leq 1. \end{cases}$$

Assuming that

$$v_k \geq 0, \quad v_k + v_{k-1} \geq 0, \quad \ldots \quad v_k + \cdots + v_2 \geq 0, \quad v_k + \cdots + v_1 = 1, \quad (2.9.13)$$

the function $\omega(x)$ is nonnegative and (2.9.12) becomes an inner product in the space of polynomials on the real line. Denote by $\{p_j(x)\}_{j=0}^{s}$ the set of polynomials orthogonal with respect to the inner product (2.9.12). These polynomials depend on $v_1, v_2, \ldots, v_k$, and can be computed from the three-term recurrence relation

$$p_0(x) = 1, \quad p_1(x) = x - \beta_0,$$
$$p_{j+1}(x) = (x - \beta_j)p_j(x) - \gamma_j p_{j-1}(x),$$

$j = 1, 2, \ldots, s - 1$, where

$$\beta_j = \frac{\langle xp_j, p_j \rangle}{\langle p_j, p_j \rangle}, \quad \gamma_j = \frac{\langle p_j, p_j \rangle}{\langle p_{j-1}, p_{j-1} \rangle}.$$

This recurrence relation is well defined since the assumption (2.9.13) implies that $\langle p_j, p_j \rangle \neq 0$ (compare [146, Lemma 9.12] or [28, Theorem 4]).

The main results of Burrage [28] can be summarized in the following theorem.

Theorem 2.9.10 *Assume that $v_j \geq 0$, $j = 1, 2, \ldots, k$, and that $\sum_{j=1}^{k} v_j = 1$. Then MRK method (2.1.13) with abscissas $c_1, c_2, \ldots, c_s$, which are the zeros of the polynomial $p_s(x)$ and with coefficients b_i, a_{ij}, and u_{ij} defined by*

$$b_i = \sum_{j=1}^{k} v_j \int_{1-j}^{1} \ell_i(x)dx, \quad i = 1, 2, \ldots, s,$$

$$a_{ij} = \frac{b_j}{b_i} \int_{c_j}^{1} \ell_i(x)dx, \qquad i, j = 1, 2, \ldots, s,$$

$$u_{ij} = \frac{v_j}{b_j} \int_{1-j}^{1} \ell(x)dx, \qquad i = 1, 2, \ldots, s, \quad j = 1, 2, \ldots, k,$$

where

$$\ell_i(x) = \prod_{l=1, l \neq i}^{s} \frac{x - c_l}{c_i - c_l},$$

has order $p = 2s$. Moreover, this method is G-stable, with matrix $\mathbf{G}$ defined by

$$\mathbf{G} = diag\big(1, v_2 + \cdots + v_k, \ldots, v_{k-1} + v_k, v_k\big).$$

In the remainder of this section we describe the important recent work of Hewitt and Hill [154], in which they reformulate the standard conditions for algebraic stability and use the concepts of method equivalence, reducibility, and order conditions for methods of high stage order to simplify the construction of such methods of order p and stage order $q = p$ or $q = p - 1$. This approach is connected to a branch of control theory concerned with the algebraic discrete Riccati equation and to the theorem of Albert [3] on block-symmetric nonnegative definite matrices, and has the potential to provide a systematic approach to the construction of algebraically stable GLMs of high order and stage order.

Following Butcher [43], we introduce the following definitions of method equivalence and reducibility.

Definition 2.9.11 *The GLMs defined by the coefficients matrices*

$$\left[\begin{array}{c|c} \widetilde{\mathbf{A}} & \widetilde{\mathbf{U}} \\ \hline \widetilde{\mathbf{B}} & \widetilde{\mathbf{V}} \end{array}\right] \quad and \quad \left[\begin{array}{c|c} \mathbf{A} & \mathbf{U} \\ \hline \mathbf{B} & \mathbf{V} \end{array}\right]$$

are equivalent if there exist a permutation matrix $\mathbf{P}$ and a nonsingular matrix $\mathbf{Q}$ such that

$$\left[\begin{array}{c|c} \widetilde{\mathbf{A}} & \widetilde{\mathbf{U}} \\ \hline \widetilde{\mathbf{B}} & \widetilde{\mathbf{V}} \end{array}\right] = \left[\begin{array}{c|c} \mathbf{P}^T \mathbf{A} \mathbf{P} & \mathbf{P}^T \mathbf{U} \mathbf{Q} \\ \hline \mathbf{Q}^{-1} \mathbf{B} \mathbf{P} & \mathbf{Q}^{-1} \mathbf{V} \mathbf{Q} \end{array}\right].$$

Definition 2.9.12 *GLM (2.1.2) is reducible if $s = s_1 + s_2$ and $r = r_1 + r_2 + r_3$ with $s_2 + r_2 + r_3 > 0$, so that the equivalent GLM method has a sparsity pattern of the form*

$$
\begin{array}{c c}
 & \begin{array}{c c c c c} s_1 & s_2 & r_1 & r_2 & r_3 \end{array} \\
\begin{array}{c} s_1 \\[1.2em] s_2 \\[1.2em] r_1 \\[1.2em] r_2 \\[1.2em] r_3 \end{array} &
\left[
\begin{array}{c c | c c c}
\mathbf{A}_{11} & \mathbf{0} & \mathbf{U}_{11} & \mathbf{0} & \mathbf{U}_{13} \\[0.6em]
\mathbf{A}_{21} & \mathbf{A}_{22} & \mathbf{U}_{21} & \mathbf{U}_{22} & \mathbf{0} \\
\hline
\mathbf{B}_{11} & \mathbf{0} & \mathbf{V}_{11} & \mathbf{0} & \mathbf{V}_{13} \\[0.6em]
\mathbf{B}_{21} & \mathbf{B}_{22} & \mathbf{V}_{21} & \mathbf{V}_{22} & \mathbf{V}_{23} \\[0.6em]
\mathbf{0} & \mathbf{0} & \mathbf{0} & \mathbf{0} & \mathbf{V}_{33}
\end{array}
\right].
\end{array}
$$

In this case the method can be reduced to a GLM with coefficient matrices

$$
\left[
\begin{array}{c | c}
\mathbf{A}_{11} & \mathbf{U}_{11} \\
\hline
\mathbf{B}_{11} & \mathbf{V}_{11}
\end{array}
\right]
$$

with s_1 internal stages and r_1 external stages. The method is said to be irreducible if it is not reducible.

Earlier definitions of algebraic stability [33, 42] required only that both matrices $\mathbf{G}$ and $\mathbf{D}$ be nonnegative definite, while the definition in [146] requires only that $\mathbf{G}$ is positive definite and $\mathbf{D}$ is nonnegative definite. In this context the result by Hewitt and Hill [154] is of interest. This result shows that all these definitions are, in fact, equivalent. We have the following theorem.

Theorem 2.9.13 (Hewitt and Hill [154]) *Assume that for an irreducible GLM with coefficients $\mathbf{A}$, $\mathbf{U}$, $\mathbf{B}$, and $\mathbf{V}$, matrix $\mathbf{M}$ of (2.9.10) is nonnegative definite for some real, symmetric, and nonnegative definite matrix $\mathbf{G}$ and a real, diagonal, and nonnegative definite matrix $\mathbf{D}$. Then $\mathbf{G}$ and $\mathbf{D}$ are positive definite.*

We now describe the reformulations of algebraic stability proposed recently by Hewitt and Hill [154], which simplify somewhat the construction of algebraically stable GLMs. These reformulations are based on the notion of Moore-Penrose pseudo-inverse and on Albert's theorem [3] on block-symmetric nonnegative definite matrices.

Suppose that $A \in \mathbb{R}^{m \times n}$ and $\mathrm{rank}(A) = r < n$. The singular value decomposition of A is defined by

$$ A = U \Sigma V^T, $$

where $U \in \mathbb{R}^{m \times m}$ and $V \in \mathbb{R}^{n \times n}$ are orthogonal matrices and

$$ \Sigma = \mathrm{diag}\big(\sigma_1, \ldots, \sigma_r, 0, \ldots, 0\big) \in \mathbb{R}^{m \times n} $$

(compare [110, 136]). Then the Moore-Penrose pseudo-inverse of A, denoted A^+, is defined by

$$A^+ = V\Sigma^+ U^T,$$

where

$$\Sigma^+ = \text{diag}\left(\frac{1}{\sigma_1}, \ldots, \frac{1}{\sigma_r}, 0, \ldots, 0\right) \in \mathbb{R}^{n \times m}.$$

It can be verified that if $\text{rank}(A) = n$, then $A^+ = (A^T A)^{-1} A^T$, while if $m = n = \text{rank}(A)$, then $A^+ = A^{-1}$. The pseudo-inverse of A can also be defined to be the unique matrix $X \in \mathbb{R}^{n \times m}$ that satisfies the Moore-Penrose conditions

$$AXA = A, \quad XAX = X, \quad (AX)^T = AX, \quad (XA)^T = XA$$

again see [110, 136]. We have the following theorem.

Theorem 2.9.14 (Albert [3]) *The matrix*

$$\mathbf{M} = \left[\begin{array}{c|c} M_{11} & M_{12} \\ \hline M_{12}^T & M_{22} \end{array}\right]$$

is nonnegative definite if and only if

$$M_{11} \geq 0 \quad and \quad M_{22} - M_{12}^T M_{11}^+ M_{12} \geq 0.$$

We recall that $A \geq 0$ means that matrix A is nonnegative definite. Applying this theorem to matrix $\mathbf{M}$ of (2.9.10) and to matrix $\mathbf{M}^*$ defined by

$$\mathbf{M}^* = \left[\begin{array}{c|c} M_{22} & M_{12}^T \\ \hline M_{12} & M_{11} \end{array}\right],$$

which is nonnegative definite if and only if $\mathbf{M}$ is nonnegative definite, we obtain the following reformulations of the criteria for algebraic stability.

Reformulation 1. GLM (2.1.2) with coefficient matrices $\mathbf{A}$, $\mathbf{U}$, $\mathbf{B}$, and $\mathbf{V}$ is algebraically stable if there exist a real, symmetric, and positive definite matrix $\mathbf{G}$ and a real, diagonal, and positive definite matrix $\mathbf{D}$ such that

$$\mathbf{DA} + \mathbf{A}^T \mathbf{D} - \mathbf{B}^T \mathbf{GB} \geq 0$$

and

$$\mathbf{G} - \mathbf{V}^T \mathbf{GV} - \left(\mathbf{U}^T \mathbf{D} - \mathbf{V}^T \mathbf{GB}\right)\left(\mathbf{DA} + \mathbf{A}^T \mathbf{D} - \mathbf{B}^T \mathbf{GB}\right)^+ \left(\mathbf{DU} - \mathbf{B}^T \mathbf{GV}\right) \geq 0.$$

In control theory, the last condition is known as the discrete Riccati equation [196].

Reformulation 2. GLM (2.1.2) with coefficient matrices $\mathbf{A}$, $\mathbf{U}$, $\mathbf{B}$, and $\mathbf{V}$ is algebraically stable if there exist a real, symmetric, and positive definite matrix $\mathbf{G}$ and a real, diagonal, and positive definite matrix $\mathbf{D}$ such that

$$\mathbf{G} - \mathbf{V}^T\mathbf{G}\mathbf{V} \geq 0$$

and

$$\mathbf{D}\mathbf{A} + \mathbf{A}^T\mathbf{D} - \mathbf{B}^T\mathbf{G}\mathbf{B} - \left(\mathbf{D}\mathbf{U} - \mathbf{B}^T\mathbf{G}\mathbf{V}\right)\left(\mathbf{G} - \mathbf{V}^T\mathbf{G}\mathbf{V}\right)^{+}\left(\mathbf{U}^T\mathbf{D} - \mathbf{V}^T\mathbf{G}\mathbf{B}\right) \geq 0.$$

Another interesting idea of Hewitt and Hill [154] in their search for algebraically stable methods is to consider the equivalent GLMs with a simple structure of matrix $\mathbf{G}$. We have the following results.

Lemma 2.9.15 ([154]) *Assume that GLMs with coefficient matrices $\mathbf{A}$, $\mathbf{U}$, $\mathbf{B}$, $\mathbf{V}$ and $\widetilde{\mathbf{A}}$, $\widetilde{\mathbf{U}}$, $\widetilde{\mathbf{B}}$, $\widetilde{\mathbf{V}}$ are equivalent (compare Definition 2.9.11). Then the GLM defined by $\mathbf{A}$, $\mathbf{U}$, $\mathbf{B}$, $\mathbf{V}$ is algebraically stable if and only if the GLM defined by $\widetilde{\mathbf{A}}$, $\widetilde{\mathbf{U}}$, $\widetilde{\mathbf{B}}$, $\widetilde{\mathbf{V}}$ is algebraically stable.*

Lemma 2.9.16 ([154]) *An algebraically stable GLM with coefficients $\mathbf{A}$, $\mathbf{U}$, $\mathbf{B}$, $\mathbf{V}$ is equivalent to an algebraically stable method with coefficients $\widetilde{\mathbf{A}}$, $\widetilde{\mathbf{U}}$, $\widetilde{\mathbf{B}}$, $\widetilde{\mathbf{V}}$ for which $\widetilde{\mathbf{G}} = \mathbf{I}$. Furthermore, if $\mathbf{D} > 0$ is such that matrix $\mathbf{M}$ of (2.9.10) satisfies $\mathbf{M} \geq 0$, then we have $\widetilde{\mathbf{M}} \geq 0$, where*

$$\widetilde{\mathbf{M}} = \left[\begin{array}{c|c} \mathbf{D}\widetilde{\mathbf{A}} + \widetilde{\mathbf{A}}^T\mathbf{D} - \widetilde{\mathbf{B}}^T\widetilde{\mathbf{B}} & \mathbf{D}\widetilde{\mathbf{U}} - \widetilde{\mathbf{B}}^T\widetilde{\mathbf{V}} \\ \hline \widetilde{\mathbf{U}}^T\mathbf{D} - \widetilde{\mathbf{V}}^T\widetilde{\mathbf{B}} & \mathbf{I} - \widetilde{\mathbf{V}}^T\widetilde{\mathbf{V}} \end{array}\right].$$

Using Lemma 2.9.16 combined with reformulations 1 and 2 and with the aid of symbolic manipulation packages, Hewitt and Hill [154] constructed new algebraically stable GLMs with $r = s = 2$, order $p = 3$ and stage order $q = 2$, and of order $p = 4$ and stage order $q = 2$ and 3. The example of method of order $p = 3$ and stage order $q = 3$ constructed using Lemma 2.9.16 and Reformulation 1 is

$$\mathbf{A} = \left[\begin{array}{cc} \frac{84-4\sqrt{141}+3y}{144} - \frac{y^3}{2592} & \frac{276-20\sqrt{141}-51y}{144} + \frac{y^3}{2592} \\[2mm] \frac{276-20\sqrt{141}+51y}{144} - \frac{y^3}{2592} & \frac{84-4\sqrt{141}-3y}{144} + \frac{y^3}{2592} \end{array}\right],$$

$$\mathbf{U} = \left[\begin{array}{cc} 1 & \frac{z}{6} \\[2mm] 1 & \frac{z(9-24y^2-2y^3)}{54} \end{array}\right],$$

$$\mathbf{B} = \left[\begin{array}{cc} \frac{1}{2} + \frac{y}{24} & \frac{1}{2} - \frac{y}{24} \\[2mm] \frac{z}{72}(12+y) & \frac{z}{72}(y-12)\left(191 - 16\sqrt{141} + \frac{2y^3}{9}\right) \end{array}\right], \quad \mathbf{V} = \left[\begin{array}{cc} 1 & 0 \\[2mm] 0 & \frac{1}{2} \end{array}\right],$$

with $\mathbf{c} = [0, 2y/3]^T$, where

$$y := \sqrt{72 - 6\sqrt{141}}, \quad z := \sqrt{4548 - 383\sqrt{141} - 380y + 32y\sqrt{141}}.$$

The decimal representation of this method is

$$
\left[\begin{array}{c|c}
\mathbf{A} & \mathbf{U} \\ \hline
\mathbf{B} & \mathbf{V}
\end{array}\right]
=
\left[
\begin{array}{cc|cc}
0.2713275450 & -0.03981838948 & 1 & 0.05494949955 \\
0.5747233653 & 0.2356534502 & 1 & -0.06352190785 \\ \hline
0.5161792288 & -0.5361792288 & 1 & 0 \\
0.05892556058 & -0.06352190785 & 0 & 0.5
\end{array}
\right],
$$

with $\mathbf{c} = [0, 0.5788676600]^T$. The example of method of order $p = 4$ and stage order $q = 3$ constructed using Lemma 2.9.16 and Reformulation 2 is

$$
\mathbf{A} =
\left[
\begin{array}{cc}
\frac{265}{864} + \frac{793y^2}{576} - \frac{5y^4}{6} - \frac{123y^6}{64} - \frac{27y^8}{16} & \frac{215}{864} - \frac{5299y^2}{576} + \frac{623y^4}{96} + \frac{915y^6}{64} + \frac{189y^8}{16} \\[2mm]
\frac{101}{432} + \frac{3821y^2}{288} - \frac{463y^4}{48} - \frac{669y^6}{32} - \frac{135y^8}{8} & \frac{67}{432} + \frac{793y^2}{288} - \frac{5y^4}{3} - \frac{123y^6}{32} - \frac{27y^8}{8}
\end{array}
\right],
$$

$$
\mathbf{B} =
\left[
\begin{array}{cc}
\frac{2}{3} & \frac{1}{3} \\[2mm]
\frac{17y - 1125y^3 + 828y^5 + 1783y^7 + 1458y^9}{24} & \frac{-11y + 1125y^3 - 828y^5 - 1782y^7 - 1458y^9}{24}
\end{array}
\right],
$$

$$
\mathbf{U} =
\left[
\begin{array}{cc}
1 & -\frac{7y + 9y^3}{16} \\[2mm]
0 & \frac{y + 9y^3}{8}
\end{array}
\right], \quad
\mathbf{V} =
\left[
\begin{array}{cc}
1 & 0 \\[1mm]
0 & \frac{1}{2}
\end{array}
\right],
$$

with

$$\mathbf{c} = \left[\begin{array}{cc} 0 & \frac{193y^2 - 129y^4 - 297y^6 - 243y^8}{8} - \frac{1}{6}\end{array}\right]^T.$$

Here $y = \pm\sqrt{z/3}$ and z is one of the two positive roots of the equation

$$9z^5 + 33z^4 + 46z^3 - 186z^2 + 9z + 1 = 0.$$

Choosing the root $z = 0.1032814360$ and $y = \sqrt{z/3}$, the decimal representation of the resulting method is

$$
\left[\begin{array}{c|c}
\mathbf{A} & \mathbf{U} \\ \hline
\mathbf{B} & \mathbf{V}
\end{array}\right]
=
\left[
\begin{array}{cc|cc}
0.3530415762 & -0.0595835887 & 1 & -0.08476931053 \\
0.6782443859 & 0.2477498188 & 1 & 0.03037947026 \\ \hline
0.6666666667 & 0.3333333333 & 1 & 0 \\
-0.1598351741 & 0.2062215576 & 0 & 0.5
\end{array}
\right],
$$

with $\mathbf{c} = [0, 0.6432188884]^T$.

Additional examples of algebraically stable GLMs with $r = s = 2$ and $p = q = 2$, $p = 3$, $q = 2$, and $p = 4$, $q = 2$, constructed using the algorithms based on results of Hewitt and Hill [154] are presented in their companion work [155].

2.10 UNDERLYING ONE-STEP METHOD

In this section we describe briefly results of Kirchgraber [191] and Stoffer [267] which establish a connection of strongly zero-stable linear multistep methods and strongly zero-stable GLMs, respectively, with a one-step method of the same order.

Let y_0 be the given initial value of the problem (2.1.1) and denote by $y_1, y_2, \ldots, y_{k-1}$ the starting values of linear multistep method (2.1.4), which are used to generate the sequence of approximations $y_k, y_{k+1}, \ldots$. These starting values, $y_1, y_2, \ldots, y_{k-1}$, will be referred to as supplementary initial values. Denote by $\widetilde{y}_0, \widetilde{y}_1, \ldots$, the sequence of numerical approximations to (2.1.1) generated by some one-step method, such as, for example, RK method (2.1.3). We have the following theorem.

Theorem 2.10.1 (Kirchgraber [191]) *There exists a one-step method and* $h_0 > 0$ *such that for* $h \in (0, h_0)$, *the following statements are true.*
(a) Let $\widetilde{y}_1, \widetilde{y}_2, \ldots$ *be the sequence generated by a one-step method with* $\widetilde{y}_0 = y_0$. *If the supplementary initial values of strongly zero-stable linear multistep method (2.1.4) are chosen as* $y_1 = \widetilde{y}_1$, $y_2 = \widetilde{y}_2$, ..., $y_{k-1} = \widetilde{y}_{k-1}$, *then*

$$y_i = \widetilde{y}_i, \quad i = k, k+1, \ldots.$$

(b) There are constants $\kappa \in (0, 1)$ *and* $k > 0$ *which do not depend on* h *such that if the supplementary initial values* $y_1, y_2, \ldots, y_{k-1}$ *are prescribed arbitrarily, there exists* $\overline{y}_0$ *such that if* $\overline{y}_1, \overline{y}_2, \ldots$ *is the sequence generated by one-step method starting from* $\overline{y}_0$, *then*

$$\|y_i - \overline{y}_i\| \le \kappa^i k (\|y_1 - \overline{y}_1\| + \cdots + \|y_{k-1} - \overline{y}_{k-1}\|), \quad i = k, k+1, \ldots.$$

As observed by Kirchgraber [191], part (a) of the theorem says that for a judicious choice of supplementary initial values, the numerical approximations generated by linear multistep and one-step-methods are the same. Unfortunately, this is not as interesting as it sounds, since the choice $y_1 = \widetilde{y}_1$ assuming that $y_0 = \widetilde{y}_0$ is equivalent to construction of the associated one-step method whose existence is claimed in Theorem 2.10.1. However, part (b) implies that for any choice of supplementary initial values the difference between the approximations generated by a linear multistep method and a one-step method is exponentially small as $h \to 0$. This associated one-step method, whose existence is guaranteed by Theorem 2.10.1, will be referred to as the underlying one-step method.

The results in Theorem 2.10.1 were generalized by Stoffer [267] to strongly zero-stable GLMs (2.1.2): that is, methods for which the coefficient matrix $\mathbf{V}$ has a simple eigenvalue equal to 1 and the remaining $r-1$ eigenvalues inside the unit circle (compare Definition 2.2.6). For such methods there exists a nonsingular matrix $\mathbf{T} \in \mathbb{R}^{r \times r}$ such that the methods can be transformed into the form

$$
\left[\begin{array}{c} Y^{[n]} \\ z^{[n]} \end{array}\right] = \left[\begin{array}{c|c} \mathbf{A} \otimes \mathbf{I} & \mathbf{U}\mathbf{T} \otimes \mathbf{I} \\ \hline \mathbf{T}^{-1}\mathbf{B} \otimes \mathbf{I} & \mathbf{T}^{-1}\mathbf{V}\mathbf{T} \otimes \mathbf{I} \end{array}\right] \left[\begin{array}{c} hF(Y^{[n]}) \\ z^{[n-1]} \end{array}\right], \qquad (2.10.1)
$$

where $z^{[n]} = (\mathbf{T}^{-1} \otimes \mathbf{I})y^{[n]}$, and the coefficient matrix $\mathbf{T}^{-1}\mathbf{V}\mathbf{T}$ assumes the form

$$
\mathbf{T}^{-1}\mathbf{V}\mathbf{T} \otimes \mathbf{I} = \left[\begin{array}{c|c} 1 & 0 \\ \hline 0 & \overline{\mathbf{V}} \end{array}\right], \qquad (2.10.2)
$$

with $\rho(\overline{\mathbf{V}}) < 1$. It follows from (2.10.2) and (2.2.1) that the preconsistency vector for the transformed method (2.10.1) is $\mathbf{q}_0 = \mathbf{e}_1$. This implies that the first subvector of $z^{[n]}$, which is denoted by y_n, approximates directly $y(t_n)$.

Denote by $\mathcal{M}_h : \mathbb{R}^{mr} \to \mathbb{R}^{mr}$ the action of the method (2.10.1)

$$
z^{[n]} = \mathcal{M}_h\big(z^{[n-1]}\big).
$$

We have the following theorem.

Theorem 2.10.2 (Stoffer [267]) *Assume that $\mathcal{M}_h$ is a strictly zero-stable GLM of order p. Then there exists a unique one-step method of order p, $y_{n+1} = \mathcal{R}_h(y_n)$, $\mathcal{R}_h : \mathbb{R}^m \to \mathbb{R}^m$, and a unique starting procedure $\mathcal{S}_h : \mathbb{R}^m \to \mathbb{R}^{mr}$ satisfying (2.4.4) such that the following diagram commutes:*

$$
\begin{array}{ccc}
& \mathcal{M}_h & \\
\mathbb{R}^{mr} & \longrightarrow & \mathbb{R}^{mr} \\
\mathcal{S}_h \uparrow & & \uparrow \mathcal{S}_h \\
\mathbb{R}^m & \longrightarrow & \mathbb{R}^m \\
& \mathcal{R}_h &
\end{array}
$$

that is,

$$
\mathcal{M}_h \circ \mathcal{S}_h = \mathcal{S}_h \circ \mathcal{R}_h. \qquad (2.10.3)
$$

Moreover,

$$
\mathcal{F}_h \circ \mathcal{S}_h = Id, \qquad (2.10.4)
$$

where $\mathcal{F}_h : \mathbb{R}^{mr} \to \mathbb{R}^m$ is a finishing procedure that selects from $z^{[n]}$ the first subvector y_n, and $Id : \mathbb{R}^m \to \mathbb{R}^m$ stands for the identity map.

It follows from (2.10.3) and (2.10.4) that the one-step method $\mathcal{R}_h$ can be represented by

$$\mathcal{R}_h = \mathcal{F}_h \circ \mathcal{M}_h \circ \mathcal{S}_h.$$

The concept of an underlying one-step method was extended by Hairer et al. [142] to nearly all GLMs, including weakly stable methods, i.e., methods for which the coefficient matrix $\mathbf{V}$ has, in addition to the simple eigenvalue equal to 1, additional simple eigenvalues with modulus 1.

It was demonstrated by Stoffer [267] that the result in Theorem 2.10.2 may be used to show that some general properties of one-step methods carry over to GLMs. An example of such a property is the existence of a hyperbolic invariant closed curve of a one-step method near the orbit of the periodic solution, assuming that the differential equation admits such a solution [20, 24, 123]. This result was generalized Eirola and Nevanlinna [124] to linear multistep methods. Another example of such a property [267] is the existence, for every small step size h, of a compact, uniformly asymptotically stable set $\Lambda(h)$ containing a compact, uniformly asymptotically stable attractor Λ of the differential system, assuming again that such an attractor exists [192] and such that $\Lambda(h)$ is convergent to Λ in the Hausdorff metric as $h \to 0$. We refer to Edgar [120] for a definition of a Hausdorff metric (distance). This result was generalized to linear multistep methods by Kloeden and Lorenz [193].

We refer to the papers [52, 75, 293] for additional discussion of underlying one-step methods in the context of GLMs, and to the paper [142] for the discussion of the dynamics of weakly stable methods.

2.11   STARTING PROCEDURES

Many modern codes for the numerical solution of ODEs are based on a family of methods which includes a method of order $p = 1$. Such codes usually start with a method of order $p = 1$ and adapt its step size and order automatically according to the smoothness of the solution. For such codes the required starting values, y_0 or y_0 and $y_0' = f(t_0, y_0)$, are readily available, and no special starting procedures are required. This is the case, for example, for classical codes based on linear multistep methods such as DIFSUB [131, 132, 133], ODE/DE/STEP [262], LSODE [242], VODE [23, 159], VODPK [25, 83], for the code dim18 for nonstiff equations, which is based on a family of type 1 DIMSIMs of order $1 \leq p \leq 8$, and for the code dim13s for stiff differential systems, which is based on A- and L-stable type 2 DIMSIMs of order $1 \leq p \leq 3$. These codes are self-starting and require only known information. The codes dim18 and dim13s are discussed briefly in Section 2.12 and in more detail in Chapter 4.

Codes based on RK methods of fixed order are also self-starting and require only the initial value y_0. The situation is different, however, for codes based on GLMs of fixed order such as DIMSIMs discussed in Chapters 3 and 4, TSRK methods discussed in Chapters 5 and 6, and GLMs with inherent

Runge-Kutta stability (IRKS) discussed in Chapters 7 and 8. The codes based on these and other GLMs of fixed order require starting procedures to compute approximations to the starting vector of external approximations $y^{[0]}$ for GLMs (2.1.2), or stage values $Y^{[0]}$ and the approximation y_1 at the point $t_1 = t_0 + h_0$ for TSRK methods (2.1.12).

The construction of starting procedures for GLMs of fixed order have been discussed by VanWieren [273, 274, 275], Wright [291, 293], and Huang [167]. VanWieren obtained starting values for his codes DIMEXx [273, 274] for nonstiff equations and DIMSTIFF2 and DIMSTIFF5 [273, 275] for stiff equations by computing approximations to the derivatives of the solution of sufficiently high order using only information available at the start of the integration. This is discussed in detail by VanWieren [273, 274], and these codes are discussed briefly in Section 2.12. Enenkel and Jackson [126, 127] introduced the class of GLMs for which the vector $y^{[0]}$ of external approximations satisfies $y_i^{[0]} \approx y(t_0 + d_i h_0)$, $i = 1, 2, \ldots, r$, where the abscissas d_i are distinct. These methods are discussed briefly in Section 2.12. They computed the required starting values $y_i^{[0]}$ using variable order solver LSODE [242] with the same order tolerance as that specified for GLM.

The construction of starting procedures for general TSRK methods (2.1.12) is discussed in Section 6.2. For reasons discussed by Hairer and Wanner [147], the construction of such procedures is quite complicated if the order p of TSRK method is at least greater by 2 than the stage order q (i.e., if $q \leq p - 2$). The starting procedures for some TSRK methods satisfying this restriction have been derived by Verner [277, 278]. The situation is much simpler for TSRK formulas such that $q = p - 1$ or $q = p$, and if this is the case, the required starting values can be computed, for example, by any continuous RK method of order p. This is the approach adopted in the codes tsrk5 [16], tsrk23 and tsrk33 [91] and in the codes discussed by Bartoszewski and Jackiewicz [17]. These codes are discussed briefly in Section 2.12 and in more detail in Chapter 6.

The construction of starting procedures for GLMs with IRKS is discussed in Section 8.2. In that section examples of such procedures are also given for methods of order $p = 2$, 3, and 4. The starting procedures for these methods are also discussed by Wright [293] and Huang [167] and the approach taken in these papers is discussed at the end of Section 8.2.

2.12 CODES BASED ON GENERAL LINEAR METHODS

Although GLMs of the form (2.1.2) for ODEs were introduced in 1980 [33], the main progress in the theory and construction of such methods with desirable stability properties was obtained only recently, mainly in the last 15 years. Various implementation issues for these methods were also investigated, such as the choice of appropriate starting procedures, Nordsieck representation of various classes of these methods, estimation of the principal part of the local discretization error, step size, and order changing strategies, construction of continuous interpolants, updating the vector of external approximations, and the efficient solution of resulting systems of nonlinear equations for stiff differential systems. Although significant progress on all these issues was obtained, the implementation aspects of GLMs are still not as well understood as is the case for RK, linear multistep, or predictor-corrector methods. Also, so far only a few codes based on some classes of GLMs have been developed for nonstiff and stiff differential systems.

Probably the first such code, written in FORTRAN, is due to Cash and Considine [88, 89] and is based on MEBDF formulas of order $1 \leq p \leq 9$ proposed by Cash [86]. These methods are also discussed by Hairer and Wanner [146] and in Section 2.1 (compare equations (2.1.7), (2.1.8), and (2.1.10)). These formulas are not linear multistep methods and are not restricted by Dahlquist's second barrier [99] for A-stable methods. MEBDF methods have good stability properties: they are A- and L-stable up to the order $p = 4$ and $A(\alpha)$-stable up to the order $p = 9$, with the angles α listed in Table 2.1.1. They form a class of GLMs. This is discussed in Section 2.1, where the coefficient matrices $\mathbf{A}$, $\mathbf{U}$, $\mathbf{B}$, and $\mathbf{V}$ for these methods are also given. It was demonstrated by Cash and Considine [88, 89] as well as by Hairer and Wanner [146] that the code based on MEBDF methods shows good performance on selected examples of stiff problems, although in some cases it was beneficial to restrict the maximal order for better performance. This was the case, for example, for the BEAM problem, where the maximal order was restricted to $p_{\max} = 4$ in the experiments presented by Hairer and Wanner [146].

Enenkel and Jackson [126, 127] introduced a subclass of GLMs which they called diagonally implicit single-eigenvalue methods (DIMSEMs) for which the coefficient matrix $\mathbf{A}$ is diagonal and all components c_i of the abscissa vector $\mathbf{c}$ are equal to zero or one. Moreover, the stability matrix $\mathbf{M}(z)$ defined by (2.6.4) has a single nonzero eigenvalue. They constructed A- and L-stable methods of order $2 \leq p \leq 6$ and developed experimental variable step size codes of fixed order $p = 3$, $p = 4$, $p = 5$, and $p = 6$ based on these methods. Although the algorithms based on these methods are not production codes, the results of numerical experiments presented by Enenkel and Jackson [126, 127] demonstrate that these methods are competitive with fixed order versions of the popular LSODE solver developed by Radhakrishnan and Hindmarsh [242] for $p = 3$, $p = 4$, and $p = 5$ on many practical test problems.

VanWieren [273, 274, 275] developed a family of research solvers based on explicit and implicit DIMSIMs of fixed order. The codes DIMEXx, based on explicit DIMSIM of fixed order x, are written as a collection of double precision FORTRAN 77 subroutines called by a driver that provides the interface to the user's calling program. These codes are based on type 1 methods which have the FASAL (first approximately same as last) property which facilitates the efficient implementation of these methods. This property is similar to the FSAL (first same as last) property introduced by Dormand and Prince [117] in the context of RK methods (compare also [116, 257, 260] and the discussion in Section 6.3). The results of some numerical experiments with DIMEX2 and DIMEX5 on some DETEST problems from Hull et al. [168] are given by VanWieren [273, 274]. The complete code for DIMEX5 appears in the appendix to Part 1 of VanWieren report [274]. The prototype codes DIMSTIFF2 and DIMSTIFF5 for stiff differential systems were developed in Part 2 of the report [275]. They are based on A- and L-stable DIMSIMs of type 2 of order $p = 2$ and $p = 5$, where a modified Newton iteration with Gaussian elimination is used to solve the nonlinear systems that arise in calculations of the internal stages. The user is asked to provide a subroutine to compute a Jacobian of the problem. These codes were tested in Part 2 of the report [275] on the Prothero-Robinson problem (1.7.1) only for $\lambda = -2$, $\lambda - 1000$, and $\lambda = -10000$.

Butcher et al. [58] developed an experimental Matlab code dim18 for nonstiff differential systems which is based on type 1 DIMSIMs of order $1 \leq p \leq 8$. This code utilizes the Nordsieck representation of DIMSIMs proposed earlier by Butcher et al. [57] in which the vector of external stages approximates the scaled derivatives of the solution. This representation facilitates a convenient way of changing the step size of the method by rescaling a vector of external approximations. Moreover, as demonstrated in [57], the methods in this formulation are zero-stable for any choice of variable mesh. The code dim18 starts with the method of order $p = 1$ and adapts its step size and order according to the smoothness of the solution. The relevant issues related to the implementation of these methods are discussed by Butcher et al. [58] and in Sections 4.2, and 4.4–4.6 of this book. The code was tested on several problems, and numerical experiments demonstrate that the error estimation employed in this code is very reliable and the step size and order changing strategy is very robust (compare [58] and Section 4.9). These experiments also indicate that dim18 outperforms the code ode45 from Matlab ODE suite [263] for moderate and stringent tolerances. The DIMSIMs of type 1 employed in this code were derived by a number of authors [44, 66, 68, 74, 291, 292]. The coefficients of these methods in Nordsieck representation are also given by Butcher et al. [58] together with the coefficients of the estimators of the local discretization errors that were used in this code.

Jackiewicz [177] developed an experimental Matlab code dim13s for stiff differential systems which is based on the Nordsieck representation of type 2 DIMSIMs, so far of order $1 \leq p \leq 3$ only. The methods employed in this code are A- and L-stable and were derived by Butcher [44] and Butcher and Jackiewicz [66]. The coefficients of these methods in Nordsieck representation are also given by Jackiewicz [177] together with the coefficients of the estimators of the local discretization errors. These estimators were designed not only to be asymptotically correct but also to be accurate for "large" step sizes compared to certain characteristics of the problem using the approach of Shampine and Baca [258] proposed in the context of RK methods for stiff differential systems. Various implementation issues related to this code are discussed by Jackiewicz [177] and in Sections 4.2–4.6 and 4.8 of this book.

The numerical experiments presented by Jackiewicz [177] and in Section 4.10 indicate that this code is especially well suited for problems whose Jacobian have eigenvalues close to the imaginary axis. An example of such a problem is the BEAM system (1.8.6).

Experimental Matlab codes based on explicit TSRK methods have been developed by Bartoszewski and Jackiewicz [16, 17] and Chollom and Jackiewicz [91]. Various issues related to the development of a new code tsrk5 for nonstiff differential systems are discussed by Bartoszewski and Jackiewicz [16]. This code is based on the explicit TSRK method of order $p = 5$ and stage order $q = 5$ constructed in [15]. The numerical experiments presented in [16] on selected test problems indicate that this code is competitive with algorithm ode45 for all tolerances. These experiments also demonstrate that the error estimation used in this code is very accurate and very reliable for variable step sizes and that the step size changing strategy is very robust for a wide range of error tolerances. Some numerical experiments with the two codes tsrk23 and tsrk33, which are based on explicit TSRK methods of order $p = 3$ and stage order $q = 2$ and $q = 3$, respectively, are presented by Chollom and Jackiewicz [91]. The coefficients of these methods are listed in their article [91] and in Section 5.6.1. The TSRK methods employed in these codes have large regions of absolute stability compared with RK methods of the same order. These regions are plotted in [91] and in Section 5.6.1. These experiments indicate a high potential of explicit TSRK methods as building blocks of software for nonstiff differential systems. Codes based on a family of explicit TSRK formulas of order $p = 3$ and stage order $q = 3$ with error constant E, which is given in advance, were developed by Bartoszewski and Jackiewicz [17]. The methods employed in these codes correspond to $E = 1/12, 1/24, 1/48$, and $1/120$, and the coefficients of the resulting formulas are listed in their article [17] and in Section 6.6. We have implemented these methods in Nordsieck form derived in [17] and discussed in Section 6.4. This representation, as in the case of DIMSIMs, facilitates a convenient way of changing step size by rescaling the vector of external approximations. The results of numerical experiments with these codes, which are presented by Bartoszewski and Jackiewicz [17] and in Section 6.8, indicate that these codes are more efficient and in many cases also more accurate than the code ode23, which is based on an embedded pair of RK methods of order $p = 2$ and $p = 3$ constructed by Bogacki and Shampine [22]. These experiments also demonstrate the high quality of error estimation used in these codes, especially codes based on methods with moderate error constants (compare [17] and Section 6.8).

Weiner and coworkers [238, 239, 251, 252, 286, 288, 289] developed Matlab and FORTRAN codes based on the class of two-step peer methods mentioned in Section 2.1 for the numerical solution of both nonstiff [286] and stiff differential systems [238, 239, 251, 252, 288]. Codes based on these methods have been implemented in Matlab by Weiner et al. [286] for nonstiff systems and compared with ode45 and in FORTRAN [289] and compared with DOPRI5 and DOP853 [143]. The codes for stiff systems have been implemented in

FORTRAN and compared by Weiner et al. [288] with the code VODPK [83] and the code TSW3B based on the two-step W-method [237], by Podhaisky et al. [238] with the codes based on Rosenbrock methods RODAS [146], RODASP [266], and ROS3P [198], by Schmitt et al. [252] with the code VODPK [83], the ROWMAP [287], and the Runge-Kutta Chebyshev code RKC [265], [170], and by Schmitt et al. [239, 250, 251] with RODAS [146]. The numerical experiments presented in these papers demonstrate a high potential of two-step peer methods for both nonstiff and stiff differential systems in sequential and parallel computing environments.

Butcher and Podhaisky [77] developed an experimental Matlab code based on stiffly accurate GLMs with IRKS of order $1 \leq p \leq 4$. Such methods are discussed in Section 7.2. The estimation of $h_n^{p+1} y^{(p+1)}(t_n)$ and $h_n^{p+2} y^{(p+2)}(t_n)$ for these methods is discussed in their paper [77] and in Section 8.10. The order changing strategy for this experimental code is also discussed in [77] with additional implementation details given in the paper by Butcher et al. [64]. Some numerical experiments with this code have also been presented for the nonstiff Brusselator equation and for the problems HIRES, OREGO, and VDPOL (see [77]).

In his Ph.D. thesis Wright [293] performed extensive numerical experiments with the fixed step size and variable step size implementations of explicit GLMs with IRKS which indicate that these methods have the potential to be the kernel of competitive codes for nonstiff differential systems. In her Ph.D. thesis Huang [167] discussed various issues related to the implementation of implicit GLMs with IRKS, such as the construction of starting procedures, efficient computation of stage values, local error estimation, rescaling the Nordsieck vector of external approximations, and interpolation. She has developed an experimental FORTRAN code based on the implicit method of order $p = 4$ and stage order $q = 4$. The coefficients of this method and the a code listing are given by Huang [167]. The results of numerical experiments with this code on selected examples of stiff differential systems such as ROBER (1.8.2) and HIRES (1.8.4) indicate the high potential of this code for the numerical solution of stiff differential equations.

Various software issues for ODEs, including the implementation of GLMs, were outlined in the panel discussion held at ANODE workshop in Auckland in January 2001. The record of this panel discussion is presented in [51].

CHAPTER 3

DIAGONALLY IMPLICIT MULTISTAGE INTEGRATION METHODS

3.1 REPRESENTATION OF DIMSIMS

Diagonally implicit multistage integration methods which we referr to as DIM-SIMs, are a subclass of GLMs (2.1.2) that is characterized by the following properties:

1. Coefficient matrix $\mathbf{A}$ is lower triangular with the same element λ on the diagonal. If $\lambda = 0$, the methods are explicit, but they will still be referred to as DIMSIMs.

2. Coefficient matrix $\mathbf{V}$ is a rank 1 matrix with nonzero eigenvalue equal to 1 to guarantee preconsistency.

3. Order p, stage order q, number of external stages r, and the number of internal stages s are related by $p = q$ or $p = q + 1$ and $r = s$ or $r = s + 1$.

DIMSIMs can be divided into four types according to the classification of GLMs introduced in Section 2.7. In the first part of this chapter we describe the construction of methods of all four types with some desirable stability properties. We aim at RK stability for methods of type 1, large intervals or regions of absolute stability for methods of type 3, and A- and L-stability

General Linear Methods for Ordinary Differential Equations. By Zdzisław Jackiewicz **131**
Copyright © 2009 John Wiley & Sons, Inc.

for methods of types 2 and 4. Moreover, for type 2 methods we also aim for SDIRK stability.

We start the discussion assuming that $p = q = r = s$ or $p - 1 = q = r = s$, the matrix $\mathbf{U} = \mathbf{I}$, and the matrix $\mathbf{V}$ has the form $\mathbf{V} = \mathbf{e}\mathbf{v}^T$, where $\mathbf{v}^T\mathbf{e} = 1$. This leads to the formulas

$$Y^{[n]} = h(\mathbf{A} \otimes \mathbf{I})F(Y^{[n]}) + (\mathbf{U} \otimes \mathbf{I})y^{[n-1]},$$

$$y^{[n]} = h(\mathbf{B} \otimes \mathbf{I})F(Y^{[n]}) + (\mathbf{V} \otimes \mathbf{I})y^{[n-1]}, \tag{3.1.1}$$

which can be represented in the form

$$\left[\begin{array}{c|c} \mathbf{A} & \mathbf{U} \\ \hline \mathbf{B} & \mathbf{V} \end{array}\right] = \left[\begin{array}{cccc|cccc} \lambda & & & & 1 & 0 & \cdots & 0 \\ a_{21} & \lambda & & & 0 & 1 & \cdots & 0 \\ \vdots & \vdots & \ddots & & \vdots & \vdots & \ddots & \vdots \\ a_{s1} & a_{s2} & \cdots & \lambda & 0 & 0 & \cdots & 1 \\ \hline b_{11} & b_{12} & \cdots & b_{1s} & v_1 & v_2 & \cdots & v_s \\ b_{21} & b_{22} & \cdots & b_{2s} & v_1 & v_2 & \cdots & v_s \\ \vdots & \vdots & \ddots & \vdots & \vdots & \vdots & \ddots & \vdots \\ b_{s1} & b_{s2} & \cdots & b_{ss} & v_1 & v_2 & \cdots & v_s \end{array}\right], \tag{3.1.2}$$

where $\sum_{i=1}^{s} v_i = 1$. After imposing the appropriate stage order and order conditions, the construction of DIMSIMs with desirable stability properties leads to the solution of large systems of polynomial equations for the remaining unknown parameters of the methods. If the order of the methods is not too high ($p \leq 4$), these systems can be generated and solved using symbolic manipulation packages such as MATHEMATICA or MAPLE. Alternatively, we could generate these systems by symbolic manipulation packages and then solve them numerically with the aid of algorithms such as PITCON [245], [246], ALCON [114], or HOMPACK [284] based on a continuation (homotopy) approach [10, 11]. The advantage of the latter approach is that we can generate entire families of methods, depending on some parameters of the methods by a judicious choice of the underlying homotopy map.

In Section 3.2 we derive the representation formula for the coefficient matrix **B**, and then in Section 3.3 we present the transformation that simplifies the derivation of DIMSIMs. In Sections 3.4–3.7 we discuss the construction of DIMSIMs of all four types and present several examples of such methods up to the order $p \leq 4$.

For higher orders ($p \geq 5$) it is no longer possible to generate the corresponding systems of polynomial equations in manageable form by symbolic manipulation packages, so a different approach is needed. We have developed an approach based on a variant of the Fourier series method in Butcher and Jackiewicz [68]. The systems that were obtained by this approach were then solved by state-of-the-art minimization software. Jackiewicz and Mittelmann [179] were also able to exploit the structure of the resulting polynomial systems of equations to speed up the search for methods of types 1 and 2 with desirable stability properties. These developments are described in Sections 3.8 and 3.9, and in Section 3.10 some examples of DIMSIMs of types 1 and 2 up to order $p \leq 6$ are given. We refer to the articles cited above [74, 179] for examples of DIMSIMs up to the order $p \leq 8$.

In Section 3.11 the alternative Nordsieck representation of DIMSIMs is discussed which, as we demonstrate in Chapter 4, facilitates an efficient and robust implementation of these methods. Then in Section 3.12 we derive the representation formulas for the coefficient matrices **P** and **G** appearing in Nordsieck representation, and in Section 3.13 we present several examples of methods of types 1 and 2 in this representation. The chapter concludes in Section 3.14 with a discussion of regularity properties of DIMSIMs.

3.2   REPRESENTATION FORMULAS FOR THE COEFFICIENT MATRIX B

The order conditions derived in Theorems 2.4.1 and 2.4.2 or formulas (2.4.8), (2.4.9), and (2.4.10) also apply to DIMSIMs (3.1.1) with coefficients defined by (3.1.2) and can be utilized in the constriction of these methods with $p = q$ or $p = q + 1$. However, for DIMSIMs with $\mathbf{U} = \mathbf{I}$, $\mathbf{V} = \mathbf{ev}^T$, $\mathbf{v}^T\mathbf{e} = 1$, and the abscissa vector $\mathbf{c} = [c_1, \ldots, c_s]^T$ with distinct components c_i, we have a very convenient representation for the coefficient matrix **B** in terms of **c**, **A**, and **V** if $q = r = s$ and $p = q$ or $p = q + 1$. These representations are derived in the following theorems.

Theorem 3.2.1 (Butcher [44]) *Let* $r = s$ *and* $\mathbf{U} = \mathbf{I}$. *Then DIMSIM* $(\mathbf{c}, \mathbf{A}, \mathbf{U}, \mathbf{B}, \mathbf{V})$ *has order* p *and stage order* q *equal to* $q = p = r = s$ *if and only if*

$$\mathbf{B} = \mathbf{B}_0 - \mathbf{A}\mathbf{B}_1 - \mathbf{V}\mathbf{B}_2 + \mathbf{V}\mathbf{A}, \tag{3.2.1}$$

where $\mathbf{B}_0$, $\mathbf{B}_1$, *and* $\mathbf{B}_2$ *are* $s \times s$ *matrices with elements defined by*

$$\frac{\displaystyle\int_0^{1+c_i} \phi_j(x)dx}{\phi_j(c_j)}, \qquad \frac{\phi_j(1 + c_i)}{\phi_j(c_j)}, \qquad \frac{\displaystyle\int_0^{c_i} \phi_j(x)dx}{\phi_j(c_j)}. \tag{3.2.2}$$

Here

$$\phi_i(x) = \prod_{j=1, j \neq i}^{s} (x - c_j),$$

$i = 1, 2, \ldots, s.$

Proof: Since $\mathbf{U} = \mathbf{I}$ it follows from equation (2.4.5) that the vector

$$\mathbf{w} = \mathbf{w}(z) = \sum_{k=0}^{s} \mathbf{q}_k z^k$$

defined in Section 2.4 takes the form

$$\mathbf{w}(z) = (\mathbf{I} - z\mathbf{A})e^{\mathbf{c}z} + O(z^{p+1}).$$

Substituting this relation into (2.4.6), we obtain

$$e^z(\mathbf{I} - z\mathbf{A})e^{\mathbf{c}z} = z\mathbf{B}e^{\mathbf{c}z} + \mathbf{V}(\mathbf{I} - z\mathbf{A})e^{\mathbf{c}z} + O(z^{p+1})$$

or

$$(\mathbf{B} - \mathbf{V}\mathbf{A})ze^{\mathbf{c}z} = e^z e^{\mathbf{c}z} - \mathbf{A}ze^z e^{\mathbf{c}z} - \mathbf{V}e^{\mathbf{c}z} + O(z^{p+1}). \qquad (3.2.3)$$

Observe that by putting $z = 0$ in this relation we get the preconsistency condition $\mathbf{V}\mathbf{e} = \mathbf{e}$. Define the differential operators $\Phi_j(D)$, $j = 1, 2, \ldots, s$, $D = d/dz$, where $\Phi_j(x)$ are polynomials of degree s given by

$$\Phi_j(x) = \int_0^x \phi_j(t)dt. \qquad (3.2.4)$$

Applying $\Phi_j(D)$ to both sides of (3.2.3) and setting $z = 0$, we get

$$\begin{aligned}
(\mathbf{B} - \mathbf{V}\mathbf{A})\Phi_j(D)(ze^{\mathbf{c}z})\Big|_{z=0} &= \Phi_j(D)(e^z e^{\mathbf{c}z})\Big|_{z=0} \\
&- \mathbf{A}\Phi_j(D)(ze^z e^{\mathbf{c}z})\Big|_{z=0} - \mathbf{V}\Phi_j(D)(e^{\mathbf{c}z})\Big|_{z=0}.
\end{aligned} \qquad (3.2.5)$$

Let us write the differential operator $\Phi_j(D)$ in the form

$$\Phi_j(D) = d_{s,j}D^s + d_{s-1,j}D^{s-1} + \cdots + d_{1,j}D + d_{0,j}I.$$

Since

$$D^k(e^{\mathbf{c}z})\Big|_{z=0} = \mathbf{c}^k, \quad D^k(e^z e^{\mathbf{c}z})\Big|_{z=0} = (\mathbf{e} + \mathbf{c})^k,$$

it follows that

$$\Phi_j(D)(e^{\mathbf{c}z})\Big|_{z=0} = \Phi_j(\mathbf{c}), \quad \Phi_j(D)(e^z e^{\mathbf{c}z})\Big|_{z=0} = \Phi_j(\mathbf{e} + \mathbf{c}).$$

We also have

$$D^k(ze^{\mathbf{c}z})\Big|_{z=0} = \sum_{j=0}^{k} \binom{k}{j} D^j(z)D^{k-j}(e^{\mathbf{c}z})\Big|_{z=0} = k\mathbf{c}^{k-1}$$

and

$$D^k(ze^z e^{\mathbf{c}z})\Big|_{z=0} = k(\mathbf{e} + \mathbf{c})^{k-1}.$$

Hence,

$$\Phi_j(D)(ze^{\mathbf{c}z})\Big|_{z=0} = \Phi'_j(\mathbf{c}) = \phi_j(\mathbf{c})$$

and

$$\Phi_j(D)(ze^z e^{\mathbf{c}z})\Big|_{z=0} = \Phi'_j(\mathbf{e}+\mathbf{c}) = \phi_j(\mathbf{e}+\mathbf{c}).$$

Substituting the relations above into (3.2.5), we obtain

$$(\mathbf{B} - \mathbf{VA})\phi_j(\mathbf{c}) = \Phi_j(\mathbf{e}+\mathbf{c}) - \mathbf{A}\phi_j(\mathbf{e}+\mathbf{c}) - \mathbf{V}\Phi_j(\mathbf{c}),$$

$j = 1, 2, \ldots, s$, or in matrix form,

$$(\mathbf{B} - \mathbf{VA})D_\phi = \widetilde{\mathbf{B}}_0 - \mathbf{A}\widetilde{\mathbf{B}}_1 - \mathbf{V}\widetilde{\mathbf{B}}_2,$$

where

$$D_\phi = \left[\begin{array}{ccc} \phi_1(\mathbf{c}) & \cdots & \phi_s(\mathbf{c}) \end{array} \right], \quad \widetilde{\mathbf{B}}_0 = \left[\begin{array}{ccc} \Phi_1(\mathbf{e}+\mathbf{c}) & \cdots & \Phi_s(\mathbf{e}+\mathbf{c}) \end{array} \right],$$

$$\widetilde{\mathbf{B}}_1 = \left[\begin{array}{ccc} \phi_1(\mathbf{e}+\mathbf{c}) & \cdots & \phi_s(\mathbf{e}+\mathbf{c}) \end{array} \right], \quad \widetilde{\mathbf{B}}_2 = \left[\begin{array}{ccc} \Phi_1(\mathbf{c}) & \cdots & \Phi_s(\mathbf{c}) \end{array} \right].$$

Since

$$\phi_j(c_i) = \begin{cases} \phi_i(c_i), & j = i, \\ 0, & j \neq i, \end{cases}$$

and $\phi_i(c_i) \neq 0$, the diagonal matrix D_ϕ is invertible and we obtain

$$\mathbf{B} = \mathbf{B}_0 - \mathbf{A}\mathbf{B}_1 - \mathbf{V}\mathbf{B}_2 + \mathbf{VA},$$

where

$$\mathbf{B}_0 = \widetilde{\mathbf{B}}_0 D_\phi^{-1}, \quad \mathbf{B}_1 = \widetilde{\mathbf{B}}_1 D_\phi^{-1}, \quad \mathbf{B}_2 = \widetilde{\mathbf{B}}_2 D_\phi^{-1}$$

are matrices whose elements are given by (3.2.2). This completes the proof. ∎

For DIMSIMs with $p = q + 1$ we have the following result.

Theorem 3.2.2 (Butcher and Jackiewicz [65]) *Let $r = s$ and $\mathbf{U} = \mathbf{I}$. Then the method $(\mathbf{c}, \mathbf{A}, \mathbf{U}, \mathbf{B}, \mathbf{V})$ has stage order $q = r = s$ and order $p = q+1$ if and only if the matrix $\mathbf{B}$ is given by (3.2.1) and there exists $\mathbf{q}_p$ such that the following condition is satisfied:*

$$\mathbf{B}\mathbf{c}^{p-1} = \frac{(\mathbf{e}+\mathbf{c})^p - \mathbf{c}^p}{p} - \mathbf{A}\left((\mathbf{e}+\mathbf{c})^{p-1} - \mathbf{c}^{p-1}\right) - (p-1)!(\mathbf{V} - \mathbf{I})\mathbf{q}_p. \quad (3.2.6)$$

Proof: Formula (3.2.1) for the coefficient matrix $\mathbf{B}$ was proved in Theorem 3.2.1. Substituting $\mathbf{w} = \mathbf{w}(z)$ computed from formula (2.4.7) with $\mathbf{U} = \mathbf{I}$ into (2.4.6), we obtain

$$(\mathbf{B} - \mathbf{VA})ze^{\mathbf{c}z} - e^z e^{\mathbf{c}z} + \mathbf{A}ze^z e^{\mathbf{c}z} + \mathbf{V}e^{\mathbf{c}z} - \mathbf{V}\mathbf{b}_p z^p + \mathbf{b}_p z^p e^z = O(z^{p+1}),$$

where

$$\mathbf{b}_p = \frac{\mathbf{c}^p}{p!} - \frac{\mathbf{A}\mathbf{c}^{p-1}}{(p-1)!} - \mathbf{q}_p. \tag{3.2.7}$$

Formula (3.2.6) now follows by equating to zero the coefficient of z^p in the Taylor expansion of the left-hand side of the relation above around $z = 0$. This completes the proof. ∎

The matrix $\mathbf{V} - \mathbf{I}$ in (3.2.6) is singular and it is not clear that there exists $\mathbf{q}_p$ such that this condition is satisfied. We prove in the next theorem that this is always the case. This theorem will also provide a simpler approach to the construction of GLMs with $r = s = q$ and $p = q + 1$ than that based on Theorem 3.2.2. This approach is based on the idea of exponential fitting. To be more precise, we have the following result.

Theorem 3.2.3 (Butcher and Jackiewicz [65]) *Let* $\mathbf{V}$ *be a matrix such that* $\mathbf{V}\mathbf{e} = \mathbf{e}$ *and such that its characteristic polynomial* P *satisfies* $P'(1) \neq 0$. *Assume that there exist vectors* $\mathbf{q}_0, \mathbf{q}_1, \ldots, \mathbf{q}_q$ *such that the DIMSIM with coefficients* $(\mathbf{c}, \mathbf{A}, \mathbf{U}, \mathbf{B}, \mathbf{V})$ *and with the starting vector* $\widetilde{y}^{[0]}$ *defined by*

$$\widetilde{y}_i^{[0]} = \sum_{k=0}^{q} q_{ik} h^k y^{(k)}(t_0) + O(h^{p+1}), \quad i = 1, 2, \ldots, r,$$

has stage order q and order p, which is at least equal to q. Suppose further that

$$\det\left(e^z \mathbf{I} - \mathbf{M}(z)\right) = O(z^{q+2}), \tag{3.2.8}$$

where $\mathbf{M}(z)$ *is the stability matrix defined by (2.6.4). Then there exist a vector* $\mathbf{q}_{q+1}$ *such that this method with starting vector* $y^{[0]}$ *given by*

$$y_i^{[0]} = \sum_{k=0}^{q+1} q_{ik} h^k y^{(k)}(t_0) + O(h^{p+1}), \quad i = 1, 2, \ldots, r,$$

has stage order q and order $p = q + 1$.

Proof: Set

$$\widetilde{\mathbf{w}}(z) = \mathbf{q}_0 + \mathbf{q}_1 z + \cdots + \mathbf{q}_q z^q.$$

Since $p \geq q$ it follows that

$$e^{\mathbf{c}z} = z\mathbf{A}e^{\mathbf{c}z} + \mathbf{U}\widetilde{\mathbf{w}}(z) + O(z^{q+1}) \quad \text{and} \quad \mathbf{M}(z)\widetilde{\mathbf{w}}(z) = O(z^{q+1}).$$

We have to show that there exists a vector $\mathbf{q}_{q+1}$ such that

$$e^{\mathbf{c}z} = z\mathbf{A}e^{\mathbf{c}z} + \mathbf{U}\mathbf{w}(z) + O(z^{q+1}) \tag{3.2.9}$$

and

$$\mathbf{M}(z)\mathbf{w}(z) = O(z^{q+2}), \tag{3.2.10}$$

where
$$\mathbf{w}(z) = \widetilde{\mathbf{w}}(z) + \mathbf{q}_{q+1} z^{q+1}.$$

Denote by $\mathbf{x}$ and $\mathbf{y} = \mathbf{e}$ the left and right eigenvectors corresponding to the single eigenvalue equal to zero of the matrix $\mathbf{M}_0 = \mathbf{M}(0) = \mathbf{I} - \mathbf{V}$ and such that $\mathbf{x}^T \mathbf{y} = 1$. We can assume without loss of generality that $\mathbf{q}_0 = \mathbf{y} = \mathbf{e}$. It follows from condition (3.2.8) that we can choose constants $R > 0$ and $\Delta > 0$ such that for $|z| < R$ the matrix $e^z \mathbf{I} - \mathbf{M}(z)$ has a single eigenvalue $\lambda(z) = O(z^{q+2})$ which satisfies $|\lambda(z)| < \Delta$ and such that all other eigenvalues of this matrix have magnitude greater than Δ. For $|z| < R$, define the function $\beta(z)$ as the unique solution to the system

$$\mathbf{M}(z)\beta(z) = \lambda(z)\beta(z), \quad \mathbf{x}^T \beta(z) = 1,$$

with $\beta(0) = \beta_0 = \mathbf{q}_0$. Since $\beta(z)$ is a rational function of $\lambda(z)$ we can conclude that this function is analytic in the disk $|z| < R$. Let

$$\beta(z) = \widetilde{\beta}(z) + O(z^{q+2}),$$

where
$$\widetilde{\beta}(z) = \beta_0 + \beta_1 z + \cdots + \beta_q z^q + \beta_{q+1} z^{q+1}.$$

Then
$$\mathbf{M}(z)\widetilde{\beta}(z) = O(z^{q+2}).$$

If $\beta_i = \mathbf{q}_i$, $i = 1, 2, \ldots, q$, we can choose $\mathbf{q}_{q+1} = \beta_{q+1}$ and the theorem follows. If $\beta_i \neq \mathbf{q}_i$ for some $0 < i \leq q$, let k be the smallest index such that $\beta_k \neq \mathbf{q}_k$. We have

$$z^k \mathbf{M}(z)(\mathbf{q}_k - \beta_k) + O(z^{k+1}) = O(z^{q+1}),$$

and it follows that $\mathbf{M}_0(\mathbf{q}_k - \beta_k) = 0$. Therefore, there exists $\theta_k \neq 0$ such that $\mathbf{q}_k - \beta_k = \theta_k \mathbf{y}$. Consider the function

$$\varphi(z) = (1 + \theta_k z^k)\widetilde{\beta}(z).$$

Then
$$
\begin{aligned}
\varphi(z) &= \beta_0 + \beta_1 z + \cdots + \beta_{k-1} z^{k-1} \\
&\quad + (\beta_k + \theta_k \beta_0) z^k + \varphi_{k+1} z^{k+1} + \cdots + \varphi_{q+1} z^{q+1} + O(z^{q+2}) \\
&= \mathbf{q}_0 + \mathbf{q}_1 z + \cdots + \mathbf{q}_{k-1} z^{k-1} + \mathbf{q}_k z^z \\
&\quad + \varphi_{k+1} z^{k+1} + \cdots + \varphi_{q+1} z^{q+1} + O(z^{q+2}),
\end{aligned}
$$

where φ_j are coefficients in the Taylor expansion of $\varphi(z)$ around $z = 0$. Hence, $\varphi(z)$ agrees with $\widetilde{\mathbf{w}}(z)$ up to the terms of order k, and this function still satisfies the relation

$$\mathbf{M}(z)\varphi(z) = O(z^{q+2}).$$

Obviously, this process can be continued until $\varphi(z)$ agrees with $\widetilde{\mathbf{w}}(z)$ up to the terms of order q. We can now choose $\mathbf{q}_{q+1} = \varphi_{q+1}$, and the resulting $\mathbf{w}(z) = \widetilde{\mathbf{w}}(z) + \mathbf{q}_{q+1}z^{q+1}$ satisfies (3.2.9) and (3.2.10). This completes the proof. ∎

We conclude this section with the representation formula for the matrix $\mathbf{B}$ if $r = s = p$ and $q = p - 1$. We have the following theorem.

Theorem 3.2.4 (Butcher and Jackiewicz [65]) *Assume that $r = s$ and that $\mathbf{U} = \mathbf{I}$. Then the method $(\mathbf{c}, \mathbf{A}, \mathbf{U}, \mathbf{B}, \mathbf{V})$ has order $p = r = s$ and stage order $q = p - 1$ if and only if*

$$\mathbf{B} = \mathbf{B}_0 - \mathbf{A}\mathbf{B}_1 - \mathbf{V}\mathbf{B}_2 + \mathbf{V}\mathbf{A} + (\mathbf{V} - \mathbf{I})\mathbf{Q}, \qquad (3.2.11)$$

where

$$\mathbf{Q} = \left(\frac{\mathbf{c}^p}{p} - \mathbf{A}\mathbf{c}^{p-1} - (p-1)!\mathbf{q}_p \right) \mathbf{e}^T D_\phi^{-1}. \qquad (3.2.12)$$

Here $\mathbf{B}_0$, $\mathbf{B}_1$, and $\mathbf{B}_2$ are defined by (3.2.2) and the matrix D_ϕ is defined in the proof of Theorem 3.2.1.

Proof: Similar to the proof of Theorem 3.2.2, substituting $\mathbf{w} = \mathbf{w}(z)$ computed from formula (2.4.7) with $\mathbf{U} = \mathbf{I}$ into (2.4.6), we obtain

$$(\mathbf{B} - \mathbf{V}\mathbf{A})ze^{\mathbf{c}z} = e^z e^{\mathbf{c}z} - \mathbf{A}ze^z e^{\mathbf{c}z} - \mathbf{V}e^{\mathbf{c}z} + \mathbf{V}\mathbf{b}_p z^p - \mathbf{b}_p z^p e^z + O(z^{p+1}),$$

where $\mathbf{b}_p$ is defined by (3.2.7). Applying the operator $\Phi_j(D)$, $j = 1, 2, \ldots, s$, $D = d/dz$, where $\Phi_j(x)$ is defined by (3.2.4), to both sides of this equation and setting $z = 0$, we obtain

$$(\mathbf{B} - \mathbf{V}\mathbf{A})\Phi_j(D)(ze^{\mathbf{c}z})\Big|_{z=0} = \Phi_j(D)(e^z e^{\mathbf{c}z})\Big|_{z=0} - \mathbf{A}\Phi_j(D)(ze^z e^{\mathbf{c}z})\Big|_{z=0}$$

$$- \mathbf{V}\Phi_j(D)(e^{\mathbf{c}z})\Big|_{z=0} + \mathbf{V}\mathbf{b}_p\Phi_j(D)z^p\Big|_{z=0} - \mathbf{b}_p\Phi_j(D)(z^p e^z)\Big|_{z=0}.$$

Similarly as in the proof of Theorem 3.2.1, this leads to the equation

$$(\mathbf{B} - \mathbf{V}\mathbf{A})\phi_j(\mathbf{c}) = \Phi_j(\mathbf{e} + \mathbf{c}) - \mathbf{A}\phi_j(\mathbf{e} + \mathbf{c}) - \mathbf{V}\Phi_j(\mathbf{c}) + (p-1)!(\mathbf{V} - \mathbf{I})\mathbf{b}_p,$$

$j = 1, 2, \ldots, s$, or in matrix form

$$(\mathbf{B} - \mathbf{V}\mathbf{A})D_\phi = \widetilde{\mathbf{B}}_0 - \mathbf{A}\widetilde{\mathbf{B}}_1 - \mathbf{V}\widetilde{\mathbf{B}}_2 + (p-1)!(\mathbf{V} - \mathbf{I})\mathbf{b}_p\mathbf{e}^T,$$

where we have also used the relations

$$\Phi_j(D)z^p\Big|_{z=0} = (p-1)! \quad \text{and} \quad \Phi_j(D)(z^p e^z)\Big|_{z=0} = (p-1)!.$$

Hence, it follows that

$$\mathbf{B} = \widetilde{\mathbf{B}}_0 D_\phi^{-1} - \mathbf{A}\widetilde{\mathbf{B}}_1 D_\phi^{-1} - \mathbf{V}\widetilde{\mathbf{B}}_2 D_\phi^{-1} + (p-1)!(\mathbf{V} - \mathbf{I})\mathbf{b}_p\mathbf{e}^T D_\phi^{-1},$$

which is equivalent to (3.2.11) with $\mathbf{Q}$ defined by (3.2.12). This completes the proof. ∎

3.3 A TRANSFORMATION FOR THE ANALYSIS OF DIMSIMS

It was demonstrated in Section 3.2 that for DIMSIMs (3.1.1) with $\mathbf{U} = \mathbf{I}$ and $p = r = s$ and $q = p$ or $q = p - 1$ there are representation formulas for the coefficient matrix $\mathbf{B}$ in terms of the abscissa vector $\mathbf{c}$ and coefficient matrices $\mathbf{A}$ and $\mathbf{V}$. These representation formulas are given by (3.2.1) if $q = p$ or (3.2.11) if $q = p - 1$, where matrices $\mathbf{B}_0$, $\mathbf{B}_1$, and $\mathbf{B}_2$, which depend on vector $\mathbf{c}$, are defined by (3.2.2) and the matrix $\mathbf{Q}$ is defined by (3.2.12).

Let $\mathbf{T} \in \mathbb{R}^{r \times r}$ be a nonsingular matrix and define the transformed matrices

$$\overline{\mathbf{A}} = \mathbf{T}^{-1} \mathbf{A} \mathbf{T}, \quad \overline{\mathbf{B}} = \mathbf{T}^{-1} \mathbf{B} \mathbf{T}, \quad \overline{\mathbf{V}} = \mathbf{T}^{-1} \mathbf{V} \mathbf{T},$$

$$\overline{\mathbf{B}}_0 = \mathbf{T}^{-1} \mathbf{B}_0 \mathbf{T}, \quad \overline{\mathbf{B}}_1 = \mathbf{T}^{-1} \mathbf{B}_1 \mathbf{T}, \quad \overline{\mathbf{B}}_2 = \mathbf{T}^{-1} \mathbf{B}_2 \mathbf{T},$$

so that (3.2.1) transform to

$$\overline{\mathbf{B}} = \overline{\mathbf{B}}_0 - \overline{\mathbf{A}} \, \overline{\mathbf{B}}_1 - \overline{\mathbf{V}} \, \overline{\mathbf{B}}_2 + \overline{\mathbf{V}} \, \overline{\mathbf{A}}$$

and (3.2.11) transforms to

$$\overline{\mathbf{B}} = \overline{\mathbf{B}}_0 - \overline{\mathbf{A}} \, \overline{\mathbf{B}}_1 - \overline{\mathbf{V}} \, \overline{\mathbf{B}}_2 + \overline{\mathbf{V}} \, \overline{\mathbf{A}} + (\overline{\mathbf{V}} - \mathbf{I}) \overline{\mathbf{Q}},$$

where $\overline{\mathbf{Q}} = \mathbf{T}^{-1} \mathbf{Q} \mathbf{T}$. In this section we describe the similarity transformation $\mathbf{T}$, first proposed by Butcher [46], which when applied to $\mathbf{A}$, $\mathbf{B}$, $\mathbf{V}$, $\mathbf{B}_0$, $\mathbf{B}_1$, and $\mathbf{B}_2$ will preserve the special triangular form of $\mathbf{A}$ but will transform $\mathbf{B}_0$, $\mathbf{B}_1$, and $\mathbf{B}_2$ into special form. This will result in significant simplifications to expressions of the stability function in terms of the free parameters of the method. This stability polynomial now takes the form

$$p(w, z) = \det\left(w\mathbf{I} - \mathbf{M}(z)\right) = \det\left(w\mathbf{I} - \overline{\mathbf{M}}(z)\right),$$

where

$$\overline{\mathbf{M}}(z) = \overline{\mathbf{V}} + z\overline{\mathbf{B}}(\mathbf{I} - z\overline{\mathbf{A}})^{-1}$$

is the stability matrix of the transformed method.

Define the matrix $\mathbf{T} = [t_{ij}] \in \mathbb{R}^{r \times r}$ by the formula

$$t_{ij} = \begin{cases} \displaystyle\prod_{k=1}^{j-1}(c_i - c_k), & j \leq i, \\ 0, & j > i. \end{cases} \tag{3.3.1}$$

The inverse of this matrix is determined by the following result [46].

Lemma 3.3.1 *The elements of* $\mathbf{T}^{-1} = [t_{ij}^{(-1)}]$ *are given by*

$$t_{ij}^{(-1)} = \begin{cases} \displaystyle\prod_{\nu=1,\nu\neq j}^{i} \frac{1}{c_j - c_\nu}, & j \leq i, \\ 0, & j > i. \end{cases} \tag{3.3.2}$$

Proof: It follows from the Newton interpolation formula that

$$p(x) = p[c_1] + p[c_1, c_2](x - c_1) + \cdots + p[c_1, c_2, \ldots, c_r](x - c_1)(x - c_2) \cdots (x - c_{r-1})$$

for all polynomials $p(x)$ of degree less than or equal to r. Here, for any function f, $f[c_1, c_2, \ldots, c_i]$ are divided differences defined recursively by

$$f[c_1] = f(c_1),$$

$$f[c_1, c_2, \ldots, c_i] = \frac{f[c_2, c_3, \ldots, c_i] - f[c_1, c_2, \ldots, c_{i-1}]}{c_i - c_1}.$$

This leads to the relation

$$
\begin{bmatrix} p(c_1) \\ p(c_2) \\ \vdots \\ p(c_r) \end{bmatrix} = \mathbf{T} \begin{bmatrix} p[c_1] \\ p[c_1, c_2] \\ \vdots \\ p[c_1, c_2, \ldots, c_r] \end{bmatrix}.
\tag{3.3.3}
$$

Denote by $\ell_j(x)$, $j = 1, 2, \ldots, r$, the Lagrange fundamental polynomials

$$\ell_j(x) = \prod_{k=1, k \neq j}^{r} \frac{x - c_k}{c_j - c_k}.
\tag{3.3.4}$$

Substituting $p = \ell_j$, $j = 1, 2, \ldots, r$, into (3.3.3) and taking into account that $\ell_j(c_k) = \delta_{jk}$, where $\delta_{jk} = 1$ if $j = k$ and $\delta_{jk} = 0$ if $j \neq k$, we can conclude that matrix $\mathbf{T}^{-1}$ takes the form

$$
\mathbf{T}^{-1} = \begin{bmatrix} \ell_1[c_1] & \ell_2[c_1] & \cdots & \ell_r[c_1] \\ \ell_1[c_1, c_2] & \ell_2[c_1, c_2] & \cdots & \ell_r[c_1, c_2] \\ \vdots & \vdots & \ddots & \vdots \\ \ell_1[c_1, c_2, \ldots, c_r] & \ell_2[c_1, c_2, \ldots, c_r] & \cdots & \ell_r[c_1, c_2, \ldots, c_r] \end{bmatrix}.
$$

Then it follows from

$$f[c_1, c_2, \ldots, c_i] = \sum_{k=1}^{i} \left(\prod_{\nu=1, \nu \neq k}^{i} \frac{1}{c_k - c_\nu} \right) f(c_k)
\tag{3.3.5}$$

that

$$\ell_j[c_1, c_2, \ldots, c_i] = \begin{cases} \displaystyle\prod_{\nu=1, \nu \neq j}^{i} \frac{1}{c_j - c_\nu}, & j \leq i, \\[2ex] 0, & j > i, \end{cases}
\tag{3.3.6}$$

which implies formula (3.3.2) for $t_{ij}^{(-1)}$. ∎

The next result specifies the structure of transformed matrices $\overline{\mathbf{A}}$ and $\overline{\mathbf{V}}$.

Lemma 3.3.2 (Butcher [46]) *The matrix $\overline{\mathbf{A}}$ is lower triangular and the matrix $\overline{\mathbf{V}}$ has the form*

$$
\overline{\mathbf{V}} =
\begin{bmatrix}
1 & \overline{v}_2 & \cdots & \overline{v}_r \\
0 & 0 & \cdots & 0 \\
\vdots & \vdots & \ddots & \vdots \\
0 & 0 & \cdots & 0
\end{bmatrix}.
$$

Proof: $\overline{\mathbf{A}}$ is clearly lower triangular as a product of three lower triangular matrices $\mathbf{T}^{-1}$, $\mathbf{A}$, and $\mathbf{T}$. To show that $\overline{\mathbf{V}}$ has the structure above observe first that it follows from the form of $\mathbf{T}^{-1}$ that

$$
\sum_{j=1}^{r} t_{ij}^{(-1)} =
\begin{cases}
1, & i = 1, \\
0, & i \neq 1.
\end{cases}
$$

Since $\mathbf{V} = \mathbf{e}\mathbf{v}^T$, $\mathbf{v}^T\mathbf{e} = 1$ (compare Section 3.1), $\overline{\mathbf{V}} = \mathbf{T}^{-1}\mathbf{e}\mathbf{v}^T\mathbf{T}$, and the first column of $\mathbf{T}$ is equal to $\mathbf{e}$, the result now follows from the relations

$$
\mathbf{T}^{-1}\mathbf{e} = \begin{bmatrix} 1 & 0 & \cdots & 0 \end{bmatrix}^T \quad \text{and} \quad \mathbf{v}^T\mathbf{T} = \begin{bmatrix} 1 & \overline{v}_2 & \cdots & \overline{v}_r \end{bmatrix}.
$$

This completes the proof. ∎

The next theorem determines the structure of matrices $\overline{\mathbf{B}}_0$, $\overline{\mathbf{B}}_1$, and $\overline{\mathbf{B}}_2$.

Theorem 3.3.3 (Butcher [46]) *Matrices $\overline{\mathbf{B}}_0$ and $\overline{\mathbf{B}}_2$ are upper Hessenberg and matrix $\overline{\mathbf{B}}_1$ is upper triangular.*

Proof: Observe first that it follows from (3.2.2) that the elements of matrices $\mathbf{B}_0$, $\mathbf{B}_1$, and $\mathbf{B}_2$ can be written in terms of the Lagrange fundamental polynomials (3.3.4) as follows:

$$
\int_0^{1+c_i} \ell_j(x)dx, \quad \ell_j(1 + c_i), \quad \int_0^{c_i} \ell_j(x)dx.
$$

Let $p_1(x), p_2(x), \ldots, p_r(x)$ be polynomials given by

$$
p_1(x) = 1, \quad p_2(x) = x - c_1, \quad \cdots \quad p_r(x) = \prod_{k=1}^{r-1}(x - c_k),
$$

so that

$$\mathbf{T} = \begin{bmatrix} p_1(c_1) & p_2(c_1) & \cdots & p_r(c_1) \\ p_1(c_2) & p_2(c_2) & \cdots & p_r(c_2) \\ \vdots & \vdots & \ddots & \vdots \\ p_1(c_r) & p_2(c_r) & \cdots & p_r(c_r) \end{bmatrix}.$$

We next compute the matrix $\mathbf{B}_0\mathbf{T}$. Since

$$(\mathbf{B}_0\mathbf{T})_{ij} = \sum_{k=1}^r \int_0^{1+c_i} \ell_k(x)dx\, p_j(c_k) = \int_0^{1+c_i} \sum_{k=1}^r p_j(c_k)\ell_k(x)dx$$

$$= \int_0^{1+c_i} p_j(x)dx = P_j(1+c_i),$$

where

$$P_j(x) = \int_0^x p_j(t)dt,$$

this matrix takes the form

$$\mathbf{B}_0\mathbf{T} = \begin{bmatrix} P_1(1+c_1) & P_2(1+c_1) & \cdots & P_r(1+c_1) \\ P_1(1+c_2) & P_2(1+c_2) & \cdots & P_r(1+c_2) \\ \vdots & \vdots & \ddots & \vdots \\ P_1(1+c_r) & P_2(1+c_r) & \cdots & P_r(1+c_r) \end{bmatrix}.$$

Set $Q_i(x) = P_i(1+x)$. To compute $\mathbf{T}^{-1}\mathbf{B}_0\mathbf{T}$, observe that

$$(\mathbf{T}^{-1}\mathbf{B}_0\mathbf{T})_{ij} = \sum_{k=1}^r \ell_k[c_1, c_2, \ldots, c_k]P_j(1+c_k)$$

$$= \sum_{k=1}^i \left(\prod_{\nu=1,\nu\neq k}^i \frac{1}{c_k - c_\nu} \right) Q_j(c_k) = Q_j[c_1, c_2, \ldots, c_i],$$

where we have used formulas (3.3.5) and (3.3.6). Hence, matrix $\overline{\mathbf{B}}_0 = \mathbf{T}^{-1}\mathbf{B}_0\mathbf{T}$ takes the form

$$\overline{\mathbf{B}}_0 = \begin{bmatrix} Q_1[c_1] & Q_2[c_1] & \cdots & Q_r[c_1] \\ Q_1[c_1, c_2] & Q_2[c_1, c_2] & \cdots & Q_r[c_1, c_2] \\ \vdots & \vdots & \ddots & \vdots \\ Q_1[c_1, c_2, \ldots, c_r] & Q_2[c_1, c_2, \ldots, c_r] & \cdots & Q_r[c_1, c_2, \ldots, c_r] \end{bmatrix}.$$

Because $Q_1(x), Q_2(x), \ldots, Q_r(x)$ are polynomials of degrees $1, 2, \ldots, r$, respectively, the divided differences $Q_j[c_1, c_2, \ldots, c_r]$ vanish if $i > j+1$. This means that $\overline{\mathbf{B}}_0$ has an upper Hessenberg form. The results for $\overline{\mathbf{B}}_1 = \mathbf{T}^{-1}\mathbf{B}_1\mathbf{T}$ and $\overline{\mathbf{B}}_2 = \mathbf{T}^{-1}\mathbf{B}_2\mathbf{T}$ follow similarly from the observations that

$$
\overline{\mathbf{B}}_1 = \begin{bmatrix} q_1[c_1] & q_2[c_1] & \cdots & q_r[c_1] \\ q_1[c_1, c_2] & q_2[c_1, c_2] & \cdots & q_r[c_1, c_2] \\ \vdots & \vdots & \ddots & \vdots \\ q_1[c_1, c_2, \ldots, c_r] & q_2[c_1, c_2, \ldots, c_r] & \cdots & q_r[c_1, c_2, \ldots, c_r] \end{bmatrix},
$$

where $q_i(x) = p_i(1 + x)$, and

$$
\overline{\mathbf{B}}_2 = \begin{bmatrix} P_1[c_1] & P_2[c_1] & \cdots & P_r[c_1] \\ P_1[c_1, c_2] & P_2[c_1, c_2] & \cdots & P_r[c_1, c_2] \\ \vdots & \vdots & \ddots & \vdots \\ P_1[c_1, c_2, \ldots, c_r] & P_2[c_1, c_2, \ldots, c_r] & \cdots & P_r[c_1, c_2, \ldots, c_r] \end{bmatrix}.
$$

This completes the proof.   ■

From the formulas derived for $\overline{\mathbf{B}}_0$, $\overline{\mathbf{B}}_1$, and $\overline{\mathbf{B}}_2$, it can be verified that the $(i, i-1)$ element of each $\overline{\mathbf{B}}_0$, and $\overline{\mathbf{B}}_2$, is $1/(i-1)$ for $i = 2, 3, \ldots, r$, and each diagonal element of $\overline{\mathbf{B}}_1$ is equal to 1.

3.4   CONSTRUCTION OF DIMSIMS OF TYPE 1

In this section we describe various approaches to the construction of methods with $p = q = r = s$, for which, by design, the stability function $p(w, z)$ defined by (2.6.5) with $\mathbf{M}(z)$ given by (2.6.4) has the simple form

$$
p(w, z) = w^{s-1}\big(w - R(z)\big), \tag{3.4.1}
$$

where

$$
R(z) = 1 + z + \frac{1}{2!}z^2 + \cdots + \frac{1}{s!}z^s \tag{3.4.2}
$$

is an approximation of order $p = s$ to the exponential function $\exp(z)$. This then implies that the corresponding DIMSIM has the same region of absolute stability as explicit RK method of order $p = s$ with stability function $R(z)$ given by (3.4.2).

We consider two choices for the abscissa vector $\mathbf{c}$. The first of these is to space out the $\mathbf{c}$ values uniformly in the interval $[0, 1]$ so that

$$
\mathbf{c} = \begin{bmatrix} 0 & \dfrac{1}{s-1} & \cdots & \dfrac{s-2}{s-1} & 1 \end{bmatrix}^T, \tag{3.4.3}
$$

in analogy to what is done for low-order explicit RK methods. A second natural choice is to space out the $\mathbf{c}$ values at step values $t_{n-s+1}, t_{n-s+2}, \ldots, t_n$, so that

$$\mathbf{c} = \begin{bmatrix} -s+2 & -s+1 & \cdots & 0 & 1 \end{bmatrix}^T. \tag{3.4.4}$$

The corresponding methods could then be regarded as generalizations of backward differentiation formulas. Other choices might have some advantages, but so far there is no evidence of this.

It can be demonstrated that the stability function of DIMSIM of type 1 is a polynomial of the form

$$p(w, z) = w^s - p_1(z)w^{s-1} + \cdots + (-1)^{s-1}p_{s-1}(z)w + (-1)^s p_s(z),$$

where

$$
\begin{aligned}
p_1(z) &= 1 + p_{11}z + p_{12}z^2 + \cdots + p_{1s}z^s, \\
p_2(z) &= p_{21}z + p_{22}z^2 + \cdots + p_{2s}z^s, \\
&\ \ \vdots \\
p_{s-1}(z) &= p_{s-1,s-2}z^{s-2} + p_{s-1,s-1}z^{s-1} + p_{s-1,s}z^s, \\
p_s(z) &= p_{s,s-1}z^{s-1} + p_{ss}z^s.
\end{aligned}
$$

In the case $p = q = r = s$ under consideration, coefficient matrix $\mathbf{B}$ can be expressed in terms of $\mathbf{c}$, $\mathbf{A}$, and $\mathbf{V}$ by formula (3.2.1) in Theorem 3.2.1. As a result, the coefficients p_{ij} of the polynomials $p_i(z)$ depend on a_{ij}, $i = 2, 3, \ldots, s$, $j = 1, 2, \ldots, i-1$, and v_i, $i = 1, 2, \ldots, s-1$ (recall that $v_s = 1 - \sum_{j=1}^{s-1} v_j$). This leads to the system of $(s-1)(s+2)/2$ nonlinear equations

$$p_{kl} = 0, \quad k = 2, 3, \ldots, s, \quad l = k-1, k, \ldots, s, \tag{3.4.5}$$

with respect to $(s-1)(s+2)/2$ unknowns a_{ij} and v_i.

For low orders ($p \leq 4$) this system can be analyzed completely with the aid of symbolic manipulation packages such as MATHEMATICA or MAPLE. The case $p = 1$ corresponds to the forward Euler method, and the case $p = 2$ was discussed in Section 2.8.1. To analyze the case $p = 3$ we use the transformation described in Section 3.3 and work with transformed matrices $\overline{\mathbf{A}}$, $\overline{\mathbf{B}}$, $\overline{\mathbf{V}}$, $\overline{\mathbf{B}}_0$, $\overline{\mathbf{B}}_1$, and $\overline{\mathbf{B}}_2$. It can be verified (compare [46]) that for $\mathbf{c} = [0, \frac{1}{2}, 1]^T$, matrices $\overline{\mathbf{B}}_0$, $\overline{\mathbf{B}}_1$, and $\overline{\mathbf{B}}_2$ are given by

$$
\overline{\mathbf{B}}_0 = \begin{bmatrix} 1 & \frac{1}{2} & \frac{1}{12} \\ 1 & \frac{5}{4} & \frac{23}{24} \\ 0 & \frac{1}{2} & \frac{5}{4} \end{bmatrix}, \quad
\overline{\mathbf{B}}_1 = \begin{bmatrix} 1 & 1 & \frac{1}{2} \\ 0 & 1 & 2 \\ 0 & 0 & 1 \end{bmatrix}, \quad
\overline{\mathbf{B}}_2 = \begin{bmatrix} 0 & 0 & 0 \\ 1 & \frac{1}{4} & -\frac{1}{24} \\ 0 & \frac{1}{2} & \frac{1}{4} \end{bmatrix},
$$

and for $\mathbf{c} = [-1, 0, 1]^T$ these matrices take the form

$$
\overline{\mathbf{B}}_0 = \begin{bmatrix} 0 & 0 & 0 \\ 1 & \frac{3}{2} & \frac{5}{6} \\ 0 & \frac{1}{2} & \frac{3}{2} \end{bmatrix}, \quad
\overline{\mathbf{B}}_1 = \begin{bmatrix} 1 & 1 & 0 \\ 0 & 1 & 2 \\ 0 & 0 & 1 \end{bmatrix}, \quad
\overline{\mathbf{B}}_2 = \begin{bmatrix} -1 & -\frac{1}{2} & \frac{1}{6} \\ 1 & \frac{1}{2} & -\frac{1}{6} \\ 0 & \frac{1}{2} & \frac{1}{2} \end{bmatrix}.
$$

Consider first the case $\mathbf{c} = [0, \frac{1}{2}, 1]^T$. Computing $\overline{\mathbf{B}}$ from the relation

$$\overline{\mathbf{B}} = \overline{\mathbf{B}}_0 - \overline{\mathbf{A}}\,\overline{\mathbf{B}}_1 - \overline{\mathbf{V}}\,\overline{\mathbf{B}}_2 + \overline{\mathbf{V}}\,\overline{\mathbf{A}},$$

we obtain the system of polynomial equations (3.4.5) with respect to $\overline{a}_{21}$, $\overline{a}_{31}$, $\overline{a}_{32}$, $\overline{v}_2$, and $\overline{v}_3$, where $\overline{a}_{ij}$ are coefficients of the matrix $\overline{\mathbf{A}}$ and $[1, \overline{v}_2, \overline{v}_3]$ is the first row of the matrix $\overline{\mathbf{V}}$. Solving first the equations $p_{32} = 0$ and $p_{33} = 0$ with respect to $\overline{a}_{21}$ and $\overline{a}_{31}$, and then the equation $p_{23} = 0$ with respect to $\overline{a}_{32}$ and substituting the resulting expressions into p_{21} and p_{22}, we obtain a system of two polynomial equations

$$p_{21} = 0, \quad p_{22} = 0$$

with respect to $\overline{v}_2$ and $\overline{v}_3$. This system has three solutions, given by

$$
\begin{aligned}
(\overline{v}_2, \overline{v}_3) &= (1, \tfrac{1}{6}), \\
(\overline{v}_2, \overline{v}_3) &\approx (0.95037130, 0.097685112), \\
(\overline{v}_2, \overline{v}_3) &\approx (-2.01509894, -0.0078712370).
\end{aligned}
$$

This leads to the following methods expressed in terms of original matrices $\mathbf{A}$, $\mathbf{U}$, $\mathbf{B}$, and $\mathbf{V} = \mathbf{e}\mathbf{v}^T$:

$$
\left[\begin{array}{c|c} \mathbf{A} & \mathbf{U} \\ \hline \mathbf{B} & \mathbf{V} \end{array}\right] =
\left[\begin{array}{ccc|ccc}
0 & 0 & 0 & 1 & 0 & 0 \\
1 & 0 & 0 & 0 & 1 & 0 \\
\frac{1}{4} & 1 & 0 & 0 & 0 & 1 \\
\hline
\frac{5}{4} & \frac{1}{3} & \frac{1}{6} & -\frac{2}{3} & \frac{4}{3} & \frac{1}{3} \\
\frac{35}{24} & -\frac{1}{3} & \frac{1}{8} & -\frac{2}{3} & \frac{4}{3} & \frac{1}{3} \\
\frac{17}{12} & 0 & \frac{1}{12} & -\frac{2}{3} & \frac{4}{3} & \frac{1}{3}
\end{array}\right],
$$

$$
\mathbf{A} = \begin{bmatrix}
0 & 0 & 0 \\
0.92418548 & 0 & 0 \\
0.46978371 & 0.91532538 & 0
\end{bmatrix},
$$

$$
\mathbf{B} = \begin{bmatrix}
1.30682500 & 0.21191312 & 0.19702172 \\
1.51515833 & -0.45475354 & 0.23116957 \\
1.55816629 & -0.37544408 & 0.14792854
\end{bmatrix},
$$

$$\mathbf{v} = \begin{bmatrix} -0.70537237 & 1.51000215 & 0.19537022 \end{bmatrix}^T$$

and

$$\mathbf{A} = \begin{bmatrix} 0 & 0 & 0 \\ 0.96002505 & 0 & 0 \\ -191.850709 & 64.8421242 & 0 \end{bmatrix},$$

$$\mathbf{B} = \begin{bmatrix} 0.18369583 & 0.98929051 & 0.00267737344 \\ 0.39202917 & 0.32262384 & 0.00098566065 \\ -63.4917617 & 192.182330 & 0.49368033 \end{bmatrix},$$

$$\mathbf{v} = \begin{bmatrix} 5.01445541 & -3.99871293 & -0.015742474 \end{bmatrix}^T.$$

Choosing next the abscissa vector $\mathbf{c} = [-1, 0, 1]^T$ the system (3.4.5) has two solutions and the resulting methods are given by

$$\mathbf{A} = \begin{bmatrix} 0 & 0 & 0 \\ -0.78936376 & 0 & 0 \\ 0.8301728 & 2.5336861 & 0 \end{bmatrix},$$

$$\mathbf{B} = \begin{bmatrix} -0.52120879 & 0.34886447 & -\frac{1}{12} \\ -0.60454212 & 1.8048949 & \frac{1}{3} \\ -0.1878545 & -1.147975 & -0.28368611 \end{bmatrix},$$

$$\mathbf{v} = \begin{bmatrix} -0.083677237 & 0.86694179 & 0.21673545 \end{bmatrix}^T$$

and

$$\mathbf{A} = \begin{bmatrix} 0 & 0 & 0 \\ 3.723271912 & 0 & 0 \\ 20.32196115 & -0.5371619499 & 0 \end{bmatrix},$$

$$\mathbf{B} = \begin{bmatrix} 0.632221303 & 0.637221992 & -\frac{1}{12} \\ 0.54888797 & -2.41938325 & \frac{1}{3} \\ 0.965554636 & -20.3514058 & 2.78716195 \end{bmatrix},$$

$$\mathbf{v} = \begin{bmatrix} 0.96755166 & 0.0259586716 & 0.00648966791 \end{bmatrix}^T.$$

The case $p = 4$ was first analyzed successfully by Wright [291, 292], who also used the transformation described in Section 3.3. For $\mathbf{c} = [0, \frac{1}{3}, \frac{2}{3}, 1]^T$ the coefficients of the method found by Wright [291, 292] are

$$
A = \begin{bmatrix}
0 & 0 & 0 & 0 \\
0.3739348246 & 0 & 0 & 0 \\
0.2949848977 & 0.4816828233 & 0 & 0 \\
-0.6903740089 & 2.2712602977 & -0.2257249932 & 0
\end{bmatrix},
$$

$$
B = \begin{bmatrix}
2.9444524372 & -6.1678663794 & 4.9955647402 & -1.0248304881 \\
2.8194524372 & -5.6539774905 & 4.1761202957 & -0.6348764238 \\
2.5372463716 & -4.6501532283 & 2.9967727914 & -0.2465466791 \\
1.1878127621 & -0.2420326723 & -1.2663733375 & 0.7127522620
\end{bmatrix},
$$

and

$$
v = \begin{bmatrix} -18.3637007103 & 47.9911902596 & -32.4937789808 & 3.8662894315 \end{bmatrix}^T.
$$

We can also construct methods with $p = q = r = s = 4$ using a continuation approach. Setting

$$
X = \begin{bmatrix} a_{21} & a_{31} & \cdots & a_{s,s-1} & v_1 & \cdots & v_{s-1} \end{bmatrix}^T
$$

and

$$
F = F(X) = \begin{bmatrix} p_{21} & p_{22} & \cdots & p_{ss} \end{bmatrix}^T
$$

we can rewrite system (3.4.5) in the compact form

$$
F(X) = 0. \tag{3.4.6}
$$

This system can be generated by MATHEMATICA or MAPLE. We can compute solutions to (3.4.6) by embedding this system into a homotopy which depends on the parameter $\xi \in [0,1]$. We considered mainly two embeddings, given by

$$
H_1(X, Y, \xi) = \xi F(X) + (1 - \xi)\big(F(X) - F(Y)\big) = 0
$$

or

$$
H_2(X, Y, \xi) = \xi F(X) + (1 - \xi)(X - Y) = 0,
$$

which for $\xi = 0$ have a known solution $X = Y$. Following the homotopy curve from $\xi = 0$ (for which $X = Y$) to $\xi = 1$ will then yield a solution to the original system (3.4.6). Continuation packages that have been used experimentally and, to a large extent, successfully in our search for methods are: (1) PCON61, version 61 of PITCON written by Rheinboldt and Burkardt [245, 246]; (2) ALCON2, written by Deufhard et al. [114]; and (3) FIXPNF, from the HOMPACK suite of subroutines written by Watson et al. [284]. Some examples of methods found in this way are given by Butcher and Jackiewicz [66].

3.5   CONSTRUCTION OF DIMSIMS OF TYPE 2

The stability function $p(w, z)$ of the DIMSIM of type 2 is a rational function which is more complicated to deal with than the stability polynomial of methods of type 1 discussed in Section 3.4. However, substituting

$$z = \frac{\widehat{z}}{1 + \lambda \widehat{z}} \qquad \text{or} \qquad \widehat{z} = \frac{z}{1 - \lambda z}$$

and

$$\mathbf{A} = \widehat{\mathbf{A}} + \lambda \mathbf{I} \qquad \text{or} \qquad \widehat{\mathbf{A}} = \mathbf{A} - \lambda \mathbf{I}$$

into the stability matrix $\mathbf{M}(z)$, we can work instead with the modified stability matrix $\widehat{\mathbf{M}}(\widehat{z})$ defined by

$$\widehat{\mathbf{M}}(\widehat{z}) \ := \ \mathbf{M}(z) = \ \mathbf{M}\left(\frac{\widehat{z}}{1 + \lambda \widehat{z}}\right)$$

$$= \ \mathbf{V} + \frac{\widehat{z}}{1 + \lambda \widehat{z}} \mathbf{B}\left(\mathbf{I} - \frac{\widehat{z}}{1 + \lambda \widehat{z}} \mathbf{A}\right)^{-1} \ = \ \mathbf{V} + \widehat{z} \mathbf{B}\left(\mathbf{I} - \widehat{z} \widehat{\mathbf{A}}\right)^{-1}$$

and the corresponding stability function

$$\widehat{p}(w, \widehat{z}) := p(w, z) = \det\left(w\mathbf{I} - \mathbf{M}(z)\right) = \det\left(\mathbf{I} - \widehat{M}(\widehat{z})\right),$$

which is a polynomial since the matrix $\widehat{\mathbf{A}}$ is strictly lower triangular. We look for methods which, by design, have the same stability properties as SDIRK methods of the same order. These methods were reviewed in Section 2.7. This means that the stability function $p(w, z)$ of these methods takes the form

$$p(w, z) = w^{s-1}\left(w - R(z)\right),$$

where

$$R(z) = \frac{P(z)}{(1 - \lambda z)^s}$$

is the stability function of an s-stage SDIRK method of order $p = s$. The stability function of SDIRK methods is defined by (2.7.5). This requirement can be reformulated in terms of $\widehat{p}(w, \widehat{z})$ as

$$\widehat{p}(w, \widehat{z}) = w^{s-1}\left(w - R(z)\right) = w^{s-1}\left(w - \widehat{R}(\widehat{z})\right),$$

where

$$\widehat{R}(\widehat{z}) \ := \ R(z) \ = \ R\left(\frac{\widehat{z}}{1 + \lambda \widehat{z}}\right)$$

$$= \ \frac{P\left(\dfrac{\widehat{z}}{1 + \lambda \widehat{z}}\right)}{\left(1 - \lambda \dfrac{\widehat{z}}{1 + \lambda \widehat{z}}\right)^s} \ = \ (1 + \lambda \widehat{z})^s P\left(\frac{\widehat{z}}{1 + \lambda \widehat{z}}\right)$$

$$= \ (-1)^s \sum_{j=0}^{s} L_s^{(s-j)}\left(\frac{1}{\lambda}\right)(1 + \lambda \widehat{z})^{s-j}(\lambda \widehat{z})^j$$

is a polynomial of degree s with respect to $\widehat{z}$. We recall that $L_s(x)$ are Laguerre polynomials defined by (2.7.6). A simpler explicit expression for this polynomial $\widehat{R}(\widehat{z})$ follows from the following lemma.

Lemma 3.5.1 (compare [79]) *Assume that $\lambda \neq 0$ and that $|\widehat{z}| < 1/\lambda$. Then*

$$\exp\left(\frac{\widehat{z}}{1 + \lambda\widehat{z}}\right) = 1 + \sum_{n=1}^{\infty} \frac{(-1)^n}{n} \lambda^{n-1} \widehat{z}^n L_n'\left(\frac{1}{\lambda}\right). \tag{3.5.1}$$

Proof: Making the substitutions $u = \lambda\widehat{z}$ and $t = 1/\lambda$, formula (3.5.1) is equivalent to

$$\exp\left(\frac{tu}{1 + u}\right) = 1 + \sum_{n=1}^{\infty} \frac{(-u)^n}{n} t L_n'(t) \tag{3.5.2}$$

for $|u| < 1$. Expanding the left-hand side of this relation into a Taylor series around $t = 0$, then substituting the formula

$$L_n'(t) = \sum_{j=1}^{n} (-1)^j \binom{n}{j} \frac{t^{j-1}}{(j-1)!}$$

into the right-hand side of (3.5.2) and reversing the order of summation in the resulting double sum, formula (3.5.2) is transformed into

$$1 + \sum_{j=1}^{\infty} \left(\frac{u}{1+u}\right)^j \frac{t^j}{j!} = 1 + \sum_{j=1}^{\infty} \sum_{n=j}^{\infty} \frac{(-u)^n}{n} (-1)^j \binom{n}{j} \frac{t^j}{(j-1)!}. \tag{3.5.3}$$

The coefficient of $t^j/j!$ on the left-hand side of the relation above is equal to

$$\left(\frac{u}{1+u}\right)^j = u^j \sum_{k=0}^{\infty} \binom{-j}{k} u^k.$$

The coefficient of $t^j/j!$ on the right-hand side of (3.5.3) is

$$\sum_{n=j}^{\infty} (-1)^{n-j} \frac{j}{n} \binom{n}{j} u^n = u^j \sum_{n=j}^{\infty} (-1)^{n-j} \binom{n-1}{j-1} u^{n-j}$$

$$= u^j \sum_{n=j}^{\infty} (-1)^{n-j} \binom{n-1}{n-j} u^{n-j} = u^j \sum_{k=0}^{\infty} (-1)^k \binom{k+j-1}{k} u^k.$$

It follows from the relation

$$(-1)^k \binom{k+j-1}{k} = \binom{-j}{k}$$

that these coefficients are equal. This completes the proof.   ∎

This lemma implies the following corollary.

Corollary 3.5.2 (compare [79]) *The polynomial $\widehat{R}(\widehat{z})$ corresponding to the type 2 DIMSIM with $p = q = r = s$ has the form*

$$\widehat{R}(\widehat{z}) = 1 + \sum_{n=1}^{s} \frac{(-1)^n}{n} \lambda^{n-1} \widehat{z}^n L_n' \left(\frac{1}{\lambda}\right). \tag{3.5.4}$$

We usually assume that the abscissa vector **c** is given by (3.4.3) or (3.4.4) as was done for DIMSIMs of type 1.

It can be demonstrated that the stability function $\widehat{p}(w, \widehat{z})$ of methods of type 2 is a polynomial of the form

$$\widehat{p}(w, \widehat{z}) = w^s - \widehat{p}_1(\widehat{z})w^{s-1} + \cdots + (-1)^{s-1}\widehat{p}_{s-1}(\widehat{z})w + (-1)^s\widehat{p}_s(\widehat{z}),$$

where

$$\begin{aligned}
\widehat{p}_1(\widehat{z}) &= 1 + \widehat{p}_{11}\widehat{z} + \widehat{p}_{12}\widehat{z}^2 + \cdots + \widehat{p}_{1s}\widehat{z}^s, \\
\widehat{p}_2(\widehat{z}) &= \widehat{p}_{21}\widehat{z} + \widehat{p}_{22}\widehat{z}^2 + \cdots + \widehat{p}_{2s}\widehat{z}^s, \\
&\vdots \\
\widehat{p}_{s-1}(\widehat{z}) &= \widehat{p}_{s-1,s-2}\widehat{z}^{s-2} + \widehat{p}_{s-1,s-1}\widehat{z}^{s-1} + \widehat{p}_{s-1,s}\widehat{z}^s, \\
\widehat{p}_s(\widehat{z}) &= \widehat{p}_{s,s-1}\widehat{z}^{s-1} + \widehat{p}_{ss}\widehat{z}^s,
\end{aligned}$$

and the coefficients $\widehat{p}_{ij}$ of the polynomials $\widehat{p}_i(\widehat{z})$ depend on $\widehat{a}_{ij} = a_{ij}$, $i = 2, 3, \ldots, s$, $j = 1, 2, \ldots, s - 1$, and v_i, $i = 1, 2, \ldots, s - 1$. As in Section 3.4, this leads to the system of $(s - 1)(s + 2)/2$ nonlinear equations

$$\widehat{p}_{kl} = 0, \quad k = 2, 3, \ldots, s, \quad l = k - 1, k, \ldots, s, \tag{3.5.5}$$

with respect to the $(s - 1)(s + 2)/2$ unknowns a_{ij} and v_i. This system also depends on the parameter λ (the diagonal element of the matrix **A**), which is usually chosen in advance in such a way that the resulting method is A- or L-stable (compare Tables 2.7.1 and 2.7.2).

As in Section 3.4, for low orders ($p \leq 4$) this system can be analyzed completely by MATHEMATICA or MAPLE. The case $p = 1$ corresponds to methods with **c** $= \lambda$ and

$$\left[\begin{array}{c|c} \mathbf{A} & \mathbf{U} \\ \hline \mathbf{B} & \mathbf{V} \end{array} \right] = \left[\begin{array}{c|c} \lambda & 1 \\ \hline 1 & 1 \end{array} \right],$$

$\lambda > 0$. (For $\lambda = \frac{1}{2}$ this method attains order $p = 2$ and corresponds to the trapezoidal rule.) These methods are A-stable for $\lambda \geq \frac{1}{2}$ and L-stable for $\lambda = 1$, which corresponds to the backward Euler method. The case $p = 2$

was discussed in Section 2.8.2. The case $p = 3$ was analyzed by Butcher and Jackiewicz [66]. Choosing $\lambda \approx 0.43586652$ (compare Table 2.7.2), which is a root of the polynomial

$$\varphi(\lambda) = \lambda^3 - 3\lambda^2 + \tfrac{3}{2}\lambda - \tfrac{1}{6},$$

and then solving system (3.5.5), we can obtain methods that are A- and L-stable. An example of such a method with $\mathbf{c} = [0, \tfrac{1}{2}, 1]^T$ is given by

$$\mathbf{A} = \begin{bmatrix} 0.43586652 & 0 & 0 \\ 0.25051488 & 0.43586652 & 0 \\ -1.2115943 & 1.0012746 & 0.43586652 \end{bmatrix},$$

$$\mathbf{B} = \begin{bmatrix} 0.83379073 & 0.64599891 & -0.31582709 \\ 0.60625754 & 1.2869318 & -0.47974168 \\ -0.30841677 & 3.8034216 & -1.1207225 \end{bmatrix},$$

and

$$\mathbf{v} = \begin{bmatrix} 0.55209096 & 0.73485666 & -0.28694762 \end{bmatrix}^T.$$

An example of A-stable and L-stable method with $\mathbf{c} = [-1, 0, 1]^T$ is

$$\mathbf{A} = \begin{bmatrix} 0.43586652 & 0 & 0 \\ 1.1720924 & 0.43586652 & 0 \\ 1.1074469 & 1.0003697 & 0.43586652 \end{bmatrix},$$

$$\mathbf{B} = \begin{bmatrix} 0.91915247 & 0.19260634 & -0.067563751 \\ 0.83581914 & 0.12304716 & -0.086763606 \\ 0.81661928 & 0.16195887 & -0.042199635 \end{bmatrix},$$

and

$$\mathbf{v} = \begin{bmatrix} 0.78498369 & 0.32691525 & -0.11189895 \end{bmatrix}^T.$$

The case $p = 4$ was first analyzed successfully by Wright [291, 292]. Choosing $\lambda \approx 0.57281606$ and solving system (3.5.5) leads to methods that are A- and L-stable. An example of such a method with $\mathbf{c} = [0, \tfrac{1}{3}, \tfrac{2}{3}, 1]^T$ derived by Wright [292] is

$$\mathbf{A} = \begin{bmatrix} 0.57281606 & 0 & 0 & 0 \\ 0.15022075 & 0.57281606 & 0 & 0 \\ 0.59515808 & -0.26632807 & 0.57281606 & 0 \\ 1.7717286 & -1.64234444 & 0.39147320 & 0.57281606 \end{bmatrix},$$

$$
\mathbf{B} =
\begin{bmatrix}
13.153119 & -25.451598 & 16.029984 & -3.7980058 \\
13.600935 & -27.228974 & 18.647436 & -4.9027858 \\
14.289167 & -29.534083 & 20.999448 & -5.4831961 \\
15.679828 & -33.750697 & 24.407110 & -5.9235938
\end{bmatrix},
$$

and

$$
\mathbf{v} = \begin{bmatrix} 15.615037 & -46.967269 & 41.290082 & -8.9378502 \end{bmatrix}^T.
$$

Similarly as in the case of DIMSIMs of type 1, methods of type 2 can also be obtained using the homotopy approach described at the end of Section 3.4. Examples of methods derived in this way are given by Butcher and Jackiewicz [66]. The attractive feature of this approach is the possibility of generating entire families of formulas, depending, for example, on the parameter λ or some of the abscissas c_i by making a judicious choice of the underlying homotopy map. This is discussed in more detail in the paper [66].

3.6 CONSTRUCTION OF DIMSIMS OF TYPE 3

A systematic approach to the construction of DIMSIMs of type 3 is described by Butcher [45]. It follows from the assumption that $\mathbf{V}$ is a rank 1 matrix that the stability polynomial of type 3 DIMSIM with $p = q = r = s$ takes the form

$$
\begin{aligned}
p(w, z) = {} & w^s - (\alpha_0 - \beta_1 z)w^{s-1} \\
& - z(\alpha_1 - \beta_2 z)w^{s-2} - \cdots - z^{s-1}(\alpha_{s-1} - \beta_s z),
\end{aligned} \tag{3.6.1}
$$

where the coefficients $\alpha_0, \alpha_1, \ldots, \alpha_{s-1}$ and $\beta_1, \beta_2, \ldots, \beta_s$ depend on the choice of the vector $\mathbf{v}$ appearing in the coefficient matrix $\mathbf{V}$. Butcher [45] discusses a choice of coefficients in (3.6.1) which allow factorization of the form

$$
p(w, z) = w^{s-\widetilde{s}}\widetilde{p}(w, z) \tag{3.6.2}
$$

for as low a value of $\widetilde{s}$ as possible, since a choice based on this principle will make the methods close to type 1 DIMSIMs.

It can be verified using (3.6.1) and the relation

$$
p(e^z, z) = O(z^{s+1}), \quad z \to 0, \tag{3.6.3}
$$

that

$$
w = \widehat{F}\left(\frac{z}{w}\right) = \frac{N\left(\dfrac{z}{w}\right)}{D\left(\dfrac{z}{w}\right)}, \tag{3.6.4}
$$

where

$$N\left(\frac{z}{w}\right) = \alpha_0 + \alpha_1\frac{z}{w} + \alpha_2\left(\frac{z}{w}\right)^2 + \cdots + \alpha_{s-1}\left(\frac{z}{w}\right)^{s-1}$$

and

$$D\left(\frac{z}{w}\right) = 1 + \beta_1\frac{z}{w} + \beta_2\left(\frac{z}{w}\right)^2 + \cdots + \beta_s\left(\frac{z}{w}\right)^s.$$

The relation (3.6.3) also implies that the rational function $\widehat{F}(z/w)$ is an approximation correct to within $O((z/w)^{s+1})$ to the function F defined in a neighborhood of $z = 0$ by the functional equation

$$e^z = F(ze^{-z}).$$

Butcher [45] studied this function with a view of finding rational approximations with numerator and denominator of degrees as low as possible for a given order of approximation. In particular, Padé approximations to F with degree m in the denominator and degree n in the numerator are studied and listed for $0 \leq m \leq 5$ and $0 \leq n \leq 5$. Unfortunately, this approach leads to stability polynomials and the corresponding methods with rather small regions of absolute stability, and an alternative approach is also proposed which is more promising [45]. In this approach the parameters $\alpha_1, \alpha_2, \ldots, \alpha_{s-1}$ are regarded as free parameters and $\beta_1, \beta_2, \ldots, \beta_r$ are chosen in terms of them to satisfy the conditions for order $p = s$. The criterion for choosing $\alpha_1, \alpha_2, \ldots, \alpha_{s-1}$ is then based on obtaining large intervals of absolute stability $[-X, 0]$ of (3.6.1). This is done by forcing the resulting stability polynomial $p(w, z)$ evaluated at $z = -X$ to have all its zeros on the unit circle $|w| = 1$. Using this approach, type 3 methods with $p = q = r = s = 3$ and $p = q = r = s = 4$ are found. The coefficients of methods of order $p = 4$ are also given by Butcher [45], corresponding to the choices of abscissa vector $\mathbf{c} = [0, \frac{1}{3}, \frac{2}{3}, 1]^T$ and $\mathbf{c} = [-2, -1, 0, 1]^T$.

In what follows we describe a somewhat different approach which is based on maximizing the area of the region of absolute stability of the polynomial $p(w, z)$, which corresponds to the method of order $p = s$. For $s = 3$ this polynomial depends on two free parameters, which are chosen as α_2 and β_3. We have plotted in Fig. 3.6.1 the area of the region of absolute stability of $p(w, z)$ versus α_2 and β_3 and the corresponding contour plots. This area achieves its maximum value approximately equal to 1.264 for the parameter values

$$\alpha_2 = \frac{279}{256}, \quad \beta_3 = -\frac{255}{182}.$$

The interval of absolute stability is $[-1.21, 0]$. The stability polynomial takes the form

$$p(w, z) = w^3 - \left(1 + \frac{59783}{34944}z\right)w^2 + z\left(\frac{24839}{34944} + \frac{90901}{69888}z\right)w - z^2\left(\frac{279}{256} + \frac{255}{182}z\right).$$

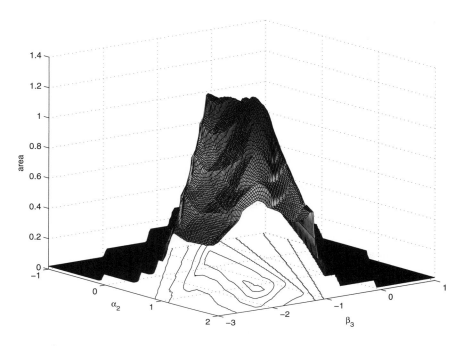

Figure 3.6.1   Area of the region of absolute stability of the polynomial $p(w, z)$ of order $p = 3$ and the corresponding contour plots

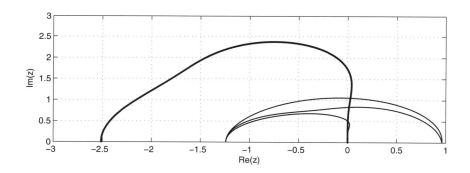

Figure 3.6.2   Stability regions of type 3 DIMSIM of order $p = 3$ and a 3-stage RK method of order $p = 3$

For $\mathbf{c} = [0, \frac{1}{2}, 1]^T$, coefficients of a type 3 method corresponding to this polynomial are

$$
\mathbf{B} = \begin{bmatrix} \dfrac{174457}{104832} & -\dfrac{27577}{52416} & -\dfrac{14431}{7488} \\[2mm] \dfrac{196297}{104832} & -\dfrac{62521}{52416} & -\dfrac{7255}{7488} \\[2mm] \dfrac{296761}{104832} & -\dfrac{202297}{52416} & \dfrac{9281}{7488} \end{bmatrix},
$$

$$\mathbf{v} = \begin{bmatrix} \frac{223417}{34944} & -\frac{125497}{8736} & \frac{104505}{11648} \end{bmatrix}^T,$$

and for $\mathbf{c} = [-1, 0, 1]$, the coefficients are

$$\mathbf{B} = \begin{bmatrix} \frac{338809}{419328} & -\frac{45049}{52416} & -\frac{519005}{419328} \\[2mm] \frac{303865}{419328} & -\frac{10105}{52416} & -\frac{344285}{419328} \\[2mm] \frac{478585}{419328} & -\frac{79993}{52416} & \frac{459427}{419328} \end{bmatrix},$$

$$\mathbf{v} = \begin{bmatrix} \frac{255}{182} & -\frac{108025}{34944} & \frac{94009}{34944} \end{bmatrix}^T.$$

A region of absolute stability of these methods is shown in Fig. 3.6.2 (thin line) together with the region of absolute stability of a 3-stage RK method of order $p = 3$ (thick line).

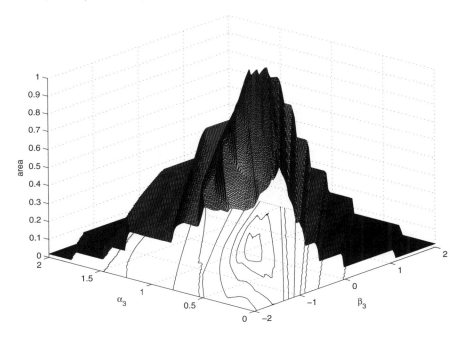

Figure 3.6.3   Area of the region of absolute stability of the polynomial $p(w, z)$ of order $p = 4$ and the corresponding contour plots

For $s = 4$ stability polynomial $p(w, z)$ of the form (3.6.1), which satisfies (3.6.3) depends on three free parameters which are chosen as β_3, α_3, and β_4. We have performed a computer search trying to maximize the area of the region of absolute stability of $p(w, z)$, and this leads to the following values:

$$\beta_3 = \frac{301}{857}, \quad \alpha_3 = \frac{339}{317}, \quad \beta_4 = -\frac{776}{409}.$$

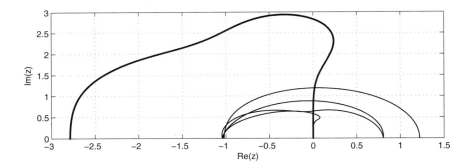

Figure 3.6.4 Stability regions of type 3 DIMSIM of order $p = 4$ and a 4-stage RK method of order $p = 4$

To illustrate that this area attains its maximum value, we have plotted on Fig. 3.6.3 the area of the region of absolute stability and the corresponding contour plots of $p(w, z)$ versus β_3 and α_3 for $\beta_4 = -\frac{301}{857}$. The stability polynomial $p(w, z)$ takes the form

$$p(w, z) = w^4 - \left(1 + \frac{1972056569}{1111126210} z\right) w^3 + z\left(\frac{860930359}{1111126210} + \frac{4758391001}{6666757260} z\right) w^2$$
$$- z^2\left(\frac{2926187477}{6666757260} - \frac{301}{857} z\right) w - z^3\left(\frac{339}{317} + \frac{776}{409} z\right).$$

For $\mathbf{c} = [0, \frac{1}{3}, \frac{2}{3}, 1]^T$ the coefficients of the type 3 method corresponding to this polynomial are

$$\mathbf{B} = \begin{bmatrix} -\frac{731642674841}{60000815340} & \frac{338127108937}{20000271780} & \frac{170879833211}{20000271780} & -\frac{898889950567}{60000815340} \\ -\frac{1478285553517}{120001630680} & \frac{696810052759}{40000543560} & \frac{102993814409}{13333514520} & -\frac{1706111988809}{120001630680} \\ -\frac{784976732921}{60000815340} & \frac{45224103773}{2222252420} & \frac{73100726731}{20000271780} & -\frac{718887504547}{60000815340} \\ -\frac{1838290445557}{120001630680} & \frac{1141260536759}{40000543560} & -\frac{273248690813}{40000543560} & -\frac{852767059529}{120001630680} \end{bmatrix},$$

$$\mathbf{v} = \begin{bmatrix} -\frac{231381282511}{3333378630} & \frac{380257759451}{1666689315} & \frac{168827831353}{666675726} & -\frac{159169149502}{1666689315} \end{bmatrix}^T,$$

and for $\mathbf{c} = [-2, -1, 0, 1]^T$ the coefficients are

$$\mathbf{B} = \begin{bmatrix} -\frac{43313516513}{40000543560} & \frac{22693257667}{13333514520} & -\frac{16941972961}{13333514520} & -\frac{82948747141}{40000543560} \\ -\frac{11245051457}{10000135890} & \frac{1246482418}{555563105} & -\frac{2429913149}{3333378630} & -\frac{10576929557}{5000067945} \\ -\frac{43313516513}{40000543560} & \frac{27137762507}{13333514520} & \frac{278682133}{4444504840} & -\frac{69615232621}{40000543560} \\ -\frac{14578430087}{10000135890} & \frac{5961699674}{1666689315} & -\frac{7985544199}{3333378630} & \frac{2756584963}{5000067945} \end{bmatrix},$$

$$\mathbf{v} = \begin{bmatrix} -\dfrac{776}{409} & \dfrac{23131515119}{3333378630} & -\dfrac{14853472073}{1666689315} & \dfrac{16233261977}{3333378630} \end{bmatrix}^T.$$

A region of absolute stability of these methods is shown in Fig. 3.6.4 (thin line) together with the region of absolute stability of a 4-stage RK method of order $p = 4$ (thick line).

3.7 CONSTRUCTION OF DIMSIMS OF TYPE 4

A systematic approach to the construction of DIMSIMs of type 4, which generalizes the approach of Butcher [45], is described by him in a subsequent article [48]. It follows from the assumption that $\mathbf{V}$ is a rank 1 matrix that the stability polynomial of type 4 DIMSIM with $p = q = r = s$ takes the form

$$\begin{aligned} p(w, z) = {} & (1 - \lambda z)^s w^s - (\alpha_0 - \beta_1 z)(1 - \lambda z)^{s-1} w^{s-1} \\ & - z(\alpha_1 - \beta_2 z)(1 - \lambda z)^{s-2} w^{s-2} - \cdots - z^{s-1}(\alpha_{s-1} - \beta_s z), \end{aligned} \tag{3.7.1}$$

with coefficients $\alpha_0, \alpha_1, \ldots, \alpha_{s-1}$ and $\beta_1, \beta_2, \ldots, \beta_s$, which depend on the choice of vector $\mathbf{v}$ in $\mathbf{V}$. Observe that for $\lambda = 0$, (3.7.1) reduces to (3.6.1).

It follows from (3.7.1) and the relation

$$p(e^z, z) = O(z^{s+1}), \quad z \to 0, \tag{3.7.2}$$

that

$$w(1 - \lambda z) = \widehat{F}\left(\frac{z}{w(1 - \lambda z)}\right) = \frac{N\left(\dfrac{z}{w(1 - \lambda z)}\right)}{D\left(\dfrac{z}{w(1 - \lambda z)}\right)}, \tag{3.7.3}$$

where

$$N\left(\frac{z}{w(1 - \lambda z)}\right) = \alpha_0 + \alpha_1\left(\frac{z}{w(1 - \lambda z)}\right) + \cdots + \alpha_{s-1}\left(\frac{z}{w(1 - \lambda z)}\right)^{s-1},$$

$$D\left(\frac{z}{w(1 - \lambda z)}\right) = 1 + \beta_1\left(\frac{z}{w(1 - \lambda z)}\right) + \cdots + \beta_s\left(\frac{z}{w(1 - \lambda z)}\right)^{s}.$$

The relation (3.7.3) reduces to (3.6.4) for $\lambda = 0$. It follows from (3.7.2) that $\widehat{F}(z/(w(1 - \lambda z)))$ is an approximation, correct to within

$$O\left(\left(\frac{z}{w(1 - \lambda z)}\right)^{s+1}\right)$$

to the function F defined in a neighborhood of $z = 0$ by the functional equation

$$(1 - \lambda z)e^z = F\left(\frac{ze^{-z}}{1 - \lambda z}\right). \tag{3.7.4}$$

Rational approximations to F have been investigated by Butcher [48]. In this section we reproduce these results for $\beta_1 = \beta_2 = \cdots = \beta_s = 0$, which corresponds to the first row of the Padé table for this function.

To formulate this approximation we have to invert the Taylor series

$$\omega = f(z) = \omega_0 + a_1(z - z_0) + \cdots + a_n(z - z_0)^n + \cdots,$$

$\omega_0 = f(z_0)$, of an analytic function $\omega = f(z)$ at the point z_0, where

$$a_1 = f'(z_0) \neq 0.$$

This can be accomplished by the Lagrange series described in the following theorem.

Theorem 3.7.1 (see [210]) *Assume that the function $\omega = f(z)$ is analytic at z_0 and single-valued in a neighborhood of $\omega_0 = f(z_0)$. Then*

$$z = f^{-1}(\omega) = z_0 + \sum_{n=1}^{\infty} \frac{1}{n!} \frac{d^{n-1}}{dz^{n-1}} \psi(z)^n \bigg|_{z=z_0} (\omega - \omega_0)^n,$$

where the function $\psi(z)$ is defined by

$$\psi(z) = \frac{z - z_0}{f(z) - \omega_0}.$$

Let

$$\omega = \frac{ze^{-z}}{1 - \lambda z}. \tag{3.7.5}$$

We have the following theorem.

Theorem 3.7.2 (Butcher [48]) *The function $F(\omega)$ can be expanded in a neighborhood of $\omega = 0$ into the power series*

$$F(\omega) = 1 + \sum_{n=1}^{\infty} (-1)^{n+1} \frac{\lambda^n}{n+1} L'_{n+1}\left(\frac{n+1}{\lambda}\right) \omega^n, \tag{3.7.6}$$

where $L_{n+1}(x)$ are the Laguerre polynomials defined by formula (2.7.6), and $L'_{n+1}(x) = (d/dx)L_{n+1}(x)$.

Proof: Consider the function

$$\omega = f(z) = \frac{z}{\psi(z)},$$

where

$$\psi(z) = (1 - \lambda z)e^z.$$

Then $f(0) = 0$, and since $f'(0) \neq 0$ the function $f(z)$ is single-valued in a neighborhood of $w_0 = 0$. Hence, it follows from Theorem 3.7.1 with $z_0 = 0$, $w_0 = 0$, that

$$z = f^{-1}(w) = \sum_{n=1}^{\infty} \frac{1}{n!} \frac{d^{n-1}}{dz^{n-1}} \psi(z)^n \bigg|_{z=0} w^n.$$

Using the relation (3.7.5), the functional equation (3.7.4) takes the form

$$F(w) = \psi(z)$$

and it follows that

$$wF(w) = w\psi(z) = z.$$

Hence,

$$
\begin{aligned}
F(w) &= \frac{z}{w} = \sum_{n=1}^{\infty} \frac{1}{n!} \frac{d^{n-1}}{dz^{n-1}} \psi(z)^n \bigg|_{z=0} w^{n-1} \\
&= 1 + \sum_{n=1}^{\infty} \frac{1}{(n+1)!} \frac{d^n}{dz^n} \psi(z)^{n+1} \bigg|_{z=0} w^n.
\end{aligned}
$$

Comparing this relation with (3.7.3) corresponding to $\beta_1 = \beta_2 = \cdots = \beta_s = 0$, the coefficients α_n are

$$
\begin{aligned}
\alpha_n &= \frac{1}{(n+1)!} \frac{d^n}{dz^n} \psi(z)^{n+1} \bigg|_{z=0} \\
&= \frac{1}{(n+1)!} \frac{d^n}{dz^n} \left((1 - \lambda z)^{n+1} e^{(n+1)z} \right) \bigg|_{z=0} \\
&= \frac{1}{(n+1)!} \sum_{k=0}^{n} \binom{n}{k} \frac{d^{n-k}}{dz^{n-k}} (1 - \lambda z)^{n+1} \bigg|_{z=0} \frac{d^k}{dz^k} e^{(n+1)z} \bigg|_{z=0} \\
&= \sum_{k=0}^{n} \binom{n}{k} \frac{(-\lambda)^{n-k}}{(k+1)!} (n+1)^k.
\end{aligned}
$$

We also have

$$L'_{n+1}(x) = -\sum_{i=1}^{n+1} \binom{n+1}{i} \frac{(-x)^{i-1}}{(i-1)!} = -\sum_{k=0}^{n} \binom{n+1}{k+1} \frac{(-x)^k}{k!}.$$

Hence,

$$
\begin{aligned}
L'_{n+1}\left(\frac{n+1}{\lambda}\right) &= -\sum_{k=0}^{n}\binom{n+1}{k+1}\frac{(n+1)^k}{(-\lambda)^k k!} \\
&= -(n+1)\sum_{k=0}^{n}\binom{n}{k}\frac{(-\lambda)^{-k}}{(k+1)!}(n+1)^k \\
&= -\frac{n+1}{(-\lambda)^n}\sum_{k=0}^{n}\binom{n}{k}\frac{(-\lambda)^{n-k}}{(k+1)!}(n+1)^k \\
&= -\frac{n+1}{(-\lambda)^n}\alpha_n,
\end{aligned}
$$

and it follows that

$$
\alpha_n = (-1)^{n+1}\frac{\lambda^n}{n+1}L'_{n+1}\left(\frac{n+1}{\lambda}\right).
$$

This shows that $F(\omega)$ satisfies (3.7.6). ∎

For $\beta_1 = \beta_2 = \cdots = \beta_s = 0$ the stability polynomial $p(w, z)$ takes the form

$$
p(w, z) = (1 - \lambda z)^s w^s - (1 - \lambda z)^{s-1} w^{s-1} - \cdots - \alpha_{s-1} z^{s-1}. \tag{3.7.7}
$$

To obtain a method of order $p = s$ we have to choose λ in such a way that $\alpha_s = 0$, which is equivalent to the condition

$$
L'_{s+1}\left(\frac{s+1}{\lambda}\right) = 0, \tag{3.7.8}
$$

and the values $\alpha_0, \alpha_1, \ldots, \alpha_{s-1}$ according to Theorem 3.7.2:

$$
\alpha_n = (-1)^{n+1}\frac{\lambda^n}{n+1}L'_{n+1}\left(\frac{n+1}{\lambda}\right), \tag{3.7.9}
$$

$n = 0, 1, \ldots, s - 1$. Since

$$
\lim_{z \to \infty}\frac{p(w, z)}{(1 - \lambda z)^s} = w^s
$$

it follows that if the stability polynomial of the form (3.7.7) is A-stable, it is also L-stable. The search for A-stable methods can be carried out numerically. Since $\lambda > 0$ it follows from the maximum principle that $p(w, z)$ is A-stable if all roots of $p(w, iy)$ are in the unit circle for all $y \in \mathbb{R}$. This can be checked by plotting the roots $w_i(y)$, $i = 1, 2, \ldots, s$, as y ranges over the real values, and noting if $|w_i(y)| \le 1$. Alternatively, we could plot the boundary locus curve

$$
p(e^{i\theta}, z) = 0
$$

for $\theta \in [0, 2\pi]$, starting from the solutions z_i, $i = 1, 2, \ldots, s$, to the polynomial equation

$$p(1, z) = 0,$$

and noting if the boundary curve stays in the right-half complex plane. If this is the case, the corresponding stability polynomial $p(w, z)$ is A-stable.

λ	α_1	α_2	A- and L-stability
0.51554560	0.48445440	0.21915046	no
1.21013832	−0.21013831	−0.66598020	yes
4.27431609	−3.27431609	6.94682974	almost

Table 3.7.1   Parameters λ that satisfy (3.7.8) and α_1, α_2 that satisfy (3.7.9) for $s = 3$

λ	α_1	α_2	α_3	A- and L-stability
0.45645867	0.54354133	0.33897851	0.17001919	no
0.87242088	0.12757912	−0.35614445	−0.41000663	almost
1.94428836	−0.94428836	−0.54260786	2.44399301	yes
6.72683210	−5.72683210	26.5697738	−84.037339	almost

Table 3.7.2   Parameters λ that satisfy (3.7.8) and α_1, α_2, α_3 that satisfy (3.7.9) for $s = 4$

We have listed in Tables 3.7.1 and 3.7.2 the values of the parameter λ that satisfy equation (3.7.8), the corresponding values of the parameters α_1, α_2 for $s = 3$ and α_1, α_2, α_3 for $s = 4$ ($\alpha_0 = 0$ for any s), and whether the stability polynomial $p(w, z)$ is A- and L-stable. The roots of $p(w, iy)$ for $y \in \mathbb{R}$ are then plotted in Fig. 3.7.1 for $s = 3$ and in Fig. 3.7.2 for $s = 4$. We have also plotted in the right lower corner of Fig. 3.7.1 the boundary locus curve $p(e^{i\theta}, z)$ for $s = 3$ and $\lambda = 0.51555$, which confirms instability. The entry "almost" in the tables means that the roots of $p(w, iy)$ corresponding to the specific values of the parameter λ extend only slightly from the unit circle (this is not apparent in Figs. 3.7.1 3.7.2), so that the corresponding methods are almost A-stable in this sense.

Following Butcher [48] next we derive formulas for the coefficients of the methods with given stability polynomials, making use of some transformation of the coefficient matrices. The transformation proposed in [48] is more

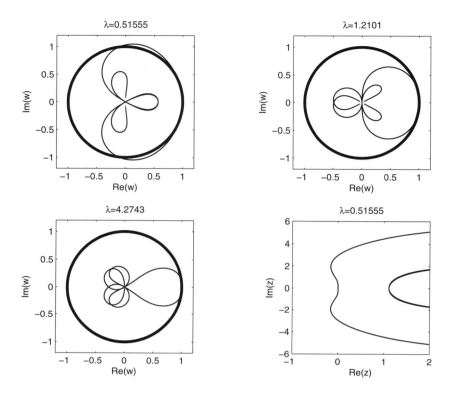

Figure 3.7.1 Roots of $p(w, iy)$ for λ which satisfy (3.7.8) and α_1, α_2 given by (3.7.9) for $s = 3$ and a boundary locus curve for $\lambda = 0.51555$

convenient than that discussed in Section 3.3 since the transformed matrices $\overline{\mathbf{B}}$ and $\overline{\mathbf{V}}$ do not depend on the abscissa vector $\mathbf{c}$. Hence, we can regard the choice of abscissas as a separate question from the choice of coefficients, which is possible only for type 3 and 4 methods. We have the following theorem.

Theorem 3.7.3 (Butcher [48]) *Consider a type 4 DIMSIM with abscissa vector $\mathbf{c}$ such that $p = q = r = s$ and $\mathbf{Ve} = \mathbf{e}$. Define the matrix $\mathbf{T}$ by*

$$
\mathbf{T} = \begin{bmatrix} P^{(s)}(c_1) & P^{(s-1)}(c_1) & \cdots & P'(c_1) \\ P^{(s)}(c_2) & P^{(s-1)}(c_2) & \cdots & P'(c_2) \\ \vdots & \vdots & \ddots & \vdots \\ P^{(s)}(c_s) & P^{(s-1)}(c_s) & \cdots & P'(c_s) \end{bmatrix},
$$

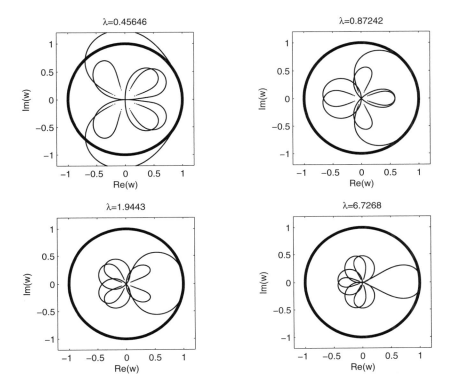

Figure 3.7.2   Roots of $p(w, iy)$ for λ which satisfy (3.7.8) and α_1, α_2, α_3 given by (3.7.9) for $s = 4$

where the polynomial $P(x)$ is defined by

$$P(x) = \frac{1}{s!} \prod_{i=1}^{s} (x - c_i).$$

Then the transformed matrices $\overline{\mathbf{B}}$ and $\overline{\mathbf{V}}$ defined by

$$\overline{\mathbf{B}} = \mathbf{T}^{-1} \mathbf{B} \mathbf{T}, \quad \overline{\mathbf{V}} = \mathbf{T}^{-1} \mathbf{V} \mathbf{T},$$

satisfy

$$\overline{\mathbf{V}} \mathbf{e}_1 = \mathbf{e}_1 \tag{3.7.10}$$

and

$$\overline{\mathbf{B}} = \widehat{\mathbf{E}} - \lambda \mathbf{E} + \overline{\mathbf{V}}(\lambda \mathbf{I} - \mathbf{J}), \tag{3.7.11}$$

where

$$\widehat{\mathbf{E}} = \begin{bmatrix} 1 & \frac{1}{2!} & \frac{1}{3!} & \cdots & \frac{1}{(s-1)!} & \frac{1}{s!} \\ 1 & 1 & \frac{1}{2!} & \cdots & \frac{1}{(s-2)!} & \frac{1}{(s-1)!} \\ 0 & 1 & 1 & \cdots & \frac{1}{(s-3)!} & \frac{1}{(s-2)!} \\ \vdots & \vdots & \vdots & \ddots & \vdots & \vdots \\ 0 & 0 & 0 & \cdots & 1 & \frac{1}{2!} \\ 0 & 0 & 0 & \cdots & 1 & 1 \end{bmatrix},$$

$$\mathbf{E} = \begin{bmatrix} 1 & 1 & \frac{1}{2!} & \cdots & \frac{1}{(s-1)!} \\ 0 & 1 & 1 & \cdots & \frac{1}{(s-2)!} \\ 0 & 0 & 1 & \cdots & \frac{1}{(s-3)!} \\ \vdots & \vdots & \vdots & \ddots & \vdots \\ 0 & 0 & 0 & \cdots & 1 \end{bmatrix}, \quad \mathbf{J} = \begin{bmatrix} 0 & 0 & \cdots & 0 & 0 \\ 1 & 0 & \cdots & 0 & 0 \\ 0 & 1 & \cdots & 0 & 0 \\ \vdots & \vdots & \ddots & \vdots & \vdots \\ 0 & 0 & \cdots & 1 & 0 \end{bmatrix}.$$

Proof: We have

$$\overline{\mathbf{V}}\mathbf{e}_1 = \mathbf{T}^{-1}\mathbf{V}\mathbf{T}\mathbf{e}_1 = \mathbf{T}^{-1}\mathbf{V}\mathbf{e} = \mathbf{T}^{-1}\mathbf{e} = \mathbf{e}_1,$$

which proves (3.7.10). This relation also implies that $\overline{\mathbf{V}}$ can be partitioned as

$$\overline{\mathbf{V}} = \begin{bmatrix} \mathbf{e}_1 \mid V_0 \end{bmatrix},$$

so that if $\widehat{\mathbf{V}}$ is defined by

$$\widehat{\mathbf{V}} = \begin{bmatrix} V_0 \mid 0 \end{bmatrix},$$

the last term in (3.7.11) takes the form

$$\overline{\mathbf{V}}(\lambda\mathbf{I} - \mathbf{J}) = \lambda\overline{\mathbf{V}} - \widehat{\mathbf{V}}.$$

Eliminating vector $\mathbf{w}(z)$ from stage order and order conditions (2.4.5) and (2.4.6) in Theorem 2.4.1, we obtain

$$e^z(1 - \lambda z)e^{\mathbf{c}z} = z\mathbf{B}e^{\mathbf{c}z} + \mathbf{V}(1 - \lambda z)e^{\mathbf{c}z} + O(z^{s+1}). \tag{3.7.12}$$

Define the sequence of numbers $\gamma_1, \gamma_2, \ldots, \gamma_s$, and the function $\phi(z)$ by

$$P(x) = \frac{x^s}{s!} + \gamma_1\frac{x^{s-1}}{(s-1)!} + \cdots + \gamma_{s-1}x + \gamma_s,$$

$$\phi(z) = 1 + \gamma_1 z + \gamma_2 z^2 + \cdots + \gamma_{s-1}z^{s-1} + \gamma_s z^s.$$

Also define matrices $\mathbf{C}$ and $\mathbf{G}$ and vector $\mathbf{Z}$ by

$$
\mathbf{C} = \begin{bmatrix} 1 & c_1 & \frac{c_1^2}{2!} & \cdots & \frac{c_1^s}{s!} \\ 1 & c_2 & \frac{c_2^2}{2!} & \cdots & \frac{c_2^s}{s!} \\ 1 & c_3 & \frac{c_3^2}{2!} & \cdots & \frac{c_3^s}{s!} \\ \vdots & \vdots & \vdots & \ddots & \vdots \\ 1 & c_s & \frac{c_s^2}{2!} & \cdots & \frac{c_s^s}{s!} \end{bmatrix}, \quad \mathbf{G} = \begin{bmatrix} 1 & \gamma_1 & \gamma_2 & \cdots & \gamma_s \\ 0 & 1 & \gamma_1 & \cdots & \gamma_{s-1} \\ 0 & 0 & 1 & \cdots & \gamma_{s-2} \\ \vdots & \vdots & \vdots & \ddots & \vdots \\ 0 & 0 & 0 & \cdots & 1 \end{bmatrix},
$$

$$
\mathbf{Z} = \begin{bmatrix} 1 & z & \cdots & z^s \end{bmatrix}^T.
$$

Then it can be verified that

$$
\mathbf{GZ} = \phi(z)\mathbf{Z} + O(z^{s+1}) \tag{3.7.13}
$$

and

$$
\mathbf{CZ} = e^{\mathbf{c}z} + O(z^{s+1}). \tag{3.7.14}
$$

It follows from (3.7.14) that we can substitute $e^{\mathbf{c}z}$ by $\mathbf{CZ}$ in (3.7.12) since this only affects terms of order $O(z^{s+1})$. Multiplying the resulting expression by $\phi(z)$, the order conditions (3.7.12) can be reformulated in the equivalent form

$$
\big(e^z(1 - \lambda z) - z\mathbf{B} - (1 - \lambda z)\mathbf{V}\big)\mathbf{CGZ} = O(z^{s+1}). \tag{3.7.15}
$$

It can be verified that the product $\mathbf{CG}$ is given by

$$
\mathbf{CG} = \begin{bmatrix} \mathbf{T} \mid P(\mathbf{c}) \end{bmatrix} = \begin{bmatrix} \mathbf{T} \mid 0 \end{bmatrix},
$$

where $P(\mathbf{c})$ is evaluated componentwise. Substituting this relation into (3.7.15), we obtain

$$
\big(e^z(1 - \lambda z) - z\mathbf{B} - (1 - \lambda z)\mathbf{V}\big)\mathbf{T}\widehat{\mathbf{Z}} = O(z^{s+1}), \tag{3.7.16}
$$

where $\widehat{\mathbf{Z}}$ is made up from the first s components of vector $\mathbf{Z}$. Multiplying (3.7.16) by $\mathbf{T}^{-1}$ from the left, the order conditions take the form

$$
e^z(1 - \lambda z)\widehat{\mathbf{Z}} = z\overline{\mathbf{B}}\widehat{\mathbf{Z}} + \widehat{\mathbf{V}}(1 - \lambda z)\widehat{\mathbf{Z}} + O(z^{s+1}). \tag{3.7.17}
$$

Subtracting $\mathbf{e}_1\mathbf{e}_1^T\widehat{\mathbf{Z}}$ from both sides of this relation, taking into account that

$$
e^z\widehat{\mathbf{Z}} - \mathbf{e}_1\mathbf{e}_1^T\widehat{\mathbf{Z}} = z\widehat{\mathbf{E}}\widehat{\mathbf{Z}} + O(z^{s+1})
$$

and

$$
\overline{\mathbf{V}}\widehat{\mathbf{Z}} - \mathbf{e}_1\mathbf{e}_1^T\widehat{\mathbf{Z}} = \widehat{\mathbf{V}}\widehat{\mathbf{Z}},
$$

and dividing both sides of the resulting relation by z, we obtain

$$
\widehat{\mathbf{E}}\widehat{\mathbf{Z}} - \lambda\mathbf{E}\widehat{\mathbf{Z}} = \overline{\mathbf{B}}\widehat{\mathbf{Z}} + \widehat{\mathbf{V}}\widehat{\mathbf{Z}} - \lambda\overline{\mathbf{V}}\widehat{\mathbf{Z}} + O(z^s). \tag{3.7.18}
$$

This equation implies that

$$\overline{\mathbf{B}} = \widehat{\mathbf{E}} - \lambda\mathbf{E} - \widehat{\mathbf{V}} + \lambda\overline{\mathbf{V}},$$

which is equivalent to (3.7.11). This completes the proof. ∎

We now consider the remaining question of how matrix $\overline{\mathbf{V}}$ should be chosen so that the method with matrix $\overline{\mathbf{B}}$ given by (3.7.11) is L-stable, that is, stability matrix $\mathbf{M}(z)$ or, equivalently, transformed stability matrix

$$\overline{\mathbf{M}}(z) = \mathbf{T}^{-1}\mathbf{M}(z)\mathbf{T}$$

has a zero spectral radius as $z \to \infty$. Using (3.7.11), matrix $\overline{\mathbf{M}}(z)$ takes the form

$$\overline{\mathbf{M}}(z) = \overline{\mathbf{V}} + \frac{z}{1 - \lambda z}\overline{\mathbf{B}} = \overline{\mathbf{V}} + \frac{z}{1 - \lambda z}\left(\widehat{\mathbf{E}} - \lambda\mathbf{E} - \widehat{\mathbf{V}} + \lambda\overline{\mathbf{V}}\right)$$

and

$$\lim_{z \to \infty} \overline{\mathbf{M}}(z) = \overline{\mathbf{M}}(\infty) = -\frac{1}{\lambda}(\widehat{\mathbf{E}} - \lambda\mathbf{E}) + \frac{1}{\lambda}\widehat{\mathbf{V}}.$$

Assuming that $\overline{\mathbf{V}}$ is a rank 1 matrix, it follows from (3.7.10) that this matrix assumes the form

$$\overline{\mathbf{V}} = \mathbf{e}_1 \begin{bmatrix} 1 & v_2 & \cdots & v_s \end{bmatrix}.$$

Hence,

$$\widehat{\mathbf{V}} = \overline{\mathbf{V}}\mathbf{J} = \mathbf{e}_1 \begin{bmatrix} v_2 & \cdots & v_s & 0 \end{bmatrix}$$

and

$$\widehat{\mathbf{M}} := -\lambda\overline{\mathbf{M}}(\infty) = \widehat{\mathbf{E}} - \lambda\mathbf{E} - \mathbf{e}_1 \begin{bmatrix} v_2 & \cdots & v_s & 0 \end{bmatrix}.$$

The conditions for matrix $\widehat{\mathbf{M}}$ to have spectral radius equal to zero were found by Butcher [48]. This result is technically very complicated and we refer the reader to the original paper [48] for a proof of the theorem that follows.

Theorem 3.7.4 (Butcher [48]) *Matrix* $\widehat{\mathbf{M}}$ *has spectral radius equal to zero if and only if*

$$v_i = (-1)^{s+1}\frac{s - i + 2}{s + 1}\lambda^{i-1}L_{s+1}^{(s-i+2)}\left(\frac{s+1}{\lambda}\right)$$

and λ *satisfies equation (3.7.8). Here* $L_{s+1}^{(s-i+2)}$ *is an* $(s-i+2)$*-fold derivative of* L_{s+1}.

We conclude this section by listing in a rational format some examples of A- and L-stable DIMSIMs of type 4 for $s = 3$ and $s = 4$ derived by Butcher [48].

Example 1. $s = 3$, $\lambda = 1.21013836$, $\mathbf{c} = [0, \frac{1}{2}, 1]^T$:

$$\mathbf{B} = \begin{bmatrix} -\frac{3713699}{575604} & \frac{13386050}{954257} & -\frac{4525336}{702095} \\ -\frac{10143085}{1360824} & \frac{9868186}{580773} & -\frac{7718629}{976125} \\ -\frac{3592443}{402940} & \frac{28151576}{1381639} & -\frac{11391581}{1221028} \end{bmatrix},$$

$$\mathbf{v} = \begin{bmatrix} -\frac{8190471}{1881082} & \frac{20278361}{1848683} & -\frac{2718914}{484227} \end{bmatrix}^T.$$

Example 2. $s = 3$, $\lambda = 1.21013836$, $\mathbf{c} = [-1, 0, 1]^T$:

$$\mathbf{B} = \begin{bmatrix} -\frac{2346473}{1639992} & \frac{3425188}{1093005} & -\frac{5551958}{5176473} \\ -\frac{827713}{546664} & \frac{6027409}{1202946} & -\frac{2563125}{1373587} \\ -\frac{2807601}{1216682} & \frac{3519979}{481686} & -\frac{285494}{120481} \end{bmatrix},$$

$$\mathbf{v} = \begin{bmatrix} -\frac{998030}{1238361} & \frac{899740}{277503} & -\frac{1181171}{822346} \end{bmatrix}^T.$$

Example 3. $s = 4$, $\lambda = 1.94428836$, $\mathbf{c} = [0, \frac{1}{3}, \frac{2}{3}, 1]^T$:

$$\mathbf{B} = \begin{bmatrix} -\frac{294809551}{907812} & \frac{628839333}{650200} & -\frac{1030246907}{1112646} & \frac{327992002}{1139573} \\ -\frac{277481113}{859266} & \frac{558142952}{581469} & -\frac{1133031719}{1238155} & \frac{450719411}{1594050} \\ -\frac{16528673}{52000} & \frac{590225249}{626947} & -\frac{352947587}{395705} & \frac{232089033}{849143} \\ -\frac{338254916}{1096701} & \frac{629824162}{693045} & -\frac{1255916163}{1470929} & \frac{107748367}{416413} \end{bmatrix},$$

$$\mathbf{v} = \begin{bmatrix} -\frac{145167427}{944816} & \frac{593469083}{1231242} & -\frac{86368735}{176174} & \frac{72595683}{445688} \end{bmatrix}^T.$$

Example 4. $s = 4$, $\lambda = 1.94428836$, $\mathbf{c} = [-2, -1, 0, 1]^T$:

$$\mathbf{B} = \begin{bmatrix} -\frac{16176817}{1365928} & \frac{30344282}{736427} & -\frac{38632409}{1120034} & \frac{6715194}{798697} \\ -\frac{6087649}{512223} & \frac{35588036}{814545} & -\frac{22670041}{631569} & \frac{11942577}{1427510} \\ -\frac{16176817}{1365928} & \frac{14780263}{339914} & -\frac{30709763}{926141} & \frac{2452519}{360838} \\ -\frac{5095141}{495935} & \frac{48597076}{1304729} & -\frac{7802303}{325755} & \frac{1994001}{612496} \end{bmatrix},$$

$$\mathbf{v} = \begin{bmatrix} -\frac{3759945}{791263} & \frac{16762402}{853779} & -\frac{34730174}{1759939} & \frac{7760482}{1326039} \end{bmatrix}^T.$$

In all these examples $\mathbf{V}$ is a rank 1 matrix of the form $\mathbf{V} = \mathbf{ev}^T$. Butcher [48] also considered construction of highly stable type 4 methods for which $\mathbf{V}$ is a matrix of rank 2.

3.8   FOURIER SERIES APPROACH TO THE CONSTRUCTION OF DIMSIMS OF HIGH ORDER

For DIMSIMs of high order ($p \geq 5$) it is no longer possible to generate systems of nonlinear equations (3.4.5) and (3.5.5) by symbolic manipulation packages, so a different approach to the construction of such methods is needed. In this section we describe an approach to generate the corresponding systems of nonlinear equations using some variant of the Fourier series method. Systems obtained in this way will then be solved by algorithms based on least-squares minimization.

A Fourier series approach can be summarized as follows. Assume that w_μ, $\mu = 1, 2, \ldots, N_1$, where N_1 is a sufficiently large integer, are complex numbers distributed uniformly on the unit circle. Multiplying the relations

$$p(w_\mu, z) = w_\mu^s - p_1(z)w_\mu^{s-1} + \cdots + (-1)^{s-1}p_{s-1}(z)w_\mu + (-1)^s p_s(z)$$

and

$$\widehat{p}(w_\mu, \widehat{z}) = w_\mu^s - \widehat{p}_1(\widehat{z})w_\mu^{s-1} + \cdots + (-1)^{s-1}\widehat{p}_{s-1}(\widehat{z})w_\mu + (-1)^s \widehat{p}_s(\widehat{z}),$$

by w_μ^{k-s}, $k = 1, 2, \ldots, s$, and summing with respect to μ, we can isolate polynomials $p_k(z)$ and $\widehat{p}_k(z)$. Here $p(w, z)$ and $\widehat{p}(w, \widehat{z})$ are stability polynomials of DIMSIMs of types 1 and 2 defined in Sections 3.4 and 3.5, respectively. This leads to

$$(-1)^k p_k(z) = \frac{1}{N_1} \sum_{\mu=1}^{N_1} w_\mu^{k-s} p(w_\mu, z)$$

and

$$(-1)^k \widehat{p}_k(z) = \frac{1}{N_1} \sum_{\mu=1}^{N_1} w_\mu^{k-s} \widehat{p}(w_\mu, z),$$

$k = 1, 2, \ldots, s$. Assume next that z_ν and $\widehat{z}_\nu$, $\nu = 1, 2, \ldots, N_2$, where N_2 is a sufficiently large integer, are complex numbers distributed uniformly on the unit circle. Repeating this process again, this time multiplying

$$p_k(z_\nu) = p_{k,k-1} z_\nu^{k-1} + p_{kk} z_\nu^k + \cdots + p_{ks} z_\nu^s$$

and

$$\widehat{p}_k(\widehat{z}_\nu) = \widehat{p}_{k,k-1} \widehat{z}_\nu^{k-1} + \widehat{p}_{kk} \widehat{z}_\nu^k + \cdots + \widehat{p}_{ks} \widehat{z}_\nu^s,$$

$k = 2, 3, \ldots, s$, by z_ν^{-l} and $\widehat{z}_\nu^{-l}$, respectively, and summing with respect to ν, we get

$$p_{kl} = \frac{1}{N_2} \sum_{\nu=1}^{N_2} z_\nu^{-l} p_k(z_\nu) = (-1)^k \frac{1}{N_1 N_2} \sum_{\mu=1}^{N_1} \sum_{\nu=1}^{N_2} w_\mu^{k-s} z_\nu^{-l} p(w_\mu, z_\nu)$$

and

$$\widehat{p}_{kl} = \frac{1}{N_2} \sum_{\nu=1}^{N_2} \widehat{z}_\nu^{-l} \widehat{p}_k(\widehat{z}_\nu) = (-1)^k \frac{1}{N_1 N_2} \sum_{\mu=1}^{N_1} \sum_{\nu=1}^{N_2} w_\mu^{k-s} \widehat{z}_\nu^{-l} \widehat{p}(w_\mu, \widehat{z}_\nu),$$

$k = 2, 3, \ldots, s$, $l = k - 1, k, \ldots, s$. We then solve numerically the systems

$$\sum_{\mu=1}^{N_1} \sum_{\nu=1}^{N_2} w_\mu^{k-s} z_\nu^{-l} p(w_\mu, z_\nu) = 0 \tag{3.8.1}$$

and

$$\sum_{\mu=1}^{N_1} \sum_{\nu=1}^{N_2} w_\mu^{k-s} \widehat{z}_\nu^{-l} \widehat{p}(w_\mu, \widehat{z}_\nu) = 0, \tag{3.8.2}$$

which are equivalent to (3.4.5) and (3.5.5), respectively, if the integers N_1 and N_2 are chosen sufficiently large.

The stability polynomial $p(w, z)$ of DIMSIMs of type 1 will be computed from the equation

$$p(w, z) = \det \big(\mathbf{Q}(w, z) \big), \tag{3.8.3}$$

where

$$\mathbf{Q}(w, z) = w(\mathbf{I} - z\mathbf{A}) - \mathbf{V} - z\mathbf{B}_0 + z\mathbf{A}\mathbf{B}_1 + z\mathbf{V}\mathbf{B}_2. \tag{3.8.4}$$

These relations follow from formula (3.2.1) and the fact that $\det(\mathbf{I} - z\mathbf{A})^{-1}$ is equal to 1. Similarly, the polynomial $\widehat{p}(w, \widehat{z})$ corresponding to DIMSIMs of type 2 will be computed from the relations

$$\widehat{p}(w, \widehat{z}) = \det \big(\widehat{\mathbf{Q}}(w, \widehat{z}) \big), \tag{3.8.5}$$

where

$$\widehat{\mathbf{Q}}(w, \widehat{z}) = w(\mathbf{I} - \widehat{z}\widehat{\mathbf{A}}) - \mathbf{V} - \widehat{z}\widehat{\mathbf{B}}_0 + \widehat{z}\widehat{\mathbf{A}}\mathbf{B}_1 + \widehat{z}\mathbf{V}\widehat{\mathbf{B}}_2, \tag{3.8.6}$$

and matrices $\widehat{\mathbf{B}}_0$ and $\widehat{\mathbf{B}}_2$ are defined by

$$\widehat{\mathbf{B}}_0 = \mathbf{B}_0 - \lambda\mathbf{B}_1 \quad \text{and} \quad \widehat{\mathbf{B}}_2 = \mathbf{B}_2 - \lambda\mathbf{I}.$$

Observe that (3.8.5) and (3.8.6) have the same form as (3.8.3) and (3.8.4), where z, $\mathbf{A}$, $\mathbf{B}_0$, and $\mathbf{B}_2$ are replaced by $\widehat{z}$, $\widehat{\mathbf{A}}$, $\widehat{\mathbf{B}}_0$, and $\widehat{\mathbf{B}}_2$, respectively.

We can use the approach described above to construct A- and L-stable methods of type 2 with a stability function of the form

$$p(w, z) = w^{s-1}\big(w - R(z)\big).$$

It follows from Table 2.7.1 that this is possible for $p = 5$, $p = 6$, and $p = 8$ by choosing the appropriate parameter λ for which the stability function $R(z)$

of the SDIRK method of the same order is A- and L-stable. However, this approach is not applicable for $p = 7$ since the stability function of the SDIRK method of this order cannot be A- and L-stable for any λ (compare Tables 2.7.1 and 2.7.2). For $p = 7$ we follow a different approach, proposed by Butcher et al. [74], and attempt to construct highly stable methods for which the stability polynomial $p(w, z)$ is equal to a given polynomial $p^*(w, z)$ of the form

$$p^*(w, z) = w^{s-2}\left(w^2 - \frac{p_1^*(z)}{(1 - \lambda z)^s}w + \frac{p_2^*(z)}{(1 - \lambda z)^s}\right),$$

where $p_1^*(z)$ and $p_2^*(z)$ are polynomials of degree less than or equal to s. These polynomials will be chosen in such a way that $p(w, z)$ is an A- and L-stable generalized approximation of order $p = s$ to the exponential function $\exp(z)$ [60]; that is,

$$p^*\big(\exp(z), z\big) = O(z^{s+1}). \tag{3.8.7}$$

Although this approach was first motivated to deal with the case $p = 7$, it is clearly also applicable to the construction of methods of any order. Moreover, in contrast to the approach described at the beginning of this section, it usually leads to an entire interval of the parameter λ, for which the corresponding methods are A-stable and L-stable.

Set

$$p_1^*(z) = 1 + p_{11}^*z + p_{12}^*z^2 + \cdots + p_{1s}^*z^s,$$

$$p_2^*(z) = p_{21}^*z + p_{22}^*z^2 + \cdots + p_{2s}^*z^s.$$

Then the order requirement (3.8.7) leads to the system of s polynomial equations for the parameter λ and the coefficients p_{kl}^* of p_k^*, $k = 1, 2$. Assuming that

$$p_{1l}^* = 0, \quad l = 5, 6, \ldots, s, \qquad p_{2l}^* = 0, \quad l = s - 3, s - 2, \ldots, s,$$

and solving the system corresponding to (3.8.7), we can express the remaining coefficients p_{1l}^*, $l = 0, 1, 2, 3, 4$, and p_{2l}^*, $l = 1, 2, \ldots, s - 4$, in terms of λ. The parameter λ can then be chosen in such a way that polynomial $p^*(w, z)$ is A-stable. Once this is done, the corresponding function $p^*(w, z)$ will also be L-stable since $p_{1s}^* = p_{2s}^* = 0$.

Polynomial $p^*(w, z)$ has a root $w = 0$ of multiplicity $s - 2$, and to assure A-stability the remaining two roots, $R_1^*(z)$ and $R_2^*(z)$, should satisfy

$$\big|R_k^*(z)\big| \leq 1, \quad k = 1, 2, \tag{3.8.8}$$

for $\mathrm{Re}(z) \leq 0$. This condition can be analyzed using the Schur theorem, Theorem 2.8.1. It can be verified, for example, that (3.8.8) is satisfied for

$$0.25864444 \leq \lambda \leq 0.27688498 \quad \text{if} \quad p = 7$$

and

$$0.19799408 \leq \lambda \leq 0.20136462 \quad \text{if} \quad p = 8.$$

Choosing, for example, $\lambda = \frac{13}{50}$, polynomials $p_1^*(z)$ and p_2^* corresponding to the generalized approximation $p^*(w, z)$ of order $p = 7$ to the exponential function are

$$p_1^*(z) = 1 - \frac{1022815846}{1708984375} z - \frac{7864050101}{68359375000} z^2 + \frac{21301028013}{136718750000} z^3 - \frac{4289969757}{136718750000} z^4$$

and

$$p_2^*(z) = \frac{757102683}{3417968750} z + \frac{469409809}{68359375000} z^2 - \frac{3769899337}{410156250000} z^3.$$

To construct appropriate methods, it is more convenient to work with the transformed variable $\widehat{z} = z/(1 - \lambda z)$, which was introduced in Section 3.5, and transformed polynomial $\widehat{p}^*(w, \widehat{z})$, defined by

$$\widehat{p}^*(w, \widehat{z}) = \frac{p^*(w, z)}{(1 - \lambda z)^s}.$$

This polynomial takes the form

$$\widehat{p}^*(w, \widehat{z}) = w^{s-2}\big(w^2 - \widehat{p}_1^*(\widehat{z})w + \widehat{p}_2^*(\widehat{z})\big),$$

where

$$\widehat{p}_1^*(\widehat{z}) = 1 + \widehat{p}_{11}^*\widehat{z} + \widehat{p}_{12}^*\widehat{z}^2 + \cdots + \widehat{p}_{1s}^*\widehat{z}^s,$$

$$\widehat{p}_2^*(\widehat{z}) = \widehat{p}_{21}^*\widehat{z} + \widehat{p}_{22}^*\widehat{z}^2 + \cdots + \widehat{p}_{2s}^*\widehat{z}^s.$$

It can be verified [74] that for given λ, the construction of DIMSIMs for which transformed stability polynomial $\widehat{p}(w, \widehat{z})$ is equal to a polynomial $\widehat{p}^*(w, \widehat{z})$ given in advance leads to a system of $(s - 1)(s + 2)/2$ polynomial equations

$$\frac{1}{N_1 N_2} \sum_{\mu=1}^{N_1} \sum_{\nu=1}^{N_2} w_\mu^{2-s}\widehat{z}_\nu^{-l}\widehat{p}(w_\mu, \widehat{z}_\nu) - \widehat{p}_{2l}^* = 0, \quad l = 1, 2, \ldots, s,$$

$$\sum_{\mu=1}^{N_1} \sum_{\nu=1}^{N_2} w_\mu^{k-s}\widehat{z}_\nu^{-l}\widehat{p}(w_\mu, \widehat{z}_\nu) = 0, \qquad\qquad l = k - 1, k, \ldots, s,$$

$$(3.8.9)$$

$k = 3, 4, \ldots, s$, for the unknown $(s-1)(s+2)/2$ coefficients a_{ij}, $i = 2, 3, \ldots, s$, $j = 1, 2, \ldots, i - 1$, and v_i, $i = 1, 2, \ldots, s - 1$, of the methods. Here, as before, w_μ, $\mu = 1, 2, \ldots, N_1$, and $\widehat{z}_\nu$, $\nu = 1, 2, \ldots, N_2$, are complex numbers distributed uniformly on the unit circle.

Stability polynomials $p(w, z)$ and $\widehat{p}(w, \widehat{z})$ of DIMSIMs of types 1 and 2 depend on coefficients a_{ij} and v_i of these methods. We conclude this section with a derivation of the exact expressions for partial derivatives $\partial p/\partial a_{ij}$, $\partial p/\partial v_i$, $\partial \widehat{p}/\partial a_{ij}$, and $\partial \widehat{p}/\partial v_i$ of these polynomials. These expressions allow for the derivation of the formula for the Jacobian matrix of systems (3.8.1), (3.8.2), and (3.8.9).

Let $\mathbf{M} = [m_{ij}]_{i,j=1}^{s}$ and define the function

$$f = f(m_{ij}) = f(m_{11}, m_{12}, \ldots, m_{ss}) = \det(\mathbf{M}).$$

We need the following lemma.

Lemma 3.8.1 *Assume that matrix $\mathbf{M}$ is nonsingular. Then*

$$\left[\frac{\partial f}{\partial m_{ij}} \right]_{i,j=1}^{s} = \det(\mathbf{M})(\mathbf{M}^{-1})^{T}.$$

Proof: Expanding $\det(\mathbf{M})$ with respect to the ith row we obtain

$$f = \sum_{k=1}^{s} (-1)^{i+k} m_{ik} \det(\mathbf{M}_{ik}),$$

where $\mathbf{M}_{ik}$ is the submatrix of $\mathbf{M}$ obtained by deleting row number i and column number k from $\mathbf{M}$. Hence,

$$\left[\frac{\partial f}{\partial m_{ij}} \right]_{i,j=1}^{s} = \left[(-1)^{i+j} \det(\mathbf{M}_{ij}) \right]_{i,j=1}^{s} = \det(\mathbf{M})(\mathbf{M}^{-1})^{T},$$

and the proof is complete. ∎

Denote by $\mathbf{e}_1, \mathbf{e}_2, \ldots, \mathbf{e}_s$ the canonical basis in $\mathbb{R}^s$, so that $\mathbf{e} = \sum_{i=1}^{s} \mathbf{e}_i$. We have the following theorem.

Theorem 3.8.2 *The partial derivatives of polynomial $p(w, z)$ with respect to a_{ij} and v_i are given by*

$$\frac{\partial p}{\partial a_{ij}} = z \det(\mathbf{Q}) \mathbf{e}_j^T (\mathbf{B}_1 - w\mathbf{I}) \mathbf{Q}^{-1} \mathbf{e}_i, \tag{3.8.10}$$

$i = 2, 3, \ldots, s, \; j = 1, 2, \ldots, i - 1, \; and$

$$\frac{\partial p}{\partial v_i} = \det(\mathbf{Q})(\mathbf{e}_i - \mathbf{e}_s)^T (z\mathbf{B}_2 - \mathbf{I})\mathbf{Q}^{-1}\mathbf{e}, \tag{3.8.11}$$

$i = 1, 2, \ldots, s - 1.$

Proof: Let

$$\mathbf{Q} = \mathbf{Q}(w, z) = \left[q_{ij} \right]_{i,j=1}^{s}, \quad \mathbf{B}_1 = \left[b_{ij}^{(1)} \right]_{i,j=1}^{s}, \quad \mathbf{B}_2 = \left[b_{ij}^{(2)} \right]_{i,j=1}^{s}, \quad \mathbf{I} = \left[\delta_{ij} \right]_{i,j=1}^{s}.$$

We have

$$\frac{\partial p}{\partial a_{ij}} = \sum_{k=1}^{s} \frac{\partial q_{ik}}{\partial a_{ij}} \frac{\partial p}{\partial q_{ik}} = \left[\begin{array}{cccc} \dfrac{\partial q_{i1}}{\partial a_{ij}} & \dfrac{\partial q_{i2}}{\partial a_{ij}} & \cdots & \dfrac{\partial q_{is}}{\partial a_{ij}} \end{array} \right] \det(\mathbf{Q})\mathbf{Q}^{-1}\mathbf{e}_i,$$

where we note that $\mathbf{Q}^{-1}\mathbf{e}_i$ is the ith column of matrix $\mathbf{Q}^{-1}$. We also have

$$\frac{\partial q_{ik}}{\partial a_{ij}} = -zw\delta_{jk} + zb_{jk}^{(1)},$$

or in matrix form,

$$\left[\frac{\partial q_{ik}}{\partial a_{ij}}\right]_{k=1}^{s} = -zw\mathbf{e}_j^T\mathbf{I} + z\mathbf{e}_j^T\mathbf{B}_1,$$

where we note that premultiplication of a matrix $\mathbf{S}$ by $\mathbf{e}_j^T$ selects the jth row of matrix $\mathbf{S}$. Substituting this relation into the expression for $\partial p/\partial a_{ij}$, we obtain (3.8.10).

To prove (3.8.11), observe that

$$\frac{\partial p}{\partial v_i} = \sum_{j,k=1}^{s} \frac{\partial p}{\partial q_{jk}} \frac{\partial q_{jk}}{\partial v_i} = \sum_{j=1}^{s}\left(\sum_{k=1}^{s} \frac{\partial p}{\partial q_{jk}} \frac{\partial q_{jk}}{\partial v_i}\right)$$

$$= \sum_{j=1}^{s}\left[\begin{array}{cccc}\dfrac{\partial q_{j1}}{\partial v_i} & \dfrac{\partial q_{j2}}{\partial v_i} & \cdots & \dfrac{\partial q_{js}}{\partial v_i}\end{array}\right]\det(\mathbf{Q})\mathbf{Q}^{-1}\mathbf{e}_j$$

and

$$\frac{\partial q_{jk}}{\partial v_i} = -\delta_{ik} + \delta_{is} + z\big(b_{ik}^{(2)} - b_{sk}^{(2)}\big),$$

or in matrix form,

$$\left[\frac{\partial q_{ik}}{\partial v_i}\right]_{k=1}^{s} = -\mathbf{e}_i^T + \mathbf{e}_s^T + z\big(\mathbf{e}_i^T\mathbf{B}_2 - \mathbf{e}_s^T\mathbf{B}_2\big).$$

Substituting the last relation into the equation for $\partial p/\partial v_i$, we find that

$$\frac{\partial p}{\partial v_i} = \det(\mathbf{Q})\Big((\mathbf{e}_s^T - \mathbf{e}_i^T)\mathbf{I} + z(\mathbf{e}_i^T - \mathbf{e}_s^T)\mathbf{B}_2\Big)\sum_{j=1}^{s}\mathbf{Q}^{-1}\mathbf{e}_j,$$

which is equivalent to (3.8.11).   ∎

Replacing $\mathbf{Q}$, $\mathbf{B}_1$, and $\mathbf{B}_2$ by $\widehat{\mathbf{Q}}$, $\widehat{\mathbf{B}}_1$, and $\widehat{\mathbf{B}}_2$ we can obtain formulas for $\partial\widehat{p}/\partial a_{ij}$ and $\partial\widehat{p}/\partial v_i$ which are analogous to equations (3.8.10) and (3.8.11). These relations allow us to compute the Jacobian matrix of systems (3.8.2) and (3.8.9) corresponding to DIMSIMs of type 2.

3.9   LEAST-SQUARES MINIMIZATION

In Section 3.8 we derived systems of nonlinear equations that define DIMSIMs of types 1 and 2 with $p = q = r = s$. These systems have been solved by minimizing the sum of squares of the objective functions defined by the right-hand sides of equation (3.8.1), (3.8.2), or (3.8.9), which were obtained using a variant of the Fourier series approach. Then the minima of this sum, which are equal to zero, are also solutions of the corresponding systems. For $p = 5$ and $p = 6$ we have used for this purpose the Levenberg-Marquardt algorithm as implemented in the FORTRAN subroutines `lmdif.f` and `lmder.f` from MINPACK.

This algorithm was first suggested by Levenberg [207] and Marquardt [211] and has been nicely summarized by Dennis and Schnabel [113]. To explain the main idea of this algorithm, we write the minimization problem in the generic form

$$f(x) = \frac{1}{2}\mathbf{R}^T(x)\mathbf{R}(x) = \frac{1}{2}\sum_{i=1}^{N} r_i^2(x) \to \min, \qquad (3.9.1)$$

where

$$\mathbf{R}(x) = \begin{bmatrix} r_1(x) & r_2(x) & \cdots & r_N(x) \end{bmatrix}^T$$

and $r_i(x)$, $i = 1, 2, \ldots, N$, $N = (s-1)(s+2)/2$, stands for the right-hand sides of equation (3.8.1), (3.8.2), or (3.8.9) and x is the vector of unknown parameters of the method

$$x = \begin{bmatrix} a_{21} & a_{31} & \cdots & a_{s,s-1} & v_1 & \cdots & v_{s-1} \end{bmatrix}^T.$$

Denote by $\mathbf{J}(x)$ the Jacobian matrix of $\mathbf{R}(x)$ defined by

$$\mathbf{J}(x) = \left[\frac{\partial r_i}{\partial x_j}\right]_{i,j=1}^{s}.$$

Then we choose the next iterate, x_+, in terms of the previous iterate, x_c, by minimizing the functional

$$\left\|\mathbf{R}(x_c) + \mathbf{J}(x_c)(x_+ - x_c)\right\|_2 \to \min$$

over $\mathbb{R}^N$ subject to

$$\|x_+ - x_c\|_2 \leq \delta_c,$$

with appropriately chosen bounds δ_c. As explained by Dennis and Schnabel [113], this leads to the formula

$$x_+ = x_c - \left(\mathbf{J}(x_c)^T\mathbf{J}(x_c) + \mu_c\mathbf{I}\right)^{-1}\mathbf{J}(x_c)^T\mathbf{R}(x_c), \qquad (3.9.2)$$

where $\mu_c = 0$ if

$$\delta_c \geq \left\|\left(\mathbf{J}(x_c)^T\mathbf{J}(x_c)\right)^{-1}\mathbf{J}(x_c)^T\mathbf{R}(x_c)\right\|_2$$

and $\mu_c > 0$ otherwise. The strategy for choosing μ_c has been described by Moré [217]. This method reduces to the Gauss-Newton algorithm for $\mu_c = 0$.

However, when the Gauss-Newton step is too long, the Levenberg-Marquardt step (3.9.2) is close to being in the steepest-descent direction.

The program `lmdif.f` uses subroutine `fcn.f` supplied by the user, which calculates the functions. The Jacobian matrix is then calculated by a forward-difference approximation. The program `lmder.f` uses subroutine `fcn.f`, which calculates both the functions and the Jacobian matrix. This matrix can be calculated with the aid of formulas (3.8.10) and (3.8.11) for $\partial p/\partial a_{ij}$ and $\partial p/\partial v_i$ (or analogous formulas for $\partial \widehat{p}/\partial a_{ij}$ and $\partial \widehat{p}/\partial v_i$) derived in Section 3.8.

To compute $\det(\mathbf{Q}(w, z))$ or $\det(\widehat{\mathbf{Q}}(w, \widehat{z}))$ appearing in the definitions of the objective functions corresponding to (3.8.1), (3.8.2), or (3.8.9), or $\det(\mathbf{Q}(w, z))$ and $\mathbf{Q}^{-1}(w, z)$ or $\det(\widehat{\mathbf{Q}}(w, \widehat{z}))$ and $\widehat{\mathbf{Q}}^{-1}(w, \widehat{z})$ appearing in the definition of the corresponding Jacobian matrices, we have used subroutines `zgefa.f` and `zgedi.f` from the LINPACK library. The program `zgefa.f` computes the PLU factorization of a COMPLEX*16 matrix which is then used by `zgedi.f` to compute the determinant, or the determinant and the inverse of the matrix, if desired, using the factors computed by `zgefa.f`.

We have used $N_1 = N_2 = p + 1$ in (3.8.1) and (3.8.2) to derive DIMSIMs of order $p = 5$ and $p = 6$. The iteration process was continued until the difference between two consecutive iterates was less than or equal to 10^{-16}. After the solution was found, we checked using MATHEMATICA or MAPLE that we computed the right solution.

We have also tried to use the Levenberg-Marquardt algorithm, (3.9.2), to construct DIMSIMs of order $p = 7$ and $p = 8$ but did not get satisfactory results. We have experimented with many different methods for least-squares minimization, and based on the experience that was gathered in this process, a choice was made to use a rather sophisticated algorithm available in DN2G and DN2GB. These are nonlinear least-squares routines available in NETLIB/PORT, the public part of the PORT library [269]. They are new versions of the original code NL2SOL [111, 112]. A description of the improvements was provided by Bunch et al. [26]. In particular, the least-squares problem is considered with additional bound constraints on its variables as proposed by Gay [129].

The least-squares solution in NL2SOL was developed as a generalization and improvement of standard approaches such as the Levenberg-Marquardt algorithm in `lmdif.f` and `lmder.f` from MINPACK which we had used for $p = 5$ and $p = 6$. Specifically, Dennis et al. [111, Sec. 8] have compared NL2SOL with algorithms from MINPACK.

Algorithm DN2G (without bound constraint) and DN2GB (with bound constraint) utilize adaptive quadratic modeling. Special problem structure is exploited by maintaining a secant approximation to the second-order part of the Hessian of the objective function $f(x)$ defined by (3.9.1). The program switches adaptively between a Gauss-Newton and an augmented Hessian approximation, where the Gauss-Newton steps are computed from a corrected seminormal approach. The Gauss-Newton model at the iterate x_c is

$$q_c^G(x) = \frac{1}{2}\mathbf{R}(x_c)^T\mathbf{R}(x_k) + (x - x_c)^T\mathbf{J}(x_c)^T\mathbf{R}(x_c)$$

$$+ \frac{1}{2}(x - x_c)^T\mathbf{J}(x_c)^T\mathbf{J}(x_c)(x - x_c).$$

In NL2SOL, one adds an approximation $\mathbf{S}_c$ to the difference between this and the standard quadratic Taylor model of the Newton method to obtain another model,

$$q_c^S(x) = \frac{1}{2}\mathbf{R}(x_c)^T\mathbf{R}(x_k) + (x - x_c)^T\mathbf{J}(x_c)^T\mathbf{R}(x_c)$$

$$+ \frac{1}{2}(x - x_c)^T\left(\mathbf{J}(x_c)^T\mathbf{J}(x_c) + \mathbf{S}_c\right)(x - x_c).$$

To update $\mathbf{S}_c$, a straightforward modification of the Oren-Luenberger self-scaling technique [226] is used. Finally, the choice of which of the models above to use is intimately related to the trust region approach utilized to pick $\Delta x_c = x_+ - x_c$, which has the form

$$\Delta x_c = \left(\mathbf{H}_c + \mu_c\mathbf{D}_c^2\right)^{-1}\nabla f(x_c).$$

Here $\mathbf{H}_c$ is the current approximation to the Hessian $\nabla^2 f(x)$ of $f(x)$,

$$\nabla^2 f(x) = \mathbf{J}(x)^T\mathbf{J}(x) + \sum_{i=1}^{N} r_i(x)\nabla^2 r_i(x),$$

$\mathbf{D}_c$ is a diagonal scaling matrix, and $\mu_c \geq 0$ is chosen by the same procedure as in the Levenberg-Marquardt algorithm [217]. The approximation to the gradient

$$\Delta f(x) = \left[\begin{array}{cccc} \Delta f_1(x) & \Delta f_2(x) & \cdots & \Delta f_N(x) \end{array}\right]^T$$

of the function $f(x)$ is computed through the following algorithm:

> **for** $i := 1(1)N$ **do**
>
> $\delta x = \delta(1 + |x_i|);$
>
> $f_1 = f(x - \delta x\mathbf{e}_i);$ $f_2 = f(x + \delta x\mathbf{e}_i);$
>
> $f_3 = f(x - 2\delta x\mathbf{e}_i);$ $f_4 = f(x + 2\delta x\mathbf{e}_i);$
>
> $f_5 = f(x - 4\delta x\mathbf{e}_i);$ $f_6 = f(x + 4\delta x\mathbf{e}_i);$
>
> $s_1 = (f_2 - f_1)/(2\delta x);$ $s_2 = (f_4 - f_3)/(4\delta x);$
>
> $s_3 = (f_6 - f_5)/(8\delta x);$
>
> $\nabla f_i = s_1 + 0.4(s_1 - s_2) + (s_1 - 2s_2 + s_3)/45;$
>
> **enddo**

Here $\delta = 0.25$ eps, eps is the machine epsilon, and $\mathbf{e}_i$ denotes the ith unit vector in $\mathbb{R}^N$. The overall code is much larger and more complex than standard least-squares algorithms but has proven to be very robust and at the same time, reasonably efficient.

To find appropriate methods, we performed an extensive computer search starting with many random points distributed uniformly in the interval $[-1, 1]$. We have used the Levenberg-Marquardt algorithm in case $p = 5$ and $p = 6$ and DN2G and DN2GB algorithms in case $p = 7$ and $p = 8$. Those points that had small f-values and not too large components were subsequently improved. This way, in all cases, solutions with sufficiently small residuals could be derived. Examples of methods found in this way are listed by Butcher and Jackiewicz [68] for $p = 5$ and $p = 6$ and by Butcher et al. [74] for $p = 7$ and $p = 8$. These examples, up to order $p = 6$, are also reproduced in Section 3.10.

The efficiency of computer searches for high order methods can be improved further by exploiting the special structure of the coefficients p_{kl} and $\widehat{p}_{kl}$ appearing in (3.8.1) and (3.8.2). This is investigated by Jackiewicz and Mittelmann [179] where the following theorem was obtained.

Theorem 3.9.1 *The coefficients p_{kl} and $\widehat{p}_{kl}$, $k = 2, 3, \ldots, s$, $l = k-1, k, \ldots, s$, of stability polynomials $p(w, z)$ and $\widehat{p}(w, \widehat{z})$ of DIMSIMs of types 1 and 2 are linear with respect to v_i, $i = 1, 2, \ldots, s-1$. Moreover, for methods with $c_1 = 0$ and $c_s = 1$, coefficients p_{ks} and $\widehat{p}_{ks}$, $k = 2, 3, \ldots, s$, corresponding to $w^{s-k} z^s$ and $w^{s-k} \widehat{z}^s$ are homogeneous with respect to v_i, $i = 1, 2, \ldots, s-1$.*

Proof :   It was demonstrated in Section 3.8 that $p(w, z)$ and $\widehat{p}(w, \widehat{z})$ can be computed from formulas (3.8.3) and (3.8.5), where matrices $\mathbf{Q}(w, z)$ and $\widehat{\mathbf{Q}}(w, \widehat{z})$ are defined by (3.8.4) and (3.8.6). To show the first part of the theorem we use the transformation for the analysis of DIMSIMs introduced by Butcher [46] and discussed in Section 3.3. Using this transformation, we work with the matrices

$$\overline{\mathbf{A}} = \mathbf{T}^{-1} \mathbf{A} \mathbf{T}, \quad \overline{\mathbf{B}} = \mathbf{T}^{-1} \mathbf{B} \mathbf{T}, \quad \overline{\mathbf{V}} = \mathbf{T}^{-1} \mathbf{V} \mathbf{T},$$

where the matrix $\mathbf{T}$ is defined by (3.3.1). It was demonstrated in Lemma 3.3.2 that the matrix $\overline{\mathbf{V}}$ has the form

$$\overline{\mathbf{V}} = \begin{bmatrix} 1 & \overline{v}_2 & \cdots & \overline{v}_s \end{bmatrix} \mathbf{e}_1$$

and it is easy to verify that $\overline{v}_i$, $i = 2, 3, \ldots, s$, are linear functions of v_j, $j = 1, 2, \ldots, s-1$. As in Section 3.3, set

$$\overline{\mathbf{B}}_0 = \mathbf{T}^{-1} \mathbf{B}_0 \mathbf{T}, \quad \overline{\mathbf{B}}_1 = \mathbf{T}^{-1} \mathbf{B}_1 \mathbf{T}, \quad \overline{\mathbf{B}}_2 = \mathbf{T}^{-1} \mathbf{B}_2 \mathbf{T},$$

and define

$$\overline{\widehat{\mathbf{B}}}_0 = \mathbf{T}^{-1} \widehat{\mathbf{B}}_0 \mathbf{T}, \quad \overline{\widehat{\mathbf{B}}}_2 = \mathbf{T}^{-1} \widehat{\mathbf{B}}_2 \mathbf{T}.$$

Then

$$p(w, z) = \det \big(\mathbf{Q}(w, z) \big) = \det \big(\overline{\mathbf{Q}}(w, z) \big),$$

$$\widehat{p}(w, \widehat{z}) = \det\left(\widehat{\mathbf{Q}}(w, \widehat{z})\right) = \det\left(\overline{\widehat{\mathbf{Q}}}(w, \widehat{z})\right),$$

where

$$\overline{\mathbf{Q}}(w, z) = w(\mathbf{I} - z\overline{\mathbf{A}}) - \overline{\mathbf{V}} - z\overline{\mathbf{B}}_0 + z\overline{\mathbf{A}}\,\overline{\mathbf{B}}_1 + z\overline{\mathbf{V}}\,\overline{\mathbf{B}}_2,$$

$$\overline{\widehat{\mathbf{Q}}}(w, \widehat{z}) = w(\mathbf{I} - \widehat{z}\overline{\widehat{\mathbf{A}}}) - \overline{\mathbf{V}} - \widehat{z}\overline{\widehat{\mathbf{B}}}_0 + \widehat{z}\overline{\widehat{\mathbf{A}}}\,\overline{\mathbf{B}}_1 + \widehat{z}\overline{\mathbf{V}}\,\overline{\widehat{\mathbf{B}}}_2.$$

The first part of the theorem now follows by expanding $\det(\overline{\mathbf{Q}}(w, z))$ and $\det(\overline{\widehat{\mathbf{Q}}}(w, \widehat{z}))$ with respect to the first row of $\overline{\mathbf{Q}}(w, z)$ and $\overline{\widehat{\mathbf{Q}}}(w, \widehat{z})$.

To prove the second part of the theorem, we must show that for methods with $c_1 = 0$ and $c_s = 1$ we have

$$p_{ks}\Big|_{v_i=0} = 0 \qquad \text{and} \qquad \widehat{p}_{ks}\Big|_{v_i=0} = 0.$$

Observe that

$$p(w, z)\Big|_{v_i=0} = \det(\mathbf{Q}_0(w, z)), \qquad \widehat{p}(w, \widehat{z})\Big|_{v_i=0} = \det(\widehat{\mathbf{Q}}_0(w, \widehat{z})),$$

where

$$\mathbf{Q}_0(w, z) = w(\mathbf{I} - z\mathbf{A}) - \mathbf{V}_0 - z\mathbf{B}_0 + z\mathbf{A}\mathbf{B}_1 + z\mathbf{V}_0\mathbf{B}_2,$$

$$\widehat{\mathbf{Q}}_0(w, \widehat{z}) = w(\mathbf{I} - \widehat{z}\widehat{\mathbf{A}}) - \mathbf{V}_0 - \widehat{z}\widehat{\mathbf{B}}_0 + \widehat{z}\widehat{\mathbf{A}}\mathbf{B}_1 + \widehat{z}\mathbf{V}_0\widehat{\mathbf{B}}_2,$$

and

$$\mathbf{V}_0 = \mathbf{V}\Big|_{v_i=0} = \mathbf{e}\begin{bmatrix} 0 & 0 & \cdots & 0 & 1 \end{bmatrix}.$$

Since $c_1 = 0$ and $c_s = 1$, it follows from formulas (3.2.2) for the elements of matrices $\mathbf{B}_0$, $\mathbf{B}_1$, and $\mathbf{B}_2$ that the first row of $\mathbf{B}_0$ is equal to the last row of $\mathbf{B}_2$. Moreover, the first row of $\mathbf{B}_1$ is equal to $[0, 0, \ldots, 0, 1]$. Hence, it is easy to verify that the first rows of of $\mathbf{B}_0 - \mathbf{V}_0\mathbf{B}_2$ and $\widehat{\mathbf{B}}_0 - \mathbf{V}_0\widehat{\mathbf{B}}_2$ are equal to zero. This implies that the first rows of $\mathbf{Q}_0(w, z)$ and $\widehat{\mathbf{Q}}_0(w, \widehat{z})$ are independent of z and $\widehat{z}$. Expanding $\det(\mathbf{Q}_0(w, z))$ and $\det(\widehat{\mathbf{Q}}_0(w, \widehat{z}))$ with respect to the first rows of $\mathbf{Q}_0(w, z)$ and $\widehat{\mathbf{Q}}_0(w, \widehat{z})$, it follows that the polynomials $p(w, z)|_{v_i=0}$ and $\widehat{p}(w, \widehat{z})|_{v_i=0}$ have at most degree $s - 1$ with respect to z and $\widehat{z}$. This is equivalent to the second part of the theorem. ∎

Set $\widetilde{\mathbf{v}} = [v_1, \ldots, v_{s-1}]^T$. It follows from this theorem that systems (3.8.1) and (3.8.2) can be written in the form

$$H(\mathbf{A})\widetilde{\mathbf{v}} = b(\mathbf{A}), \quad G(\mathbf{A})\widetilde{\mathbf{v}} = 0, \tag{3.9.3}$$

and

$$\widehat{H}(\widehat{\mathbf{A}})\widetilde{\mathbf{v}} = \widehat{b}(\widehat{\mathbf{A}}), \quad \widehat{G}(\widehat{\mathbf{A}})\widetilde{\mathbf{v}} = 0, \tag{3.9.4}$$

where $H(\mathbf{A})$, $\widehat{H}(\widehat{\mathbf{A}})$ and $G(\mathbf{A})$, $\widehat{G}(\widehat{\mathbf{A}})$ are matrices of dimensions $(s - 1) \times s(s - 1)/2$ and $(s - 1) \times (s - 1)$, respectively, and $b(\mathbf{A})$, $\widehat{b}(\widehat{\mathbf{A}})$ are vectors of dimension $s(s - 1)/2$. Subsystems $G(\mathbf{A})\widetilde{\mathbf{v}} = 0$ and $\widehat{G}(\widehat{\mathbf{A}})\widetilde{\mathbf{v}} = 0$ in (3.9.3) and

(3.9.4) correspond to equations $p_{ks} = 0$ and $\widehat{p}_{ks} = 0$, $k = 2, 3, \ldots, s$, in (3.4.5) and (3.5.5), or to equations $2, 5, 9, 13, \ldots, (s-1)(s+2)/2$ in systems (3.8.1) and (3.8.2). For given $\mathbf{A}$ or $\widehat{\mathbf{A}}$, matrices $H(\mathbf{A})$ and $G(\mathbf{A})$ or $\widehat{H}(\widehat{\mathbf{A}})$ and $\widehat{G}(\widehat{\mathbf{A}})$ can easily be assembled by computing p_{kl} or $\widehat{p}_{kl}$ with $v_i = 1$, $v_j = 0$, $j \neq i$, $i = 1, 2, \ldots, s-1$. Then the ith columns of G and $\widehat{G}$ correspond to p_{ks} or $\widehat{p}_{ks}$, $k = 2, 3, \ldots, s$, computed with v_j given above, and the ith columns of H or $\widehat{H}$ correspond to the remaining p_{kl} or $\widehat{p}_{kl}$.

If $(\mathbf{A}, \widetilde{\mathbf{v}})$ and $(\widehat{\mathbf{A}}, \widetilde{\mathbf{v}})$ are solutions to (3.9.3) and (3.9.4) (with $c_1 = 0$ and $c_s = 1$), the vector $\widetilde{\mathbf{v}}$ necessarily belongs to the null space of matrices $G(\mathbf{A})$ and $\widehat{G}(\widehat{\mathbf{A}})$. Augmenting these systems by the equations

$$\det\big(G(\mathbf{A})\big) = 0 \quad \text{and} \quad \det(\widehat{G}(\widehat{\mathbf{A}})) = 0,$$

and applying least-squares minimization methods to these augmented systems, we can improve the accuracy and reliability of searches for high order DIMSIMs. Examples of methods found in this way have been given by Jackiewicz and Mittelmann [179].

3.10 EXAMPLES OF DIMSIMS OF TYPES 1 AND 2

In this section we present examples of DIMSIMs of types 1 and 2 of order $p = 5$ and $p = 6$ with the abscissa vector $\mathbf{c}$ of dimension $s = p$ given by (3.4.3). Coefficient matrices $\mathbf{A}$, $\widehat{\mathbf{A}}$, and $\mathbf{V}$ were obtained by least-squares minimization techniques described in Section 3.9, which were applied to systems (3.8.1), (3.8.2), and (3.8.9). Coefficient matrix $\mathbf{B}$ was then computed using formula (3.2.1), derived in Theorem 3.2.1. Matrices $\mathbf{B}_0$, $\mathbf{B}_1$, and $\mathbf{B}_2$ in (3.2.1) are determined uniquely by the vector $\mathbf{C}$ with the aid of formulas (3.2.2). As a consequence, the derived methods have the required order p equal to the stage order q.

We have found many solutions to systems (3.8.1), (3.8.2), and (3.8.9). Inspecting these solutions we could see that some of them have advantages in terms of the sizes of the elements in coefficient matrices $\mathbf{A}$, $\widehat{\mathbf{A}}$, $\mathbf{V}$, and $\mathbf{B}$. Exceedingly large values are likely to cause round-off problems because of cancellation of significant digits. We display Below we cite examples of methods with moderately sized coefficients. All methods of type 2 are A- and L-stable.

Example 1. DIMSIM of type 1 with $p = q = r = s = 5$:

$$\mathbf{A} = \begin{bmatrix} 0 & 0 & 0 & 0 & 0 \\ 1.176528 & 0 & 0 & 0 & 0 \\ 1.980579 & 0.401718 & 0 & 0 & 0 \\ 3.053284 & 0.734996 & 0.267236 & 0 & 0 \\ 1.419332 & 2.653490 & -2.277853 & 1.197891 & 0 \end{bmatrix},$$

$$\mathbf{v} = \begin{bmatrix} -0.240696 & 1.260495 & -2.481269 & 1.919907 & 0.541563 \end{bmatrix}^T.$$

Example 2. DIMSIM of type 2 with $p = q = r = s = 5$:

$$\mathbf{A} = \begin{bmatrix} 0.278054 & 0 & 0 & 0 & 0 \\ 0.220452 & 0.278054 & 0 & 0 & 0 \\ 2.294820 & -0.602367 & 0.278054 & 0 & 0 \\ 5.054621 & -1.529876 & 0.097119 & 0.278054 & 0 \\ 9.345168 & -1.412134 & -1.883402 & 0.782534 & 0.278054 \end{bmatrix},$$

$$\mathbf{v} = \begin{bmatrix} -0.0793855 & 0.554318 & -1.569590 & 2.332075 & -0.237418 \end{bmatrix}^T.$$

Example 3. DIMSIM of type 1 with $p = q = r = s = 6$:

$$\mathbf{A} = \begin{bmatrix} 0 & 0 & 0 & 0 & 0 & 0 \\ -0.378732 & 0 & 0 & 0 & 0 & 0 \\ 3.415475 & 1.075853 & 0 & 0 & 0 & 0 \\ 3.517453 & 0.493910 & 0.483207 & 0 & 0 & 0 \\ -0.398097 & -1.988293 & 1.043842 & 0.253273 & 0 & 0 \\ 1.348865 & -2.741611 & 1.388314 & 0.0327580 & 0.364587 & 0 \end{bmatrix},$$

$$\mathbf{v} = \begin{bmatrix} -3.112709 & 3.675653 & -1.938159 & 2.859978 & -2.336351 & 1.85159 \end{bmatrix}^T.$$

Example 4. DIMSIM of type 2 with $p = q = r = s = 6$:

$$\mathbf{A} = \begin{bmatrix} 0.334142 & 0 & 0 & 0 & 0 & 0 \\ 0.044043 & 0.334142 & 0 & 0 & 0 & 0 \\ 0.431528 & -0.685118 & 0.334142 & 0 & 0 & 0 \\ 0.108586 & -0.267762 & -0.383403 & 0.334142 & 0 & 0 \\ -1.833853 & 2.855822 & -1.573479 & 0.039950 & 0.334142 & 0 \\ -5.629166 & 11.996111 & -10.377635 & 4.983245 & -1.166032 & 0.334142 \end{bmatrix},$$

$$\mathbf{v} = \begin{bmatrix} 2.134887 & -9.944380 & 13.934564 & 0.329620 & -14.682606 & 9.22792 \end{bmatrix}^T.$$

In the four examples above we have displayed the coefficients of $\mathbf{A}$ and $\mathbf{v}$ truncated to about six significant digits. We refer readers to the papers [74, 179] for examples of types 1 and 2 DIMSIMs with $p = q = r = s = 7$ and $p = q = r = s = 8$.

3.11   NORDSIECK REPRESENTATION OF DIMSIMS

In the representation of DIMSIMs (3.1.1), coefficient matrix $\mathbf{U}$ was chosen as identity matrix $\mathbf{I}$ of appropriate dimension. Since the purpose of the r vectors making up the input data $y^{[n-1]}$ and the corresponding output data $y^{[n]}$ is merely to carry information from step to step, it is possible to rearrange these data by taking r independent linear combinations of the r subvectors and using the resulting combinations instead of the subvectors themselves. This is equivalent to choosing a nonsingular matrix $\mathbf{T}$ and replacing the partitioned matrix that characterizes the method

$$\begin{bmatrix} \mathbf{A} & \mathbf{U} \\ \mathbf{B} & \mathbf{V} \end{bmatrix}$$

by the "transformed method"

$$\begin{bmatrix} \mathbf{A} & \mathbf{UT} \\ \mathbf{T}^{-1}\mathbf{B} & \mathbf{T}^{-1}\mathbf{VT} \end{bmatrix}.$$

If $\mathbf{U}$ is square and nonsingular, transforming it back to the case $\mathbf{U} = \mathbf{I}$ is easily achieved by the choice $\mathbf{T} = \mathbf{U}^{-1}$.

In this section we derive a new representation of DIMSIMs which is more convenient to implement in a variable step size variable order environment than representation (3.1.1). This representation was proposed by Nordsieck [222, 223] in the context of Adams methods and its use was later promoted and used with advantage by Gear [131, 132, 133] in the code DIFSUB for nonstiff and stiff differential systems. This representation in the context of DIMSIMs was first proposed by Butcher et al. [57]. In this Nordsieck representation of DIMSIMs we have $p = q = s$, $r = s + 1$, and the vector $y^{[n]}$ of external approximations approximates directly the Nordsieck vector $z(t_n, h)$, which is defined by

$$z(t, h) = \begin{bmatrix} y(t) \\ hy'(t) \\ \vdots \\ h^p y^{(p)}(t) \end{bmatrix}. \tag{3.11.1}$$

We demonstrate in Section 4.1 that this new representation will be zero-stable for any step size pattern, and changing the step size will be accomplished by a simple rescaling and modifying of the vector of external approximations.

Following Butcher et al. [57], our starting point is the representation (3.1.1) such that $p = q = r = s$ and with $\mathbf{U} = \mathbf{I}$. In addition to $y^{[n]}$ we consider an approximation $\eta^{[n]} \in \mathbb{R}^m$ such that

$$\eta^{[n]} = \sum_{k=0}^{p} t_k h^k y^{(k)}(t_n) + O(h^{p+1}),$$

which has the form

$$\eta^{[n]} = h(\mathbf{b}^T \otimes \mathbf{I})F(Y^{[n]}) + (\mathbf{v}^T \otimes \mathbf{I})y^{[n-1]}. \tag{3.11.2}$$

Here, t_k, $k = 0, 1, \ldots, p$, are some scalars and $\mathbf{b} \in \mathbb{R}^s$, $\mathbf{v} \in \mathbb{R}^s$ are some vectors. To avoid the possibility that $\eta^{[n]}$ can be written as a linear combination of the output quantities $y_i^{[n]}$, $i = 1, 2, \ldots, s$, we assume that the matrix

$$\begin{bmatrix} \mathbf{B} & \mathbf{e} \\ \mathbf{b}^T & 1 \end{bmatrix}$$

is nonsingular. Define the vector

$$\mathbf{t} = \begin{bmatrix} t_0 & t_1 & \cdots & t_p \end{bmatrix}^T \in \mathbb{R}^{p+1}$$

and the matrix

$$\widetilde{\mathbf{W}} = \begin{bmatrix} \mathbf{W} \\ \mathbf{t}^T \end{bmatrix} = \begin{bmatrix} \mathbf{q}_0 & \mathbf{q}_1 & \cdots & \mathbf{q}_p \\ t_0 & t_1 & \cdots & t_p \end{bmatrix} \in \mathbb{R}^{(p+1) \times (p+1)}.$$

The independence of the vectors $y_i^{[n]}$, $i = 1, 2, \ldots, s$, and $\eta^{[n]}$ defined by (3.11.2) guarantees that $\widetilde{\mathbf{W}}$ is nonsingular. The relationship between vectors $\mathbf{t}$ and $\mathbf{b}$ is given by the following result.

Theorem 3.11.1 (Butcher et al. [57]) *The first component of* $\mathbf{t}$ *is given by* $t_0 = 1$, *and for given* $t_1, t_2, \ldots, t_p$, *vector* $\mathbf{b}$ *is equal to*

$$\mathbf{b}^T = \mathbf{1}^T \mathbf{C}^{-1}, \tag{3.11.3}$$

where $\mathbf{C}$ *is the Vandermonde matrix*

$$\mathbf{C} = \begin{bmatrix} \mathbf{e} & \mathbf{c} & \cdots & \mathbf{c}^{p-1} \end{bmatrix}$$

and the vector $\mathbf{l} = [l_1, \ldots, l_p]^T$ *is defined by*

$$l_k = (k-1)! \left(\sum_{j=0}^{k} \frac{t_{k-j}}{j!} - \mathbf{v}^T \mathbf{q}_k \right), \quad k = 1, 2, \ldots, s.$$

Proof: It follows from Theorem 2.4.1, formula (2.4.6), that

$$e^z \xi = z \mathbf{b}^T e^{\mathbf{c}z} + \mathbf{v}^T \mathbf{w}(z) + O(z^{p+1}),$$

where

$$\xi = \sum_{k=0}^{p} t_k z^k \quad \text{and} \quad \mathbf{w}(z) = \sum_{k=0}^{p} \mathbf{q}_k z^k.$$

Expanding both sides of this equation into a Taylor series around $z = 0$ and comparing the corresponding terms, we obtain

$$\mathbf{v}^T \mathbf{q}_0 = t_0 \tag{3.11.4}$$

and

$$\frac{\mathbf{b}^T \mathbf{c}^k}{k!} = \sum_{j=0}^{k+1} \frac{t_{k+1-j}}{j!} - \mathbf{v}^T \mathbf{q}_{k+1}, \quad k = 0, 1, \ldots, s-1. \tag{3.11.5}$$

It follows from (3.11.4) that $t_0 = 1$ and it can easily be verified that equation (3.11.5) is equivalent to (3.11.3). This completes the proof. $\blacksquare$

Define the vector

$$\widetilde{y}^{[n]} = \begin{bmatrix} y^{[n]} \\ \eta^{[n]} \end{bmatrix}$$

and consider the method

$$Y^{[n]} = h(\mathbf{A} \otimes \mathbf{I})F(Y^{[n]}) + (\widetilde{\mathbf{U}} \otimes \mathbf{I})\widetilde{y}^{[n-1]},$$

$$\widetilde{y}^{[n]} = h(\widetilde{\mathbf{B}} \otimes \mathbf{I})F(Y^{[n]}) + (\widetilde{\mathbf{V}} \otimes \mathbf{I})\widetilde{y}^{[n-1]}, \tag{3.11.6}$$

$n = 1, 2, \ldots, N$, where

$$\widetilde{\mathbf{U}} = \begin{bmatrix} \mathbf{U} & \mathbf{0} \end{bmatrix}, \quad \widetilde{\mathbf{B}} = \begin{bmatrix} \mathbf{B} \\ \mathbf{b}^T \end{bmatrix}, \quad \widetilde{\mathbf{V}} = \begin{bmatrix} \mathbf{V} & \mathbf{0} \\ \mathbf{v}^T & \mathbf{0} \end{bmatrix}.$$

Since

$$\widetilde{y}^{[n]} = (\widetilde{\mathbf{W}} \otimes \mathbf{I})z(t_n, h) + O(h^{p+1}),$$

where $z(t_n, h)$ is a Nordsieck vector given by (3.11.1), we define the vector $z^{[n]}$ by the relation

$$\widetilde{y}^{[n]} = (\widetilde{\mathbf{W}} \otimes \mathbf{I})z^{[n]}. \tag{3.11.7}$$

Observing that

$$\widetilde{\mathbf{U}}\widetilde{\mathbf{W}} = \begin{bmatrix} \mathbf{U} & \mathbf{0} \end{bmatrix} \begin{bmatrix} \mathbf{W} \\ \mathbf{t}^T \end{bmatrix} = \mathbf{U}\mathbf{W}$$

and substituting (3.11.7) into (3.11.6), we obtain

$$Y^{[n]} = h(\mathbf{A} \otimes \mathbf{I})F(Y^{[n]}) + (\mathbf{P} \otimes \mathbf{I})z^{[n-1]},$$

$$z^{[n]} = h(\mathbf{G} \otimes \mathbf{I})F(Y^{[n]}) + (\mathbf{Q} \otimes \mathbf{I})z^{[n-1]}, \tag{3.11.8}$$

$n = 1, 2, \ldots, N$, where the new coefficient matrices $\mathbf{P}$, $\mathbf{G}$, and $\mathbf{Q}$ are defined by

$$\mathbf{P} = \mathbf{UW}, \quad \mathbf{G} = \widetilde{\mathbf{W}}^{-1}\mathbf{B}, \quad \mathbf{Q} = \widetilde{\mathbf{W}}^{-1}\widetilde{\mathbf{V}}\widetilde{\mathbf{W}}.$$

Since

$$z^{[n]} = (\widetilde{\mathbf{W}}^{-1} \otimes \mathbf{I})\widetilde{y}^{[n]} = z(t_n, h) + O(h^{p+1}), \tag{3.11.9}$$

(3.11.8) is the desired Nordsieck representation of DIMSIMs.

Next we investigate linear stability properties of the augmented method (3.11.8). If $\mathbf{M}(z) = \mathbf{V} + z\mathbf{B}(\mathbf{I} - z\mathbf{A})^{-1}\mathbf{U}$ denotes a stability matrix of a method (3.1.1) (this matrix is defined by (2.6.4)), the stability matrix $\widetilde{\mathbf{M}}(z)$ of the "extended" method (3.11.6) is given by

$$
\begin{aligned}
\widetilde{\mathbf{M}}(z) &= \begin{bmatrix} \mathbf{V} & \mathbf{0} \\ \mathbf{v}^T & 0 \end{bmatrix} + z \begin{bmatrix} \mathbf{B} \\ \mathbf{b}^T \end{bmatrix} (\mathbf{I} - z\mathbf{A})^{-1} \begin{bmatrix} \mathbf{U} & \mathbf{0} \end{bmatrix} \\
&= \begin{bmatrix} \mathbf{M}(z) & \mathbf{0} \\ \mathbf{v}^T + z\mathbf{b}^T(\mathbf{I} - z\mathbf{A})^{-1}\mathbf{U} & 0 \end{bmatrix}.
\end{aligned}
$$

As a consequence, $\widetilde{\mathbf{M}}(z)$ has one more eigenvalue than matrix $\mathbf{M}(z)$, and this extra eigenvalue is equal to zero. As a result, the linear stability properties of method (3.1.1) remain unchanged for constant step sizes through the augmentation process (3.11.6), which leads to the Nordsieck representation of DIMSIMs given by (3.11.8).

Now we derive now the formulas for the computation of the coefficient matrices $\mathbf{Q}$ and $\mathbf{G}$. To simplify the matrix $\mathbf{Q}$, observe that

$$\widetilde{\mathbf{W}} \begin{bmatrix} \mathbf{e} \\ 1 \end{bmatrix} = \mathbf{e}_1,$$

where $\mathbf{e}_1 = [1, 0, \ldots, 0]^T \in \mathbb{R}^{s+1}$. This implies that

$$\mathbf{Q} = \widetilde{\mathbf{W}}^{-1} \begin{bmatrix} \mathbf{e} \\ 1 \end{bmatrix} \begin{bmatrix} \mathbf{v}^T & 0 \end{bmatrix} \begin{bmatrix} \mathbf{W} \\ \mathbf{t}^T \end{bmatrix} = \mathbf{e}_1 \begin{bmatrix} 1 & \mathbf{v}^T\mathbf{q}_1 & \cdots & \mathbf{v}^T\mathbf{q}_p \end{bmatrix}.$$

The matrix $\mathbf{G}$ is characterized by the following result.

Theorem 3.11.2 (Butcher et al. [57]) *The matrix $\mathbf{G}$ is determined by the formula*

$$\mathbf{G} = \mathbf{L}\,\mathbf{C}^{-1}, \tag{3.11.10}$$

where $\mathbf{C}$ is the Vandermonde matrix defined in Theorem 3.11.1 and $\mathbf{L}$ is a matrix with columns L_k, given by

$$L_k = (k-1)! \left(\sum_{j=0}^{k} \frac{\mathbf{e}_{j+1}}{(k-j)!} - \mathbf{Q}\mathbf{e}_{k+1} \right), \quad k = 1, 2, \ldots, s.$$

Here $\mathbf{e}_i$, $i = 1, 2, \ldots, p + 1$, *is the canonical basis in* $\mathbb{R}^{p+1}$. *In particular, matrix* **G** *is independent of* $t_1, t_2, \ldots, t_p$.

Proof: It follows from Theorem 2.4.1, formula (2.4.6), that

$$e^z \widetilde{\mathbf{w}} = z\mathbf{G}e^{\mathbf{c}z} + \mathbf{Q}\widetilde{\mathbf{w}}(z) + O(z^{p+1}), \tag{3.11.11}$$

where

$$\widetilde{\mathbf{w}}(z) = \sum_{k=0}^{p} \mathbf{e}_{k+1} z^k.$$

Expanding both sides of (3.11.11) into a Taylor series around $z = 0$ and equating the corresponding coefficients of z^k, we obtain

$$\mathbf{Q}\mathbf{e}_1 = \mathbf{e}_1 \tag{3.11.12}$$

and

$$\frac{\mathbf{G}\mathbf{c}^k}{k!} = \sum_{j=0}^{k+1} \frac{\mathbf{e}_{j+1}}{(k+1-j)!} - \mathbf{Q}\mathbf{e}_{k+2}, \quad k = 0, 1, \ldots, p - 1. \tag{3.11.13}$$

It follows from the form of the matrix **Q** that condition (3.11.12) is satisfied automatically and (3.11.13) is equivalent to (3.11.10). This completes the proof. ∎

3.12 REPRESENTATION FORMULAS FOR COEFFICIENT MATRICES **P** AND **G**

In Section 3.11 we derived the Nordsieck representation of DIMSIMs by the augmentation process of the vector of external approximation $y^{[n]}$ corresponding to representation (3.1.1) with coefficient matrices given by (3.1.2). In this section we demonstrate that we can obtain directly the representation formulas for coefficient matrices **P** and **G** of the Nordsieck representation (3.11.8). This direct approach is based on use of the modified form of the order and stage order conditions reformulated specifically for methods of the form (3.11.8) with $p = q = s$ and $r = s + 1$, where the vector $z^{[n]}$ is an approximation of order p to the Nordsieck vector $z(t_n)$ defined by (3.11.1).

Define the matrices $\mathbf{K}_p \in \mathbb{R}^{(p+1)\times(p+1)}$ and $\mathbf{E}_p \in \mathbb{R}^{(p+1)\times(p+1)}$ by

$$\mathbf{K}_p = \begin{bmatrix} 0 & 1 & 0 & \cdots & 0 \\ 0 & 0 & 1 & \cdots & 0 \\ \vdots & \vdots & \vdots & \ddots & \vdots \\ 0 & 0 & 0 & \cdots & 1 \\ 0 & 0 & 0 & \cdots & 0 \end{bmatrix}, \quad \mathbf{E}_p = \exp(\mathbf{K}_p) = \begin{bmatrix} 1 & 1 & \frac{1}{2!} & \cdots & \frac{1}{p!} \\ 0 & 1 & 1 & \cdots & \frac{1}{(p-1)!} \\ 0 & 0 & 1 & \cdots & \frac{1}{(p-2)!} \\ \vdots & \vdots & \vdots & \ddots & \vdots \\ 0 & 0 & 0 & \cdots & 1 \end{bmatrix}.$$

We have the following theorem.

Theorem 3.12.1 ([177], [293]) *Assume that $r = s+1$. Then method (3.11.8) with abscissa vector $\mathbf{c} = [c_1, \ldots, c_s]^T$ has order $p = s$; that is,*

$$z_i^{[n-1]} = h^{i-1} y^{(i-1)}(t_{n-1}) + O(h^{s+1})$$

implies that

$$z_i^{[n]} = h^{i-1} y^{(i-1)}(t_n) + O(h^{s+1}),$$

$i = 1, 2, \ldots, r$, *and stage order* $q = p$,

$$Y_i^{[n]} = y(t_{n-1} + c_i h) + O(h^{s+1}), \quad i = 1, 2, \ldots, s$$

if and only if

$$e^{\mathbf{c}z} - z\mathbf{A}e^{\mathbf{c}z} - \mathbf{PZ} = O(z^{s+1}) \tag{3.12.1}$$

and

$$\mathbf{E}_p\mathbf{Z} - z\mathbf{G}e^{\mathbf{c}z} - \mathbf{QZ} = O(z^{s+1}) \tag{3.12.2}$$

as $z \to 0$, *where*

$$\mathbf{Z} = \begin{bmatrix} 1 & z & \cdots & z^s \end{bmatrix}^T \quad and \quad e^{\mathbf{c}z} = \begin{bmatrix} e^{c_1 z} & e^{c_2 z} & \cdots & e^{c_s z} \end{bmatrix}^T.$$

Proof: Expanding $y(t_{n-1}+c_i h)$ and $y'(t_{n-1}+c_i h)$ into a Taylor series around t_{n-1}, we obtain

$$Y^{[n]} = \sum_{k=0}^{s} \frac{c_i^k}{k!} h^k y^{(k)}(t_{n-1}) + O(h^{s+1})$$

and

$$hf(Y_i^{[n]}) = \sum_{k=1}^{s} \frac{c_i^{k-1}}{(k-1)!} h^k y^{(k)}(t_{n-1}) + O(h^{s+1}),$$

$i = 1, 2, \ldots, s$. We also have

$$z_i^{[n-1]} = h^{i-1} y^{(i-1)}(t_{n-1}) + O(h^{s+1})$$

and

$$
\begin{aligned}
z_i^{[n]} &= h^{i-1} y^{(i-1)}(t_n) + O(h^{s+1}) \\
&= \sum_{k=i-1}^{s} \frac{1}{(k-i+1)!} h^k y^{(k)}(t_{n-1}) + O(h^{s+1}) \\
&= \sum_{k=0}^{s} e_{i,k+1} h^k y^{(k)}(t_{n-1}) + O(h^{s+1}),
\end{aligned}
$$

$i = 1, 2, \ldots, s + 1$. Here e_{ij} are the elements of the matrix $\mathbf{E}_p$. Substituting the relations above into (3.11.8) yields

$$\sum_{k=0}^{s} \left(c_i^k - k \sum_{j=1}^{s} a_{ij} c_j^{k-1} - k! p_{i,k+1} \right) \frac{h^k}{k!} y^{(k)}(t_{n-1}) = O(h^{s+1}), \qquad (3.12.3)$$

$i = 1, 2, \ldots, s$, and

$$\sum_{k=0}^{s} \left(k! e_{i,k+1} - k \sum_{j=1}^{s} g_{ij} c_j^{k-1} - k! q_{i,k+1} \right) \frac{h^k}{k!} y^{(k)}(t_{n-1}) + O(h^{s+1}), \quad (3.12.4)$$

$i = 1, 2, \ldots, s + 1$. Here a_{ij}, p_{ij}, and q_{ij} are elements of matrices **A**, **P**, and **Q**, respectively. Equating the coefficients of $h^k y^{(k)}(t_{n-1})/k!$ to zero and then multiplying these coefficients by $z^k/k!$ and adding them from $k = 0$ to $k = s$, we obtain

$$e^{c_i z} - z \sum_{j=1}^{s} a_{ij} e^{c_j z} - \sum_{k=0}^{s} p_{i,k+1} z^k = O(z^{s+1}),$$

$i = 1, 2, \ldots, s$, and

$$\sum_{k=0}^{s} e_{i,k+1} z^k - z \sum_{j=1}^{s} g_{ij} e^{c_j z} - \sum_{k=0}^{s} q_{i,k+1} z^k = O(z^{s+1}),$$

$i = 1, 2, \ldots, s + 1$. These relations are equivalent to (3.12.1) and (3.12.2), respectively. This completes the proof.   ∎

Since $\mathbf{E}_p \mathbf{Z} = e^z \mathbf{Z} + O(z^{s+1})$ the order condition (3.12.2) can be formulated in the equivalent form

$$e^z \mathbf{Z} - z \mathbf{G} e^{\mathbf{c} z} - \mathbf{Q} \mathbf{Z} = O(z^{s+1}),$$

proposed by Wright [293].

Define matrix $\mathbf{C}_p$ by

$$\mathbf{C}_p = \left[\begin{array}{ccccc} \mathbf{e} & \mathbf{c} & \dfrac{\mathbf{c}^2}{2!} & \cdots & \dfrac{\mathbf{c}^p}{p!} \end{array} \right] \in \mathbb{R}^{p \times (p+1)},$$

where $\mathbf{c}^i$ stands for componentwise exponentiation. It follows from the proof of Theorem 3.12.1 that coefficient matrices **P** and **Q** are determined completely by abscissa vector **c** and coefficient matrices **A** and **G**. To be more precise, we have the following corollary.

Corollary 3.12.2 (compare [177], [293]) *Assume that $p = q = s$ and $r = s + 1$ in (3.11.8). Then*

$$\mathbf{P} = \mathbf{C}_p - \mathbf{A} \mathbf{C}_p \mathbf{K}_p \qquad (3.12.5)$$

and

$$\mathbf{Q} = \mathbf{E}_p - \mathbf{G} \mathbf{C}_p \mathbf{K}_p. \qquad (3.12.6)$$

Proof: It follows from (3.12.3) and (3.12.4) that

$$
p_{i1} = 1, \quad p_{i\nu} = \frac{c_i^{\nu-1}}{(\nu-1)!} + \sum_{j=1}^{s} a_{ij} \frac{c_j^{\nu-2}}{(\nu-2)!},
$$

$i = 1, 2, \ldots, s$, $\nu = 2, 3, \ldots, s+1$, and

$$
q_{i1} = e_{i1}, \quad q_{i\nu} = e_{i\nu} + \sum_{j=1}^{s} g_{ij} \frac{c_j^{\nu-2}}{(\nu-2)!},
$$

$i = 1, 2, \ldots, s+1$, $\nu = 2, 3, \ldots, s+1$. Taking into account that

$$
\mathbf{C}_p \mathbf{K}_p = \begin{bmatrix} \mathbf{0} & \mathbf{e} & \mathbf{c} & \dfrac{\mathbf{c}^2}{2!} & \dfrac{\mathbf{c}^{p-1}}{(p-1)!} \end{bmatrix},
$$

these relations for p_{ij} and q_{ij} are equivalent to (3.12.5) and (3.12.6). ∎

Constructing DIMSIMs in Nordsieck form, we usually assume that matrix $\mathbf{Q}$ has the form

$$
\mathbf{Q} = \mathbf{e}_1 \mathbf{q}^T, \quad \mathbf{q} = \begin{bmatrix} 1 \\ \widetilde{\mathbf{q}} \end{bmatrix},
$$

and then we are interested in expressing the coefficient matrix $\mathbf{G}$ in terms of $\mathbf{c}$ and $\mathbf{Q}$. This can be accomplished by "inversion" of formula (3.12.6). Let us partition matrices $\mathbf{G}$, $\mathbf{Q}$, and $\mathbf{E}_p$ as follows:

$$
\mathbf{G} = \left[\begin{array}{c} \mathbf{g}^T \\ \hline \widetilde{\mathbf{G}} \end{array} \right], \quad
\mathbf{Q} = \left[\begin{array}{c|c} 1 & \widetilde{\mathbf{q}}^T \\ \hline 0 & 0 \end{array} \right], \quad
\mathbf{E}_p = \left[\begin{array}{c|c} 1 & \mathbf{e}_{p-1}^T \\ \hline 0 & \mathbf{E}_{p-1} \end{array} \right].
$$

where $\mathbf{g}^T$ stands for the first row of $\mathbf{G}$ and $\mathbf{0}$ stands for the zero vector or matrix of appropriate dimension. We have the following result.

Theorem 3.12.3 (Butcher and Jackiewicz [69]) *Assume that $p = q = s$ and $r = s + 1$ in (3.11.8). Assume also that the components of the abscissa vector $\mathbf{c}$ are distinct. Then*

$$
\mathbf{g}^T = (\mathbf{e}_{p-1} - \widetilde{\mathbf{q}})\mathbf{C}_{p-1}^{-1} \tag{3.12.7}
$$

and

$$
\widetilde{\mathbf{G}} = \mathbf{E}_{p-1}\mathbf{C}_{p-1}^{-1}. \tag{3.12.8}
$$

Proof: It follows from (3.12.6) and the relation $\mathbf{C}_p \mathbf{K}_p = [\,\mathbf{0} \,|\, \mathbf{C}_{p-1}\,]$ that

$$
\left[\begin{array}{c|c} 1 & \widetilde{\mathbf{q}}^T \\ \hline 0 & 0 \end{array} \right] =
\left[\begin{array}{c|c} 1 & \mathbf{e}_{p-1}^T \\ \hline 0 & \mathbf{E}_{p-1} \end{array} \right] -
\left[\begin{array}{c|c} 0 & \mathbf{g}^T \mathbf{C}_{p-1} \\ \hline 0 & \widetilde{\mathbf{G}}\mathbf{C}_{p-1} \end{array} \right],
$$

and comparing the corresponding elements on both sides of this matrix equation, we obtain

$$\mathbf{e}_{p-1} - \mathbf{g}^T \mathbf{C}_{p-1} = \widetilde{\mathbf{q}}^T \quad \text{and} \quad \mathbf{E}_{p-1} - \widetilde{\mathbf{G}} \mathbf{C}_{p-1} = \mathbf{0}.$$

It follows from the assumptions of the theorem that the matrix $\mathbf{C}_{p-1}$ is invertible and the relations above are equivalent to (3.12.7) and (3.12.8), respectively. ∎

3.13 EXAMPLES OF DIMSIMS IN NORDSIECK FORM

In this section we list examples of types 1 and 2 DIMSIMs in Nordsieck form up to the order $p = 3$. These methods can be derived starting from representation (3.1.1) with coefficients given by (3.1.2) using the approach described in Section 3.11. Alternatively, we can use representation formulas (3.12.5) for matrix $\mathbf{P}$ and (3.12.7) and (3.12.8) for matrix $\mathbf{G}$ and then compute coefficient matrices $\mathbf{A}$ and $\mathbf{Q}$ so that the resulting method has some desirable stability properties. As in Sections 3.4 and 3.5, we aim at RK stability for explicit type 1 methods and A- and L-stability for implicit type 2 formulas. This leads again to systems of polynomial equations for unknown coefficients a_{ij} and $\widetilde{q}_i$ of the methods. These systems can be solved by symbolic manipulation packages if $1 \le p \le 4$ or using the approach based on a variant of the Fourier series method described in Section 3.8 and least-squares minimization discussed in Section 3.9 if $5 < p \le 8$. The coefficients of explicit methods for $5 < p \le 8$ are given by Butcher et al. [58].

Example 1. Type 1 method with $p = q = s = 1$, $r = 2$, and $\mathbf{c} = c_1 = 0$:

$$\left[\begin{array}{c|c} \mathbf{A} & \mathbf{P} \\ \hline \mathbf{G} & \mathbf{Q} \end{array} \right] = \left[\begin{array}{c|cc} 0 & 1 & 0 \\ \hline 1 & 1 & 0 \\ 1 & 0 & 0 \end{array} \right].$$

This method corresponds to the forward Euler formula

$$\left[\begin{array}{c|c} \mathbf{A} & \mathbf{U} \\ \hline \mathbf{B} & \mathbf{V} \end{array} \right] = \left[\begin{array}{c|c} 0 & 1 \\ \hline 1 & 1 \end{array} \right]$$

with $\mathbf{c} = c_1 = 0$.

Example 2. Type 2 method with $p = q = s = 1$, $r = 2$, and $\mathbf{c} = c_1 = 1$:

$$\left[\begin{array}{c|c} \mathbf{A} & \mathbf{P} \\ \hline \mathbf{G} & \mathbf{Q} \end{array} \right] = \left[\begin{array}{c|cc} 1 & 1 & 0 \\ \hline 1 & 1 & 0 \\ 1 & 0 & 0 \end{array} \right].$$

This method corresponds to the backward Euler formula

$$
\left[\begin{array}{c|c} \mathbf{A} & \mathbf{U} \\ \hline \mathbf{B} & \mathbf{V} \end{array}\right] = \left[\begin{array}{c|c} 1 & 1 \\ \hline 1 & 1 \end{array}\right]
$$

with $\mathbf{c} = c_1 = 1$.

Example 3. Type 1 method with $p = q = s = 2$, $r = 3$, and $\mathbf{c} = [0, 1]^T$:

$$
\left[\begin{array}{c|c} \mathbf{A} & \mathbf{P} \\ \hline \mathbf{G} & \mathbf{Q} \end{array}\right] = \left[\begin{array}{cc|ccc} 0 & 0 & 1 & 0 & 0 \\ 2 & 0 & 1 & -1 & \frac{1}{2} \\ \hline \frac{5}{4} & \frac{1}{4} & 1 & -\frac{1}{2} & \frac{1}{4} \\ 0 & 1 & 0 & 0 & 0 \\ -1 & 1 & 0 & 0 & 0 \end{array}\right].
$$

This method corresponds to the DIMSIM of type 1 of order $p = 2$ with $\mathbf{c} = [0, 1]^T$ given by Butcher [44] and also in Section 2.8.1. The details of the derivation of this method are given by Jackiewicz [178].

Example 4. Type 2 method with $p = q = s = 2$, $r = 3$, and $\mathbf{c} = [0, 1]^T$:

$$
\left[\begin{array}{c|c} \mathbf{A} & \mathbf{P} \\ \hline \mathbf{G} & \mathbf{Q} \end{array}\right] = \left[\begin{array}{cc|ccc} \frac{2-\sqrt{2}}{2} & 0 & 1 & \frac{\sqrt{2}-2}{2} & 0 \\ \frac{6+2\sqrt{2}}{7} & \frac{2-\sqrt{2}}{2} & 1 & \frac{3(\sqrt{2}-4)}{14} & \frac{\sqrt{2}-1}{2} \\ \hline \frac{73-34\sqrt{2}}{28} & \frac{2\sqrt{2}-1}{4} & 1 & \frac{10\sqrt{2}-19}{14} & \frac{3-2\sqrt{2}}{4} \\ 0 & 1 & 0 & 0 & 0 \\ -1 & 1 & 0 & 0 & 0 \end{array}\right].
$$

This method corresponds to DIMSIM of type 2 of order $p = 2$ with $\mathbf{c} = [0, 1]^T$ given by Butcher [44] and listed also in Section 2.8.2 for $\lambda = (2 - \sqrt{2})/2$. The details of the derivation of this method are given by Jackiewicz [178].

Example 5. Type 1 method with $p = q = s = 3$, $r = 4$, and $\mathbf{c} = [0, \frac{1}{2}, 1]^T$:

$$
\left[\begin{array}{c|c} \mathbf{A} & \mathbf{P} \\ \hline \mathbf{G} & \mathbf{Q} \end{array}\right] = \left[\begin{array}{ccc|cccc} 0 & 0 & 0 & 1 & 0 & 0 & 0 \\ 1 & 0 & 0 & 1 & -\frac{1}{2} & \frac{1}{8} & \frac{1}{48} \\ \frac{1}{4} & 1 & 0 & 1 & -\frac{1}{4} & 0 & \frac{1}{24} \\ \hline \frac{5}{4} & \frac{1}{3} & \frac{1}{6} & 1 & -\frac{3}{4} & \frac{1}{6} & \frac{1}{24} \\ 0 & 0 & 1 & 0 & 0 & 0 & 0 \\ 1 & -4 & 3 & 0 & 0 & 0 & 0 \\ 4 & -8 & 4 & 0 & 0 & 0 & 0 \end{array}\right].
$$

This method corresponds to the DIMSIM of type 1 of order $p = 3$ with $\mathbf{c} = [0, \frac{1}{2}, 1]^T$ given by Butcher and Jackiewicz [66] and also in Section 3.4.

Example 6. Type 2 method with $p = q = s = 3$, $r = 4$, and $\mathbf{c} = [-1, 0, 1]^T$:

$$
\mathbf{A} = \begin{bmatrix} 0.43586652 & 0 & 0 \\ 1.1720924 & 0.43586652 & 0 \\ 1.1074469 & 1.0003697 & 0.43586652 \end{bmatrix},
$$

$$
\mathbf{P} = \begin{bmatrix} 1 & -1.43586652 & 0.93586652 & -0.38459993 \\ 1 & -1.60795889 & 1.17209237 & -0.58604618 \\ 1 & -1.54368307 & 1.17158037 & -0.60499004 \end{bmatrix},
$$

$$
\mathbf{G} = \begin{bmatrix} 0.83581914 & 1.29513953 & 0.34910292 \\ 0 & 0 & 1 \\ 0.5 & -2 & 1.5 \\ 1 & -2 & 1 \end{bmatrix},
$$

$$
\mathbf{q}^T = \begin{bmatrix} 1 & -1.48006158 & 0.98671622 & -0.42579436 \end{bmatrix}.
$$

This method corresponds to the DIMSIM of type 2 of order $p = 3$ with $\mathbf{c} = [-1, 0, 1]^T$ given in Section 3.5.

3.14 REGULARITY PROPERTIES OF DIMSIMS

In this section we investigate if the GLM (3.1.1) with coefficients defined by (3.1.2) has the same set of finite asymptotic values as the underlying differential system (2.1.1). We follow the investigation of this question by Jackiewicz et al. [188]

Setting $z^{[n-1]} := \sum_{j=1}^{r} v_j y_j^{[n-1]}$, the GLM (3.1.1) can be written in the form

$$
\begin{aligned}
Y_i^{[n]} &= h \sum_{j=1}^{s} a_{ij} f(Y_j^{[n]}) + \sum_{j=1}^{r} u_{ij} y_j^{[n-1]}, \quad i = 1, 2, \ldots, s, \\
y_i^{[n]} &= h \sum_{j=1}^{s} b_{ij} f(Y_j^{[n]}) + z^{[n-1]}, \qquad\qquad i = 1, 2, \ldots, r,
\end{aligned} \tag{3.14.1}
$$

$n = 0, 1, \ldots, N$. This formulation reflects the fact that to compute the external stages $y_i^{[n]}$, $i = 1, 2, \ldots, r$, the linear combination $z^{[n-1]}$ of $y_j^{[n-1]}$ needs to be propagated from the current step, t_{n-1}, to the next step, t_n.

Following Hairer et al. [141] and Iserles [171], we denote by $\mathcal{F}$ the set of all possible bounded asymptotic values of (2.1.1), that is,

$$\mathcal{F} = \{ y \in \mathbb{R}^m : f(y) = 0 \},$$

and define

$$\widetilde{\mathcal{F}} = \{ (y, \dots, y) \in \mathbb{R}^{(s+r)m} : y \in \mathcal{F} \}.$$

Then we consider the system

$$\widehat{Y}_i = h \sum_{j=1}^{s} a_{ij} f(\widehat{Y}_j) + \sum_{j=1}^{r} u_{ij} \widehat{y}_j, \quad i = 1, 2, \dots, s,$$

$$\widehat{y}_i = h \sum_{j=1}^{s} b_{ij} f(\widehat{Y}_j) + \widehat{z}, \qquad i = 1, 2, \dots, r, \tag{3.14.2}$$

with $\widehat{z} = \sum_{j=1}^{r} v_j \widehat{y}_j$, which is obtained formally from (3.14.1) by passing with n to infinity and setting

$$\widehat{Y}_i = \lim_{n \to \infty} Y_i^{[n]}, \quad \widehat{y}_i = \lim_{n \to \infty} y_i^{[n]},$$

assuming that these limits exist. Also define the sets

$$\mathcal{F}_h = \left\{ \widehat{z} = \sum_{j=1}^{r} v_j \widehat{y}_j : \ (\widehat{y}_1, \dots, \widehat{y}_r, \widehat{Y}_1, \dots, \widehat{Y}_s) \text{ satisfy (3.14.2)} \right\}$$

and

$$\widetilde{\mathcal{F}}_h = \left\{ (\widehat{y}_1, \dots, \widehat{y}_r, \widehat{Y}_1, \dots, \widehat{Y}_s) : \ (\widehat{y}_1, \dots, \widehat{y}_r, \widehat{Y}_1, \dots, \widehat{Y}_s) \text{ satisfy (3.14.2)} \right\}.$$

It will be demonstrated that $\mathcal{F} \subset \mathcal{F}_h$ and $\widetilde{\mathcal{F}} \subset \widetilde{\mathcal{F}}_h$ for any $h > 0$, but in general, sets $\mathcal{F}_h - \mathcal{F}$ and $\widetilde{\mathcal{F}}_h - \widetilde{\mathcal{F}}$ may be nonempty. GLM (3.14.1) is said to be regular if and only if $\mathcal{F}_h = \mathcal{F}$ and strongly regular if and only if $\widetilde{\mathcal{F}}_h = \widetilde{\mathcal{F}}$ for any $h > 0$.

In the case of DIMSIMs, the motivation behind these definitions is as follows. If $y(t) \to y^*$ as $t \to \infty$, then, necessarily, $y^* \in \mathcal{F}$, and if y is sufficiently smooth, we also have $y^{(i)}(t) \to 0$ as $t \to \infty$ for $i = 1, 2, \dots, p$. Assume now that $y_i^{[n]} \to \widehat{y}_i$, $i = 1, 2, \dots, r$, and $Y_i^{[n]} \to \widehat{Y}_i$, $i = 1, 2, \dots, s$, as $n \to \infty$. Since $y^{[n]}$ approximates $\sum_{k=0}^{p} q_{ik} h^k y^{(k)}(t_n)$, $Y_i^{[n]}$ approximates $y(t_{n-1} + c_i h)$, and $q_{i0} = 1$ (this follows from $\mathbf{U} = \mathbf{I}$), it would be desirable if all these limits would equal y^*, which is the limit of the solution $y(t)$ to (2.1.1) as $t \to \infty$. This is obviously the case if $\widetilde{\mathcal{F}}_h = \widetilde{\mathcal{F}}$. Moreover, since for DIMSIMs we have $\mathbf{v}^T \mathbf{e} = 1$, it follows that if $y_i^{[n]} \to y^*$, then $z^{[n]} \to y^*$ as $n \to \infty$, which motivates the weaker requirement that $\mathcal{F}_h = \mathcal{F}$.

Setting $r = 1$, we have $\mathbf{U} = \mathbf{e}$, $\mathbf{B} = \mathbf{b}^T$, $\mathbf{V} = \mathbf{v} = 1$, and (3.14.1) reduces to the RK method (2.1.3). Since $z^{[n-1]} = y_1^{[n-1]} = y_{n-1}$, regularity now means that the system

$$\widehat{Y}_i = \widehat{y} + h \sum_{j=1}^{s} a_{ij} f(\widehat{Y}_j), \quad i = 1, 2, \ldots, s,$$

$$\sum_{j=1}^{s} b_j f(\widehat{Y}_j) = 0$$

(3.14.3)

has only solutions $(\widehat{y}, \widehat{Y}_1, \ldots, \widehat{Y}_s)$ such that $f(\widehat{y}) = 0$. Consequently, the RK method (2.1.3) is regular in the sense introduced in [141], [171], and the definition of regularity of DIMSIMs is a generalization of the concept of regularity for RK methods. Strong regularity means instead that the system (3.14.3) has only the solution of the form $(\widehat{y}, \ldots, \widehat{y}) \in \mathbb{R}^{(s+1)m}$ such that $f(\widehat{y}) = 0$, which is a stronger requirement than regularity as introduced by Hairer et al. [141] and Iserles [171]. However, it was proved by Jackiewicz et al. [187] that these two concepts are equivalent for a class of essentially one stage RK methods with $s = 2$.

To investigate regularity properties of DIMSIMs, we use vector formulation (3.1.1) which will be rewritten in the form

$$K^{[n]} = f\big(h(\mathbf{A} \otimes \mathbf{I})K^{[n]} + (\mathbf{U} \otimes \mathbf{I})y^{[n-1]}\big),$$

$$y^{[n]} = h(\mathbf{B} \otimes \mathbf{I})K^{[n]} + (\mathbf{V} \otimes \mathbf{I})y^{[n]},$$

(3.14.4)

$n = 1, 2, \ldots, N$. This is analogous to the k-notation for RK schemes (see [109]). Consider also the system

$$\widehat{K} = f\big(h(\mathbf{A} \otimes \mathbf{I})\widehat{K} + (\mathbf{U} \otimes \mathbf{I})\widehat{y}\big),$$

$$\widehat{y} = h(\mathbf{B} \otimes \mathbf{I})\widehat{K} + (\mathbf{V} \otimes \mathbf{I})\widehat{y},$$

(3.14.5)

which corresponds to system (3.14.2) in k-notation with $\widehat{y} = [\widehat{y}_1, \ldots, \widehat{y}_r]^T$. Let $\mathbf{J}(y^*) = (\partial f/\partial y)(y^*)$ be the Jacobian matrix of $f(y)$ and define the map

$$F(\widehat{y}) := h(\mathbf{B} \otimes \mathbf{I})\widehat{K} + (\mathbf{V} \otimes \mathbf{I})\widehat{y},$$

where $\widehat{K}$ is defined implicitly by (3.14.5). Denote by $\sigma(A)$ the spectrum of matrix A and by $\mathcal{A}$ the region of absolute stability of method (3.14.4) (compare Definition 2.6.2). Denote by $\overline{\mathcal{A}}$ the closure of $\mathcal{A}$. We have the following result, which is an analog of Theorem 3 in [171].

Theorem 3.14.1 (Jackiewicz at al. [188]) *For any step size $h > 0$ we have $\mathcal{F} \subset \mathcal{F}_h$ and $\widetilde{\mathcal{F}} \subset \widetilde{\mathcal{F}}_h$. Moreover, y^* is attractive as an asymptotic value of $F([y^*, \ldots, y^*]^T)$ if $\sigma(h\mathbf{J}(y^*)) \subset \mathcal{A}$ and only if $\sigma(h\mathbf{J}(y^*)) \subset \overline{\mathcal{A}}$.*

Proof: The inclusions $\mathcal{F} \subset \mathcal{F}_h$ and $\widetilde{\mathcal{F}} \subset \widetilde{\mathcal{F}}_h$ follow from the fact that since $\mathbf{U}\mathbf{e} = \mathbf{e}$ and $\mathbf{V}\mathbf{e} = \mathbf{e}$, system (3.14.5) obviously has the solution $\widehat{y} = [y^*, \ldots, y^*]^T$ and $\widehat{K} = [0, \ldots, 0]^T$ if $f(y^*) = 0$ and that $\widehat{z} = (\mathbf{v}^T \otimes \mathbf{I})\widehat{y} = y^*$.

To investigate if $y^* \in \mathcal{F}$ is attractive, we first establish the relationship between the spectrum of $\mathbf{J} = \mathbf{J}(y^*)$ and the spectrum of $\partial F/\partial \widehat{y} = (\partial F/\partial \widehat{y})(y^*, \ldots, y^*)$. Let $\partial \widehat{K}/\partial \widehat{y} = (\partial \widehat{K}/\partial \widehat{y})(y^*, \ldots, y^*)$. It can be verified that

$$\frac{\partial \widehat{K}}{\partial \widehat{y}} = h(\mathbf{A} \otimes \mathbf{J})\frac{\partial \widehat{K}}{\partial \widehat{y}} + \mathbf{U} \otimes \mathbf{J}, \quad \frac{\partial F}{\partial \widehat{y}} = h(\mathbf{B} \otimes \mathbf{I})\frac{\partial \widehat{K}}{\partial \widehat{y}} + \mathbf{V} \otimes \mathbf{I}.$$

Computing $\partial \widehat{K}/\partial \widehat{y}$ from the first of the equations above and substituting the result into the second equation, we obtain

$$\frac{\partial F}{\partial \widehat{y}} = h(\mathbf{B} \otimes \mathbf{I})\big(\mathbf{I} - h(\mathbf{A} \otimes \mathbf{J})\big)^{-1}(\mathbf{U} \otimes \mathbf{J}) + \mathbf{V} \otimes \mathbf{I},$$

where $\mathbf{I}$ stand for the identity matrices of appropriate dimensions. Assume that $\mathbf{J}x = \lambda x$. We will show that for any $z \in \mathbb{R}^m$,

$$\frac{\partial F}{\partial \widehat{y}}(z \otimes x) = \big(\mathbf{M}(h\lambda) \otimes \mathbf{I}\big)(z \otimes x), \tag{3.14.6}$$

where $\mathbf{M}(z) = \mathbf{V} + z\mathbf{B}(\mathbf{I} - z\mathbf{A})^{-1}\mathbf{U}$ is the stability matrix of the method (3.14.4). We have

$$\frac{\partial F}{\partial \widehat{y}}(z \otimes x) = h\lambda(\mathbf{B} \otimes \mathbf{I})\big(\mathbf{I} - h(\mathbf{A} \otimes \mathbf{J})\big)^{-1}(\mathbf{U}z \otimes \mathbf{e}) + \mathbf{V}z \otimes \mathbf{e}.$$

It can be verified that

$$\big(\mathbf{I} - h(\mathbf{A} \otimes \mathbf{J})\big)^{-1}(\mathbf{U}z \otimes x) = (\mathbf{I} - h\lambda\mathbf{A})^{-1}\mathbf{U}z \otimes x.$$

Hence,

$$\begin{aligned}
\frac{\partial F}{\partial \widehat{y}}(z \otimes x) &= h\lambda(\mathbf{B} \otimes \mathbf{I})\big((\mathbf{I} - h\lambda\mathbf{A})^{-1}\mathbf{U}z \otimes x\big) + \mathbf{V} \otimes \mathbf{e} \\
&= h\lambda\mathbf{B}(\mathbf{I} - h\lambda\mathbf{A})^{-1}\mathbf{U}z \otimes x + \mathbf{V} \otimes x \\
&= \big(h\lambda\mathbf{B}(\mathbf{I} - h\lambda\mathbf{A})^{-1}\mathbf{U} \otimes \mathbf{I} + \mathbf{V} \otimes \mathbf{I}\big)(z \otimes x) \\
&= \big(\mathbf{M}(h\lambda) \otimes \mathbf{I}\big)(z \otimes x),
\end{aligned}$$

which proves (3.14.6). Assume now that $\mathbf{M}(h\lambda)z = \mu z$. Then

$$\frac{\partial F}{\partial \widehat{y}}(z \otimes x) = \mu(z \otimes x)$$

and assuming that the eigenvalues of $\mathbf{J}$ are distinct and that the eigenvalues of $\mathbf{M}(h\lambda)$, $\lambda \in \sigma(\mathbf{J})$, are distinct, it follows that

$$\sigma\left(\frac{\partial F}{\partial \widehat{y}}\right) = \bigcup_{\lambda \in \sigma(\mathbf{J})} \sigma\big(\mathbf{M}(h\lambda)\big).$$

Using the perturbation arguments, it follows that these sets are also equal without the foregoing assumptions on the eigenvalues of $\mathbf{J}$ and $\mathbf{M}(h\lambda)$. The theorem follows from this relation between the spectra of $\partial F/\partial \widehat{y}$ and $\mathbf{J}$ and the definition of the region of absolute stability $\mathcal{A}$.   ∎

The next result provides the conditions for strong regularity of the method (3.14.4). We have the following theorem.

Theorem 3.14.2 (Jackiewicz et al. [188]) *Assume that $p \geq 1$, $q \geq 1$, and that*

$$(\mathbf{A} + \mathbf{U}\mathbf{B})x = \xi e, \quad \xi \in \mathbb{R}, \tag{3.14.7}$$

for any $x \in S$, where $S = \{x : \mathbf{v}^T\mathbf{B}x = 0\}$. Then method (3.14.4) is strongly regular.

Proof : Observe first that it follows from the definition of the tensor product that (3.14.7) is equivalent to

$$\big((\mathbf{A} + \mathbf{U}\mathbf{B}) \otimes I\big)x = e \otimes \eta, \quad \eta \in \mathbb{R}^m,$$

for any x such that $(\mathbf{v}^T\mathbf{B} \otimes I)x = 0$. Multiplying the second equation of (3.14.5) by $\mathbf{v}^T \otimes I$, it follows that

$$(\mathbf{v}^T \otimes I)(\mathbf{B} \otimes I)\widehat{K} = (\mathbf{v}^T\mathbf{B} \otimes I)\widehat{K} = 0,$$

and setting $\vartheta = (\mathbf{v}^T \otimes I)\widehat{y} \in \mathbb{R}^m$, we obtain

$$\widehat{y} = h(\mathbf{B} \otimes I)\widehat{K} + e \otimes \vartheta.$$

Substituting this relation into the first equation of (3.14.5), we get

$$\widehat{K} = f\big(h((\mathbf{A} + \mathbf{U}\mathbf{B}) \otimes I)\widehat{K} + e \otimes \vartheta\big).$$

It follows from condition (3.14.7) that $\widehat{K}$ has the form $\widehat{K} = e \otimes \rho$ for some $\rho \in \mathbb{R}^m$. Denote by $\widehat{K}_{i,j}$ the jth component of the vector $\widehat{K}_i \in \mathbb{R}^m$ and set $\widehat{K}_{:,j} = [\widehat{K}_{1,j}, \ldots, \widehat{K}_{s,j}]^T \in \mathbb{R}^s$. Then $\widehat{K}_{:,j} = \rho_j e$, where ρ_j is the jth component of the vector ρ. Since for DIMSIM (3.14.4) we have $q_0 = e$, $\mathbf{V} = e\mathbf{v}^T$, $\mathbf{v}^T e = 1$, it follows from the consistency condition $\mathbf{B}e + \mathbf{V}q_1 = q_0 + q_1$ that $\mathbf{v}^T\mathbf{B}e = 1$. Hence,

$$\rho_j = \rho_j \mathbf{v}^T\mathbf{B}e = \mathbf{v}^T\mathbf{B}\widehat{K}_{:,j} = 0.$$

This implies that $\widehat{K}_{:,j} = 0$, $\widehat{K} = 0$, $\widehat{Y} = \widehat{y} = \mathbf{e} \otimes \vartheta$, and $f(\widehat{y}) = 0$, which proves the strong regularity of (3.14.4).   ∎

Set $\mathbf{Q} = \mathbf{A} + \mathbf{UB}$ and denote by $r_k(\mathbf{Q})$ the kth row of $\mathbf{Q}$. Then it is easy to check that condition (3.14.7) is equivalent to the requirement that $r_k(\mathbf{Q}) - r_l(\mathbf{Q})$ is proportional to vector $\mathbf{v}^T\mathbf{B}$ for any pair of indices k and l such that $k \neq l$. In what follows we show that the necessary condition for both regularity and strong regularity of (3.14.4) is the existence of two distinct indices k and l such that $r_k(\mathbf{Q}) - r_l(\mathbf{Q})$ is proportional to $\mathbf{v}^T\mathbf{B}$.

Consider the system

$$h(\mathbf{A} \otimes \mathbf{I})\widehat{K} + (\mathbf{U} \otimes \mathbf{I})\widehat{y} = \widehat{Y},$$

$$((\mathbf{I} - \mathbf{V}) \otimes \mathbf{I})\widehat{y} - h(\mathbf{B} \otimes \mathbf{I})\widehat{K} = 0,$$

(3.14.8)

where

$$\widehat{y} = \begin{bmatrix} \widehat{y}_1 & \cdots & \widehat{y}_s \end{bmatrix}^T, \quad \widehat{Y} = \begin{bmatrix} \widehat{Y}_1 & \cdots & \widehat{Y}_s \end{bmatrix}^T, \quad \widehat{K} = \begin{bmatrix} \widehat{K}_1 & \cdots & \widehat{K}_s \end{bmatrix}^T,$$

and $\widehat{y}_i$, $\widehat{Y}_i$, and $\widehat{K}_i$ are vectors from $\mathbb{R}^m$. The solution $(\widehat{y}, \widehat{Y}, \widehat{K})$ to (3.14.8) is said to be admissible if $\widehat{K}_i = \widehat{K}_j$ whenever $\widehat{Y}_i = \widehat{Y}_j$. We have the following result.

Lemma 3.14.3 (Jackiewicz et al. [188]) *Method (3.14.4) is regular if and only if for every $m \geq 1$ and $h > 0$ every admissible solution to (3.14.8) admits an index ν such that $\widehat{K}_\nu = 0$ and $\widehat{Y}_\nu = \widehat{z}$, where $\widehat{z} = \sum_{j=1}^r v_j\widehat{y}_j$.*

Proof: The sufficiency is obvious. The necessity follows from the fact that we can construct a continuous function $f(z)$ such that $f(\widehat{z}) \neq 0$ and $\widehat{K}_i = f(\widehat{Y}_i)$, $i = 1, 2, \ldots, s$, whenever the assumptions of the lemma do not hold.   ∎

The next lemma provides a characterization of strong regularity. The straightforward proof of this result is omitted.

Lemma 3.14.4 (Jackiewicz et al. [188]) *Method (3.14.4) is strongly regular if and only if for every $m \geq 1$ and $h > 0$, system (3.14.8) has an admissible solution only if $\widehat{Y}_i = \widehat{Y}_j$ for all $i, j = 1, 2, \ldots, s$, and for all such solutions we have $\widehat{y}_i = \widehat{y}_j$, $i, j = 1, 2, \ldots, r$, and $\widehat{K}_i = 0$, $i = 1, 2, \ldots, s$.*

For the remaining part of this section we consider only the scalar case $m = 1$ to simplify the presentation of the results about regularity and strong regularity of DIMSIMs. These results can easily be generalized to the vector case $m > 1$ by using the tensor product notation.

We introduce first the generalization to DIMSIMs of the concept of essentially one stage (EOS) RK methods which was introduced by Hairer et al. [141]. A DIMSIM (3.14.4) is said to be EOS if there exists an index $\nu \in \{1, 2, \ldots, s\}$ such that

$$r_\nu(\mathbf{U})\mathbf{B} = \mathbf{v}^T\mathbf{B} = \mathbf{e}_\nu^T, \quad r_\nu(\mathbf{A}) = c_\nu\mathbf{e}_\nu^T,$$

for some $c_\nu \in \mathbb{R}$. It is easy to verify that whenever $r_\nu(\mathbf{U})y^{[0]} = \mathbf{v}^T y^{[0]}$, for an EOS method the quantity $z^{[n]}$ is propagated by using one stage according to the formulas

$$Y_\nu^{[n]} = h c_\nu K_\nu^{[n]} + z^{[n-1]}, \quad z^{[n]} = h K_\nu^{[n]} + z^{[n-1]},$$

whereas the other stages are nonessential. For technical reasons we must also consider a wider class of methods for which there exists an index $\nu \in \{1, 2, \ldots, s\}$ such that

$$\mathbf{v}^T \mathbf{B} = \mathbf{e}_\nu^T, \quad r_\nu(\mathbf{U})\mathbf{B} + r_\nu(\mathbf{A}) = \gamma_\nu \mathbf{e}_\nu^T,$$

for some $\gamma_\nu \in \mathbb{R}$. Observe that as $n \to \infty$, such DIMSIMs have an EOS structure, and for this reason they will be called asymptotically essentially one stage (AEOS) methods. Clearly, the EOS method is also AEOS. We have the following lemma.

Lemma 3.14.5 (Jackiewicz et al. [188]) *An AEOS method is regular.*

Proof: Since method (3.14.4) is AEOS, multiplying the second equation of (3.14.8) by $r_\nu(\mathbf{U})$ and then by $\mathbf{v}^T$, we get $r_\nu(\mathbf{U})\widehat{y} = h r_\nu(\mathbf{U})\mathbf{B}\widehat{K} + \widehat{z}$, where $\widehat{z} = \mathbf{v}^T \widehat{y}$, and $\widehat{K} = 0$. Substituting this into the first equation of (3.14.8), we obtain

$$\widehat{Y}_\nu = h\big(r_\nu(\mathbf{A}) + r_\nu(\mathbf{U})\mathbf{B}\big)\widehat{K} + \widehat{z} = \widehat{z},$$

and the conclusion follows from Lemma 3.14.3. ∎

AEOS methods are clearly exceptional and not very interesting. The next theorem gives a necessary condition for the regularity of DIMSIMs that are not AEOS.

Theorem 3.14.6 (Jackiewicz et al. [188]) *Assume that method (3.14.4) is not AEOS and regular. Then there exists distinct k and l, $1 \le k, l \le s$, such that $r_k(\mathbf{Q}) - r_l(\mathbf{Q})$ is proportional to $\mathbf{v}^T \mathbf{B}$, where $\mathbf{Q} = \mathbf{A} + \mathbf{UB}$, and $r_k(\mathbf{Q})$ stands for the kth row of $\mathbf{Q}$.*

Proof: The proof of this theorem is based on the ideas of the proof of Theorem 3 in Hairer et al. [141]. Assume first that $\mathbf{B}^T \mathbf{v} \ne \mathbf{e}_k$ for all $k = 1, 2, \ldots, s$. Then there exists a set $H = \{\xi \in \mathbb{R}^m : \sigma(l)\xi_l > 0, \ l = 1, 2, \ldots, s\}$ with $\sigma(l) = \pm 1$ such that $H^* = H \cap (\mathbf{B}^T \mathbf{v})^\perp \ne \emptyset$. Here, $(\mathbf{B}^T \mathbf{v})^\perp$ is the orthogonal complement of $\mathbf{B}^T \mathbf{v}$ in $\mathbb{R}^s$. We will prove that

$$h\mathbf{Q}H^* \subset \bigcup_{i \ne j} \{Y \in \mathbb{R}^s : Y_i = Y_j\}, \tag{3.14.9}$$

where $\mathbf{Q}H^* = \{\mathbf{Q}\xi : \xi \in H^*\}$. Assume to the contrary that (3.14.9) is false. Then there exists $K \in H^*$ such that $Y = h\mathbf{Q}K$ has pairwise distinct components. Since $K \in (\mathbf{B}^T \mathbf{v})^\perp$, it follows that $y = h\mathbf{B}K$ is a solution to

$(\mathbf{I} - \mathbf{V})y = h\mathbf{B}K$. Hence, $Y = h\mathbf{A}K + \mathbf{U}y = h\mathbf{Q}K$ and we can conclude that $Y_i \neq Y_j$ for all distinct i and j. Thus, (y, Y, K) is an admissible solution to (3.14.8). Moreover, since $K \in H^*$, it follows from Lemma 3.14.3 that method (3.14.4) is not regular, contradicting our assumption and thus proving (3.14.9).

Since (3.14.9) is valid, it follows from the convexity of $\mathbf{Q}H^*$ that there exist distinct k and l such that $h\mathbf{Q}H^* \subset \{Y \in \mathbb{R}^s : Y_k = Y_l\}$. Similarly as in Hairer et al. [141], this implies that $r_k(\mathbf{Q}) - r_l(\mathbf{Q})$ is proportional to $\mathbf{v}^T \mathbf{B}$.

Assume next that $\mathbf{B}^T \mathbf{v} = \mathbf{e}_1$. Since the method (3.14.4) is not AEOS, it follows that $r_1(\mathbf{Q}) \neq \gamma \mathbf{e}_1$ and $H_1 = \{\xi \in \mathbb{R}^s : \xi_1 = 0, \ r_1(\mathbf{Q})\xi > 0\} \neq \emptyset$. Moreover, we can find a set $H_2 = \{\xi \in \mathbb{R}^s : \xi_1 = 0, \ (-1)^{\sigma(i)}\xi_i > 0, \ i = 2, 3, \dots, s\}$ with $\sigma(i) = \pm 1$ such that $H = H_1 \cap H_2 \neq \emptyset$. Similarly as before, we can prove that

$$h\mathbf{Q}H \subset \bigcup_{i \neq j} \{Y \in \mathbb{R}^s : Y_i = Y_j\}. \tag{3.14.10}$$

Indeed, assuming that (3.14.10) is false, there exists $K \in H$ such that $Y = h\mathbf{Q}K$ has pairwise distinct components. Since $K \in \mathbf{e}_1^\perp$ (i.e., $K_1 = 0$), it follows that $y = h\mathbf{B}K$ is a solution to the system $(\mathbf{I} - \mathbf{V})y = h\mathbf{B}K$. Hence, $Y = h\mathbf{A}K + \mathbf{U}y = h\mathbf{Q}K$ and it follows that (y, Y, K) is an admissible solution to (3.14.8), which contradicts the regularity of (3.14.4) since

$$K_1 = 0, \quad Y_1 = hr_1(\mathbf{Q})K > 0, \quad \mathbf{v}^T y = h\mathbf{v}^T \mathbf{B}K = 0,$$

and $K_\nu \neq 0$, $\nu = 2, 3, \dots, s$.

This establishes (3.14.10), and the rest of the proof is the same as in the case of $\mathbf{B}^T \mathbf{v}$ not being a coordinate vector.   ∎

This theorem gives, in particular, the necessary condition for strong regularity of DIMSIMs (3.14.4). We formulate this as the following corollary.

Corollary 3.14.7 (Jackiewicz et al. [188]) *Assume that method (3.14.4) is strongly regular. Then there exist distinct k and l, $1 \leq k, l \leq s$, such that $r_k(\mathbf{Q}) - r_l(\mathbf{Q})$ is proportional to $\mathbf{v}^T \mathbf{B}$.*

It is also possible to prove this result in a much simpler way than the proof of Theorem 3.14.6. This direct proof is given by Jackiewicz et al. [188].

The next theorem illustrates that it is possible to reduce the question of regularity or strong regularity of the original method (3.14.4) with s internal stages and r external stages to the question of regularity or strong regularity of the folded method with $s - 1$ internal stages and $r - 1$ external stages. Following Hairer et al. [141], the coefficient matrices of the folded method are defined by

$$
\left[
\begin{array}{c|c}
\mathbf{A}^* & \mathbf{U}^* \\
\hline
\mathbf{B}^* & \mathbf{V}^*
\end{array}
\right]
=
\left[
\begin{array}{c|c}
\left[\begin{array}{cc} \mathbf{I} & 0 \end{array}\right] \mathbf{A} \left[\begin{array}{c} \mathbf{I} \\ \mathbf{e}_1^T \end{array}\right] & \left[\begin{array}{cc} \mathbf{I} & 0 \end{array}\right] \mathbf{U} \\[4mm]
\hline
\mathbf{B} \left[\begin{array}{c} \mathbf{I} \\ \mathbf{e}_1^T \end{array}\right] & \mathbf{V}
\end{array}
\right],
\tag{3.14.11}
$$

where $\mathbf{I}$ stands for the identity matrix of dimension $s - 1$. We have the following theorem.

Theorem 3.14.8 (Jackiewicz et al. [188]) *Assume that $r_k(\mathbf{Q}) - r_l(\mathbf{Q}) = \xi \mathbf{v}^T \mathbf{B}$ for some $\xi \in \mathbb{R}$ and distinct k and l, $1 \le k, l \le s$. After reordering the coefficient matrices so that $k \to 1$ and $l \to s$, we have that method (3.14.4) is regular or strongly regular if and only if folded method (3.14.11) is regular or strongly regular.*

Proof: We can assume, without loss of generality, that $k = 1$ and $l = s$. Consider system (3.14.8), whose solution is given by $\widehat{y} = h\mathbf{B}\widehat{K} + \lambda\mathbf{e}$, $\lambda \in \mathbb{R}$, where $\widehat{K}$ satisfies the system

$$
\begin{aligned}
\mathbf{v}^T \mathbf{B}\widehat{K} &= 0, \\
h\mathbf{Q}\widehat{K} &= \widehat{Y} - \lambda\mathbf{e}.
\end{aligned}
\tag{3.14.12}
$$

Consider also the system corresponding to folded method (3.14.11)

$$
\begin{aligned}
h\mathbf{A}^* K^* + \mathbf{U}^* y^* &= Y^*, \\
(\mathbf{I} - \mathbf{V}^*) y^* - h\mathbf{B}^* K^* &= 0.
\end{aligned}
\tag{3.14.13}
$$

Using the same arguments as in the case of the original method (3.14.4), the solution to (3.14.13) is $y^* = h\mathbf{B}^* K^* + \lambda^*\mathbf{e}$, for some $\lambda^* \in \mathbb{R}$, where K^* satisfies the system

$$
\begin{aligned}
\mathbf{v}^T \mathbf{B}^* K^* &= 0, \\
h\mathbf{Q}^* K^* &= Y^* - \lambda^*\mathbf{e},
\end{aligned}
\tag{3.14.14}
$$

with

$$
\begin{aligned}
\mathbf{Q}^* &= \mathbf{A}^* + \mathbf{U}^* \mathbf{B}^* = \left[\begin{array}{cc} \mathbf{I} & 0 \end{array}\right] \mathbf{A} \left[\begin{array}{c} \mathbf{I} \\ \mathbf{e}_1^T \end{array}\right] + \left[\begin{array}{cc} \mathbf{I} & 0 \end{array}\right] \mathbf{U}\mathbf{B} \left[\begin{array}{c} \mathbf{I} \\ \mathbf{e}_1^T \end{array}\right] \\[3mm]
&= \left[\begin{array}{cc} \mathbf{I} & 0 \end{array}\right] (\mathbf{A} + \mathbf{U}\mathbf{B}) \left[\begin{array}{c} \mathbf{I} \\ \mathbf{e}_1^T \end{array}\right] = \left[\begin{array}{cc} \mathbf{I} & 0 \end{array}\right] \mathbf{Q} \left[\begin{array}{c} \mathbf{I} \\ \mathbf{e}_1^T \end{array}\right].
\end{aligned}
$$

Since $r_1(\mathbf{Q}) - r_s(\mathbf{Q}) = \xi \mathbf{v}^T \mathbf{B}$ it follows that system (3.14.12) can have solutions only if $\widehat{Y}_1 = \widehat{Y}_s$. Setting $\widehat{K}_1 = \widehat{K}_s$, $K_i^* = \widehat{K}_i$, $i = 1, 2, \ldots, s - 1$, and $Y_i^* = \widehat{Y}_i$, $i = 1, 2, \ldots, s - 1$, it is easy to see that system (3.14.12) is equivalent to (3.14.14). Thus, Lemmas 3.14.3 and 3.14.4 imply that the original method (3.14.4) is regular or strongly regular if and only if folded method (3.14.11) is regular or strongly regular, and the proof is complete. ■

Theorems 3.14.2 and 3.14.6 provide a complete characterization of regular and strongly regular DIMSIMs (3.14.4) with two stages. We formulate this as the following corollary.

Corollary 3.14.9 (Jackiewicz et al. [188]) *Method (3.14.4) with $s = 2$ $p \geq 1$ and $q \geq 1$, which is not AEOS, is regular if and only if*

$$r_1(\mathbf{Q}) - r_2(\mathbf{Q}) = \xi \mathbf{v}^T \mathbf{B} \tag{3.14.15}$$

for some $\xi \in \mathbb{R}$. Condition (3.14.15) is also sufficient and necessary for strong regularity of such methods.

Corollary 3.14.9 was used by Jackiewicz et al. [188] to describe DIMSIMs (3.14.4) of type 1, 2, 3, and 4 with $s = r = q = 2$ and $p \geq 2$, and we refer readers to that paper for details.

CHAPTER 4

IMPLEMENTATION OF DIMSIMS

4.1 VARIABLE STEP SIZE FORMULATION OF DIMSIMS

In this chapter we discuss various practical issues related to the implementation of DIMSIMs, such as variable step size formulation, the choice of initial step size and initial order of integration, error propagation and estimation of the principal part of the local discretization errors for small and large step sizes, construction of continuous interpolants, step size and order changing strategies using the Nordsieck representation derived in Section 3.11, updating vectors of external approximations, step-control stability, and the solution of nonlinear systems of equations resulting in implicit formulas by some variants of Newton method. We also describe the codes for nonstiff and stiff differential systems, which are based on DIMSIMs of types 1 and 2, respectively, constructed in Chapter 3. We conclude the chapter with the results of numerical experiments with these codes and comparison with codes from the Matlab ODE suite [263].

On a nonuniform grid

$$t_0 < t_1 < \cdots < t_N, \quad t_N \geq T,$$

General Linear Methods for Ordinary Differential Equations. By Zdzisław Jackiewicz   **201**
Copyright © 2009 John Wiley & Sons, Inc.

the Nordsieck representation of DIMSIMs (3.11.8) takes the form

$$Y^{[n]} = h_n(\mathbf{A} \otimes \mathbf{I})F(Y^{[n]}) + (\mathbf{P} \otimes \mathbf{I})\widehat{z}^{[n-1]},$$

$$z^{[n]} = h_n(\mathbf{G} \otimes \mathbf{I})F(Y^{[n]}) + (\mathbf{Q} \otimes \mathbf{I})\widehat{z}^{[n-1]},$$

(4.1.1)

$n = 1, 2, \ldots, N$, where $h_n = t_n - t_{n-1}$, and $\widehat{z}^{[n-1]}$ is an approximation of order p to the vector

$$\widehat{z}(t_{n-1}, h_n) = \begin{bmatrix} y(t_{n-1}) \\ h_n y'(t_{n-1}) \\ \vdots \\ h_n^p y^{(p)}(t_{n-1}) \end{bmatrix}.$$

(4.1.2)

Since

$$\begin{bmatrix} y(t_{n-1}) \\ h_n y'(t_{n-1}) \\ \vdots \\ h_n^p y^{(p)}(t_{n-1}) \end{bmatrix} = \begin{bmatrix} \mathbf{I} & \mathbf{0} & \cdots & \mathbf{0} \\ \mathbf{0} & \delta_n \mathbf{I} & \cdots & \mathbf{0} \\ \vdots & \vdots & \ddots & \vdots \\ \mathbf{0} & \mathbf{0} & \cdots & \delta_n^p \mathbf{I} \end{bmatrix} \begin{bmatrix} y(t_{n-1}) \\ h_{n-1} y'(t_{n-1}) \\ \vdots \\ h_{n-1}^p y^{(p)}(t_{n-1}) \end{bmatrix},$$

where $\delta_n = h_n/h_{n-1}$, the vector $\widehat{z}^{[n-1]}$ appearing in (4.1.1) is defined by the formula

$$\widehat{z}^{[n-1]} = (\mathbf{D}(\delta_n) \otimes \mathbf{I})z^{[n-1]},$$

(4.1.3)

where $\mathbf{D}(\delta)$ is the rescaling matrix given by

$$\mathbf{D}(\delta) = \mathrm{diag}(1, \delta, \ldots, \delta^p).$$

(4.1.4)

It follows from (4.1.1) and (4.1.3) that the zero-stability properties of method (4.1.1) are determined by the eigenvalues of the matrix $\mathbf{QD}(\delta_n) \otimes \mathbf{I}$. Since the matrix $\mathbf{QD}(\delta_n)$ has a simple eigenvalue equal to 1 and eigenvalue zero of multiplicity p for any step size ratio δ_n, it follows that the method (4.1.1) is zero-stable for any variable step size pattern. This is in contrast to the strategy proposed before by Butcher and Jackiewicz [67] for DIMSIMs of the form (3.1.1), where the desirable zero-stability properties were enforced by a suitable choice of some free parameters associated with a matrix that affects the computation of some rescaled quantities corresponding to the input vector $y^{[n-1]}$.

Substituting (4.1.3) with $\mathbf{D}(\delta_n)$ defined by (4.1.4) into (4.1.1), we obtain the following representation of DIMSIMs:

$$Y^{[n]} = h_n(\mathbf{A} \otimes \mathbf{I})F(Y^{[n]}) + (\mathbf{PD}(\delta_n) \otimes \mathbf{I})z^{[n-1]},$$

$$z^{[n]} = h_n(\mathbf{G} \otimes \mathbf{I})F(Y^{[n]}) + (\mathbf{QD}(\delta_n) \otimes \mathbf{I})z^{[n-1]},$$

(4.1.5)

$n = 1, 2, \ldots, N$.

Butcher et al. [57] and Butcher and Jackiewicz [67] (see also Section 2.5) defined the local discretization error $\Gamma(t_n)$ of method (4.1.5) at the grid point t_n by

$$\Gamma(t_n) = z(t_n, h_n) - h_n(\mathbf{G} \otimes \mathbf{I})F(\widehat{Y}^{[n]}) - \big(\mathbf{Q}\mathbf{D}(\delta_n) \otimes \mathbf{I}\big)z(t_{n-1}, h_n),$$

where

$$\widehat{Y}^{[n]} = h_n(\mathbf{A} \otimes \mathbf{I})F(\widehat{Y}^{[n]}) + \big(\mathbf{P}\mathbf{D}(\delta_n) \otimes \mathbf{I}\big)z(t_{n-1}, h_n),$$

and $z(t_n, h_n)$ and $z(t_{n-1}, h_n)$ are Nordsieck vectors defined by (3.11.1). It was demonstrated [57, 67] that $\Gamma(t_n)$ is given by

$$\Gamma(t_n) = \delta_n^{p+1} h_{n-1}^{p+1}(\varphi_p \otimes \mathbf{I})y^{(p+1)}(t_{n-1}) + O(h_n^{p+2}),$$

where

$$\varphi_p = \mathbf{a} - \frac{\mathbf{G}\mathbf{c}^p}{p!} \quad \text{and} \quad \mathbf{a} = \left[\begin{array}{ccccc} \dfrac{1}{(p+1)!} & \dfrac{1}{p!} & \cdots & \dfrac{1}{2!} & 1 \end{array} \right]^T.$$

Here $\mathbf{c}^p = [c_1^p, \ldots, c_s^p]^T$. Since $(\mathbf{Q}\mathbf{D}(\delta_n) \otimes \mathbf{I})z^{[n-1]}$ propagates to the next step, we have tried to estimate and control the quantity

$$\delta_n^{p+1} h_{n-1}^{p+1} \mathbf{q}^T \varphi_p y^{(p+1)}(t_{n-1}),$$

where $\mathbf{q}^T$ stands for the first row of $\mathbf{Q}$. This quantity was estimated [57] by

$$\delta_n^{p+1} h_{n-1}^{p+1} \mathbf{q}^T \varphi_p y^{(p+1)}(t_{n-1})$$

$$= \delta_n h_{n-1} \left(\big(\beta^T(\delta_n) \otimes \mathbf{I}\big) F(Y^{[n]}) + \frac{1}{\delta_n}\big(\gamma^T(\delta_n) \otimes \mathbf{I}\big) F(Y^{[n-1]}) \right)$$

$$+ O(h_n^{p+2}),$$

where the vectors $\beta(\delta_n)$ and $\gamma(\delta_n)$ were computed from a system of equations resulting from expanding both sides of the relation above into a Taylor series around the point t_{n-1} and comparing the coefficients of corresponding powers of h_{n-1} up to the order $p+1$. In [58, 67] $\Gamma(t_n)$ was estimated by formula that is more convenient in a variable order setting,

$$\delta_n^{p+1} h_{n-1}^{p+1} \mathbf{q}^T \varphi_p y^{(p+1)}(t_{n-1})$$

$$= \delta_n h_{n-1}\big(\beta^T(\delta_n) \otimes \mathbf{I}\big) F(Y^{[n]}) + \big(\gamma^T(\delta_n) \otimes \mathbf{I}\big)z^{[n-1]} + O(h_n^{p+2}),$$

where $\beta(\delta_n)$ and $\gamma(\delta_n)$ again satisfy the appropriate system of equations. It was demonstrated [57, 58, 67] that these formulas are quite accurate and reliable if the step size ratio δ_n is not too large. However, for large values of δ_n, the accuracy of these formulas deteriorates significantly (see the graphs in [57, 58, 67]). In the next section we describe an approach, first proposed by Butcher and Jackiewicz [69], which is very accurate and very reliable for any step size pattern. This approach is based on isolating and controlling the error of the first component $z_1^{[n]}$ only of the vector of external approximations $z^{[n]}$.

4.2 LOCAL ERROR ESTIMATION

As in Section 3.12, denote by $\mathbf{g}^T$ and $\mathbf{q}^T$ the first rows of the coefficient matrices $\mathbf{G}$ and $\mathbf{Q}$, respectively. Then

$$z_1^{[n-1]} = h_{n-1}(\mathbf{g}^T \otimes \mathbf{I})F(Y^{[n-1]}) + (\mathbf{q}^T \mathbf{D}(\delta_{n-1}) \otimes \mathbf{I})z^{[n-2]}$$

and

$$(\mathbf{e}_1 \otimes \mathbf{I})z_1^{[n-1]} = h_{n-1}(\mathbf{e}_1 \mathbf{g}^T \otimes \mathbf{I})F(Y^{[n-1]}) + (\mathbf{Q}\mathbf{D}(\delta_{n-1}) \otimes \mathbf{I})z_1^{[n-1]}$$

since $\mathbf{Q} = \mathbf{e}_1 \mathbf{q}^T$. Computing $(\mathbf{Q}\mathbf{D}(\delta_{n-1}) \otimes \mathbf{I})z^{[n-2]}$ from this relation and substituting it into $z^{[n-1]}$, we obtain

$$z^{[n-1]} = h_{n-1}((\mathbf{G} - \mathbf{e}_1 \mathbf{g}^T) \otimes \mathbf{I})F(Y^{[n-1]}) + (\mathbf{e}_1 \otimes \mathbf{I})z_1^{[n-1]}. \qquad (4.2.1)$$

We also have

$$z_1^{[n]} = h_n(\mathbf{g}^T \otimes \mathbf{I})F(Y^{[n]}) + (\mathbf{q}^T \mathbf{D}(\delta_n) \otimes \mathbf{I})z^{[n-1]}.$$

Substituting into this relation the formula for $z^{[n-1]}$ obtained above and taking into account that $\mathbf{q}^T \mathbf{D}(\delta_n)\mathbf{e}_1 = 1$ yields

$$
\begin{aligned}
z_1^{[n]} =\ & h_n(\mathbf{g}^T \otimes \mathbf{I})F(Y^{[n]}) \\
& + h_{n-1}(\mathbf{q}^T \mathbf{D}(\delta_n)(\mathbf{G} - \mathbf{e}_1 \mathbf{g}^T) \otimes \mathbf{I})F(Y^{[n-1]}) + z_1^{[n-1]}.
\end{aligned}
\qquad (4.2.2)
$$

We define a local discretization error $le(t_n)$ of the first component of $z^{[n]}$ as the residual obtained by replacing in (4.2.2) $z_1^{[n]}$ by $y(t_n)$, $z_1^{[n-1]}$ by $y(t_{n-1})$, $F(Y^{[n]})$ by $y'(t_{n-1}+ch_n)$ and $F(Y^{[n-1]})$ by $y'(t_{n-1}+(\mathbf{c}-\mathbf{e})h_{n-1})$. This leads to

$$
\begin{aligned}
le(t_n) =\ & y(t_n) - y(t_{n-1}) - h_n(\mathbf{g}^T \otimes \mathbf{I})F(Y^{[n]}) \\
& + h_{n-1}(\mathbf{q}^T \mathbf{D}(\delta_n)(\mathbf{G} - \mathbf{e}_1 \mathbf{g}^T) \otimes \mathbf{I})F(Y^{[n-1]}).
\end{aligned}
\qquad (4.2.3)
$$

We have the following theorem.

Theorem 4.2.1 (Butcher and Jackiewicz [69]) *Assume that the function f appearing in (2.1.1) is sufficiently smooth. Then the local discretization error $le(t_n)$ of method (4.1.5) takes the form*

$$le(t_n) = \vartheta(\delta_n)h_n^{p+1}y^{(p+1)}(t_{n-1}) + O(h_n^{p+2}), \qquad (4.2.4)$$

where the error constant $\vartheta_p(\delta)$ is given by

$$\vartheta_p(\delta) = \frac{1}{(p+1)!}\left(1-(p+1)\mathbf{g}^T\mathbf{c}^p - \frac{p+1}{\delta^{p+1}}\mathbf{q}^T\mathbf{D}(\delta)(\mathbf{G}-\mathbf{e}_1\mathbf{g}^T)(\mathbf{c}-\mathbf{e})^p\right). \quad (4.2.5)$$

Proof: Expanding $y(t_n)$, $y'(t_{n-1}+\mathbf{c}h_n)$, and $y'(t_{n-1}+(\mathbf{c}-\mathbf{e})h_{n-1})$ into a Taylor series around t_{n-1}, we obtain

$$\mathrm{le}(t_n) = \sum_{k=1}^{p+1}\left(1 - k\mathbf{g}^T\mathbf{c}^{k-1} - \frac{k}{\delta_n^k}\mathbf{q}^T\mathbf{D}(\delta_n)(\mathbf{G}-\mathbf{e}_1\mathbf{g}^T)(\mathbf{c}-\mathbf{e})^{k-1}\right)$$
$$\times \; y^{(k)}(t_{n-1})\frac{h_n^k}{k!} + O(h_n^{p+2}).$$

Since method (4.2.5) has order p and stage order $q = p$, the terms corresponding to $k = 1, 2, \ldots, p$ are equal to zero and we obtain formula (4.2.4) with error constant $\vartheta_p(\delta_n)$ given by (4.2.5). This completes the proof. ∎

Observe that the error constant $\vartheta_p(\delta_n)$ and, as a result, the principal part of the local discretization error $\mathrm{le}(t_n)$ depend on the ratio of step sizes δ_n. This is in contrast to the quantity $h_n^{p+1}\mathbf{q}^T\varphi_p y^{(p+1)}(t_{n-1})$, whose estimates were used in [57, 58, 67] to control the step size of the underlying numerical methods. This makes possible the reliable estimation of $\mathrm{le}(t_n)$ for any step size pattern.

The scaled derivative $h_n^{p+1}y^{(p+1)}(t_{n-1})$ appearing in $\mathrm{le}(t_n)$ given by (4.2.4) can be estimated by formulas that depend on $F(Y^{[n]})$ and $F(Y^{[n-1]})$ or $F(Y^{[n]})$ and $z^{[n-1]}$, respectively. As noted in Section 4.1, an estimate that depends on $F(Y^{[n]})$ and $z^{[n-1]}$ is more convenient in a variable order setting. This estimate of $h_n^{p+1}y^{(p+1)}(t_{n-1})$ is given in the following result.

Theorem 4.2.2 (Butcher and Jackiewicz [69]) *Assume that the function f appearing in (2.1.1) is sufficiently smooth. Then*

$$\delta_n^{p+1}h_{n-1}^{p+1}y^{(p+1)}(t_{n-1}) = \delta_n h_{n-1}\left(\beta^T(\delta_n)\otimes\mathbf{I}\right)F(Y^{[n]})$$
$$+ \left(\gamma^T(\delta_n)\otimes\mathbf{I}\right)z^{[n-1]} + O(h_n^{p+2}), \qquad (4.2.6)$$

where

$$\gamma(\delta_n) = \begin{bmatrix} 0 \\ \widetilde{\gamma}(\delta_n) \end{bmatrix} \in \mathbb{R}^{s+1}$$

and the vectors $\beta(\delta_n) \in \mathbb{R}^s$ and $\widetilde{\gamma}(\delta_n) \in \mathbb{R}^s$ satisfy the system of equations

$$\beta^T(\delta_n)\mathbf{C}_p\mathbf{K}_p\mathbf{D}(\delta_n)\mathbf{E}_p + \widetilde{\gamma}^T(\delta_n)\widetilde{\mathbf{G}}\mathbf{C}_p\mathbf{K}_p = 0,$$
$$\beta^T(\delta_n)\left(\mathbf{C}_p\mathbf{K}_p\mathbf{D}(\delta_n)\mathbf{a} + \delta_n^{p+1}\frac{\mathbf{c}^p}{p!}\right) + \widetilde{\gamma}^T(\delta_n)\widetilde{\mathbf{G}}\frac{\mathbf{c}^p}{p!} = \delta_n^{p+1}, \qquad (4.2.7)$$
$$\left(\delta_n^{p+1}\beta^T(\delta_n) + \widetilde{\gamma}^T(\delta_n)\widetilde{\mathbf{G}}\right)\mathbf{e} = 0.$$

Here vector **a** *is defined in Section 4.1, the rescaling matrix* $\mathbf{D}(\delta_n)$ *is defined by (4.1.4), and matrices* $\mathbf{C}_p$, $\mathbf{K}_p$, $\mathbf{E}_p$ *and* $\widetilde{\mathbf{G}}$ *are defined in Section 3.12.*

Proof: Since method (4.1.5) has order p and stage order $q = p$, there exists a function $\nu(t)$ such that

$$Y^{[n]} = y(t_{n-1} + \mathbf{c}h_n) + \nu(t_{n-1} + \mathbf{c}h_n)h_n^p + O(h_n^{p+1})$$

as $h_n \to 0$ (compare [57, Theorem 10]). Hence,

$$h_n F(Y^{[n]}) = h_n y'(t_{n-1} + \mathbf{c}h_n) + \frac{\partial f}{\partial y}\big(y(t_{n-1} + \mathbf{c}h_n)\big)\nu(t_{n-1} + \mathbf{c}h_n)h_n^{p+1}$$
$$+ O(h_n^{p+2}).$$

Expanding $y'(t_{n-1} + \mathbf{c}h_n)$ into a Taylor series around the point t_{n-1}, we obtain

$$h_n F(Y^{[n]}) = h_n\big(\mathbf{e} \otimes y'(t_{n-1})\big) + h_n^2\big(\mathbf{c} \otimes y''(t_{n-1})\big) + \cdots$$
$$+ h_n^p\left(\frac{\mathbf{c}^{p-1}}{(p-1)!} \otimes y^{(p)}(t_{n-1})\right) + h_n^{p+1}\left(\frac{\mathbf{c}^p}{p!} \otimes y^{(p+1)}(t_{n-1})\right)$$
$$+ h_n^{p+1}\left(\mathbf{e} \otimes \frac{\partial f}{\partial y}\big(y(t_{n-1})\big)\nu(t_{n-1})\right) + O(h_n^{p+2}).$$

This relation can be written in a more compact form as

$$h_n F(Y^{[n]}) = \big(\mathbf{C}_p\mathbf{K}_p \otimes \mathbf{I}\big)\widehat{z}(t_{n-1}, h_n) + h_n^{p+1}\left(\frac{\mathbf{c}^p}{p!} \otimes y^{(p+1)}(t_{n-1})\right)$$
$$+ h_n^{p+1}\left(\mathbf{e} \otimes \frac{\partial f}{\partial y}\big(y(t_{n-1})\big)\nu(t_{n-1})\right) + O(h_n^{p+2}),$$

where $\widehat{z}(t_{n-1}, h_n)$ is defined by (4.1.2). We also have

$$\widehat{z}(t_{n-1}, h_n) = \big(\mathbf{D}(\delta_n)\mathbf{E}_p \otimes \mathbf{I}\big)\widehat{z}(t_{n-2}, h_{n-1}) + h_{n-1}^{p+1}\big(\mathbf{D}(\delta_n)\mathbf{a} \otimes y^{(p+1)}(t_{n-2})\big)$$
$$+ O(h_n^{p+2})$$

and

$$z^{[n-1]} = h_{n-1}(\mathbf{G} \otimes \mathbf{I})F(Y^{[n-1]}) + (\mathbf{Q} \otimes \mathbf{I})\widehat{z}(t_{n-2}, h_{n-1}) + O(h_{n-1}^{p+2})$$
$$= \big((\mathbf{G}\mathbf{C}_p\mathbf{K}_p + \mathbf{Q}) \otimes \mathbf{I}\big)\widehat{z}(t_{n-2}, h_{n-1}) + h_{n-1}^{p+1}\left(\frac{\mathbf{c}^p}{p!} \otimes y^{(p+1)}(t_{n-2})\right)$$
$$+ h_{n-1}^{p+1}\left(\mathbf{e} \otimes \frac{\partial f}{\partial y}\big(y(t_{n-2})\big)\nu(t_{n-2})\right) + O(h_{n-1}^{p+2}).$$

Substituting the relations above for $h_n F(Y^{[n]})$, $\widehat{z}(t_{n-1}, h_n)$, and $z^{[n-1]}$ into (4.2.6) and taking into account that

$$y^{(p+1)}(t_{n-1}) = y^{(p+1)}(t_{n-2}) + O(h_{n-1}),$$

$$\frac{\partial f}{\partial y}\big(y(t_{n-1})\big) = \frac{\partial f}{\partial y}\big(y(t_{n-2})\big) + O(h_{n-1})$$

and

$$\nu(t_{n-1}) = \nu(t_{n-2}) + O(h_{n-1})$$

we obtain

$$\delta_n^{p+1} h_{n-1}^{p+1} y^{(p+1)}(t_{n-2})$$

$$= \Big(\big(\beta^T(\delta_n)\mathbf{C}_p\mathbf{K}_p\mathbf{D}(\delta_n)\mathbf{E}_p + \gamma^T(\delta_n)(\mathbf{G}\mathbf{C}_p\mathbf{K}_p + \mathbf{Q})\big) \otimes \mathbf{I}\Big)\widehat{z}(t_{n-2}, h_{n-1})$$

$$+ \left(\beta^T(\delta_n)\Big(\mathbf{C}_p\mathbf{K}_p\mathbf{D}(\delta_n)\mathbf{a} + \delta_n^{p+1}\frac{\mathbf{c}^p}{p!}\Big) + \gamma^T(\delta_n)\frac{\mathbf{c}^p}{p!}\right)h_{n-1}^{p+1}y^{(p+1)}(t_{n-2})$$

$$+ \left(\delta_n^{p+1}\beta^T(\delta_n) + \gamma^T(\delta_n)\mathbf{G}\right)\mathbf{e} + O(h_{n-1}^{p+2}).$$

Comparing the corresponding terms in the relation above we obtain the following system of equations:

$$\beta^T(\delta_n)\mathbf{C}_p\mathbf{K}_p\mathbf{D}(\delta_n)\mathbf{E}_p + \gamma^T(\delta_n)(\mathbf{G}\mathbf{C}_p\mathbf{K}_p + \mathbf{Q}) = 0,$$

$$\beta^T(\delta_n)\left(\mathbf{C}_p\mathbf{K}_p\mathbf{D}(\delta_n)\mathbf{a} + \delta_n^{p+1}\frac{\mathbf{c}^p}{p!}\right) + \gamma^T(\delta_n)\frac{\mathbf{c}^p}{p!} = \delta_n^{p+1},$$

$$\left(\delta_n^{p+1}\beta^T(\delta_n) + \gamma^T(\delta_n)\mathbf{G}\right)\mathbf{e} = 0.$$

It is easy to verify that $\gamma_1(\delta_n) = 0$. Since $\mathbf{Q}$ with the first row omitted is equal to the zero matrix and

$$\gamma^T(\delta_n)\mathbf{G} = \widetilde{\gamma}^T(\delta_n)\widetilde{\mathbf{G}},$$

the system above is equivalent to (4.2.7). This completes the proof.   ∎

It follows from Theorem 3.12.3 that the matrix $\widetilde{\mathbf{G}}$ depends only on the abscissa vector $\mathbf{c}$. As a result, vectors $\beta(\delta)$ and $\widetilde{\gamma}(\delta)$ also depend only on vector $\mathbf{c}$.

For explicit and implicit formulas of order $p = 1$ listed in Examples 1 and 2 in Section 3.13, the error constants are $\vartheta_1(\delta) = \frac{1}{2}$ and $\vartheta_1(\delta) = -\frac{1}{2}$, respectively (independent of δ). Solving the first two equations of system (4.2.7), we obtain

$$\beta(\delta) = \delta, \quad \gamma(\delta) = \begin{bmatrix} 0 & -\delta^2 \end{bmatrix}^T$$

for the explicit method, and

$$\beta(\delta) = 1, \quad \gamma(\delta) = \begin{bmatrix} 0 & -\delta \end{bmatrix}^T$$

for the implicit method. For both methods the third equation of (4.2.7) now takes the form

$$\delta^2(1 - \delta) = 0$$

and is satisfied only for $\delta = 1$ (i.e., if the step size of integration is kept constant). However, in practice, these methods are only used to start or restart the variable order codes where δ is usually close to 1, and as a result, this does not have a large impact on the reliability of error estimation.

For the explicit and implicit methods listed in Examples 3 and 4 in Section 3.13, the error constants are

$$\vartheta_2(\delta) = \frac{3 + \delta}{24\delta} \quad \text{and} \quad \vartheta_2(\delta) = \frac{9 + 7\delta - 6\sqrt{2}(1 + \delta)}{24\delta},$$

respectively. For both methods, $\mathbf{c} = [0, 1]^T$, and solving the system (4.2.7) we obtain

$$\beta(\delta) = \left[\begin{array}{cc} -\dfrac{2\delta}{1 + \delta} & \dfrac{2\delta}{1 + \delta} \end{array} \right]^T, \quad \gamma(\delta) = \left[\begin{array}{ccc} 0 & 0 & -\dfrac{2\delta^3}{1 + \delta} \end{array} \right]^T.$$

For the explicit and implicit methods listed in Examples 5 and 6 in Section 3.13, the error constants are

$$\vartheta_3(\delta) = \frac{\delta^2 + 3\delta + 2}{144\delta^2}, \quad \vartheta_3(\delta) = \frac{0.122786\delta^2 - 0.425794\delta + 0.328905}{\delta^2},$$

respectively. Solving system (4.2.7) with $\mathbf{c} = [0, \frac{1}{2}, 1]^T$, we obtain

$$\beta(\delta) = \left[\begin{array}{c} \dfrac{24\delta^4 + \gamma_3(7\delta^2 + 9\delta + 2)}{3\delta^3(1 + \delta)} \\[3mm] -\dfrac{4\left(12\delta^4 + \gamma_3(2\delta^2 + 3\delta + 1)\right)}{3\delta^3(1 + \delta)} \\[3mm] \dfrac{24\delta^4 + \gamma_3(\delta^2 + 3\delta + 2)}{3\delta^3(1 + \delta)} \end{array} \right], \quad \gamma(\delta) = \left[\begin{array}{c} 0 \\[3mm] 0 \\[3mm] \gamma_3 \\[3mm] \dfrac{-12\delta^4 + \gamma_3(\delta^2 - 1)}{6(1 + \delta)} \end{array} \right]$$

for the explicit method, where γ_3 is a free parameter that may depend on δ. Similarly, solving (4.2.7) with $\mathbf{c} = [-1, 0, 1]^T$ leads to

$$\beta(\delta) = \left[\begin{array}{c} \dfrac{6\delta^4 + \gamma_3(\delta^2 + 3\delta + 2)}{6\delta^3} \\[3mm] -\dfrac{6\delta^4 + \gamma_3(\delta^2 + 2)}{3\delta^3} \\[3mm] \dfrac{6\delta^4 + \gamma_3(\delta^2 - 3\delta + 2)}{6\delta^3} \end{array} \right], \quad \gamma(\delta) = \left[\begin{array}{c} 0 \\[3mm] 0 \\[3mm] \gamma_3 \\[3mm] -\dfrac{6\delta^4 + \gamma_3(\delta^2 + 2)}{6} \end{array} \right]$$

for the implicit method.

In general, for $p > 2$ the solution to system (4.2.7) leads to a family of error estimators that depend on free parameters $\gamma_3, \ldots, \gamma_p$. The choice

$$\gamma_3 = \cdots = \gamma_p = 0$$

leads to the estimate of $h_n^{p+1} y^{(p+1)}(t_{n-1})$ described in the following result.

Theorem 4.2.3 *Assume that $p > 2$ and $\gamma_3 = \cdots = \gamma_p = 0$. Then*

$$h_n^{p+1} y^{(p+1)}(t_{n-1}) = \frac{2\delta^n}{1 + \delta_n} \left(z_{p+1}^{[n]} - \delta_n^p z_{p+1}^{[n-1]} \right) + O(h_n^{p+2}), \qquad (4.2.8)$$

where $z_{p+1}^{[n]}$ is subvector number $p + 1$ of vector $z^{[n]}$.

Proof: It follows from the assumption $\gamma_3 = \cdots = \gamma_p = 0$ that the estimate of $h_n^{p+1} y^{(p+1)}(t_{n-1})$ takes the form

$$h_n^{p+1} y^{(p+1)}(t_{n-1}) = h_n \sum_{i=1}^{p} \beta_i(\delta_n) f(Y_i^{[n]}) + \gamma_{p+1}(\delta_n) z_{p+1}^{[n-1]} + O(h_n^{p+2}).$$

Since $z_{p+1}^{[n-1]} = O(h_n^p)$, the same must be true for the term

$$h_n \sum_{i=1}^{p} \beta_i(\delta_n) f(Y_i^{[n]}),$$

and, accordingly, the only choice for this is a scalar multiple of the same expression as that used as the value of $z_{p+1}^{[n]}$. Hence, the approximation to $h_n^{p+1} y^{(p+1)}(t_{n-1})$ that we need to use has the form $\alpha z_{p+1}^{[n]} - \alpha' z_{p+1}^{[n-1]}$ for some constants α and α'. Because

$$z_{p+1}^{[n]} = \delta_n^p z_{p+1}^{[n-1]} + O(h_n^{[p+1]}),$$

α' must be chosen as $\delta_n^p \alpha$. To find the correct choice for α, we note that for $p > 2$ we have

$$z_{p+1}^{[n]} = h_n^p y^{(p)} \left(t_{n-1} + \frac{h_n}{2} \right) + O(h_n^{p+2})$$

$$= h_n^p y^{(p)}(t_{n-1}) + \frac{h_n^{p+1}}{2} y^{(p+1)}(t_{n-1}) + O(h_n^{p+2})$$

and similarly, that

$$z_{p+1}^{[n-1]} = h_{n-1}^p y^{(p)}(t_{n-1}) - \frac{h_{n-1}^{p+1}}{2} y^{(p+1)}(t_{n-1}) + O(h_{n-1}^{p+2}).$$

Hence,

$$z_{p+1}^{[n]} - \delta_n^p z_{p+1}^{[n-1]} = \left(\frac{1}{2} + \frac{1}{2\delta_n} \right) h_n^{p+1} y^{(p+1)}(t_{n-1}) + O(h_n^{p+2}),$$

and it follows that α should be chosen as $2\delta_n/(1 + \delta_n)$. This completes the proof.   ∎

It follows from Theorems 4.2.1 and 4.2.2 that the local discretization error $\mathrm{le}(t_n)$ can be estimated by the formula

$$\mathrm{est}(t_n) = \vartheta_p(\delta_n)\Big(h_n\big(\beta^T(\delta_n) \otimes \mathbf{I}\big)F(Y^{[n]}) + \big(\gamma^T(\delta_n) \otimes \mathbf{I}\big)z^{[n-1]}\Big), \qquad (4.2.9)$$

where the error constant $\vartheta_p(\delta_n)$ is given by (4.2.5) and the vectors $\beta(\delta_n)$ and $\gamma(\delta_n)$ are defined by (4.2.7). It follows from Theorem 4.2.3 that for $p > 2$ the estimate corresponding to $\gamma_3 = \cdots = \gamma_p$ is given by

$$\mathrm{est}(t_n) = \vartheta_p(\delta_n)\frac{2\delta_n}{1 + \delta_n}\left(z_{p+1}^{[n]} - \delta_n^p z_{p+1}^{[n-1]}\right). \qquad (4.2.10)$$

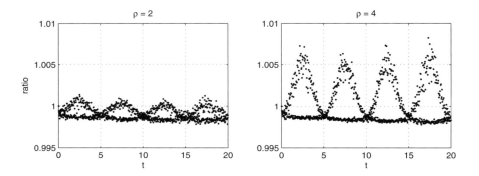

Figure 4.2.1   Ratios $\mathrm{le}(\delta_n)/\mathrm{est}(\delta_n)$ versus t for the type 1 method of order $p = 3$ given in Example 5 in Section 3.13

To illustrate the quality of these estimators we use the standard test problem

$$y'(t) = \lambda(y - e^{\mu t}) + \mu e^{\mu t}, \quad t \in [t_0, T],$$

$$y(t_0) = y_0, \qquad (4.2.11)$$

$\lambda, \mu \in \mathbb{R}$, which was used elsewhere [57, 69, 70, 71, 73]. We test these estimates under quite demanding conditions, where the step size pattern changes rapidly according to the formula

$$h_{n+1} = \rho^{(-1)^n \sin(4\pi n/(T-t_0))} h_n,$$

$n = 0, 1, \ldots, N - 1$, with $h_0 = (T - t_0)/N$, $t_0 = 0$, $T = 20$, $y_0 = 1$, $N = 1000$, and $\rho = 2$ or $\rho = 4$. This step size pattern is more demanding than that

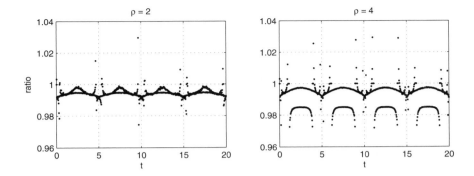

Figure 4.2.2   Ratios $\mathrm{le}(\delta_n)/\mathrm{est}(\delta_n)$ versus t for the type 2 method of order $p = 3$ given in Example 6 in Section 3.13

usually encountered in real codes, and we believe that satisfactory behavior of error estimators in such extreme conditions should be a good indicator of the quality of these estimators.

We have plotted in Figures 4.2.1 and 4.2.2 the ratios

$$\mathrm{ratio}(\delta_n) = \frac{\mathrm{le}(\delta_n)}{\mathrm{est}(\delta_n)}$$

for the type 1 and 2 methods of order $p = 3$ given in Examples 5 and 6 in Section 3.13 for $\lambda = -0.1$ and $\mu = 0.1$. These figures indicate that the error estimators are quite accurate and reliable under very demanding conditions imposed on the step size pattern. The error estimate stays within less than 1% of true local discretization error for the explicit method and with a few exceptions within less than 2% of the true local discretization error for the implicit method. The quality of error estimators for implicit methods for stiff systems of differential equations can be improved further by the technique of filtering error estimators. This is discussed in the next section.

4.3 LOCAL ERROR ESTIMATION FOR LARGE STEP SIZES

In this section we follow Jackiewicz [177] to derive error estimates for implicit methods suitable for the numerical solution of stiff differential systems. As in the case of explicit methods for nonstiff differential equations, it is necessary that these estimates be asymptotically correct as $h_n \to 0$. However, this is not sufficient in the stiff case, where as pointed out by Shampine and Baca [258], one must work with step sizes that are large compared to certain characteristics of the problem. To assess the quality of error estimators for large h_n they proposed, as in the classical theory of absolute stability, considering a restricted class of problems of the form $y' = Jy$, where J is a constant matrix that can be diagonalized by a similarity transformation. Then it is sufficient to consider the scalar problem

$$y' = \xi y, \quad t \geq 0,$$

$$y(0) = 1,$$

(4.3.1)

$\xi \in \mathbb{C}$, where ξ is the eigenvalue of J and $\mathbb{C}$ stands for the set of complex numbers. In this section we adopt this approach to the case of error estimation for type 2 DIMSIMs.

The local discretization error $\mathrm{le}(t_n)$ of method (4.1.5) is defined by (4.2.3). Replacing $y(t_n)$ and $y(t_{n-1})$ in this formula by $e^{\xi t_n}$ and $e^{\xi t_{n-1}}$, where $y(t) = e^{\xi t}$ is the solution to (4.3.1), and taking into account that

$$Y^{[n]} = e^{\xi(t_{n-1}+\mathbf{c}h_n)} + O(h_n^{p+1})$$

and

$$Y^{[n-1]} = e^{\xi(t_{n-1}+(\mathbf{c}-\mathbf{e})h_{n-1})} + O(h_n^{p+1}),$$

we obtain

$$
\begin{aligned}
\mathrm{le}(t_n) &= e^{\xi t_n} - e^{\xi t_{n-1}} - h_n \xi \mathbf{g}^T e^{\xi(t_{n-1}+\mathbf{c}h_n)} \\
&\quad - h_{n-1}\xi \mathbf{q}^T \mathbf{D}(\delta_n)(\mathbf{G}-\mathbf{e}_1\mathbf{g}^T)e^{\xi(t_{n-1}+(\mathbf{c}-\mathbf{e})h_{n-1})} + O(h_n^{p+2}) \\
&= \left(e^{\delta_n z} - 1 - \mathbf{g}^T \delta_n z e^{\mathbf{c}\delta_n z} - z\mathbf{q}^T \mathbf{D}(\delta_n)(\mathbf{G}-\mathbf{e}_1\mathbf{g}^T)e^{(\mathbf{c}-\mathbf{e})z} \right) e^{\xi t_{n-1}} \\
&\quad + O(z^{p+2}).
\end{aligned}
$$

Here $z = h_{n-1}\xi$.

We now develop a similar expression for the error estimate $\mathrm{est}(t_n)$. Substituting (4.2.1) into (4.2.9), we obtain

$$
\begin{aligned}
\mathrm{est}(t_n) &= \vartheta_p(\delta_n)\Big(h_n\big(\beta^T(\delta_n)\otimes \mathbf{I}\big)F(Y^{[n]}) \\
&\quad + \big(\gamma^T(\delta_n)\otimes \mathbf{I}\big)\Big(h_{n-1}\big((\mathbf{G}-\mathbf{e}_1\mathbf{g}^T)\otimes \mathbf{I}\big)F(Y^{[n-1]}) + (\mathbf{e}_1\otimes \mathbf{I})z_1^{[n-1]}\Big)\Big).
\end{aligned}
$$

Assume that $z_1^{[n-1]} = e^{\xi t_{n-1}}$. Proceeding as earlier, we obtain

$$
\begin{aligned}
\mathrm{est}(t_n) &= \vartheta_p(\delta_n)\Big(h_{n-1}\xi\delta_n\beta^T(\delta_n)e^{\xi(t_{n-1}+\mathbf{c}h_{n-1}\delta_n)} \\
&\quad + \gamma^T(\delta_n)\Big(h_{n-1}\xi(\mathbf{G}-\mathbf{e}_1\mathbf{g}^T)e^{\xi(t_{n-1}+(\mathbf{c}-\mathbf{e})h_{n-1})} + \mathbf{e}_1 e^{\xi t_{n-1}}\Big)\Big) + O(h_n^{p+2}) \\
&= \vartheta_p(\delta_n)\Big(z\delta_n\beta^T(\delta_n)e^{\mathbf{c}\delta_n z} + z\gamma^T(\delta_n)(\mathbf{G}-\mathbf{e}_1\mathbf{g})e^{(\mathbf{c}-\mathbf{e})z}\Big)e^{\xi t_{n-1}} \\
&\quad + O(z^{p+2}),
\end{aligned}
$$

where we have taken into account that $\gamma^T(\delta_n)\mathbf{e}_1 = 0$. To investigate the behavior of error estimates for large $z = h_{n-1}\xi$, we define the functions $R_{\mathrm{le}}(z,\delta)$ and $R_{\mathrm{est}}(z,\delta)$ by the formulas

$$R_{\mathrm{le}}(z,\delta) := e^{\delta z} - 1 - \delta z \mathbf{g}^T e^{\mathbf{c}\delta z} - z \mathbf{q}^T \mathbf{D}(\delta)\big(\mathbf{G} - \mathbf{e}_1 \mathbf{g}^T\big)e^{(\mathbf{c}-\mathbf{e})z}$$

and

$$R_{\mathrm{est}}(z,\delta) := \vartheta_p(\delta)\Big(z\delta\beta^T(\delta)e^{\mathbf{c}\delta z} + z\gamma^T(\delta)\big(\mathbf{G} - \mathbf{e}_1 \mathbf{g}\big)e^{(\mathbf{c}-\mathbf{e})z}\Big),$$

corresponding to $\mathrm{le}(t_n)$ and $\mathrm{est}(t_n)$. To assess the quality of $\mathrm{est}(t_n)$ for large step sizes, we examine the ratio

$$r(z,\delta) := \frac{R_{\mathrm{est}}(z,\delta)}{R_{\mathrm{le}}(z,\delta)}. \tag{4.3.2}$$

If

$$r(z,\delta) \sim \mathrm{const}\, z^\mu$$

for $\mathrm{Re}(z) < 0$ as $|z| \to \infty$ with a positive integer μ, the error is grossly overestimated for large z. To compensate for this, Shampine and Baca [258] suggested, in the context of RK methods, premultiplying $\mathrm{est}(t_n)$ by the filter matrix, which damps the large, stiff error components. This new error estimate $\mathrm{est}^*(t_n)$ is defined according to the formula

$$\mathrm{est}^*(t_n) = \big(\mathbf{I} - h_n \lambda \mathbf{J}(t_n)\big)^{-\mu} \mathrm{est}(t_n), \tag{4.3.3}$$

where λ is the diagonal element of the coefficient matrix $\mathbf{A}$ and $\mathbf{J}(t_n)$ is an approximation to the Jacobian of problem (2.1.1) at the point t_n. We also follow this approach in the context of DIMSIMs. Since the behavior of (4.3.3) is described by the function

$$R_{\mathrm{est}}^*(z,\delta) = \frac{R_{\mathrm{est}}(z,\delta)}{(1 - \lambda z)^\mu},$$

it follows that the modified estimate $\mathrm{est}^*(t_n)$ is asymptotically correct as $z \to 0$ and also corrects the order as $|z| \to \infty$, $\mathrm{Re}(z) < 0$.

We illustrate this process for the type 2 DIMSIMs listed in Section 3.13. For the method of order $p = 1$ given in Example 2 of Section 3.13, the function $r(z,\delta)$ defined by (4.3.2) takes the form

$$r(z,\delta) = \frac{\delta z(e^{\delta z} - 1)}{2(\delta z e^{\delta z} - e^{\delta z} + 1)}.$$

This implies that

$$r(z,\delta) \sim -\frac{\delta}{2} z, \quad |z| \to \infty, \quad \mathrm{Re}(z) < 0,$$

and it is recommended that the estimate

$$\text{est}^*(t_n) = \left(\mathbf{I} - h_n \mathbf{J}(t_n)\right)^{-1} \text{est}(t_n)$$

be used instead of $\text{est}(t_n)$ for large step sizes.

For the method of order $p = 2$ given in Example 4 in Section 3.13, the function $r(z, \delta)$ takes the form

$$r(z, \delta) = \frac{P(z, \delta)}{Q(z, \delta)},$$

with

$$P(z, \delta) = \delta\left(9 + 7\delta - 6\sqrt{2}(\delta + 1)\right)\left(\delta - e^z(1 + \delta) + e^{(1+\delta)z}\right)z$$

and

$$Q(z, \delta) = 3(1 + \delta)\left(4e^z(e^{\delta z} - 1) + \left(3\delta^2 - \delta e^z(5 + 3\delta) + \delta e^{(1+\delta)z}\right)\right)$$
$$- 3\sqrt{2}(1 + \delta)\left(2\delta^2 - 2\delta e^z(1 + \delta) + 2\delta e^{(1+\delta)z}\right)z.$$

This implies that

$$r(z, \delta) \sim \frac{9 + 7\delta - 6\sqrt{2}(1 + \delta)}{3(3 - 2\sqrt{2})(1 + \delta)}, \quad |z| \to \infty, \quad \text{Re}(z) < 0,$$

and we define $\text{est}^*(t_n) = \text{est}(t_n)$.

For the method of order $p = 3$ given in Example 6 in Section 3.13, the ratio $r(z, \delta)$ is quite complicated and is not reproduced here. Assuming that $\gamma_3 = 0$, we have

$$r(z, \delta) \sim \frac{3.96382 - 5.13148\,\delta + 1.47976\,\delta^2}{5.94572 - 5.13148\,\delta}, \quad |z| \to \infty, \quad \text{Re}(z) < 0,$$

for $\delta \in (0, 2)$, and this suggests that we use the original estimate $\text{est}(t_n)$ for all step sizes. However, the denominator of the expression above has a positive root $\tilde{\delta} \approx 1.15868$ and we have verified experimentally that, in most cases, the estimate

$$\text{est}^*(t_n) = \left(\mathbf{I} - \lambda h_n \mathbf{J}(t_n)\right)^{-1} \text{est}(t_n)$$

is more reliable than $\text{est}(t_n)$ and leads to a smaller number of rejected steps in actual implementations of this method.

A different approach to error estimation for implicit RK methods, which is based on construction of implicit error estimators, has been proposed by Swart and Söderlind [268].

4.4   CONSTRUCTION OF CONTINUOUS INTERPOLANTS

Continuous interpolants for DIMSIMs have been investigated by Butcher and Jackiewicz [67] and Jackiewicz [177]. Define the vectors

$$\mathbf{q}(\theta) = \left[\begin{array}{cccc} q_1(\theta) & q_2(\theta) & \cdots & q_{s+1}(\theta) \end{array} \right], \quad \theta \in [0, 1],$$

where $q_1(\theta) \equiv 1$, and

$$\mathbf{G}_1(\theta) = \left[\begin{array}{cccc} g_{11}(\theta) & g_{12}(\theta) & \cdots & g_{1s}(\theta) \end{array} \right], \quad \theta \in [0, 1].$$

We consider the following variable step size continuous DIMSIMs in Nordsieck form:

$$Y_i^{[n]} = h_n \sum_{j=1}^{s} a_{ij} f(Y_j^{[n]}) + \sum_{j=1}^{s+1} p_{ij} \delta_n^{j-1} z_j^{[n-1]}, \quad i = 1, 2, \ldots, s,$$

$$z_1^{[n]}(t_{n-1} + \theta h_n) = h_n \sum_{j=1}^{s} g_{1j}(\theta) f(Y_j^{[n]}) + \sum_{j=1}^{s+1} q_j(\theta) z_j^{[n-1]}, \qquad (4.4.1)$$

$$z_i^{[n]} = h_n \sum_{j=1}^{s} g_{ij} f(Y_j^{[n]}), \quad i = 2, 3, \ldots, s+1,$$

$\theta \in [0, 1]$. We say that method (4.4.1) has uniform order p if

$$z_i^{[n-1]} = h_{n-1}^{i-1} y^{(i-1)}(t_{n-1}) + O(h_n^{p+1})$$

implies that

$$z_1^{[n]}(t_{n-1} + \theta h_n) = y(t_{n-1} + \theta h_n) + O(h_n^{p+1})$$

uniformly with respect to $\theta \in [0, 1]$. We have the following theorem.

Theorem 4.4.1 (Jackiewicz [177]) *Method (4.4.1) satisfying order conditions (3.12.1) and (3.12.2) for $\theta = 1$ has a uniform order $p = s$ if and only if*

$$\mathbf{E}_1(\theta)\mathbf{Z} - z\mathbf{G}_1(\theta)e^{\mathbf{c}z} - \mathbf{q}(\theta)\mathbf{Z} = O(z^{s+1}) \qquad (4.4.2)$$

as $z \to 0$, where

$$\mathbf{E}_1(\theta) = \left[\begin{array}{cccc} 1 & \dfrac{\theta}{1!} & \cdots & \dfrac{\theta^s}{s!} \end{array} \right], \quad \theta \in [0, 1],$$

and $\mathbf{Z}$ and $e^{\mathbf{c}z}$ are defined as in Theorem 3.12.1.

Proof: Expanding $z_1^{[n]}(t_{n-1} + \theta h_n)$ into a Taylor series around the point t_{n-1}, we obtain

$$z_1^{[n]}(t_{n-1} + \theta h_n) = \sum_{k=0}^{s} \frac{\theta^k}{k!} h_n^k y^{(k)}(t_{n-1}) + O(h_n^{s+1}),$$

$h_n \to 0$. Substituting the relations for $z_1^{[n]}(t_{n-1} + \theta h_n)$, $z_i^{[n-1]}$, and

$$h_n f(Y_i^{[n]}) = \sum_{k=1}^{s} \frac{c_i^{k-1}}{(k-1)!} h_n^k y^{(k)}(t_{n-1}) + O(h_n^{s+1}),$$

$i = 1, 2, \ldots, s$ (compare the proof of Theorem 3.12.1) into a second equation of (4.4.1) yields

$$\sum_{k=0}^{s} \frac{1}{k!} \left(\theta^k - k \sum_{j=1}^{s} g_{1j}(\theta) c_j^{k-1} - k! q_{k+1}(\theta) \right) h_n y^{(k)}(t_{n-1}) = O(h_n^{s+1}).$$

Equating the coefficients of $h_n^k y^{(k)}(t_{n-1})/k!$ to zero and then multiplying these coefficients by $z^k/k!$ and adding them from $k = 0$ to $k = s$ leads to

$$\sum_{k=0}^{s} \left(\frac{\theta^k z^k}{k!} - z \sum_{j=1}^{s} g_{1j}(\theta) \frac{c_j^{k-1} z^{k-1}}{(k-1)!} - q_{k+1}(\theta) z^k \right) = 0,$$

which is equivalent to (4.4.2). This completes the proof. ∎

Relation (4.4.2) leads to s equations for the unknown functions $g_{1j}(\theta)$, $j = 1, 2, \ldots, s$, and $q_j(\theta)$, $j = 2, 3, \ldots, s+1$. These equations have the form

$$\frac{\theta^k}{k!} - \sum_{j=1}^{s} g_{1j}(\theta) \frac{c_j^{k-1}}{(k-1)!} - q_{k+1}(\theta) = 0, \qquad (4.4.3)$$

$k = 1, 2, \ldots, s$ (the constant term in (4.4.2) is equal to zero). Assuming that $q_j(\theta) \equiv q_j$, $j = 2, 3, \ldots, s+1$, where q_j are coefficients of the underlying discrete method, and solving the system (4.4.3) for $g_{1j}(\theta)$, $j = 1, 2, \ldots, s$, we obtain the following examples of continuous DIMSIMs in Nordsieck form, corresponding to the methods of order $p = 1$, $p = 2$, and $p = 3$ given in Section 3.13.

Continuous weights for types 1 and 2 methods of order $p = 1$ given in Examples 1 and 2 in Section 3.13:

$$g_{11}(\theta) = \theta, \quad \theta \in [0, 1].$$

Continuous weights for the type 1 method of order $p = 2$ given in Example 3:

$$g_{11}(\theta) = \frac{3 + 4\theta - 2\theta^2}{4}, \quad g_{12}(\theta) = \frac{2\theta^2 - 1}{4}, \quad \theta \in [0, 1].$$

Continuous weights for the type 2 method of order $p = 2$ given in Example 4:

$$g_{11}(\theta) = \frac{59 - 34\sqrt{2} + 14(2 - \theta)\,\theta}{28}, \quad g_{12}(\theta) = \frac{2\sqrt{2} - 3 + 2\,\theta^2}{4}, \quad \theta \in [0, 1].$$

Continuous weights for the type 1 method of order $p = 3$ given in Example 5:

$$g_{11}(\theta) = \frac{13 + 12\,\theta - 18\,\theta^2 + 8\,\theta^3}{12}, \quad g_{12}(\theta) = -\frac{1 - 6\,\theta^2 + 4\,\theta^3}{3},$$

$$g_{13}(\theta) = \frac{(4\,\theta - 3)\,\theta^2}{6}, \quad \theta \in [0, 1].$$

Continuous weights for the the type 2 method of order $p = 3$ given in Example 6:

$$g_{11}(\theta) = 0.919152 - \frac{\theta^2}{4} + \frac{\theta^3}{6}, \quad g_{12}(\theta) = 0.628473 + \theta - \frac{\theta^3}{3},$$

$$g_{13}(\theta) = -0.0675638 + \frac{\theta^2}{4} + \frac{\theta^3}{6}, \quad \theta \in [0, 1].$$

Observe that $g_{1j}(1) = g_{1j}$, $j = 1, 2, \ldots, s$, as should be the case.

A somewhat different approach to the construction of continuous interpolants of DIMSIMs has been given by Jackiewicz et al. [186].

4.5   STEP SIZE AND ORDER CHANGING STRATEGY

In the implementation of explicit and implicit DIMSIMs we follow closely approaches developed in the classical codes for nonstiff and stiff differential systems, such as STEP due to Shampine and Gordon [262], and RADAU5 and RADAU due to Hairer and Wanner [146, 148]. The code STEP [262] for nonstiff systems is based on the family of Adams-Bashforth Adams-Moulton methods in predictor-corrector mode with variable step size and variable order, ranging between $p = 1$ and $p = 12$. The fixed order code RADAU5 [146] for stiff systems is based on the A- and L-stable RadauIIA process of Ehle [121, 122] with $s = 3$ stages and order $p = 5$. The variable order code RADAU [148] is based on the A- and L-stable RadauIIA methods [121, 122] with $s = 3$, $s = 5$, and $s = 7$ stages of orders $p = 5$, $p = 9$, and $p = 13$. We adopt a similar approach in the implementation of DIMSIMs.

The initial order p_1 will be chosen to be equal to 1. The initial step size h_1 is computed using a modification of the approach taken by Gladwell et al. [134] (compare [143, p. 169] and [257, p. 379]). This approach consists of the following. Set

$$sc_i = \text{Atol}_i + |y_i(t_0)|\text{Rtol}_i, \quad i = 1, 2, \ldots, m,$$

where Atol_i and Rtol_i are absolute and relative error tolerances corresponding to the ith component of the solution $y_i(t)$ to (2.1.1) and, following Hairer et al. [143], define the norm $\|\cdot\|_{\text{sc}}$ by

$$\|y\|_{\text{sc}} = \sqrt{\frac{1}{m}\sum_{i=1}^{m}\frac{y_i^2}{\text{sc}_i^2}}.$$

The initial local discretization errors $\text{le}_i(t_0)$ of type 1 or 2 methods of order $p = 1$ given in Examples 1 and 2 of Section 3.13 are

$$\frac{1}{2}h_1^2 y_i''(t_0) \quad \text{and} \quad -\frac{1}{2}h_1^2 y_i''(t_0),$$

respectively, and this suggests the formula

$$h_1 = \sqrt{\frac{2}{\|d_2\|_{\text{sc}}}},$$

where d_2 is an approximation to the second derivative $y''(t_0)$. We compute this approximation from the formula

$$d_2 = \frac{f\big(y_0 + h_0 f(y_0)\big) - f(y_0)}{h_0},$$

where the step size h_0 corresponds to the method of order zero, $y_n = y_{n-1}$, and is given by

$$h_0 = \frac{1}{\|f(y_0)\|_{\text{sc}}}.$$

We start the integration with the input vector $z^{[0]}$ defined by

$$z^{[0]} = \begin{bmatrix} y_0 \\ h_1 f(y_0) \end{bmatrix}.$$

To describe step size and order changing strategy, assume that we have completed a step from t_{n-1} to t_n with a step size h_n and order p_n which resulted in the computation of quantities $Y^{[n]}$ and $z^{[n]}$. We then compute the estimate of the local discretization error (4.2.9) or (4.3.3), which is now denoted by $\text{est}(t_n, p_n)$ or $\text{est}^*(t_n, p_n)$, to indicate their dependence on the current order p_n. We also compute the measure $\text{err}(t_n, p_n)$ or $\text{err}^*(t_n, p_n)$ of the local discretization error by

$$\text{err}(t_n, p_n) = \big\|\text{est}(t_n, p_n)\big\|_{\text{sc}}$$

or

$$\text{err}^*(t_n, p_n) = \big\|\text{est}^*(t_n, p_n)\big\|_{\text{sc}}$$

with

$$\text{sc}_i = \text{Atol}_i + \max\left\{\big\|z^{[n-1]}(1:m)\big\|, \big\|z^{[n]}(1:m)\big\|\right\}\text{Rtol}_i,$$

$i = 1, 2, \ldots, m$, where $z(1 : m)$ stands for the first m components of the vector z, and compare it to 1 to find an optimal step size for the next step. This optimal step size is computed according to the formula

$$h_{\text{opt}} = h_n \cdot r_{\text{opt}} \quad \text{with} \quad r_{\text{opt}} = \left(\frac{1}{\text{err}(t_n, p_n)} \right)^{1/(p_n+1)}.$$

If

$$\text{err}(t_n, p_n) \leq 1 \quad \text{or} \quad \text{err}^*(t_n, p_n) \leq 1,$$

the step is accepted and the new step size h_{n+1} is computed from the formula

$$h_{n+1} = h_n \cdot \min \left\{ \text{facmax}, \text{fac} \cdot r_{\text{opt}} \right\}$$

(compare [143]). Here facmax and fac are safety factors build into a code to prevent the step from increasing too rapidly and, thus to avoid an excessive number of rejected steps. We have chosen facmax = 2 and fac = 0.9.

The choice of the new order p_{n+1} is based on monitoring $\text{est}(t_n, p_n)$ or $\text{est}^*(t_n, p_n)$ as well as the estimate $\text{est}(t_n, p_n-1)$ or $\text{est}^*(t_n, p_n-1)$ (for $p_n > 1$), which correspond to the method of order $p_n - 1$. For $p_n = 1$ we compare the error estimate with the error estimated in the order zero result defined by $y_n = y_{n-1}$. Since the last m components $z^{[n]}(p_n m + 1 : (p_n + 1)m)$ of vector $z^{[n]}$ approximate $h_n^{p_n} y^{(p_n)}(t_n)$, this estimate can easily be obtained from the expression

$$\text{est}(t_n, p_n - 1) = \vartheta_{p_n}(\delta_n) z^{[n]}(p_n m + 1 : (p_n + 1)m),$$

where $\vartheta_{p_n}(\delta_n)$ is the error constant defined by (4.2.5) of method (4.1.5) of order $p_n - 1$ if $p_n > 1$ and $\vartheta_0(\delta_n) = 1$ if $p_n = 1$. The decision about the new order p_{n+1} is then based on the ratio or ratio* defined by

$$\text{ratio} = \frac{\|\text{est}(t_n, p_n)\|}{\|\text{est}(t_n, p_n - 1)\|} \quad \text{or} \quad \text{ratio}^* = \frac{\|\text{est}^*(t_n, p_n)\|}{\|\text{est}^*(t_n, p_n - 1)\|}.$$

If

$$\text{ratio} < r_{\min} \quad \text{or} \quad \text{ratio}^* < r_{\min}, \quad p_n < p_{\max},$$

and the previous step was not rejected, the new order is chosen as $p_{n+1} = p_n + 1$. If

$$\text{ratio} > r_{\max} \quad \text{or} \quad \text{ratio}^* > r_{\max}, \quad p_n > 1,$$

the new order is $p_{n+1} = p_n - 1$. Otherwise, the order is not changed. We have chosen $r_{\min} = 0.9$ and $r_{\max} = 1.1$.

If

$$\text{err}(t_n, p_n) > 1 \quad \text{or} \quad \text{err}^*(t_n, p_n) > 1,$$

the step is rejected and the computations are repeated with a new step size $\widetilde{h}_n$ chosen according to the formula

$$\widetilde{h}_n = h_n \cdot \min \left\{ \text{facmin}, \text{fac} \cdot r_{\text{opt}} \right\},$$

where facmin is a safety factor built into a code to prevent the step size from decreasing too rapidly. We have chosen facmin $= 0.5$.

After a rejected step a new order $\tilde{p}_n$ is never increased and it is chosen according to the following rules. After the first rejection the order is not changed. After the second rejection the order is reduced by 1 (if $p_n > 1$). After the third rejection the order is dropped to $\tilde{p}_n = 1$.

4.6   UPDATING THE VECTOR OF EXTERNAL APPROXIMATIONS

After a successful step from t_{n-1} to t_n is completed, we have to update the vector $z^{[n]} = z^{[n]}(p_n)$ of external approximations so that it corresponds to the new order, p_{n+1}. If $p_{n+1} = p_n$ the updated vector $z^{[n]}(p_{n+1})$ is equal to $z^{[n]}(p_n)$. If $p_{n+1} = p_n + 1$, $z^{[n]}(p_{n+1})$ is computed according to the formula

$$
z^{[n]}(p_{n+1}) = \begin{bmatrix} z^{[n]}(p_n) \\ \mathrm{est}(t_n, p_n) \\ \vartheta_{p_n}(\delta_n) \end{bmatrix} \quad \text{or} \quad z^{[n]}(p_{n+1}) = \begin{bmatrix} z^{[n]}(p_n) \\ \mathrm{est}^*(t_n, p_n) \\ \vartheta_{p_n}(\delta_n) \end{bmatrix},
$$

where $\mathrm{est}(t_n, p_n)/\vartheta_{p_n}(\delta_n)$ or $\mathrm{est}^*(t_n, p_n)/\vartheta_{p_n}(\delta_n)$ is an approximation to the scaled derivative $h_{n+1}^{p+1} y^{(p+1)}(t_n)$ and $\vartheta_{p_n}(\delta_n)$ is the error constant defined by (4.2.5) of method (4.1.5) of order p_n. In this case we also want to find constants $\epsilon_1, \epsilon_2, \ldots, \epsilon_{p_n}$ so that it is appropriate to add corrections of order $O(h_n^{p+1})$ to the vector $z^{[n]}$ when the order is increased, resulting in a modified vector:

$$
\begin{bmatrix} z_2^{[n]}(p_{n+1}) \\ z_3^{[n]}(p_{n+1}) \\ \vdots \\ z_{p_{n+1}}^{[n]}(p_{n+1}) \end{bmatrix} \rightarrow \begin{bmatrix} z_2^{[n]}(p_{n+1}) + \epsilon_1 z_{p_{n+1}+1}^{[n]}(p_{n+1}) \\ z_3^{[n]}(p_{n+1}) + \epsilon_2 z_{p_{n+1}+1}^{[n]}(p_{n+1}) \\ \vdots \\ z_{p_{n+1}}^{[n]}(p_{n+1}) + \epsilon_{p_n} z_{p_{n+1}+1}^{[n]}(p_{n+1}) \end{bmatrix},
$$

where $z_i^{[n]}(p_{n+1})$ stands for the $(i-1)m + 1 : im$ components of the vector $z^{[n]}(p_{n+1})$. We choose ϵ_k so that

$$
\sum_{j=1}^{p_n} g_{k+1,j} h_n y'\big(t_n + (c_j - 1)h_n\big) + \epsilon_k h_n^{p_n+1} y^{(p_n+1)}(t_n) = h_n^k y^{(k)}(t_n) + O(h_n^{p_n+2}),
$$

$k = 1, 2, \ldots, p_n$. If y is a polynomial of degree $p_n + 1$, the two sides of this equation will be equal with the $O(h_n^{p_n+2})$ term missing. Hence, substituting $y'(t_n + \eta h_n) = \varphi(\eta)$, where $\varphi(\eta)$ is a polynomial of degree p_n, we find that

$$
\sum_{j=1}^{p_n} g_{k+1,j} \varphi(c_j - 1) + \epsilon_k \varphi^{(p_n)}(0) = \varphi^{(k-1)}(0).
$$

Choosing

$$\varphi(\eta) = \prod_{j=1}^{p_n}(\eta + 1 - c_j)$$

so that the g terms vanish, it follows that

$$\epsilon_k = \frac{\varphi^{(k-1)}(0)}{p!}.$$

It is convenient to calculate ϵ_k, $k = 1, 2, \ldots, p_n$, from the coefficients in the expansion of $\varphi(\eta)$:

$$\prod_{j=1}^{p_n}(\eta + 1 - c_j) = \eta^{p_n} + \alpha_{p_n}\eta^{p_n - 1} + \cdots + \alpha_2\eta + \alpha_1,$$

and we find that

$$\epsilon_k = \frac{(k-1)!\alpha_k}{p_n!}.$$

If the order is reduced by 1 (i.e., $p_{n+1} = p_n - 1 \geq 1$), the new vector $z^{[n]} = z^{[n]}(p_{n+1})$ is given by

$$z^{[n]}(p_{n+1}) = z^{[n]}(1 : p_n m).$$

If the step from t_{n-1} to t_n is rejected, the computations are repeated with a new step size $\widetilde{h}_n$ defined in Section 4.5 and a new order $\widetilde{p}_n$ equal to $\widetilde{p}_n = p_n$, $\widetilde{p}_n = p_n - 1$, or $\widetilde{p}_n = 1$. The vector $z^{[n-1]}(\widetilde{p}_n)$ is then updated as follows:

$$z^{[n-1]}(\widetilde{p}_n) = z^{[n-1]} \quad \text{if} \quad \widetilde{p}_n = p_n,$$

$$z^{[n-1]}(\widetilde{p}_n) = z^{[n-1]}(1 : p_n m) \quad \text{if} \quad \widetilde{p}_n = p_n - 1,$$

or

$$z^{[n-1]}(\widetilde{p}_n) = z^{[n-1]}(1 : 2m) \quad \text{if} \quad \widetilde{p}_n = 1.$$

4.7 STEP-CONTROL STABILITY OF DIMSIMS

Step-control stability of RK methods was investigated by Hall [150, 151] and by Hall and Higham [152, 157]. A very nice summary of these results can be found in the book by Hairer and Wanner [146]. In this section we follow Jackiewicz [176] and extend these ideas to DIMSIMs (4.1.5).

Assume that the step size of method (4.1.5) of order p is chosen according to the strategy

$$h_{n+1} = h_n \left(\frac{\theta\,\mathrm{Tol}}{\|\mathrm{est}(t_n)\|}\right)^{1/(p+1)}, \tag{4.7.1}$$

where θ is a safety parameter build into the scheme to reduce the number of rejected steps, Tol is a given error tolerance, and $\mathrm{est}(t_n)$ is an estimate of local discretization error discussed in Section 4.2 (compare (4.2.9)). This strategy is equivalent, in principle, to the step size changing strategy discussed in Section 4.5. To study the behavior of (4.7.1), we again use the basic test equation

$$y' = \xi y, \quad t \geq 0, \tag{4.7.2}$$

where ξ is a complex parameter. Applying (4.1.5) to (4.7.2), then eliminating $Y^{[n]}$ from the resulting equations, we obtain the following dynamical system:

$$
\begin{aligned}
z^{[n]} &= \Big(\mathbf{Q} + h_n\xi\mathbf{G}\big(\mathbf{I} - h_n\xi\mathbf{A}\big)^{-1}\mathbf{P}\Big)\mathbf{D}(\delta_n)z^{[n-1]}, \\
h_{n+1} &= h_n\left(\frac{\theta\,\mathrm{Tol}}{|\mathrm{est}(t_n)|}\right)^{1/(p+1)}, \\
\mathrm{est}(t_n) &= \vartheta_p(\delta_n)\Omega(h_n\xi, \delta_n)z^{[n-1]}
\end{aligned}
\tag{4.7.3}
$$

with

$$\Omega(h_n\xi, \delta_n) := \gamma^T(\delta_n) + h_n\xi\beta^T(\delta_n)(\mathbf{I} - h_n\xi\mathbf{A})^{-1}\mathbf{P}\mathbf{D}(\delta_n).$$

This dynamical system defines a map

$$\big(z^{[n-1]}, z_n, z_{n-1}\big) \to \big(z^{[n]}, z_{n+1}, z_n\big)$$

from $\mathbb{C}^{s+3}$ to $\mathbb{C}^{s+3}$, $z_n = h_n\xi$. Its fixed points for $\delta_n = 1$, $h_n = h$, satisfy the system

$$
\begin{aligned}
\big(\mathbf{Q} + z\mathbf{G}(\mathbf{I} + z\mathbf{A})^{-1}\mathbf{P}\big)\overline{z} &= \overline{z}, \\
\Big|\vartheta_p(1)\big(\gamma^T(1) + z\beta^T(1)(\mathbf{I} - z\mathbf{A})^{-1}\mathbf{P}\big)\overline{z}\Big| &= \theta\,\mathrm{Tol},
\end{aligned}
\tag{4.7.4}
$$

$z = h\xi$, $\overline{z} \in \mathbb{C}^{s+1}$. We recognize $\mathbf{M}(z) = \mathbf{Q} + z\mathbf{G}(\mathbf{I} - z\mathbf{A})^{-1}\mathbf{P}$ as the stability matrix of (4.1.5). This matrix, in the case of DIMSIMs of type 1, have one eigenvalue $w_1(z)$ equal to the stability function $R(z)$ of the explicit RK method of the same order with remaining eigenvalues $w_i(z)$, $i = 2, 3, \ldots, s+1$, equal to zero.

It follows from the first relation of (4.7.4) that $(\overline{z}, z)$ is a fixed point of (4.7.3) only if $\overline{z}$ is an eigenvector of $\mathbf{M}(z)$ corresponding to the eigenvalue equal to 1. This is always the case for a discrete number of points z on the boundary $\partial\mathcal{A}$ of the region of absolute stability $\mathcal{A}$ of method (4.1.5), in particular for $z = 0$. To extend the analysis to all values of $z \in \partial\mathcal{A}$, we have to relax the condition given by the first relation of (4.7.4) and consider the eigenvalues of $\mathbf{M}(z)$ of modulus equal to 1:

$$\big|w_1(z)\big| = \big|R(z)\big| = 1 \tag{4.7.5}$$

and

$$w_i(z) = 0, \quad i = 2, 3, \ldots, s+1. \tag{4.7.6}$$

(In fact, this was also done in the case of RK methods examined by Hairer and Wanner [146], where system (2.28) was replaced by (2.30)). Let $\mathbf{S}(z)$ be a nonsingular matrix such that

$$
\mathbf{S}^{-1}(z)\mathbf{M}(z)\mathbf{S}(z) =
\left[
\begin{array}{c|c}
w_1(z) & \mathbf{0} \\
\hline
\mathbf{0} & \widetilde{\mathbf{M}}(z)
\end{array}
\right].
$$

We consider now instead of the first relation of (4.7.4) the system

$$
\big|w_1(z)\big|\,|\eta_1| = |\eta_1|,
$$

$$
\widetilde{\mathbf{M}}(z)
\left[\begin{array}{ccc} \eta_2 & \cdots & \eta_{s+1} \end{array}\right]^T
=
\left[\begin{array}{ccc} \eta_2 & \cdots & \eta_{s+1} \end{array}\right]^T,
\tag{4.7.7}
$$

where $\eta = \mathbf{S}^{-1}(z)\overline{z}$. For $z \in \partial\mathcal{A}$ it follows from (4.7.5) that the first relation is satisfied automatically, and since (4.7.6) implies that $\rho(\widetilde{\mathbf{M}}(z)) = 0$, the second relation of (4.7.7) reduces to

$$
\eta_i = 0, \quad i = 2, 3, \ldots, s+1.
\tag{4.7.8}
$$

System (4.7.8) can be used together with the second relation of (4.7.4) to determine $\overline{z}$ for $z \in \partial\mathcal{A}$.

Since the stability of the map defined by (4.7.3) at the fixed point $(\overline{z}, z)$ defined as described above depends on the spectral radius of the Jacobian matrix $\mathbf{J}_{\mathrm{SC}}(\overline{z}, z)$ of this map computed at the fixed point $(\overline{z}, z)$, we can conclude that the step size control mechanism based on (4.7.1) with error estimate $\mathrm{est}(t_n)$ given by (4.2.9) is stable if

$$
\rho(\mathbf{J}_{\mathrm{SC}}(\overline{z}, z)) < 1.
$$

In such a case method (4.1.5) is said to be SC-stable for this value of $z \in \partial\mathcal{A}$. We also define the region of SC-stability as a subset of $\partial\mathcal{A}$ given be

$$
\partial\mathcal{A}_{\mathrm{SC}} = \Big\{z \in \partial\mathcal{A}: \ \rho(\mathbf{J}_{\mathrm{SC}}(\overline{z}, z)) < 1\Big\}.
$$

To illustrate the theory discussed above consider the type 1 method of order $p = 2$ given in Example 3 in Section 3.13. For this method the error constant is given by $\vartheta_2(\delta) = (3+\delta)/(24\delta)$, and the vectors $\beta(\delta)$ and $\gamma(\delta)$ appearing in (4.2.9) are

$$
\beta(\delta) = \left[\begin{array}{cc} -\dfrac{2\delta}{1+\delta} & \dfrac{2\delta}{1+\delta} \end{array}\right]^T, \quad
\gamma(\delta) = \left[\begin{array}{ccc} 0 & 0 & -\dfrac{2\delta^3}{1+\delta} \end{array}\right]^T
$$

(compare Section 4.2). We have applied this method to the system considered by Hairer and Wanner [146]:

$$
\begin{aligned}
y_1' &= -2000\big(\cos(t)y_1 + \sin(t)y_2 + 1\big), & y_1(0) &= 1, \\
y_2' &= -2000\big(-\sin(t)y_1 + \cos(t)y_2 + 1\big), & y_2(0) &= 0,
\end{aligned}
\tag{4.7.9}
$$

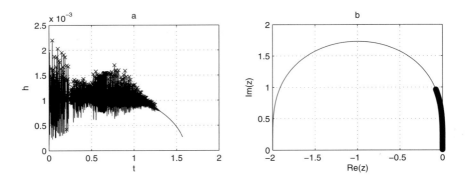

Figure 4.7.1   (a) Step sizes versus t for the method of order $p = 2$ applied to (4.7.2) with Tol $= 10^{-2}$ and $\theta = 5/6$ with step size changing strategy given by (4.7.1). (b) Region of absolute stability and SC-stability of the method of order $p = 2$

whose eigenvalues move on the circle from -2000 to $-2000 \pm 2000\,i$ as t goes from 0 to $\pi/2$. The resulting step size pattern for Tol $= 10^{-2}$ and $\theta = 5/6$ is plotted in Fig. 4.7.1a, where we have marked by the symbol "$\times$" all rejected steps. There were 835 steps rejected out of the total number of 1877.

To explain this behavior we have plotted in Fig. 4.7.1b the region of absolute stability of the method used to generate this step size pattern with the region of SC-stability represented by the thick line. For this method stability matrix $\mathbf{M}(z)$ takes the form

$$\mathbf{M}(z) = \begin{bmatrix} \dfrac{2 + 3z + z^2}{2} & -\dfrac{2 + z}{4} & \dfrac{2 + z}{8} \\[2ex] z(1 + 2z) & -z & \dfrac{z}{2} \\[2ex] 2z^2 & -z & \dfrac{z}{2} \end{bmatrix}$$

and it can be reduced to the Jordan canonical form

$$\mathbf{S}^{-1}(z)\mathbf{M}(z)\mathbf{S}(z) = \left[\begin{array}{c|cc} \dfrac{2 + 2z + z^2}{2} & 0 & 0 \\[2ex] \hline 0 & 0 & 1 \\[1ex] 0 & 0 & 0 \end{array} \right]$$

with

$$\mathbf{S}(z) = \begin{bmatrix} \dfrac{(2+z)(2+2z+z^2)}{4z^2(1+z)} & 0 & -\dfrac{1}{2z} \\[2ex] \dfrac{2+3z+2z^2}{2z(1+z)} & \dfrac{1}{2} & -\dfrac{2+z}{2z} \\[2ex] 1 & 1 & 1 \end{bmatrix}.$$

It can be verified that the system to determine fixed points of (4.7.3) consisting of (4.7.8) and the second equation of (4.7.4) now takes the form

$$4z^2(4+6z+3z^2)\bar{z}_1 - 2z(4+8z+6z^2+z^3)\bar{z}_2$$

$$-(8+12z+8z^2+2z^3+z^4)\bar{z}_3 = 0,$$

$$4z\bar{z}_1 - 2(2+z)\bar{z}_2 + (2+z)\bar{z}_3 = 0,$$

$$\left|4z^2\bar{z}_1 - 2z\bar{z}_2 - (2-z)\bar{z}_3\right| = 12\,\theta\,\text{Tol}.$$

The solution to this system is given by

$$\bar{z}_1 = \frac{3\,\theta\,\text{Tol}\,(2+z)(2+2z+z^2)}{z^2(1+z)\left|z(2+z)\right|},$$

$$\bar{z}_2 = \frac{6\,\theta\,\text{Tol}\,(2+3z+2z^2)}{z(1+z)\left|z(2+z)\right|}, \quad \bar{z}_3 = \frac{12\,\theta\,\text{Tol}}{\left|z(2+z)\right|}.$$

We can observe that the smooth behavior of the step size changing mechanism for t greater than about 1.25 corresponds to the region of SC-stability plotted in Fig. 4.7.1b as predicted approximately by this theory.

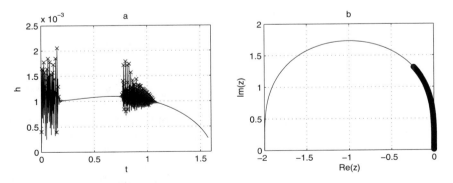

Figure 4.7.2   (a) Step sizes versus t for the method of order $p = 2$ applied to (4.7.2) with Tol $= 10^{-2}$ and $\theta = 5/6$ with $textrmPI$ step size control. (b) Region of absolute stability and PISC-stability of the method of order $p = 2$

It was demonstrated by Gustafson et al. [140] that the instability of the step size changing strategy based on (4.7.1) can usually be resolved in the case of RK methods by using PI step size control motivated by the concepts from control theory instead of (4.7.1). To analyze this algorithm in the context of DIMSIMs, we again apply (4.1.5) to (4.7.2) and replace (4.7.1) by PI step size control to obtain the dynamical system:

$$z^{[n]} = \Big(\mathbf{Q} + h_n \xi \mathbf{G} \big(\mathbf{I} - h_n \xi \mathbf{A}\big)^{-1} \mathbf{P}\Big) \mathbf{D}(\delta_n) z^{[n-1]},$$

$$h_{n+1} = h_n \left(\frac{\theta \, \mathrm{Tol}}{\big|\mathrm{est}(t_n)\big|}\right)^\alpha \left(\frac{\big|\mathrm{est}(t_{n-1})\big|}{\theta \, \mathrm{Tol}}\right)^\beta, \tag{4.7.10}$$

$$\mathrm{est}(t_n) = \vartheta_p(\delta_n) \Omega(h_n \xi, \delta_n) z^{[n-1]}$$

with $\Omega(h_n \xi, \delta_n)$ defined as in formula (4.7.3). Here α and β are constants. Observe that (4.7.10) now defines a map

$$\big(z^{[n-1]}, z_n, z^{[n-2]}, z_{n-1}, z_{n-2}\big) \to \big(z^{[n]}, z_{n+1}, z^{[n-1]}, z_n, z_{n-1}\big)$$

from $\mathbb{C}^{2s+5}$ to $\mathbb{C}^{2s+5}$, since $\mathrm{est}(t_{n-1})$ depends on $\delta_{n-1} = h_{n-1}/h_{n-2} = z_{n-1}/z_{n-2}$. The fixed points of (4.7.10) satisfy system (4.7.4), where, as before, the first equation of (4.7.4) can be replaced by (4.7.8). Denote by $\mathbf{J}_{\mathrm{PISC}}(\overline{z}, z)$ the Jacobian of the map (4.7.10) at the fixed point $(\overline{z}, z)$ defined by (4.7.4). Method (4.1.5) is said to be PISC-stable at $z \in \partial \mathcal{A}$ if

$$\rho\big(\mathbf{J}_{\mathrm{PISC}}(\overline{z}, z)\big) < 1.$$

We also define the region of PISC-stability as a subset of $\partial \mathcal{A}$ given by

$$\partial \mathcal{A}_{\mathrm{PISC}} = \Big\{z \in \partial \mathcal{A} : \ \rho\big(\mathbf{J}_{\mathrm{PISC}}(\overline{z}, z)\big) < 1\Big\}.$$

To illustrate the new approach we have applied the method given in Example 3 in Section 3.13 to the system (4.7.9), where we have replaced the step size control given by (4.7.1) by PI step size control defined by the second equation of (4.7.10). We have plotted in Fig. 4.7.2a the resulting step size pattern for Tol $= 10^{-2}$ and $\theta = 5/6$ for the parameters $\alpha = 0.175$ and $\beta = 0.089$, where we have again marked all rejected steps by the symbol "×". There were now 391 rejected steps out of total number of 1806. We have also plotted in Fig. 4.7.2b the region of PISC-stability of this method and we can observe that the actual behavior of the PI step size control mechanism is somewhat better than that predicted by this region. Fig. 4.7.2a suggests that this region of PISC-stability should contain another segment of the boundary of the region of absolute stability $\partial \mathcal{A}$ somewhere between $-2 < \mathrm{Re}(z) < -1$, which was not detected by this analysis. The possible explanation of this is that the Jacobian matrix $\mathbf{J}_{\mathrm{PISC}}$ of the map (4.7.10) controls the behavior of all components of the vector $z^{[n]}$ of external approximations, while only the quantity $\mathbf{Q}\mathbf{D}(\delta_n) z^{[n]}$ propagates to the next step.

The step-control stability of DIMSIMs for estimators of local discretization error different from those examined in this section was investigated by Jackiewicz [176].

4.8 SIMPLIFIED NEWTON ITERATIONS FOR IMPLICIT METHODS

If the function f appearing in (2.1.1) is nonlinear, system (4.1.5) for $Y^{[n]}$ has to be solved iteratively. In the context of stiff differential systems it is advantageous to use for this purpose some variant of the Newton method. This method will not be applied directly to (4.1.5), but to reduce the influence of round-off errors we adopt the suggestion of Hairer and Wanner [146], which was made in the context of implicit RK methods, and introduce the quantities

$$\eta^{[n]} = Y^{[n]} - \big(\mathbf{PD}(\delta_n) \otimes \mathbf{I}\big) z^{[n-1]},$$

which are, in general, smaller than $Y^{[n]}$. Then method (4.1.5) reformulated in terms of $\eta^{[n]}$ becomes

$$
\begin{aligned}
\eta^{[n]} &= h_n(\mathbf{A} \otimes \mathbf{I}) F\big(\eta^{[n]} + \big(\mathbf{PD}(\delta_n) \otimes \mathbf{I}\big) z^{[n-1]}\big), \\
z^{[n]} &= h_n(\mathbf{G} \otimes \mathbf{I}) F\big(\eta^{[n]} + \big(\mathbf{PD}(\delta_n) \otimes \mathbf{I}\big) z^{[n-1]}\big) \\
&+ \big(\mathbf{QD}(\delta_n) \otimes \mathbf{I}\big) z^{[n-1]},
\end{aligned}
\tag{4.8.1}
$$

$n = 1, 2, \ldots, N$. Since the coefficient matrix $\mathbf{A}$ corresponding to type 2 DIMSIMs is nonsingular, we have

$$h_n F\big(\eta^{[n]} + \big(\mathbf{PD}(\delta_n) \otimes \mathbf{I}\big) z^{[n-1]}\big) = (\mathbf{A}^{-1} \otimes \mathbf{I})\eta^{[n]} \tag{4.8.2}$$

and method (4.8.1) can be reformulated further as follows:

$$
\begin{aligned}
\eta^{[n]} &= h_n(\mathbf{A} \otimes \mathbf{I}) F\big(\eta^{[n]} + \big(\mathbf{PD}(\delta_n) \otimes \mathbf{I}\big) z^{[n-1]}\big), \\
z^{[n]} &= (\mathbf{GA}^{-1} \otimes \mathbf{I})\eta^{[n]} + \big(\mathbf{QD}(\delta_n) \otimes \mathbf{I}\big) z^{[n-1]},
\end{aligned}
\tag{4.8.3}
$$

$n = 1, 2, \ldots, N$. As observed by Shampine [254], again in the context of implicit RK methods, this formulation does not amplify iteration errors that affect the computation of $\eta^{[n]}$.

Using (4.8.2), we can also reformulate the error estimate (4.2.9) of the local discretization error. This leads to the formula

$$\mathrm{est}(t_n, p_n) = \big(\beta^T(\delta_n)\mathbf{A}^{-1} \otimes \mathbf{I}\big)\eta^{[n]} + \big(\gamma^T(\delta_n) \otimes \mathbf{I}\big) z^{[n-1]}, \tag{4.8.4}$$

where we again indicated the dependence of the error estimate on the order of the method p_n.

The vector $\eta^{[n]}$ will be computed by simplified Newton iterations. Setting $\eta = \eta^{[n]}$ and denoting by λ the diagonal element of the coefficient matrix $\mathbf{A}$, the first relation of (4.8.3) can be written as

$$\eta_i - h_n \lambda f\left(\eta_i + \sum_{k=1}^{s+1} p_{ik}\delta_n^{k-1} z_k^{[n-1]}\right) - h_n \sum_{j=1}^{i-1} a_{ij} f\left(\eta_j + \sum_{k=1}^{s+1} p_{jk}\delta_n^{k-1} z_k^{[n-1]}\right) = 0,$$

$i = 1, 2, \ldots, s$. The Newton iterations take the form

$$(\mathbf{I} - h_n \lambda \mathbf{J}_i)\Delta\eta_i^\nu = -\eta_i^\nu + h_n \lambda f\left(\eta_i^\nu + \sum_{k=1}^{s+1} p_{ik}\delta_n^{k-1} z_k^{[n-1]}\right)$$

$$+ h_n \sum_{j=1}^{i-1} a_{ij} f\left(\eta_j + \sum_{k=1}^{s+1} p_{jk}\delta_n^{k-1} z_k^{[n-1]}\right), \tag{4.8.5}$$

$$\eta_i^{\nu+1} = \eta_i^\nu + \Delta\eta_i^\nu,$$

$\nu = 0, 1, \ldots$, $i = 1, 2, \ldots, s$, where η_i^0 is the given initial guess and $\mathbf{J}_i$ is an approximation to the Jacobian matrix

$$\frac{\partial f}{\partial y}\left(\eta_i^\nu + \sum_{k=1}^{s+1} p_{ik}\delta_n^{k-1} z_k^{[n-1]}\right) = \frac{\partial f}{\partial y}\left(Y_i^{[n]}\right).$$

Since

$$Y_i^{[n]} \approx y(t_{n-1} + c_i h_n) = \sum_{k=1}^{s+1} \frac{c_i^{k-1}\delta_n^{k-1}}{(k-1)!} z_k^{[n-1]} + O(h_n^{s+1}),$$

we define

$$\mathbf{J}_i = \frac{\partial f}{\partial y}\left(\sum_{k=1}^{s+1} \frac{c_i^{k-1}\delta_n^{k-1}}{(k-1)!} z_k^{[n-1]}\right),$$

$i = 1, 2, \ldots, s$. To simplify the iteration process (4.8.5) further and to make computations more efficient, we replace all Jacobians $\mathbf{J}_i$, $i = 1, 2, \ldots, s$, by the matrix

$$\mathbf{J} = \frac{\partial f}{\partial y}\left(\sum_{k=1}^{s+1} \frac{\bar{c}^{k-1}\delta_n^{k-1}}{(k-1)!} z_k^{[n-1]}\right),$$

where $\bar{c}$ is defined by

$$\bar{c} = \frac{1}{s}\sum_{i=1}^{s} c_i.$$

Then the matrix $\mathbf{I} - h_n \lambda \mathbf{J}$ is the same for all iterations $\nu = 0, 1, \ldots$, and all stages $i = 1, 2, \ldots, s$, and its LU decomposition has to be computed only once. This matrix will also be used to compute the modified error estimates $\text{est}^*(t_n, p_n)$ defined by (4.3.3) and discussed in Section 4.3. Since

$$\eta_i \approx y(t_{n-1} + c_i h_n) - \sum_{k=1}^{s+1} p_{ik} \delta_n^{k-1} z_k^{[n-1]},$$

we define the initial guess to start the iterations as

$$\eta_i^0 = \sum_{k=1}^{s+1} \left(\frac{c_i^{k-1} \delta_n^{k-1}}{(k-1)!} - p_{ik} \delta_n^{k-1} \right) z_k^{[n-1]},$$

$i = 1, 2, \ldots, s$. Following Hairer and Wanner [146], we stop the simplified Newton iterations when

$$\rho_\nu \big\| \Delta \eta_i^\nu \big\|_{\text{SC}} \le \kappa \quad \text{or} \quad \nu > \nu_{\max} = 10,$$

where

$$\rho_\nu = \frac{\theta_\nu}{1 - \theta_\nu}, \quad \theta_\nu = \frac{\big\| \Delta \eta_i^\nu \big\|_{\text{SC}}}{\big\| \Delta \eta_i^{\nu-1} \big\|_{\text{SC}}},$$

$\nu \ge 1$, $\kappa = 0.01$, where $\| \cdot \|_{\text{SC}}$ is the norm defined in Section 4.5. For $\nu = 0$ we define

$$\rho_0 = \Big(\max \{ \rho_{\text{old}}, \text{eps} \} \Big)^{0.8},$$

where ρ_{old} is the last ρ_ν from the preceding step and eps is the machine epsilon. If there is no convergence of simplified Newton iterations, we restart the computations with the halved step size $\widetilde{h}_n = h_n/2$. Other variants of stopping criteria are discussed by Hairer and Wanner [146] and by Shampine [254].

4.9   NUMERICAL EXPERIMENTS WITH TYPE 1 DIMSIMS

To test type 1 DIMSIMs we have written a variable step size variable order experimental code `dim18` based on explicit methods of order $1 \le p \le 8$ in Nordsieck representation. The coefficients of these methods of order $1 \le p \le 3$ are given in Section 3.13, and the coefficients of methods of order $4 \le p \le 8$ are given by Butcher et al. [58]. The implementation details are discussed in Sections 4.2, 4.4, 4.5, and 4.6 as well as by Butcher et al. [58]. The local discretization error of these methods was estimated by the formula (4.2.9) for $p = 1$ and $p = 2$ and by the formula (4.2.10) for $3 \le p \le 8$. These estimates differ from the error estimates employed in the earlier version of this code, discussed by Butcher et al. [58]. This code was applied to many problems to test its accuracy, efficiency, the reliability of the local error estimation, and robustness of step size and order changing strategy. We present in this section a selection of numerical results on the problems SCALAR (1.2.1),

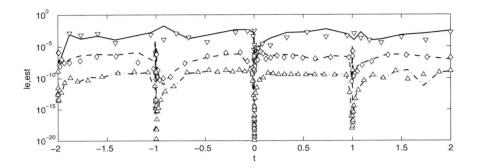

Figure 4.9.1   Local errors and local error estimates versus t for the SCALAR problem (1.2.1)

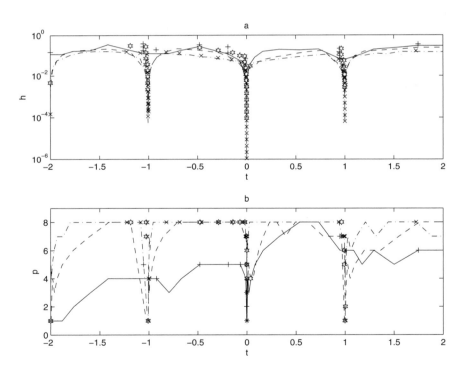

Figure 4.9.2   (a) Step size versus t and (b) order versus t for the SCALAR problem (1.2.1)

AREN (1.2.3), LRNZ (1.2.4), ROPE defined by (1.2.7) or (1.2.8), and BRUS (1.2.10). These problems are presented in Section 1.2.

As observed in Section 1.2, the solution to the SCALAR problem (1.2.1) has a discontinuity in the first derivative y' at the point $t = 0$ and discontinuities in the second derivative y'' at $t = -1$ and $t = 1$, and the step size and order changing mechanism adjusts its step size and order accordingly in their neighborhoods so that the approximation to the solution is computed to a sufficient accuracy. We have plotted in Fig. 4.9.1 the local errors and the local error estimates for Atol = Rtol = Tol = 10^{-3} (solid line, downward triangle), 10^{-6} (dashed line, diamond), and 10^{-9} (dashed-dotted line, upward triangle). The corresponding step size and order patterns are plotted in Fig. 4.9.2, where we have used a solid line, a dashed line, and a dashed-dotted line and the symbols "+", five-pointed star, and "×" to indicate rejected steps for Tol = 10^{-3}, Tol = 10^{-6}, and Tol = 10^{-9}, respectively.

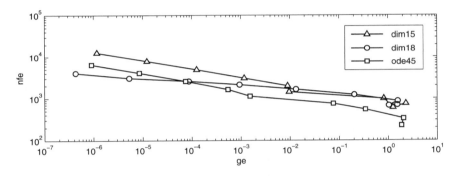

Figure 4.9.3 Number of function evaluations nfe versus global error ge for the AREN problem (1.2.2)

The AREN problem (1.2.3) was solved for $\mu = 0.012277471$ and the initial conditions

$$y_1(0) = 0.994, \quad y_1'(0) = 0, \quad y_2(0) = 0, \quad y_2'(0) = -2.001585106379.$$

As remarked in Section 1.2, this corresponds to the periodic solution of the earth-moon system with the period of motion T_1 given by $T_1 = 17.06522$. This problem was solved on the interval $[0, T_1]$, and a selection of numerical results is presented in Fig. 4.9.3. On this figure we have plotted nfe, the number of evaluations of the function f, versus ge, the global error at the end of the interval of integration for our experimental code `dim18` based on explicit DIMSIMs of order $1 \leq p \leq p_{\max} = 8$, and for the `ode45` code from the Matlab ODE suite [263]. This code is based on explicit RK pair DOPRI5 of order 4 and 5 constructed by Dormand and Prince [117]. This code uses local extrapolation, so it is effectively of order $p = 5$. For this reason we have

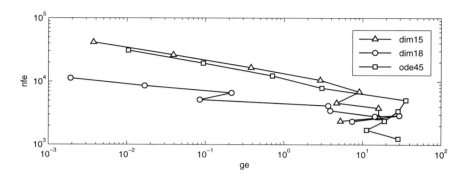

Figure 4.9.4   Number of function evaluations nfe versus global error ge for the LRNZ problem (1.2.3)

also plotted for comparison the numerical results obtained for `dim15`, which correspond to our DIMSIM code where the maximum order is restricted to $p_{max} = 5$.

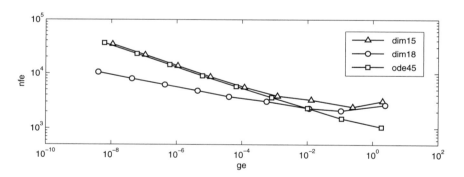

Figure 4.9.5   Number of function evaluations nfe versus global error ge for the ROPE problem defined by (1.2.7) or (1.2.8)

The LRNZ problem (1.2.4) was solved on the interval $[0, 16]$ for the parameters $b = 8/3$, $\sigma = 10$, and $r = 28$, which correspond to aperiodic solution and the initial values given by

$$y_1(0) = -8, \quad y_2(0) = 8, \quad y_3(0) = 27.$$

This solution is very sensitive to round-off errors, and it is very difficult to compute accurate approximations (compare the discussion of this topic in [143]). As $t \to \infty$ the solution approaches the famous Lorenz strange attrac-

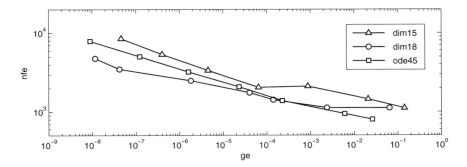

Figure 4.9.6   Number of function evaluations nfe versus global error ge for the BRUS problem (1.2.10)

tor. A selection of the results of numerical experiments with dim18, ode45, and dim15 is given in Fig. 4.9.4.

The results of numerical experiments on the ROPE problem defined by (1.2.7) or (1.2.8) are presented in Fig. 4.9.5.

The BRUS (1.2.10) was solved for $N = 21$, so the dimension of the system is 882. This problem was solved on the interval $[0, 7.5]$. The results of numerical experiments are presented in Fig. 4.9.6.

These work-precision graphs presented in Figs. 4.9.3–4.9.6 clearly indicate the advantage of high order DIMSIMs of type 1 and illustrate the high potential of these methods. The code dim18 based on explicit DIMSIMs of order $1 \leq p \leq p_{\max} = 8$ is more efficient than both dim15 and ode45 for moderate and stringent tolerances. We can also observe that the code dim15 is somewhat less efficient than ode45.

4.10   NUMERICAL EXPERIMENTS WITH TYPE 2 DIMSIMS

Although A- and L-stable DIMSIMs up to order $p = 8$ have already been constructed in the literature (compare [66, 68, 74, 179], and Sections 3.10 and 3.13), so far only methods of order $p = 1$, 2, and 3 have been implemented in the experimental variable step size variable order Matlab code dim13s for the numerical solution of stiff differential systems. This code is described by Jackiewicz [177], and various implementation issues related to this code are also discussed in this chapter. One reason for this is that the development of the code based on methods of higher orders requires, among other things, the derivation of suitable error estimators for small and large step sizes as described in Sections 4.2 and 4.3. This work, which requires a significant power of symbolic manipulation packages, has not yet been undertaken for type 2 DIMSIMs of order $p \geq 4$.

The numerical evidence to date indicates that the code dim13s is not as effective as the code ode15s from the Matlab ODE suite [263], even if the maximal order of ode15s is restricted to $p = 3$. However, there is a restricted class of problems for which the eigenvalues of the Jacobian matrix are purely imaginary, for which dim13s outperforms ode15s by a few orders of magnitude. An example of such problem is BEAM [146], which is also reproduced in Section 1.8.

In what follows we present the results of numerical experiments on two problems: the PLATE problem (1.8.5), and the BEAM problem (1.8.6). These problems are described by Hairer and Wanner [146] and in Section 1.8. They were solved by the experimental code dim13s described by Jackiewicz [177]. The selection of numerical results for the PLATE problem is given in Table 4.10.1 and for the BEAM problem in Table 4.10.2. For comparison we have also presented in Tables 4.10.3 and 4.10.4 the numerical results obtained by the code ode15s with the order restricted to $p = 3$. This code will be referred to as ode13s. In these tables Tol is the required tolerance, ns is the number of steps, nrs is the number of rejected steps, nfe is the number of functions evaluations, npd is the number of partial derivatives, nlu is the number of LU decompositions, nls is the number of linear solves, and ge is the global error at the end of the interval of integration. The selection of numerical results is also presented in Figs. 4.10.1 and 4.10.2, where we have listed nfe versus ge, and where in addition to the results corresponding to dim13s and ode13s, we have also plotted the results corresponding to ode15s.

Tol	ns	nrs	nfe	npd	nlu	nls	ge
10^{-4}	30	7	432	3	29	154	$1.15 \cdot 10^{-3}$
10^{-6}	108	6	834	3	54	479	$3.78 \cdot 10^{-5}$
10^{-8}	391	6	2377	3	178	1739	$2.15 \cdot 10^{-7}$
10^{-10}	1315	6	7349	3	527	5787	$3.42 \cdot 10^{-9}$

Table 4.10.1   Numerical results for the PLATE problem solved by dim13s

Comparing the results presented in Tables 4.10.1 and 4.10.3 and in Fig. 4.10.1, we can see that dim13s is not competitive with ode13s for the PLATE problem. This can be explained by the fact that the eigenvalues of the Jacobian of this problem lie in the sector

$$S_\alpha = \left\{ z \in \mathbb{C} : \left| \arg(-z) \right| < \alpha \right\}$$

for $\alpha \approx 71°$ (compare [146, p. 146]). Hence, they fall easily into the region of absolute stability of the Klopfenstein-Shampine numerical differentiation formula of order $p = 3$, NDF3, employed in ode13s when scaled by any step

Tol	ns	nrs	nfe	npd	nlu	nls	ge
10^{-4}	89	15	5286	56	115	667	$5.64 \cdot 10^{-2}$
10^{-6}	406	47	7418	55	276	2540	$1.38 \cdot 10^{-3}$
10^{-8}	2079	85	16703	43	785	11100	$1.71 \cdot 10^{-4}$
10^{-10}	11315	24	68161	26	1692	54770	$5.22 \cdot 10^{-6}$

Table 4.10.2   Numerical results for the BEAM problem solved by `dim13s`

Tol	ns	nrs	nfe	npd	nlu	nls	ge
10^{-4}	54	6	197	1	18	115	$1.84 \cdot 10^{-4}$
10^{-6}	157	13	392	1	38	310	$3.45 \cdot 10^{-6}$
10^{-8}	465	21	844	1	64	762	$9.33 \cdot 10^{-8}$
10^{-10}	1429	26	2145	1	120	2063	$3.42 \cdot 10^{-9}$

Table 4.10.3   Numerical results for the PLATE problem solved by `ode13s`

Tol	ns	nrs	nfe	npd	nlu	nls	ge
10^{-4}	165160	5567	210865	1	18504	210783	$1.02 \cdot 10^{+2}$
10^{-6}	295464	2787	312413	1	14079	312331	$6.02 \cdot 10^{-1}$
10^{-8}	315169	2501	330435	1	12649	330353	$8.01 \cdot 10^{-3}$
10^{-10}	314672	2446	473518	1	12522	473436	$8.96 \cdot 10^{-5}$

Table 4.10.4   Numerical results for the BEAM problem solved by `ode13s`

size h. This formula is $A(\alpha)$-stable for $\alpha \approx 80°$ (compare [263]). The code `dim13s` requires a smaller number of steps than does `ode13s` for this problem. However, all numerical differentiation methods require only one function evaluation per step, whereas the DIMSIM of order p requires p function evaluations per step, and this leads to a higher overall cost for `dim13s` than for `ode13s`.

Comparing the results in Tables 4.10.2 and 4.10.4 and in Fig. 4.10.2, we can see, however, that `dim13s` outperforms `ode13s` on the BEAM problem by a few orders of magnitude. This can be explained by the fact that the eigenvalues of the Jacobian matrix for this problem are purely imaginary and vary between $-6400i$ and $6400i$ for $n = 40$ (compare [146, p 201]). As a result,

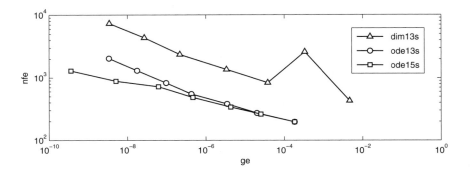

Figure 4.10.1   Number of function evaluations nfe versus global error ge for the discretization of the PLATE problem (1.8.5)

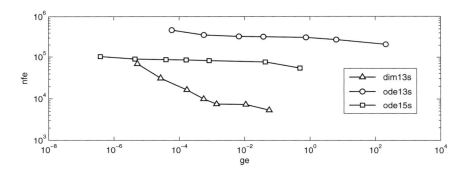

Figure 4.10.2   Number of function evaluations nfe versus global error ge for the BEAM problem (1.8.6)

they never fall into the region of absolute stability of the NDF3 method when scaled by any step size h. On the other hand, since the DIMSIM method of order $p = 3$ employed in dim13s is A-stable, these eigenvalues fall into the region of absolute stability when scaled by any h. We can also observe that the dim13s code is more accurate for this problem than ode13s for all tolerances listed in Tables 4.10.2 and 4.10.4. Moreover, the error for ode13s corresponding to Tol = 10^{-4} is very large.

So far no codes have been implemented based on DIMSIMs of type 3 discussed in Section 3.6 or type 4 discussed in Section 3.7 for nonstiff and stiff differential systems in a parallel computing environment, but this is the subject of current work.

CHAPTER 5

TWO-STEP RUNGE-KUTTA METHODS

5.1 REPRESENTATION OF TWO-STEP RUNGE-KUTTA METHODS

In this chapter we discuss the general class of two-step Runge-Kutta (TSRK) methods, which depend on stage values at two consecutive steps. These methods were introduced by Jackiewicz and Tracogna [183] (compare also [270]), and were mentioned in Section 2.1, formula (2.1.12), as an example of GLMs. This formula is reproduced below:

$$
\begin{aligned}
Y_i^{[n]} &= (1 - u_i)y_{n-1} + u_i y_{n-2} \\
&\quad + h \sum_{j=1}^{s} \Big(a_{ij} f(Y_j^{[n]}) + b_{ij} f(Y_j^{[n-1]}) \Big), \\
y_n &= (1 - \vartheta)y_{n-1} + \vartheta y_{n-2} \\
&\quad + h \sum_{j=1}^{s} \Big(v_j f(Y_j^{[n]}) + w_j f(Y_j^{[n-1]}) \Big),
\end{aligned}
\tag{5.1.1}
$$

General Linear Methods for Ordinary Differential Equations. By Zdzisław Jackiewicz **237**
Copyright © 2009 John Wiley & Sons, Inc.

$i = 1, 2, \ldots, s$, $n = 2, 3, \ldots, N$. Here the stage values $Y_i^{[n]}$ are approximations to $y(t_{n-1} + c_i h)$, $i = 1, 2, \ldots, s$, where $y(t)$ is the solution to (2.1.1).

Introducing the standard notation as in Section 2.1,

$$
Y^{[n]} = \begin{bmatrix} Y_1^{[n]} \\ \vdots \\ Y_s^{[n]} \end{bmatrix}, \quad F(Y^{[n]}) = \begin{bmatrix} f(Y_1^{[n]}) \\ \vdots \\ f(Y_s^{[n]}) \end{bmatrix},
$$

these methods can be written in the following vector form:

$$
\begin{aligned}
Y^{[n]} &= \big((\mathbf{e} - \mathbf{u}) \otimes \mathbf{I})\big)y_{n-1} + (\mathbf{u} \otimes \mathbf{I})y_{n-2} \\
&\quad + h\Big((\mathbf{A} \otimes \mathbf{I})F(Y^{[n]}) + (\mathbf{B} \otimes \mathbf{I})F(Y^{[n-1]})\Big), \\
y_n &= (1 - \vartheta)y_{n-1} + \vartheta y_{n-2} \\
&\quad + h\Big((\mathbf{v} \otimes \mathbf{I})F(Y^{[n]}) + (\mathbf{w} \otimes \mathbf{I})F(Y^{[n-1]})\Big),
\end{aligned}
\tag{5.1.2}
$$

$i = 1, 2, \ldots, s$, $n = 2, 3, \ldots, N$, $\mathbf{e} = [1, 1, \ldots, 1]^T \in \mathbb{R}^s$. These methods can be represented by the abscissa vector $\mathbf{c} = [c_1, \ldots, c_s]^T$ and the following tableaux of its coefficients:

$$
\begin{array}{c|c|c}
\mathbf{u} & \mathbf{A} & \mathbf{B} \\
\hline
\vartheta & \mathbf{v}^T & \mathbf{w}^T
\end{array}
=
\begin{array}{c|ccc|ccc}
u_1 & a_{11} & \cdots & a_{1s} & b_{11} & \cdots & b_{1s} \\
\vdots & \vdots & \ddots & \vdots & \vdots & \ddots & \vdots \\
u_s & a_{s1} & \cdots & a_{ss} & b_{s1} & \cdots & b_{ss} \\
\hline
\vartheta & v_1 & \cdots & v_s & w_1 & \cdots & w_s
\end{array}.
\tag{5.1.3}
$$

Consistent with the classification of GLMs introduced in Section 2.7, TSRK methods can be divided into four types depending on the nature of the problem to be solved (nonstiff or stiff) and the type of computer architecture at hand (sequential or parallel). For type 1 or 2 methods, the coefficient matrix $\mathbf{A}$ has the form

$$
\mathbf{A} = \begin{bmatrix} \lambda & & & \\ a_{21} & \lambda & & \\ \vdots & \vdots & \ddots & \\ a_{s1} & a_{s2} & \cdots & \lambda \end{bmatrix},
$$

where $\lambda = 0$ or $\lambda > 0$, respectively. Such methods are appropriate for nonstiff or stiff differential systems in a sequential computing environment. For type 3 or 4 methods, the matrix $\mathbf{A}$ takes the form

$$
\mathbf{A} = \mathrm{diag}\big(\lambda, \lambda, \ldots, \lambda\big) = \lambda \mathbf{I},
$$

where $\lambda = 0$ or $\lambda > 0$, respectively. Such methods are appropriate for nonstiff or stiff differential systems in a parallel computing environment.

The TSRK methods (5.1.1) generalize special cases of TSRK methods investigated by Byrne and Lambert [84], Byrne [82], Caira et al. [90], Renaut [243, 244], and Jackiewicz et al. [181, 182]. For example, the methods investigated by Jackiewicz et al. [181] have the form

$$Y_i^{[n]} = y_{n-1} + h\sum_{j=1}^{s} a_{ij} f(Y_j^{[n]}),$$

$$y_n = (1 - \vartheta)y_{n-1} + \vartheta y_{n-2} + h\sum_{j=1}^{s} \left(v_j f(Y_j^{[n]}) + w_j f(Y_j^{[n-1]}) \right), \tag{5.1.4}$$

$i = 1, 2, \ldots, s$, $n = 2, 3, \ldots, N$, which corresponds to (5.1.1) with $\mathbf{B} = \mathbf{0}$ and $\mathbf{u} = \mathbf{0}$. Verwer [279, 280, 281] considered two- and three-step explicit RK methods for the numerical integration of differential systems resulting from parabolic partial differential equations by applying the method of lines. We refer also to van der Houwen and Sommeijer [165, 166], and van der Houwen [164] for related results concerning explicit k-step m-stage RK methods. Jackiewicz and Zennaro [190] investigated variable step size TSRK of the form (5.1.4) and demonstrated that these methods can be used to estimate the local discretization error of continuous RK methods without extra evaluations of the right-hand side of the differential equation. Their approach is based on the construction of embedded pairs of continuous RK methods of order p and variable step size TSRK methods of order $p + 1$, and Jackiewicz and Zennaro [190] give examples of such pairs up to the order $p = 4$. Bellen et al. [19] used a similar approach to construct embedded pairs of singly-implicit RK (SIRK) methods used in the differential equation solver STRIDE (see [34]) and variable step size TSRK formulas. Pan [234] constructed embedded pairs of multi-implicit RK methods investigated by Orel [225] and TSRK formulas.

The presence of extra parameters in (5.1.1) as compared to (5.1.4) and RK methods makes it possible to construct high order methods with relatively few stages. For example, as demonstrated later in the chapter, there exists explicit TSRK method (5.1.1) with stage order $q = 2$ and order $p = 3$ with $s = 1$ only, and stage order $q = 4$ and order $p = 5$ with $s = 2$ only. This big gain in efficiency makes these methods very attractive for the solution of large systems of ODEs and for Volterra integral and integro-differential equations, where there is a need to evaluate the kernel at many points of the interval of integration.

5.2   ORDER CONDITIONS FOR TSRK METHODS

In this section we follow the presentation by Jackiewicz and Tracogna [183] and Tracogna [270] to derive the order conditions for the general class of TSRK methods (5.1.1). This approach is based on the ideas of Albrecht for RK methods [5, 7] and for composite and linear cyclic methods [6, 8]. A readable summary of this approach for RK methods is given by Lambert [195]. This approach was also used by Hosea [161] to compute truncation error coefficients of RK methods and by Hosea and Shampine [162] for efficiency comparisons of methods for numerical solution of ODEs.

In what follows we derive order conditions for the scalar case (i.e., where $m = 1$ in (2.1.1)). The general case of order conditions for systems of equations is technically more complicated and is discussed by Jackiewicz and Tracogna [183].

Let $\mathcal{A}$ and $\mathcal{B}$ be $(2m + 2) \times (2m + 2)$ matrices defined by

$$\mathcal{A} = \begin{bmatrix} \mathbf{0} & \mathbf{I} & \mathbf{0} & \mathbf{0} \\ \mathbf{0} & \mathbf{0} & \mathbf{u} & \mathbf{e} - \mathbf{u} \\ \mathbf{0} & \mathbf{0} & 0 & 1 \\ \mathbf{0} & \mathbf{0} & \vartheta & 1 - \vartheta \end{bmatrix}, \quad \mathcal{B} = \begin{bmatrix} \mathbf{0} & \mathbf{0} & \mathbf{0} & \mathbf{0} \\ \mathbf{B} & \mathbf{A} & \mathbf{0} & \mathbf{0} \\ \mathbf{0} & \mathbf{0} & \mathbf{0} & \mathbf{0} \\ \mathbf{w}^T & \mathbf{v}^T & 0 & 0 \end{bmatrix},$$

where $\mathbf{0}$ stands for the zero vector or the zero matrix of appropriate dimensions and $\mathbf{I}$ stands for the m-dimensional identity matrix. Also define

$$Z_n = \begin{bmatrix} Y^{[n-1]} \\ Y^{[n]} \\ y_{n-1} \\ y_n \end{bmatrix}, \quad Z_1 = \begin{bmatrix} \mathbf{0} \\ Y^{[1]} \\ y_0 \\ y_1 \end{bmatrix}, \quad F(Z_n) = \begin{bmatrix} F(Y^{[n-1]}) \\ F(Y^{[n]}) \\ f(y_{n-1}) \\ f(y_n) \end{bmatrix},$$

$n = 2, 3, \ldots, N$, where $Y^{[1]}$ is obtained by some starting procedure. Starting procedures for TSRK methods (5.1.1) are discussed in Section 5.6. Then method (5.1.1) can be represented in the following form of an A-method as defined by Albrecht [4]:

$$Z_n = \mathcal{A}Z_{n-1} + h\mathcal{B}F(Z_n), \tag{5.2.1}$$

$n = 2, 3, \ldots, N$. Since the matrix $\mathcal{A}$ has eigenvalues 1 and $-\vartheta$ and eigenvalue 0 of multiplicity $2m$, this method is zero-stable if and only if $-1 < \vartheta \leq 1$ (compare with [181, 285]).

Assume that the stage values $Y_j^{[n]}$ are approximations (possibly of low order) to $y(t_{n-1} + c_j h)$, and set

$$y(t + \mathbf{c}h) = \begin{bmatrix} y(t + c_1 h) \\ \vdots \\ y(t + c_s h) \end{bmatrix}, \quad F\big(y(t + \mathbf{c}h)\big) = \begin{bmatrix} f\big(y(t + c_1 h)\big) \\ \vdots \\ f\big(y(t + c_s h)\big) \end{bmatrix},$$

$$
z(t_n) = \begin{bmatrix} y(t_{n-2} + \mathbf{c}h) \\ y(t_{n-1} + \mathbf{c}h) \\ y(t_{n-1}) \\ y(t_n) \end{bmatrix}, \quad F\big(z(t_n)\big) = \begin{bmatrix} F\big(y(t_{n-2} + \mathbf{c}h)\big) \\ F\big(y(t_{n-1} + \mathbf{c}h)\big) \\ f\big(y(t_{n-1})\big) \\ f\big(y(t_n)\big) \end{bmatrix}.
$$

We will always require that at least $y(t_{n-1} + c_j h) = Y_j^{[n]} + O(h)$, $j = 1, 2, \ldots, s$, as $h \to 0$. Then using (5.1.1), this leads to the stage-consistency condition

$$
\mathbf{c} = (\mathbf{A} + \mathbf{B})\mathbf{e} - \mathbf{u}
$$

discussed in Section 2.2.

We define the local discretization error $hd(t_n)$ of method (5.1.1) as the residual obtained by replacing Z_n by $z(t_n)$, Z_{n-1} by $z(t_{n-1})$, and $F(Z_n)$ by $F\big(z(t_n)\big)$ in (5.2.1):

$$
z(t_n) = \mathcal{A}z(t_{n-1}) + h\mathcal{B}F\big(z(t_n)\big) + hd(t_n), \tag{5.2.2}
$$

$n = 2, 3, \ldots, N$, where $z(t_1) = [\mathbf{0}^T, y(t_0 + \mathbf{c}h)^T, y(t_0), y(t_1)]^T$. Partitioning the vector $hd(t_n)$ as

$$
hd(t_n) = h \begin{bmatrix} d^{(1)}(t_n) \\ d^{(2)}(t_n) \\ \widehat{d}^{(1)}(t_n) \\ \widehat{d}^{(2)}(t_n) \end{bmatrix},
$$

we obtain

$$
\begin{aligned}
hd^{(1)}(t_n) &= y\big(t_{n-1} + (\mathbf{c} - \mathbf{e})h\big) - y\big(t_{n-1} + (\mathbf{c} - \mathbf{e})h\big) = 0, \\
hd^{(2)}(t_n) &= y(t_{n-1} + \mathbf{c}h) - (\mathbf{e} - \mathbf{u})\,y(t_{n-1}) - \mathbf{u}\,y(t_{n-2}) \\
&\quad - h\mathbf{A}\,y'(t_{n-1} + \mathbf{c}h) - h\mathbf{B}\,y'\big(t_{n-1} + (\mathbf{c} - \mathbf{e})h\big), \\
h\widehat{d}^{(1)}(t_n) &= y(t_{n-1}) - y(t_{n-1}) = 0, \\
h\widehat{d}^{(2)}(t_n) &= y(t_n) - (1 - \vartheta)\,y(t_{n-1}) - \vartheta\,y(t_{n-2}) \\
&\quad - h\mathbf{v}^T y'(t_{n-1} + \mathbf{c}h) - h\mathbf{w}^T y'\big(t_{n-1} + (\mathbf{c} - \mathbf{e})h\big).
\end{aligned}
$$

Expanding $y(t_{n-2})$, $y(t_n)$, $y(t_{n-1} + \mathbf{c}h)$, $y'\big(t_{n-1} + (\mathbf{c} - \mathbf{e})h\big)$, and $y'(t_{n-1} + \mathbf{c}h)$ into a Taylor series around t_{n-1} leads to

$$
h \begin{bmatrix} d^{(1)}(t_n) \\ d^{(2)}(t_n) \\ \widehat{d}^{(1)}(t_n) \\ \widehat{d}^{(2)}(t_n) \end{bmatrix} = \begin{bmatrix} \mathbf{0} \\ C_1 h\,y'(t_{n-1}) + C_2 h^2 y''(t_{n-1}) + C_3 h^3 y'''(t_{n-1}) + \cdots \\ 0 \\ \widehat{C}_1 h\,y'(t_{n-1}) + \widehat{C}_2 h^2 y''(t_{n-1}) + \widehat{C}_3 h^3 y'''(t_{n-1}) + \cdots \end{bmatrix},
$$

where the error vectors C_ν and error constants $\widehat{C}_\nu$ are defined by

$$C_\nu := \frac{\mathbf{c}^\nu}{\nu!} - \frac{(-1)^\nu}{\nu!}\mathbf{u} - \frac{\mathbf{A}\mathbf{c}^{\nu-1}}{(\nu-1)!} - \frac{\mathbf{B}(\mathbf{c}-\mathbf{e})^{\nu-1}}{(\nu-1)!}, \qquad (5.2.3)$$

$$\widehat{C}_\nu := \frac{1}{\nu!} - \frac{(-1)^\nu}{\nu!}\vartheta - \frac{\mathbf{v}^T\mathbf{c}^{\nu-1}}{(\nu-1)!} - \frac{\mathbf{w}^T(\mathbf{c}-\mathbf{e})^{\nu-1}}{(\nu-1)!}, \qquad (5.2.4)$$

where $\mathbf{c}^\nu$ indicates componentwise multiplication. Observe that $C_1 = 0$ is equivalent to the stage-consistency condition

$$\mathbf{c} = (\mathbf{A} + \mathbf{B})\mathbf{e} - \mathbf{u}$$

obtained before, and $\widehat{C}_1 = 0$ is equivalent to the consistency condition

$$(\mathbf{v}^T + \mathbf{w}^T)\mathbf{e} = 1 + \vartheta$$

obtained in Section 2.2.

Subtracting (5.2.1) from (5.2.2), we obtain the linear recurrence relation

$$q(t_n) = \mathcal{A}\,q(t_{n-1}) + h\mathcal{B}\,r(t_n) + hd(t_n), \qquad (5.2.5)$$

$n = 2, 3, \ldots, N$, where

$$q(t_n) = \begin{bmatrix} q^{(1)}(t_n) \\ q^{(2)}(t_n) \\ \widehat{q}^{(1)}(t_n) \\ \widehat{q}^{(2)}(t_n) \end{bmatrix} = \begin{bmatrix} y\big(t_{n-1} + (\mathbf{c}-\mathbf{e})h\big) - Y^{[n-1]} \\ y(t_{n-1} + \mathbf{c}h) - Y^{[n]} \\ y(t_{n-1}) - y_{n-1} \\ y(t_n) - y_n \end{bmatrix},$$

$$r(t_n) = \begin{bmatrix} r^{(1)}(t_n) \\ r^{(2)}(t_n) \\ \widehat{r}^{(1)}(t_n) \\ \widehat{r}^{(2)}(t_n) \end{bmatrix} = \begin{bmatrix} F\big(y(t_{n-1} + (\mathbf{c}-\mathbf{e})h)\big) - F(Y^{[n-1]}) \\ F\big(y(t_{n-1} + \mathbf{c}h)\big) - F(Y^{[n]}) \\ f\big(y(t_{n-1})\big) - f(y_{n-1}) \\ f\big(y(t_n)\big) - f(y_n) \end{bmatrix},$$

and

$$q(t_1) = \begin{bmatrix} q^{(1)}(t_1) \\ q^{(2)}(t_1) \\ \widehat{q}^{(1)}(t_1) \\ \widehat{q}^{(2)}(t_1) \end{bmatrix} = \begin{bmatrix} \mathbf{0} \\ y(t_0 + \mathbf{c}h) - Y^{[1]} \\ y(t_0) - y_0 \\ y(t_1) - y_1 \end{bmatrix}.$$

The solution to (5.2.5) is

$$q(t_n) = \mathcal{A}^{n-1}q(t_1) + h\sum_{l=1}^{n-1} \mathcal{A}^{n-1-l}d(t_{l+1}) + h\sum_{l=1}^{n-1} \mathcal{A}^{n-1-l}\mathcal{B}\,r(t_{l+1}), \qquad (5.2.6)$$

$n = 2, 3, \ldots, N$. It can be verified that matrices $\mathcal{A}^\mu$ for $\mu \geq 2$ have the form

$$
\mathcal{A}^\mu = \begin{bmatrix}
\mathbf{0} & \mathbf{0} & \alpha_{11}^{(\mu)} & \alpha_{12}^{(\mu)} \\
\mathbf{0} & \mathbf{0} & \alpha_{21}^{(\mu)} & \alpha_{22}^{(\mu)} \\
\mathbf{0} & \mathbf{0} & \beta_{11}^{(\mu)} & \beta_{12}^{(\mu)} \\
\mathbf{0} & \mathbf{0} & \beta_{21}^{(\mu)} & \beta_{22}^{(\mu)}
\end{bmatrix},
$$

where the vectors $\alpha_{kl}^{(\mu)} \in \mathbb{R}^s$ and the scalars $\beta_{kl}^{(\mu)}$ are uniformly bounded if method (5.1.1) is zero-stable. For $n \geq 3$ we have

$$
\begin{aligned}
q(t_n) &= \mathcal{A}^{n-1} q(t_1) + h d(t_n) + h \mathcal{A}\, d(t_{n-1}) \\
&\quad + h \sum_{l=1}^{n-3} \mathcal{A}^{n-1-l} d(t_{l+1}) + h\mathcal{B}\, r(t_n) + h\mathcal{A}\mathcal{B}\, r(t_{n-1}) \\
&\quad + h \sum_{l=1}^{n-3} \mathcal{A}^{n-1-l} \mathcal{B}\, r(t_{l+1}).
\end{aligned}
\tag{5.2.7}
$$

Rewriting the equation (5.2.7) componentwise leads to the relations

$$
\begin{aligned}
q_n^{(1)} &= \alpha_{11}^{(n-1)} \widehat{q}_1^{(1)} + \alpha_{12}^{(n-1)} \widehat{q}_1^{(2)} + h d_n^{(1)} + h d_{n-1}^{(2)} \\
&\quad + h \sum_{l=1}^{n-3} \left(\alpha_{11}^{(n-1-l)} \widehat{d}_{l+1}^{(1)} + \alpha_{12}^{(n-1-l)} \widehat{d}_{l+1}^{(2)} \right) + h\mathbf{B}\, r_{n-1}^{(1)} + h\mathbf{A}\, r_{n-1}^{(2)} \\
&\quad + h \sum_{l=1}^{n-3} \alpha_{12}^{(n-1-l)} \left(\mathbf{w}^T r_{l+1}^{(1)} + \mathbf{v}^T r_{l+1}^{(2)} \right), \\
q_n^{(2)} &= \alpha_{21}^{(n-1)} \widehat{q}_1^{(1)} + \alpha_{22}^{(n-1)} \widehat{q}_1^{(2)} + h d_n^{(2)} + h\mathbf{u}\widehat{d}_{n-1}^{(1)} + h(\mathbf{e} - \mathbf{u})\widehat{d}_{n-1}^{(2)} \\
&\quad + h \sum_{l=1}^{n-3} \left(\alpha_{21}^{(n-1-l)} \widehat{d}_{l+1}^{(1)} + \alpha_{22}^{(n-1-l)} \widehat{d}_{l+1}^{(2)} \right) + h\mathbf{B}\, r_n^{(1)} + h\mathbf{A}\, r_n^{(2)} \\
&\quad + h(\mathbf{e} - \mathbf{u})\mathbf{w}^T r_{n-1}^{(1)} + h(\mathbf{e} - \mathbf{u})\mathbf{v}^T r_{n-1}^{(2)} \\
&\quad + h \sum_{l=1}^{n-3} \alpha_{22}^{(n-1-l)} \left(\mathbf{w}^T r_{l+1}^{(1)} + \mathbf{v}^T r_{l+1}^{(2)} \right), \\
\widehat{q}_n^{(1)} &= \beta_{11}^{(n-1)} \widehat{q}_1^{(1)} + \beta_{12}^{(n-1)} \widehat{q}_1^{(2)} + h \widehat{d}_n^{(1)} + h \widehat{d}_{n-1}^{(2)} \\
&\quad + h \sum_{l=1}^{n-3} \left(\beta_{11}^{(n-1-l)} \widehat{d}_{l+1}^{(1)} + \beta_{12}^{(n-1-l)} \widehat{d}_{l+1}^{(2)} \right) + h\mathbf{w}^T r_{n-1}^{(1)} + h\mathbf{v}^T r_{n-1}^{(2)} \\
&\quad + h \sum_{l=1}^{n-3} \beta_{12}^{(n-1-l)} \left(\mathbf{w}^T r_{l+1}^{(1)} + \mathbf{v}^T r_{l+1}^{(2)} \right),
\end{aligned}
$$

$$\widehat{q}_n^{(2)} = \beta_{21}^{(n-1)}\widehat{q}_1^{(1)} + \beta_{22}^{(n-1)}\widehat{q}_1^{(2)} + h\widehat{d}_n^{(2)} + h\vartheta\,\widehat{d}_{n-1}^{(1)} + h(1-\vartheta)\widehat{d}_{n-1}^{(2)}$$

$$+ \; h\sum_{l=1}^{n-3}\left(\beta_{21}^{(n-1-l)}\widehat{d}_{l+1}^{(1)} + \beta_{22}^{(n-1-l)}\widehat{d}_{l+1}^{(2)}\right) + h\mathbf{w}^T r_n^{(1)} + h\mathbf{v}^T r_n^{(2)}$$

$$+ \; h(1-\vartheta)\mathbf{w}^T r_{n-1}^{(1)} + h(1-\vartheta)\mathbf{v}^T r_{n-1}^{(2)}$$

$$+ \; h\sum_{l=1}^{n-3}\beta_{22}^{(n-1-l)}\left(\mathbf{w}^T r_{l+1}^{(1)} + \mathbf{v}^T r_{l+1}^{(2)}\right),$$

where to simplify the notation we have written $q_l^{(\nu)}$, $\widehat{q}_l^{(\nu)}$, $r_l^{(\nu)}$, $\widehat{r}_l^{(\nu)}$, $d_l^{(\nu)}$, and $\widehat{d}_l^{(\nu)}$ instead of $q^{(\nu)}(t_l)$, $\widehat{q}^{(\nu)}(t_l)$, $r^{(\nu)}(t_l)$, $\widehat{r}^{(\nu)}(t_l)$, $d^{(\nu)}(t_l)$, and $\widehat{d}^{(\nu)}(t_l)$.

Assume that the starting values have order p, that is,

$$\widehat{q}_1^{(1)} = O(h^p), \quad \widehat{q}_1^{(2)} = O(h^p), \quad h \to 0,$$

and that the last two stages of the method (5.2.1) have order of consistency p, that is,

$$\widehat{d}_l^{(1)} = O(h^p), \quad \widehat{d}_l^{(2)} = O(h^p), \quad h \to 0$$

(in fact, for this method, $\widehat{d}_l^{(1)} \equiv 0$). Recall also that $d_l^{(1)} \equiv 0$. Then

$$q_n^{(1)} = hd_{n-1}^{(2)} + h\mathbf{B}\,r_{n-1}^{(1)} + h\mathbf{A}\,r_{n-1}^{(2)} + h\sum_{l=1}^{n-3}\alpha_{12}^{(n-1-l)}\left(\mathbf{w}^T r_{l+1}^{(1)} + \mathbf{v}^T r_{l+1}^{(2)}\right) + O(h^p),$$

$$q_n^{(2)} = hd_n^{(2)} + h\mathbf{B}\,r_n^{(1)} + h\mathbf{A}\,r_n^{(2)} + h(\mathbf{e}-\mathbf{u})\mathbf{w}^T r_{n-1}^{(1)} + h(\mathbf{e}-\mathbf{u})\mathbf{v}^T r_{n-1}^{(2)}$$

$$+ \; h\sum_{l=1}^{n-3}\alpha_{22}^{(n-1-l)}\left(\mathbf{w}^T r_{l+1}^{(1)} + \mathbf{v}^T r_{l+1}^{(2)}\right) + O(h^p),$$

$$\widehat{q}_n^{(1)} = h\mathbf{w}^T r_{n-1}^{(1)} + h\mathbf{v}^T r_{n-1}^{(2)} + h\sum_{l=1}^{n-3}\beta_{12}^{(n-1-l)}\left(\mathbf{w}^T r_{l+1}^{(1)} + \mathbf{v}^T r_{l+1}^{(2)}\right) + O(h^p),$$

$$\widehat{q}_n^{(2)} = h\mathbf{w}^T r_n^{(1)} + h\mathbf{v}^T r_n^{(2)} + h(1-\vartheta)\mathbf{w}^T r_{n-1}^{(1)} + h(1-\vartheta)\mathbf{v}^T r_{n-1}^{(2)}$$

$$+ \; h\sum_{l=1}^{n-3}\beta_{22}^{(n-1-l)}\left(\mathbf{w}^T r_{l+1}^{(1)} + \mathbf{v}^T r_{l+1}^{(2)}\right) + O(h^p).$$

Taking into account that $q_n^{(1)} = q_{n-1}^{(2)}$ and $r_n^{(1)} = r_{n-1}^{(2)}$, it follows that

$$q_n^{(2)} = hd_n^{(2)} + h\mathbf{B}\,r_{n-1}^{(2)} + h\mathbf{A}\,r_n^{(2)}$$

$$+ \; h(\mathbf{e}-\mathbf{u})\mathbf{w}^T r_{n-2}^{(2)} + h(\mathbf{e}-\mathbf{u})\mathbf{v}^T r_{n-1}^{(2)}$$

$$+ \; h\sum_{l=1}^{n-3}\alpha_{22}^{(n-1-l)}\left(\mathbf{w}^T r_l^{(2)} + \mathbf{v}^T r_{l+1}^{(2)}\right) + O(h^p),$$

(5.2.8)

$$\widehat{q}_n^{(2)} = h\mathbf{w}^T r_{n-1}^{(2)} + h\mathbf{v}^T r_n^{(2)}$$

$$+ h(1-\vartheta)\mathbf{w}^T r_{n-2}^{(2)} + h(1-\vartheta)\mathbf{v}^T r_{n-1}^{(2)} \tag{5.2.9}$$

$$+ h\sum_{l=1}^{n-3} \beta_{22}^{(n-1-l)}\left(\mathbf{w}^T r_l^{(2)} + \mathbf{v}^T r_{l+1}^{(2)}\right) + O(h^p).$$

In what follows we write r_n, q_n, $\widehat{q}_n$, d_n, and $\widehat{d}_n$ instead of $r_n^{(2)}$, $q_n^{(2)}$, $\widehat{q}_n^{(2)}$, $d_n^{(2)}$, and $\widehat{d}_n^{(2)}$, respectively. Then formulas (5.2.8) and (5.2.9) lead to the following theorem.

Theorem 5.2.1 ([183, 270]) *Assume that the TSRK method (5.2.1) is zero-stable and that $\widehat{d}_l = O(h^p)$, $l = 1, 2, \ldots, N$, as $h \to 0$. Then*

$$\widehat{q}_n = O(h^p), \quad h \to 0,$$

if and only if

$$\mathbf{w}^T r_l + \mathbf{v}^T r_{l+1} = O(h^p), \quad h \to 0, \tag{5.2.10}$$

$l = 1, 2, \ldots, N-1$. *Moreover, the errors q_n of stage values $Y^{[n]}$ are given by*

$$q_n = h\mathbf{B}r_{n-1} + h\mathbf{A}r_n + hd_n + O(h^p), \quad h \to 0, \tag{5.2.11}$$

where

$$hd_n = \sum_{\nu=1}^{p} C_\nu h^\nu y^{(\nu)}(t_{n-1}) + O(h^{p+1})$$

with the error vectors C_ν defined by (5.2.3).

The condition $d_l = O(h^p)$ and equation (5.2.10) represent the general form of order conditions for TSRK methods (5.2.1). It follows from this theorem that if the method is zero-stable and satisfies $d_l = O(h^p)$ and (5.2.10), the last stage is convergent with order p (i.e., $y_n - y(t_n) = O(h^p)$) and the errors $y(t_{n-1} + \mathbf{c}h) - Y^{[n]}$ of the stages $Y^{[n]}$ are given by (5.2.11).

To reformulate (5.2.10) in a more convenient form, we first need to express r_n in terms of q_n. Denoting by $r_{j,n}$ and $q_{j,n}$ the jth components of r_n and q_n, respectively, we obtain

$$r_{j,n} = f\left(y(t_{n-1} + c_j h)\right) - f(Y_j^{[n]})$$

$$= -\sum_{\nu=1}^{p-1} D_y^\nu f\left(y(t_{n-1} + c_j h)\right)\frac{\left(Y_j^{[n]} - y(t_{n-1} + c_j h)\right)^\nu}{\nu!} + O(h^p)$$

$$= \sum_{\nu=1}^{p-1} \frac{(-1)^{\nu+1}}{\nu!}D_y^\nu f\left(y(t_{n-1} + c_j h)\right)(q_{j,n})^\nu + O(h^p),$$

where D_y^ν is a differential operator of order ν with respect to y. Define the functions $g_\nu(t_{n-1} + c_j h)$ by

$$g_\nu(t_{n-1} + c_j h) := \frac{(-1)^{\nu+1}}{\nu!} D_y^\nu f\big(y(t_{n-1} + c_j h)\big).$$

Then

$$g_\nu(t_{n-1} + c_j h) = \sum_{\mu=0}^{p-1} g_\nu^{(\mu)}(t_{n-1}) \frac{c_j^\mu h^\mu}{\mu!} + O(h^p),$$

where $g_\nu^{(\mu)}$ stands for the derivative of order μ. Setting

$$G_\nu = \mathrm{diag}\Big(g_\nu(t_{n-1} + c_1 h), g_\nu(t_{n-1} + c_2 h), \ldots, g_\nu(t_{n-1} + c_s h)\Big),$$

we obtain

$$G_\nu = \sum_{\mu=0}^{p-1} \frac{h^\mu}{\mu!} g_\nu^{(\mu)}(t_{n-1}) \Gamma_{\mathbf{c}}^\mu + O(h^p),$$

with the diagonal matrix $\Gamma_{\mathbf{c}}$ defined by

$$\Gamma_{\mathbf{c}} = \mathrm{diag}(c_1, c_2, \ldots, c_s).$$

Using the foregoing notation $r_{j,n}$, can be written in the form

$$r_{j,n} = \sum_{\nu=1}^{p-1} g_\nu(t_{n-1} + c_j h)(q_{j,n})^\nu + O(h^p)$$

or

$$r_n = G_1 q_n + G_2(q_n)^2 + G_3(q_n)^3 + \cdots + O(h^p) \tag{5.2.12}$$

as $h \to 0$. Following Albrecht [5, 7] we can argue using the implicit function theorem [212] that the functions $r_n = r_n(h)$ and $q_n = q_n(h)$ are unique in a neighborhood of $h = 0$ and that they have the following Taylor series expansions:

$$q_n = \xi_2(t_{n-1})h^2 + \xi_3(t_{n-1})h^3 + \cdots + \xi_{p-1}(t_{n-1})h^{p-1} + O(h^p), \tag{5.2.13}$$

$$r_n = \eta_2(t_{n-1})h^2 + \eta_3(t_{n-1})h^3 + \cdots + \eta_{p-1}(t_{n-1})h^{p-1} + O(h^p), \tag{5.2.14}$$

$h \to 0$. We can generate the functions $\xi_j(t_{n-1})$ and $\eta_j(t_{n-1})$ by recursively substituting (5.2.13) and (5.2.14) into (5.2.11) and (5.2.12). This leads to the following theorem.

Theorem 5.2.2 ([183, 270]) *The functions $\xi_j(t_{n-1})$ and $\eta_j(t_{n-1})$ satisfy the relations*

$$\xi_j(t_{n-1}) = C_j y^{(j)}(t_{n-1}) + \mathbf{A}\,\eta_{j-1}(t_{n-1}) + \mathbf{B} \sum_{\nu=0}^{j-3} \frac{(-1)^\nu}{\nu!} \eta_{j-\nu-1}^{(\nu)}(t_{n-1})$$

and

$$
\eta_j(t_{n-1}) = \sum_{l=0}^{j-2} \frac{1}{l!} \Gamma_{\mathbf{c}}^l \Big(g_1^{(l)}(t_{n-1}) \xi_{j-l}(t_{n-1})
$$

$$
+ \sum_{\mu,\nu \ge 2,\, \mu+\nu=j-l} g_2^{(l)}(t_{n-1}) \big(\xi_\mu(t_{n-1}) \cdot \xi_\nu(t_{n-1}) \big)
$$

$$
+ \sum_{\mu,\nu,\tau \ge 2,\, \mu+\nu+\tau=j-l} g_3^{(l)}(t_{n-1}) \big(\xi_\mu(t_{n-1}) \cdot \xi_\nu(t_{n-1}) \cdot \xi_\tau(t_{n-1}) \big) + \cdots \Big)
$$

for $j = 2, 3, \ldots, p-1$, $n = 1, 2, \ldots$, *with* $\eta_1(t_{n-1}) \equiv 0$. *Here* "$\cdot$" *denotes componentwise products.*

Proof: Using (5.2.11) and (5.2.13) and expanding $\eta_\nu(t_{n-2})$ into a Taylor series around t_{n-1}, we obtain

$$
\sum_{\nu=2}^{j} \xi_\nu(t_{n-1}) h^\nu = \sum_{\nu=1}^{j} C_\nu y^{(\nu)}(t_{n-1}) h^\nu + \mathbf{B} \sum_{\nu=2}^{j-1} \eta_\nu(t_{n-2}) h^{\nu+1}
$$

$$
+ \mathbf{A} \sum_{\nu=2}^{j-1} \eta_\nu(t_{n-1}) h^{\nu+1} + O(h^{j+1})
$$

$$
= \sum_{\nu=1}^{j} C_\nu y^{(\nu)}(t_{n-1}) h^\nu + \mathbf{B} \sum_{\nu=2}^{j-1} \sum_{\tau=0}^{j-\nu-1} \frac{(-1)^\tau}{\tau!} \eta_\nu^{(\tau)}(t_{n-1}) h^{\tau+\nu+1}
$$

$$
+ \mathbf{A} \sum_{\nu=2}^{j-1} \eta_\nu(t_{n-1}) h^{\nu+1} + O(h^{j+1})
$$

$$
= \sum_{\nu=1}^{j} C_\nu y^{(\nu)}(t_{n-1}) h^\nu + \mathbf{B} \sum_{\nu=2}^{j-1} \sum_{\tau=\nu+1}^{j} \frac{(-1)^{\tau-\nu-1}}{(\tau-\nu-1)!} \eta_\nu^{(\tau-\nu-1)}(t_{n-1}) h^\tau
$$

$$
+ \mathbf{A} \sum_{\nu=2}^{j-1} \eta_\nu(t_{n-1}) h^{\nu+1} + O(h^{j+1}).
$$

A comparison of the h^j terms yields

$$
\xi_j(t_{n-1}) = C_j y^{(j)}(t_{n-1}) + \mathbf{A}\eta_{j-1}(t_{n-1}) + \mathbf{B} \sum_{\nu=2}^{j-1} \frac{(-1)^{j-\nu-1}}{(j-\nu-1)!} \eta_\nu^{(j-\nu-1)}(t_{n-1})
$$

$$
= C_j y^{(j)}(t_{n-1}) + \mathbf{A}\eta_{j-1}(t_{n-1}) + \mathbf{B} \sum_{\nu=0}^{j-3} \frac{(-1)^\nu}{\nu!} \eta_{j-\nu-1}^{(\nu)}(t_{n-1}),
$$

which is the formula required for the function $\xi_j(t_{n-1})$.

Define

$$
q_n = s_j(t_{n-1}) + O(h^j),
$$

where
$$s_j(t_{n-1}) := \xi_2(t_{n-1})h^2 + \cdots \xi_j(t_{n-1})h^j.$$
We next use (5.2.12), (5.2.13), and (5.2.14) to obtain

$$\sum_{\nu=2}^{j} \eta_\nu(t_{n-1})h^\nu \;=\; G_1 s_j + G_2(s_j)^2 + G_3(s_j)^3 + \cdots + O(h^{j+1})$$

$$= \left(g_1(t_{n-1})\mathbf{I} + hg_1^{(1)}(t_{n-1})\Gamma_\mathbf{c} + \frac{h^2}{2!}g_1^{(2)}(t_{n-1})\Gamma_\mathbf{c}^2 + \cdots\right)$$

$$\times \sum_{\mu \geq 2, \mu \leq j} h^\mu \xi_\mu(t_{n-1})$$

$$+ \left(g_2(t_{n-1})\mathbf{I} + hg_2^{(1)}(t_{n-1})\Gamma_\mathbf{c} + \frac{h^2}{2!}g_2^{(2)}(t_{n-1})\Gamma_\mathbf{c}^2 + \cdots\right)$$

$$\times \sum_{\mu,\nu \geq 2, \mu+\nu \leq j} h^{\mu+\nu}\left(\xi_\mu(t_{n-1}) \cdot \xi_\nu(t_{n-1})\right)$$

$$+ \left(g_3(t_{n-1})\mathbf{I} + hg_3^{(1)}(t_{n-1})\Gamma_\mathbf{c} + \frac{h^2}{2!}g_3^{(2)}(t_{n-1})\Gamma_\mathbf{c}^2 + \cdots\right)$$

$$\times \sum_{\mu,\nu,\tau \geq 2, \mu+\nu+\tau \leq j} h^{\mu+\nu+\tau}\left(\xi_\mu(t_{n-1}) \cdot \xi_\nu(t_{n-1}) \cdot \xi_\tau(t_{n-1})\right) + \cdots.$$

A comparison of the h^j terms in the expression above yields

$$\eta_j(t_{n-1}) \;=\; g_1(t_{n-1})\xi_j(t_{n-1}) + g_1^{(1)}(t_{n-1})\Gamma_\mathbf{c}\xi_{j-1}(t_{n-1}) + \cdots$$

$$+ \frac{1}{(j-2)!}g_1^{(j-2)}(t_{n-1})\Gamma_\mathbf{c}^{j-2}\xi_2(t_{n-1})$$

$$+ g_2(t_{n-1}) \sum_{\mu,\nu \geq 2, \mu+\nu=j} \left(\xi_\mu(t_{n-1}) \cdot \xi_\nu(t_{n-1})\right)$$

$$+ g_2^{(1)}(t_{n-1})\Gamma_\mathbf{c} \sum_{\mu,\nu \geq 2, \mu+\nu=j-1} \left(\xi_\mu(t_{n-1}) \cdot \xi_\nu(t_{n-1})\right) + \cdots$$

$$+ \frac{1}{(j-2)!}g_2^{(j-2)}(t_{n-1})\Gamma_\mathbf{c}^{j-2} \sum_{\mu,\nu \geq 2, \mu+\nu=2} \left(\xi_\mu(t_{n-1}) \cdot \xi_\nu(t_{n-1})\right)$$

$$+ g_3(t_{n-1}) \sum_{\mu,\nu,\tau \geq 2, \mu+\nu+\tau=j} \left(\xi_\mu(t_{n-1}) \cdot \xi_\nu(t_{n-1}) \cdot \xi_\tau(t_{n-1})\right)$$

$$+ g_3^{(1)}(t_{n-1})\Gamma_\mathbf{c} \sum_{\mu,\nu,\tau \geq 2, \mu+\nu+\tau=j-1} \left(\xi_\mu(t_{n-1}) \cdot \xi_\nu(t_{n-1}) \cdot \xi_\tau(t_{n-1})\right) + \cdots$$

$$+ \frac{1}{(j-2)!}g_3^{(j-2)}(t_{n-1})\Gamma_\mathbf{c}^{j-2} \sum_{\mu,\nu,\tau \geq 2, \mu+\nu+\tau=2} \left(\xi_\mu(t_{n-1}) \cdot \xi_\nu(t_{n-1}) \cdot \xi_\tau(t_{n-1})\right)$$

$$+ \cdots.$$

Hence,

$$
\eta_j(t_{n-1}) = \sum_{l=0}^{j-2} \frac{1}{l!} \Gamma_{\mathbf{c}}^l \Big(g_1^{(l)}(t_{n-1}) \xi_{j-l}(t_{n-1})
$$

$$
+ \sum_{\mu,\nu \geq 2,\, \mu+\nu=j-l} g_2^{(l)}(t_{n-1}) \big(\xi_\mu(t_{n-1}) \cdot \xi_\nu(t_{n-1}) \big)
$$

$$
+ \sum_{\mu,\nu,\tau \geq 2,\, \mu+\nu+\tau=j-l} g_3^{(l)}(t_{n-1}) \big(\xi_\mu(t_{n-1}) \cdot \xi_\nu(t_{n-1}) \cdot \xi_\tau(t_{n-1}) \big) + \cdots \Big),
$$

which is the formula required for $\eta_j(t_{n-1})$. This completes the proof.   ∎

It follows from Theorem 5.2.1 and formula (5.2.13) that method (5.2.1) has order p if

$$
\widehat{C}_q = 0
$$

for $q = 1, 2, \ldots, p$, where $\widehat{C}_q$ are defined by (5.2.4), and if

$$
\mathbf{w}^T \eta_q(t_{n-2}) + \mathbf{v}^T \eta_q(t_{n-1}) = 0 \tag{5.2.15}
$$

for $q = 2, 3, \ldots, p-1$, where the functions η_q can be generated using Theorem 5.2.2.

Condition (5.2.15) is still difficult to apply since it involves the values of the functions η_q at two consecutive points t_{n-2} and t_{n-1}. However, the more convenient form of this condition is given by the following result.

Theorem 5.2.3 ([183, 270]) *Zero-stable method (5.2.1) has order p if and only if $\widehat{C}_q = 0$, $q = 1, 2, \ldots, p$, and*

$$
(\mathbf{v}^T + \mathbf{w}^T)\eta_q(t_{n-1}) + \mathbf{w}^T \sum_{\nu=1}^{q-2} \frac{(-1)^\nu}{\nu!} \eta_{q-\nu}^{(\nu)}(t_{n-1}) = 0, \tag{5.2.16}
$$

$q = 2, 3, \ldots, p-1$, $n = 1, 2, \ldots$.

Proof: Condition (5.2.15) is equivalent to

$$
\sum_{\nu=2}^{p-1} h^{\nu-2} \big(\mathbf{w}^T \eta_\nu(t_{n-2}) + \mathbf{v}^T \eta_\nu(t_{n-1}) \big) = O(h^{p-2}),
$$

$h \to 0$, or

$$
\mathbf{w}^T \sum_{\nu=2}^{p-1} \sum_{\mu=0}^{p-\nu-1} \frac{(-1)^\mu}{\mu!} h^{\nu+\mu-2} \eta_\nu^{(\mu)}(t_{n-1}) + \mathbf{v}^T \sum_{\nu=2}^{p-1} h^{\nu-2} \eta_\nu(t_{n-1}).
$$

Substituting $\mu = q - \nu$ and changing the order of summation in the resulting double sum, we obtain

$$
\sum_{q=2}^{p-1} h^{q-2} \left(\mathbf{w}^T \sum_{\nu=2}^{q} \frac{(-1)^{q-\nu}}{(q-\nu)!} \eta_\nu^{(q-\nu)}(t_{n-1}) + \mathbf{v}^T \eta_q(t_{n-1}) \right) = 0
$$

or

$$\sum_{q=2}^{p-1} h^{q-2} \left(\mathbf{w}^T \sum_{\mu=0}^{q-2} \frac{(-1)^\mu}{\mu!} \eta_{q-\mu}^{(\mu)}(t_{n-1}) + \mathbf{v}^T \eta_q(t_{n-1}) \right) = 0,$$

which is equivalent to (5.2.16). This completes the proof. ∎

In this section we were striving at self-contained derivation of general order conditions for TSRK methods (5.2.1), and this resulted in a rather lengthy presentation leading to Theorems 5.2.1, 5.2.2, and 5.2.3. The reader familiar with the theory of B-series [143] can find much shorter derivation of order conditions for TSRK methods (5.1.1) in the work of Hairer and Wanner [147] and Tracogna and Welfert [272]. The order conditions for TSRK methods (5.1.1) were also obtained by Butcher and Tracogna [78] using the algebraic approach of Butcher [41] and by Jackiewicz and Vermiglio [185] using the theory of GLMs with external stages of different orders. The order conditions for these methods can be also derived using a suitable modification of the approach by Burrage and Moss [35] and Burrage [29].

5.3   DERIVATION OF ORDER CONDITIONS UP TO ORDER 6

We illustrate the application of Theorems 5.2.2 and 5.2.3 by the derivation of order conditions for TSRK method (5.2.1) (equivalent to (5.1.1)) of order $p = 6$. These order conditions are

$$\widehat{C}_q = 0, \quad q = 1, 2, \dots, 6,$$

and

$$(\mathbf{v}^T + \mathbf{w}^T)\eta_2(t_{n-1}) = 0,$$

$$(\mathbf{v}^T + \mathbf{w}^T)\eta_3(t_{n-1}) - \mathbf{w}^T \eta_2^{(1)}(t_{n-1}) = 0,$$

$$(\mathbf{v}^T + \mathbf{w}^T)\eta_4(t_{n-1}) - \mathbf{w}^T \left(\eta_3^{(1)}(t_{n-1}) - \frac{1}{2}\eta_2^{(2)}(t_{n-1}) \right) = 0,$$

$$(\mathbf{v}^T + \mathbf{w}^T)\eta_5(t_{n-1}) - \mathbf{w}^T \left(\eta_4^{(1)}(t_{n-1}) - \frac{1}{2}\eta_3^{(2)}(t_{n-1}) + \frac{1}{6}\eta_2^{(3)}(t_{n-1}) \right) = 0.$$

The expressions above contain various combinations of $g_\nu^{(\mu)} = g_\nu^{(\mu)}(t_{n-1})$ and $y^{(\mu)} = y^{(\mu)}(t_{n-1})$, and we obtain order conditions for TSRK methods (5.1.1) by equating the coefficients of these combinations to zero. These combinations play the role of elementary differentials in the theory of RK methods and are given by

$p = 3$:

$$g_1 y'';$$

$p = 4$:

$$g_1 y''', \ g_1^2 y'', \ g_1^{(1)} y'';$$

$p = 5$:

$$g_1 y^{(4)}, \ g_1 g_1^{(1)} y'', \ g_1^2 y''', \ g_1^3 y'', \ g_1^{(1)} y''', \ g_1^{(1)} g_1 y'', \ g_1^{(2)} y'', \ g_2 (y'')^2;$$

$p = 6$:

$$g_1 y^{(5)}, \ g_1 g_1^{(2)} y'', \ g_1 g_1^{(1)} y''', \ g_1^{(2)} y^{(4)}, \ g_1 g_1^{(1)} g_1 y'', \ g_1^2 g_1^{(1)} y'', \ g_1^3 y''',$$
$$g_1^4 y'', \ g_1 g_2 (y'')^2, \ g_1^{(1)} y^{(4)}, \ (g_1^{(1)})^2 y'', \ g_1^{(1)} g_1 y''', \ g_1^{(1)} g_1^2 y'', \ g_1^{(2)} y''',$$
$$g_1^{(2)} g_1 y'', \ g_1^{(3)} y'', \ g_2 y'' y''', \ g_1 y'' g_1 y'', \ g_2^{(1)} (y'')^2.$$

The order conditions up to $p = 6$ are listed below, where we always assume the stage-consistency condition

$$C_1 = \mathbf{c} + \mathbf{u} - (\mathbf{A} + \mathbf{B}) \mathbf{e} = \mathbf{0},$$

and the error vectors C_ν and error constants $\widehat{C}_\nu$ are defined by (5.2.3) and (5.2.4), respectively.

$p = 1$:
$$\widehat{C}_1 = 0;$$

$p = 2$:
$$\widehat{C}_2 = 0;$$

$p = 3$:
$$\widehat{C}_3 = 0,$$
$$(\mathbf{v}^T + \mathbf{w}^T) C_2 = 0;$$

$p = 4$:
$$\widehat{C}_4 = 0,$$
$$(\mathbf{v}^T + \mathbf{w}^T) C_3 - \mathbf{w}^T C_2 = 0,$$
$$(\mathbf{v}^T + \mathbf{w}^T)(\mathbf{A} + \mathbf{B}) C_2 = 0,$$
$$(\mathbf{v}^T + \mathbf{w}^T) \Gamma_{\mathbf{c}} C_2 - \mathbf{w}^T C_2 = 0;$$

$p = 5$:
$$\widehat{C}_5 = 0,$$
$$2(\mathbf{v}^T + \mathbf{w}^T) C_4 + \mathbf{w}^T (C_2 - 2C_3) = 0,$$
$$(\mathbf{v}^T + \mathbf{w}^T)\big((\mathbf{A} + \mathbf{B}) \Gamma_{\mathbf{c}} C_2 - \mathbf{B} C_2\big) - \mathbf{w}^T (\mathbf{A} + \mathbf{B}) C_2 = 0,$$
$$(\mathbf{v}^T + \mathbf{w}^T)\big((\mathbf{A} + \mathbf{B}) C_3 - \mathbf{B} C_2\big) - \mathbf{w}^T (\mathbf{A} + \mathbf{B}) C_2 = 0,$$
$$(\mathbf{v}^T + \mathbf{w}^T)(\mathbf{A} + \mathbf{B})^2 C_2 = 0,$$
$$(\mathbf{v}^T + \mathbf{w}^T) \Gamma_{\mathbf{c}} C_3 + \mathbf{w}^T (C_2 - C_3 - \Gamma_{\mathbf{c}} C_2) = 0,$$
$$(\mathbf{v}^T + \mathbf{w}^T) \Gamma_{\mathbf{c}} (\mathbf{A} + \mathbf{B}) C_2 - \mathbf{w}^T (\mathbf{A} + \mathbf{B}) C_2 = 0,$$
$$(\mathbf{v}^T + \mathbf{w}^T) \Gamma_{\mathbf{c}}^2 C_2 + \mathbf{w}^T (C_2 - 2\Gamma_{\mathbf{c}} C_2) = 0,$$
$$(\mathbf{v}^T + \mathbf{w}^T) C_2^2 = 0;$$

$p = 6$:

$$\widehat{C}_6 = 0,$$

$$6(\mathbf{v}^T + \mathbf{w}^T)C_5 - \mathbf{w}^T(C_2 - 3C_3 + 6C_4) = 0,$$

$$(\mathbf{v}^T + \mathbf{w}^T)\big(\mathbf{B}(C_2 - 2\Gamma_{\mathbf{c}}C_2) + (\mathbf{A} + \mathbf{B})\Gamma_{\mathbf{c}}^2 C_2\big)$$
$$+ \mathbf{w}^T\big((\mathbf{A} + \mathbf{B})(C_2 - 2\Gamma_{\mathbf{c}}C_2) + 2\mathbf{B}C_2\big) = 0,$$

$$(\mathbf{v}^T + \mathbf{w}^T)\big(\mathbf{B}(C_2 - C_3 - \Gamma_{\mathbf{c}}C_2) + (\mathbf{A} + \mathbf{B})\Gamma_{\mathbf{c}}C_3\big)$$
$$+ \mathbf{w}^T\big((\mathbf{A} + \mathbf{B})(C_2 - C_3 - \Gamma_{\mathbf{c}}C_2) + 2\mathbf{B}C_2\big) = 0$$

$$(\mathbf{v}^T + \mathbf{w}^T)\big(\mathbf{A}(C_2 - 2C_3) + 2(\mathbf{A} + \mathbf{B})C_4\big)$$
$$+ \mathbf{w}^T\big((\mathbf{A} + \mathbf{B})(C_2 - 2C_3) + 2\mathbf{B}C_2\big) = 0,$$

$$(\mathbf{v}^T + \mathbf{w}^T)\big((\mathbf{A} + \mathbf{B})\Gamma_{\mathbf{c}} - \mathbf{B}\big)(\mathbf{A} + \mathbf{B})C_2 - \mathbf{w}^T(\mathbf{A} + \mathbf{B})^2 C_2 = 0,$$

$$(\mathbf{v}^T + \mathbf{w}^T)\big((\mathbf{A} + \mathbf{B})((\mathbf{A} + \mathbf{B})\Gamma_{\mathbf{c}}C_2 - \mathbf{B}C_2) - \mathbf{B}(\mathbf{A} + \mathbf{B})C_2\big)$$
$$- \mathbf{w}^T(\mathbf{A} + \mathbf{B})^2 C_2 = 0,$$

$$(\mathbf{v}^T + \mathbf{w}^T)\big((\mathbf{A} + \mathbf{B})^2 C_3 - \mathbf{B}(\mathbf{A} + \mathbf{B})C_2 - (\mathbf{A} + \mathbf{B})\mathbf{B}C_2\big)$$
$$- \mathbf{w}^T(\mathbf{A} + \mathbf{B})^2 C_2 = 0,$$

$$(\mathbf{v}^T + \mathbf{w}^T)(\mathbf{A} + \mathbf{B})^3 C_2 = 0,$$

$$(\mathbf{v}^T + \mathbf{w}^T)\big((\mathbf{A} + \mathbf{B})C_2^2 + C_2(\mathbf{A} + \mathbf{B})C_2\big) = 0,$$

$$2(\mathbf{v}^T + \mathbf{w}^T)\Gamma_{\mathbf{c}}C_4 - \mathbf{w}^T(C_2 - 2C_3 - \Gamma_{\mathbf{c}}C_2 + 2C_4 + 2\Gamma_{\mathbf{c}}C_3) = 0,$$

$$(\mathbf{v}^T + \mathbf{w}^T)\big(\Gamma_{\mathbf{c}}(\mathbf{A} + \mathbf{B})\Gamma_{\mathbf{c}}C_2 - \Gamma_{\mathbf{c}}\mathbf{B}C_2\big)$$
$$+ \mathbf{w}^T\big((\mathbf{A} + \mathbf{B})(C_2 - \Gamma_{\mathbf{c}}C_2) + \mathbf{B}C_2 - \Gamma_{\mathbf{c}}(\mathbf{A} + \mathbf{B})C_2\big) = 0,$$

$$(\mathbf{v}^T + \mathbf{w}^T)\big(\Gamma_{\mathbf{c}}(\mathbf{A} + \mathbf{B})C_3 - \Gamma_{\mathbf{c}}\mathbf{B}C_2\big)$$
$$+ \mathbf{w}^T\big((\mathbf{A} + \mathbf{B})(C_2 - C_3) + \mathbf{B}C_2 - \Gamma_{\mathbf{c}}(\mathbf{A} + \mathbf{B})C_2\big) = 0,$$

$$(\mathbf{v}^T + \mathbf{w}^T)\Gamma_{\mathbf{c}}(\mathbf{A} + \mathbf{B})^2 C_2 - \mathbf{w}^T(\mathbf{A} + \mathbf{B})^2 C_2 = 0,$$

$$(\mathbf{v}^T + \mathbf{w}^T)\Gamma_{\mathbf{c}}^2 C_3 - \mathbf{w}^T(C_2 - 2\Gamma_{\mathbf{c}}C_2 - C_3 + 2\Gamma_{\mathbf{c}}C_3 + \Gamma_{\mathbf{c}}^2 C_2) = 0,$$

$$(\mathbf{v}^T + \mathbf{w}^T)\Gamma_{\mathbf{c}}^2(\mathbf{A} + \mathbf{B})C_2 + \mathbf{w}^T\big((\mathbf{A} + \mathbf{B})C_2 - 2\Gamma_{\mathbf{c}}(\mathbf{A} + \mathbf{B})C_2\big) = 0,$$

$$(\mathbf{v}^T + \mathbf{w}^T)\Gamma_{\mathbf{c}}^3 C_2 - \mathbf{w}^T(C_2 - 3\Gamma_{\mathbf{c}}C_2 + 3\Gamma_{\mathbf{c}}^2 C_2) = 0,$$

$$(\mathbf{v}^T + \mathbf{w}^T)C_2 C_3 - \mathbf{w}^T C_2^2 = 0,$$

$$(\mathbf{v}^T + \mathbf{w}^T)(\mathbf{A} + \mathbf{B})C_2^2 = 0,$$

$$(\mathbf{v}^T + \mathbf{w}^T)\Gamma_{\mathbf{c}}C_2^2 - \mathbf{w}^T C_2^2 = 0.$$

Setting $\mathbf{B} = \mathbf{0}$, the equations above reduce to the order conditions obtained by Jackiewicz et al. [181, 182] using the approach of Hairer and Wanner [144].

These order conditions in the scalar case $m = 1$ are sufficient but not necessary since some elementary differentials are equal to each other. For example, for $p = 5$, $g_1 g_1^{(1)} y'' = g_1^{(1)} g_1 y''$ and the conditions corresponding to these elementary differentials could be replaced by one condition. However, in vector case $m > 1$, $G_1(G_1^{(1)}(y'')) \neq G_1^{(1)}(G_1(y''))$, where G_1 and $G_1^{(1)}$ are vector analogs of g_1 and $g_1^{(1)}$, and these conditions are also necessary. We refer to Jackiewicz and Tracogna [183] for a derivation of order conditions in the vector case.

It follows from Theorems 5.2.1 and 5.2.2 (compare also the structure of order conditions displayed above) that TSRK method (5.1.2) has order p and stage order $q = p - 1$ if

$$C_\nu = 0, \quad \nu = 1, 2, \ldots, p - 1, \tag{5.3.1}$$

and

$$\widehat{C}_\nu = 0, \quad \nu = 1, 2, \ldots, p, \tag{5.3.2}$$

where C_ν and $\widehat{C}_\nu$ are given by (5.2.3) and (5.2.4). These conditions imply that method (5.1.2) is convergent with order p; that is,

$$\sup \left\{ \|y(t_n) - y_n\| : \ 0 \leq n \leq N \right\} = O(h^p)$$

as $h \to 0$. If, in addition to (5.3.1) and (5.3.2), we have

$$C_p = 0; \tag{5.3.3}$$

then the TSRK method has order p and stage order $q = p$.

In the next three sections we illustrate the solution of order conditions for TSRK methods with $s = 1$, $s = 2$, and $s = 3$ stages. We are interested primarily in methods of order p and stage order $q = p - 1$ or $q = p$.

5.4   ANALYSIS OF TSRK METHODS WITH ONE STAGE

5.4.1   Explicit TSRK methods: $s = 1$, $p = 2$ or 3

Explicit TSRK methods with $s = 1$ are given by the abscissa c and the table of coefficients

$$\begin{array}{c|c|c} u & 0 & b \\ \hline \vartheta & v & w \end{array}, \tag{5.4.1}$$

where u, ϑ, b, v, and w are real parameters. These methods have been investigated by Chollom and Jackiewicz [91]. Solving the system of order and stage order conditions $\widehat{C}_1 = 0$, $\widehat{C}_2 = 0$, and $C_1 = 0$ with respect to b, v, and w, we obtain a three parameter family of methods of order $p = 2$ and stage order $q = 1$. The coefficients of these methods are given by

$$b = c + u, \quad v = \frac{3 + \vartheta - 2c(1 + \vartheta)}{2}, \quad w = \frac{\vartheta - 1 + 2c(1 + \vartheta)}{2}.$$

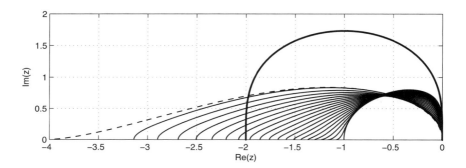

Figure 5.4.1   Stability regions of TSRK methods with $s = 1$ and $p = q + 1 = 2$ and RK method of order 2 (thick line). Stability regions of TSRK methods correspond to the values of $\vartheta = -1$ (dashed line), and $\vartheta = -0.9 + 0.1(i - 1)$, $i = 1, 2, \ldots, 20$ (thin lines, from left to right)

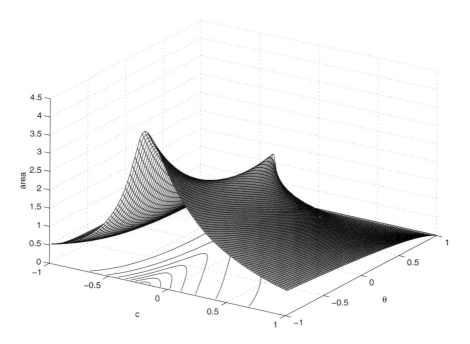

Figure 5.4.2   Area of stability regions of TSRK methods with $s = 1$ and $p = q = 2$ and contours corresponding to areas equal to 1, 1.5, 2, 2.5, 3, 3.5, and 4

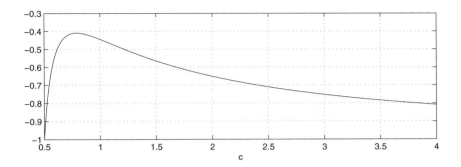

Figure 5.4.3   Stability interval versus c for $c \in [1/2, 4]$ for TSRK methods with $s = 1$, $p = 3$, and $q = 2$

To analyze the stability properties of (5.4.1), these methods are first reformulated as GLMs with coefficients given by

$$
\left[\begin{array}{c|c} \mathbf{A} & \mathbf{U} \\ \hline \mathbf{B} & \mathbf{V} \end{array}\right] = \left[\begin{array}{c|ccc} 0 & 1-u & u & b \\ \hline v & 1-\vartheta & \vartheta & w \\ 0 & 1 & 0 & 0 \\ 1 & 0 & 0 & 0 \end{array}\right]
$$

(compare Section 2.1). It follows from the linear stability theory of GLMs (compare Section 2.6) that the stability function $p(\eta, z)$ of (5.4.1) is given by (2.6.5), that is,

$$
p(\eta, z) = \det\left(\eta \mathbf{I} - \mathbf{M}(z)\right),
$$

where the stability matrix $\mathbf{M}(z)$ is defined by (2.6.4). It can be verified that stability polynomial $p(\eta, z)$ of these methods is

$$
p(\eta, z) = \eta^3 - p_2(z)\eta^2 + p_1(z)\eta - p_0(z), \tag{5.4.2}
$$

where

$$
p_2(z) = 1 - \vartheta + \frac{3 + \vartheta - 2c\vartheta - u + 2cu - \vartheta u + 2c\vartheta u}{2} z,
$$

$$
p_1(z) = -\vartheta + \frac{1 - \vartheta - 4c\vartheta - 2u + 4cu - 2\vartheta u + 4c\vartheta u}{2} z,
$$

and

$$
p_0(z) = \frac{2cu - 2c\vartheta - u - \vartheta u + 2c\vartheta u}{2} z.
$$

For fixed values of ϑ, $-1 < \vartheta \leq 1$, we have performed a numerical search in the parameter space (c, u) trying to maximize the area of the region of absolute stability $\mathcal{A}$ of the corresponding TSRK methods. These stability regions are plotted in Fig. 5.4.1 for $\vartheta = -1$ (dashed line), $\vartheta = -0.9 + 0.1(i-1)$, $i = 1, 2, \ldots, 20$ (thin lines). For comparison, we also shown, by a thick line, the stability region of RK method of order $p = 2$ with $s = 2$.

Solving in addition to $\widehat{C}_1 = 0$, $\widehat{C}_2 = 0$, and $C_1 = 0$ the stage order condition $C_2 = 0$ with respect to u, we obtain a two-parameter family of TSRK methods given by

$$u = \frac{c(2-c)}{2c-1}, \quad b = \frac{c(c+1)}{2c-1}, \quad c \neq \frac{1}{2}, \tag{5.4.3}$$

where v and w are defined as before. The area of the region of absolute stability of these methods is plotted in Fig. 5.4.2 for $-1 \leq c \leq 1$ and $-1 \leq \vartheta \leq 1$ together with the contours corresponding to the areas equal to 1, 1.5, 2, 2.5, 3, 3.5, and 4.

Solving in addition to the previous order and stage order conditions the equation $\widehat{C}_3 = 0$ with respect to ϑ, we obtain a one-parameter family of TSRK methods of order $p = 3$ and stage order $q = 2$. The coefficients of these methods are

$$\vartheta = \frac{6c^2 - 12c + 5}{1 - 6c^2}, \quad v = \frac{2(2 - 6c + 3c^2)}{1 - 6c^2}, \quad w = \frac{2(1 - 3c^2)}{1 - 6c^2}, \tag{5.4.4}$$

$c \neq \pm\sqrt{6}/6$, with u and b defined as before. We have to impose the condition $-1 < \vartheta \leq 1$ for zero-stability, which corresponds to $c > 1/2$. The stability polynomial of these methods is given by (5.4.2) with $p_2(z)$, $p_1(z)$, and $p_0(z)$ defined by

$$p_2(z) = \frac{4 - 12c + 12c^2 - (4 - 5c - 9c^2)z}{6c^2 - 1},$$

$$p_1(z) = \frac{5 - 12c + 6c^2 + 2(1 - 4c + 6c^2)z}{6c^2 - 1},$$

and

$$p_0(z) = \frac{c(3c-1)z}{6c^2 - 1},$$

$c \neq \pm\sqrt{6}/6$. We can verify using the Schur criterion given in Theorem 2.8.1 that the interval of absolute stability of these methods is

$$\left[\frac{4(1 - 3c + 3c^2)}{1 + 2c - 12c^2}, 0 \right].$$

The left-hand side of this interval is plotted in Fig. 5.4.3 versus c for $c \in [1/2, 4]$. For $c \to 1/2$ and $c \to \infty$ the interval of absolute stability attains its maximum size $[-1, 0]$. However, for these values of c we have $\vartheta = -1$, which violates the zero-stability condition of the method. The choice $c = 1$ leads to the method of the form

$$\begin{array}{c|c|c} 1 & 0 & 2 \\ \hline \frac{1}{5} & \frac{2}{5} & \frac{4}{5} \end{array},$$

which has the interval of absolute stability $[-4/9, 0]$.

Solving instead the equation $\widehat{C}_3 = 0$ with respect to c, we obtain a one-parameter family of TSRK methods of order $p = 3$ and stage order $q = 2$ with coefficients of the form

$$c = \frac{6 \pm \sqrt{6(\vartheta^2 - 4\vartheta + 1)}}{6(1 + \vartheta)}, \tag{5.4.5}$$

$$u = \frac{3(5 - 3\vartheta) \mp 5\sqrt{6(\vartheta^2 - 4\vartheta + 1)}}{6(1 + \vartheta)},$$

$$b = \frac{3(7 - 3\vartheta) \mp 4\sqrt{6(\vartheta^2 - 4\vartheta + 1)}}{6(1 + \vartheta)} \tag{5.4.6}$$

and

$$v = \frac{3(1 + \vartheta) \mp \sqrt{6(\vartheta^2 - 4\vartheta + 1)}}{6},$$

$$w = \frac{3(1 + \vartheta) \pm \sqrt{6(\vartheta^2 - 4\vartheta + 1)}}{6}. \tag{5.4.7}$$

We have to assume that $\vartheta \in (-1, 2 - \sqrt{3}]$, so that the coefficients of these methods are real. The choice $\vartheta = 0$ leads to methods with $c = (6 \pm \sqrt{6})/6$ and coefficients u, b, v, and w given by

$$\begin{array}{c|c|c} \frac{5(3\mp\sqrt{6})}{6} & 0 & \frac{21\mp4\sqrt{6}}{6} \\ \hline 0 & \frac{3\mp\sqrt{6}}{6} & \frac{3\pm\sqrt{6}}{6} \end{array},$$

with an interval of absolute stability equal to $[-6/11, 0]$.

5.4.2   Implicit TSRK methods: $s = 1$, $p = 2$ or 3

Implicit TSRK methods with $s = 1$ are given by the abscissa c and the table of coefficients

$$\begin{array}{c|c|c} u & \lambda & b \\ \hline \vartheta & v & w \end{array}, \tag{5.4.8}$$

where u, ϑ, λ, b, v, and w are real parameters. These methods have been investigated in [91]. Solving $C_1 = 0$ and order conditions up to $p = 3$ with respect to u, b, ϑ, v, and w, we obtain a two-parameter family of methods of order $p = 3$ with respect to c and λ. The coefficients of these methods are given by

$$u = \frac{2c - c^2 - 2\lambda}{2c - 1}, \quad b = \frac{c + c^2 - \lambda - 2c\lambda}{2c - 1}, \quad c \neq \frac{1}{2},$$

with ϑ, v, and w defined by (5.4.4). Observe that for $\lambda = 0$ the coefficients u and b reduce to those defined by (5.4.3). It can be verified that these methods are necessarily of stage order $q = 2$ (i.e., $C_2 = 0$).

To investigate the stability properties of these methods, it is more convenient to reformulate the formulas for the coefficients of these methods in terms of ϑ and λ instead of c and λ. In such a case, the abscissa c is given by the formula (5.4.5) with the coefficients u and b taking the form

$$
u = \frac{3(5 - 3\vartheta) \mp 5\sqrt{6(\vartheta^2 - 4\vartheta + 1)}}{6(1 + \vartheta)} - \frac{2\big(3(1 - \vartheta) \mp \sqrt{6(1 - 4\vartheta + \vartheta^2)}\big)}{1 + \vartheta}\,\lambda,
$$

$$
b = \frac{3(7 - 3\vartheta) \mp 4\sqrt{6(\vartheta^2 - 4\vartheta + 1)}}{6(1 + \vartheta)} - \frac{7 - 6\vartheta \mp 2\sqrt{6(1 - 4\vartheta + \vartheta^2)}}{1 + \vartheta}\,\lambda,
$$

and v and w given by (5.4.7). For $\lambda = 0$ the formulas for u and b reduce to (5.4.6). It is also more convenient to work with the polynomial

$$
\widetilde{p}(\eta, z) = (1 - \lambda z)\,p(\eta, z)
$$

instead of $p(\eta, z)$, where $p(\eta, z) = \det(z\mathbf{I} - \mathbf{M}(z))$ is the stability function of method (5.4.8) and $\mathbf{M}(z)$ is the stability matrix. It can be verified that $\widetilde{p}(\eta, z)$ takes the form

$$
\widetilde{p}(\eta, z) = (1 - \lambda z)\,\eta^3 - \widetilde{p}_2(z)\,\eta^2 + \widetilde{p}_1(z)\,\eta - \widetilde{p}_0(z), \tag{5.4.9}
$$

with

$$
\widetilde{p}_2(z) = 1 - \vartheta + \frac{23 - 36\,\lambda + 5\,\vartheta}{12}\,z,
$$

$$
\widetilde{p}_1(z) = -\vartheta + \frac{1 - 9\lambda - 2\vartheta}{3}\,z, \quad \widetilde{p}_0(z) = \frac{5 - 12\lambda - \vartheta}{12}\,z.
$$

We demonstrate first that there are no A-stable methods in the class of methods (5.4.8) of order $p = 3$ and stage order $q = 2$. This follows from the following arguments. We must have $\lambda > 0$ and it follows from the maximum principle that the method is A-stable if and only if $\widetilde{p}_3(\eta, z)$ is a Schur polynomial with respect to η for all $y \in \mathbb{R}$, where $\widetilde{p}_3(\eta, y) := \widetilde{p}(\eta, iy)$. Let

$$
\widehat{p}_3(\eta, y) = \eta^3\,\widetilde{p}_3(1/\eta, -y),
$$

$$
\widetilde{p}_2(\eta, y) = \big(\widehat{p}_3(0, y)\widetilde{p}_3(\eta, y) - \widetilde{p}_3(0, y)\widehat{p}_3(\eta, y)\big)/\eta, \quad \widehat{p}_2(\eta, y) = \eta^2\,\widetilde{p}_2(1/\eta, -y),
$$

$$
\widetilde{p}_1(\eta, y) = \big(\widehat{p}_2(0, y)\widetilde{p}_2(\eta, y) - \widetilde{p}_2(0, y)\widehat{p}_2(\eta, y)\big)/\eta, \quad \widehat{p}_1(\eta, y) = \eta\,\widetilde{p}_1(1/\eta, -y),
$$

and

$$
\widetilde{p}_0(y) = \big(\widehat{p}_1(0, y)\widetilde{p}_1(\eta, y) - \widetilde{p}_1(0, y)\widehat{p}_1(\eta, y)\big)/\eta.
$$

It follows from the Schur criterion (see Theorem 2.8.1) that $\widetilde{p}_3(\eta, y)$ has all roots η inside the unit disk for all $y \in \mathbb{R}$ if and only if

$$
\widetilde{p}_0(y) > 0, \quad \big|\widehat{p}_3(0, y)\big| - \big|\widetilde{p}_3(0, y)\big| > 0, \quad \text{and} \quad \big|\widehat{p}_2(0, y)\big| - \big|\widetilde{p}_2(0, y)\big| > 0
$$

for all $y \in \mathbb{R}$. These conditions take the form

$$
q_0(\lambda, \vartheta)y^4 + q_1(\lambda, \vartheta)y^6 + q_2(\lambda, \vartheta)y^8 > 0, \quad 1 + q_3(\lambda, \vartheta)y^2 > 0,
$$

and

$$q_4(\lambda, \vartheta) + q_5(\lambda, \vartheta)y^2 + q_6(\lambda, \vartheta)y^4 > 0,$$

where $q_i(\lambda, \vartheta)$, $i = 1, 2, \ldots, 6$, are polynomials in λ and ϑ. These polynomials are not reproduced here. It can be verified that the conditions above are satisfied for all $y \in \mathbb{R}$ if and only if

$$q_i(\lambda, \vartheta) > 0, \quad i = 1, 2, \ldots, 6.$$

This in turn implies, among other things, the inequalities

$$24\lambda + \vartheta - 9 < 0, \quad 24 + \vartheta - 5 > 0, \quad 24\lambda + \vartheta - 11 > 0,$$

which cannot be satisfied simultaneously for any λ and ϑ. This justifies our claim that there are no A-stable methods (5.4.8) with $p = 3$ and $q = 2$.

Next we investigate the existence of methods of the form (5.4.8) that are A_0-stable (i.e., stable on the negative real axis). To this end we have to verify that $\widetilde{p}_3(\eta, x)$ is a Schur polynomial with respect to η for all $x \in \mathbb{R}$, $x < 0$, where we now define $\widetilde{p}_3(\eta, x) := \widetilde{p}(\eta, x)$. Let

$$\widehat{p}_3(\eta, x) = \eta^3 \, \widetilde{p}_3(1/\eta, x),$$

$$\widetilde{p}_2(\eta, x) = \big(\widehat{p}_3(0, x)\widetilde{p}_3(\eta, x) - \widetilde{p}_3(0, x)\widehat{p}_3(\eta, x)\big)/\eta, \quad \widehat{p}_2(\eta, x) = \eta^2 \, \widetilde{p}_2(1/\eta, x),$$

$$\widetilde{p}_1(\eta, x) = \big(\widehat{p}_2(0, x)\widetilde{p}_2(\eta, x) - \widetilde{p}_2(0, x)\widehat{p}_2(\eta, x)\big)/\eta, \quad \widehat{p}_1(\eta, x) = \eta \, \widetilde{p}_1(1/\eta, x),$$

and

$$\widetilde{p}_0(x) = \big(\widehat{p}_1(0, x)\widetilde{p}_1(\eta, x) - \widetilde{p}_1(0, x)\widehat{p}_1(\eta, x)\big)/\eta.$$

It follows again from the Schur criterion that $\widetilde{p}_3(\eta, x)$ has all roots η inside the unit disk for $x \in \mathbb{R}$, $x < 0$, if and only if

$$\widetilde{p}_0(x) > 0, \quad \big|\widehat{p}_3(0, x)\big| - \big|\widetilde{p}_3(0, x)\big| > 0, \quad \text{and} \quad \big|\widehat{p}_2(0, x)\big| - \big|\widetilde{p}_2(0, x)\big| > 0$$

for $x \in \mathbb{R}$, $x < 0$. Careful analysis reveals that this is the case if and only if all nonzero roots of the polynomial $\widetilde{p}_0(x)$ are real and positive. These roots are

$$r_1 = \frac{6(\vartheta - 1)}{11 - 24\lambda - \vartheta}, \quad r_2 = \frac{12}{5 - \vartheta}, \quad r_3 = -\frac{12}{5 - 24\lambda - \vartheta},$$

and

$$r_{4,5} = r_{6,7} = \frac{11 - 2\vartheta - \vartheta^2 \pm \sqrt{\Delta}}{15 - 24\lambda + 2\vartheta - 24\lambda\vartheta - \vartheta^2},$$

where

$$\Delta(\lambda, \vartheta) := -239 + 576\lambda - 452\vartheta + 1151\lambda\vartheta - 42\vartheta^2 + 576\lambda\vartheta^2 + 28\vartheta^3 + \vartheta^4.$$

It is clear that for $-1 < \vartheta \leq 1$ (the condition required for zero-stability) these roots are real and positive if and only if

$$f_1(\lambda, \vartheta) := 11 - 24\lambda - \vartheta < 0, \quad f_2(\lambda, \vartheta) := 5 - 24\lambda - \vartheta < 0,$$

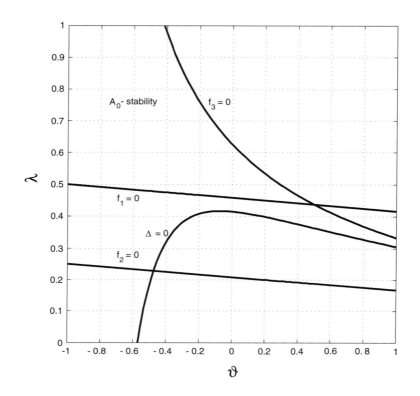

Figure 5.4.4 A_0-stable methods (5.4.8) of order $p = 3$ and stage order $q = 2$

$$f_3(\lambda, \vartheta) := 15 - 24\lambda + 2\vartheta - 24\lambda\vartheta - \vartheta^2 > 0 \quad \text{and} \quad \Delta(\lambda, \vartheta) > 0.$$

We have plotted in Fig. 5.4.4 the curves $f_i(\lambda, \vartheta) = 0$ for $i = 1, 2, 3$, and $\Delta(\lambda, \vartheta) = 0$. Methods (5.4.8) are A_0-stable for the values of (λ, ϑ) belonging to the region bounded by the curves $f_1(\lambda, \vartheta) = 0$ and $f_3(\lambda, \vartheta) = 0$ (above $f_1(\lambda, \vartheta) = 0$ and below $f_3(\lambda, \vartheta) = 0$). Observe that this conclusion differs somewhat from the results of Jackiewicz and Tracogna [183] and Tracogna [270].

Setting $\lambda = 1/2$, $\vartheta = 0$, we have $c = (6 \pm \sqrt{6})/6$, and this leads to the A_0-stable TSRK methods of order $p = 3$ and stage order $q = 2$ with coefficients given by

$$\begin{array}{c|c|c} \frac{-3\pm\sqrt{6}}{6} & \frac{1}{2} & \pm\frac{\sqrt{6}}{6} \\ \hline 0 & \frac{3\mp\sqrt{6}}{6} & \frac{3\pm\sqrt{6}}{6} \end{array}.$$

The stability polynomial of this method is

$$\widetilde{p}(\eta, z) = \left(1 - \frac{1}{2}z\right)\eta^3 - \left(1 + \frac{5}{12}z\right)\eta^2 - \frac{1}{6}z\eta + \frac{1}{12}z.$$

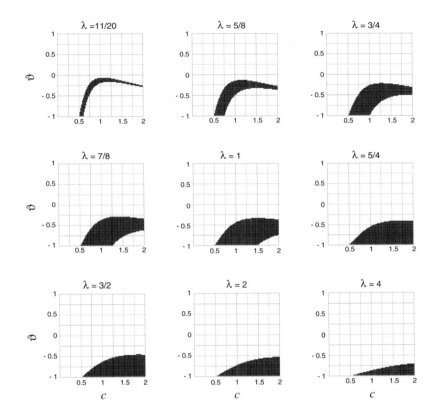

Figure 5.4.5 A-stable methods (5.4.10) of order $p = 2$ and stage order $q = 2$

We conclude this section by investigating A-stable methods of the form (5.4.8) with $p = q = 2$. Solving the system of order and stage order conditions

$$\widehat{C}_1 = 0, \quad \widehat{C}_2 = 0, \quad C_1 = 0, \quad C_2 = 0,$$

we obtain a three-parameter family of methods depending on c, λ, and ϑ, with coefficients given by

$$
\begin{array}{c|c|c}
\frac{2c - c^2 - 2\lambda}{2c - 1} & \lambda & \frac{c + c^2 - \lambda - 2c\lambda}{2c - 1} \\
\hline
\vartheta & \frac{3 - 2c + \vartheta - 2c\vartheta}{2} & \frac{2c + \vartheta - 2c\vartheta - 1}{2}
\end{array},
\tag{5.4.10}
$$

$c \neq 1/2$. The stability polynomial $\widetilde{p}(\eta, z)$ takes the form (5.4.9) with

$$\widetilde{p}_2(z) = 1 - \vartheta + \frac{3 + 2c - c^2 - 6\lambda - \vartheta - 2c^2\vartheta}{2} \, z,$$

$$\widetilde{p}_1(z) = -\vartheta + \frac{1 + 4c - 2c^2 - 6\lambda - \vartheta - 2c^2\vartheta}{2} \, z,$$

and

$$\widetilde{p}_0(z) = \frac{2c - c^2 - 2\lambda - c^2\vartheta}{2} \, z.$$

We have performed an extensive computer search in the three-dimensional space (λ, c, ϑ) looking for A-stable methods using the Schur criterion in Theorem 2.8.1 [183, 270]. Selection of the results of this search in the plane (c, ϑ) is displayed in Fig. 5.4.5 for selected values of the parameter λ equal to 11/20, 5/8, 3/4, 7/8, 1, 5/4, 3/2, 2, and 4.

It can be verified that for the parameters (c, ϑ) inside the shaded regions on Fig 5.4.5, all roots of the polynomial

$$\varphi(\eta) = \lim_{z \to \infty} \left(\widetilde{p}(\eta, z) / (1 - \lambda z) \right)$$

have modulus less than 1.

5.5   ANALYSIS OF TSRK METHODS WITH TWO STAGES

5.5.1   Explicit TSRK methods: $s = 2$, $p = 2$, $q = 1$ or 2

This section closely follows the presentation by Chollom and Jackiewicz [91]. Explicit TSRK methods with two stages are specified by the abscissa vector $\mathbf{c} = [c_1, c_2]^T$ and the table of coefficients

$$
\frac{\mathbf{u} \ \Big| \ \mathbf{A} \ \Big| \ \mathbf{B}}{\vartheta \ \Big| \ \mathbf{v}^T \ \Big| \ \mathbf{w}^T} =
\begin{array}{c|c|cc}
u_1 & & b_{11} & b_{12} \\
u_2 & a_{21} & b_{21} & b_{22} \\
\hline
\vartheta & v_1 \quad v_2 & w_1 & w_2
\end{array}, \qquad (5.5.1)
$$

where u_1, u_2, ϑ, a_{21}, b_{11}, b_{12}, b_{21}, b_{22}, v_1, v_2, w_1, and w_2 are real parameters. Assuming that $\mathbf{c} = [0, 1]^T$, $\vartheta = 0$, and solving the system of order conditions $\widehat{C}_1 = 0$, $\widehat{C}_2 = 0$, with respect to w_1 and w_2, and stage order condition $C_1 = 0$ with respect to b_{11} and b_{21}, we obtain a seven-parameter family of TSRK methods (5.5.1) of order $p = 2$ and stage order $q = 1$. The coefficients of these methods are given by

$$b_{11} = u_1 - b_{12}, \quad b_{21} = 1 - a_{21} - b_{22} + u_2,$$

$$w_1 = \frac{2v_1 - 1}{2}, \quad w_2 = \frac{3 - 2v_1 - 4v_2}{2}.$$

We choose these free parameters of the method trying to maximize the area of the intersection of the region of absolute stability of the TSRK method with a negative half-plane. This area can be approximated using numerical integration in polar coordinates. The resulting objective function for the "negative area" is then minimized using the subroutine `fminsearch` from Matlab, starting from some random initial guesses. This computer search leads to the method with coefficients

0.911557		0.692385	0.219172
0.601892	0.730872	0.635887	0.235132
0	0.892098 0.336848	−0.163152	−0.0657949

The stability region of this method is shown by a thin line in Fig. 5.5.1.

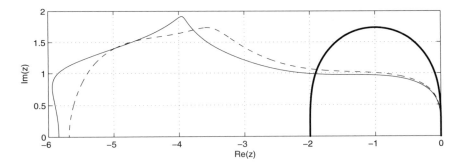

Figure 5.5.1   Stability region of TSRK method with $s = 2$ and $p = q + 1 = 2$ (thin line), TSRK method with $s = 2$ and $p = q = 2$ (thin dashed line), and RK method of order 2 (thick line) .

Next we look for methods with $p = q = 2$. Assuming again that $\mathbf{c} = [0, 1]^T$, $\vartheta = 0$, and solving the order and stage order conditions $\widehat{C}_1 = 0$, $\widehat{C}_2 = 0$, $C_1 = 0$, and $C_2 = 0$, we obtain a five-parameter family of TSRK methods (5.5.1) depending on u_1, u_2, a_{21}, v_1, and v_2. The coefficients of these methods are given by

$$b_{11} = b_{12} = \frac{u_1}{2}, \quad b_{21} = \frac{u_2 - 1}{2}, \quad b_{22} = \frac{3 - 2\,a_{21} + u_2}{2},$$

$$w_1 = \frac{2v_1 - 1}{2}, \quad w_2 = \frac{3 - 2v_1 - 4v_2}{2}.$$

A computer search in the parameter space u_1, u_2, a_{21}, v_1, and v_2 leads to a method with coefficients

0.308343			0.154172	0.154172
−0.113988	1.43462		−0.556994	0.00838564
0	1.47417	−0.0264581	−0.526458	0.0787465

for which the region of absolute stability is plotted in Fig. 5.5.1 by a thin dashed line.

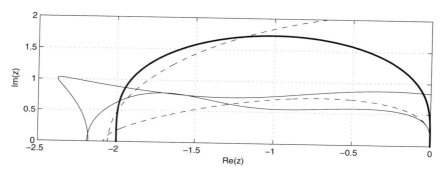

Figure 5.5.2   Stability region of TSRK methods of type 3 with $s = 2$ and $p = q + 1 = 2$ (thin line), TSRK method with $s = 2$ and $p = q = 2$ (thin dashed line), and RK method of order 2 (thick line)

We consider next type 3 methods with $p = 2$ and $q = 1$ (i.e., methods for which $a_{21} = 0$). Assuming as before that $\mathbf{c} = [0, 1]^T$ and $\vartheta = 0$, a computer search for methods with a large region of absolute stability leads to the formula with coefficients

0.860064			−0.177218	1.03728
−1.14359			0.206108	−0.349695
0	0.20866	0.366951	−0.133049	0.557438

This search was performed in a parameter space u_1, u_2, b_{12}, b_{22}, v_1, and v_2. The stability region of this formula is plotted in Fig. 5.5.2 by a thin line.

Consider next type 3 methods with $p = q = 2$. Assuming again that $\mathbf{c} = [0, 1]^T$ and $\vartheta = 0$, a computer search in a parameter space u_1, u_2, v_1, and

v_2 leads to a TSRK formula with coefficients given by

$$
\begin{array}{cc|cc}
0.303461 & & 0.151731 & 0.151731 \\
0.210821 & & -0.394589 & 1.60541 \\
\hline
0 & 0.123103 \quad -0.763431 & -1.26343 & 2.90376
\end{array}
$$

whose region of absolute stability is shown by a thin dashed line in Fig 5.5.2.

5.5.2   Implicit TSRK methods: $s = 2$, $p = 2$, $q = 1$ or 2

Implicit TSRK methods of type 2 with two stages are specified by the abscissa
vector $\mathbf{c} = [c_1, c_2]^T$ and the table of coefficients

$$
\frac{\mathbf{u} \;\big|\; \mathbf{A} \;\big|\; \mathbf{B}}{\vartheta \;\big|\; \mathbf{v}^T \;\big|\; \mathbf{w}^T}
=
\begin{array}{c|cc|cc}
u_1 & \lambda & & b_{11} & b_{12} \\
u_2 & a_{21} & \lambda & b_{21} & b_{22} \\
\hline
\vartheta & v_1 & v_2 & w_1 & w_2
\end{array}
\,, \tag{5.5.2}
$$

where u_1, u_2, ϑ, λ, a_{21}, b_{11}, b_{12}, b_{21}, b_{22}, v_1, v_2, w_1, and w_2 are real parameters.
Assume that $\mathbf{c} = [0, 1]^T$, $\vartheta = 0$, and $\mathbf{u} = [0, 0]^T$. As in Section 5.5.1, solving
the system of order conditions $\widehat{C}_1 = 0$, $\widehat{C}_2 = 0$ with respect to w_1 and w_2
and stage order condition $C_1 = 0$ with respect to b_{11} and b_{21}, we obtain a
six-parameter family of formulas (5.5.2) of order $p = 2$ and stage order $q = 1$.
The coefficients of these methods are not listed here. Solving next the stage
order condition $C_2 = 0$ with respect to b_{12} and b_{22}, we obtain a four-parameter
family of methods of order $p = 2$ and stage order $q = 2$ which depend on the
parameters λ, a_{21}, v_1, and v_2. The coefficients of these methods are

$$
\begin{array}{c|cc|cc}
0 & \lambda & & 0 & -\lambda \\
0 & a_{21} & \lambda & \frac{-1+2\lambda}{3} & \frac{3-2a_{21}-4\lambda}{2} \\
\hline
0 & v_1 & v_2 & \frac{-1+2v_2}{2} & \frac{3-2v_1-4v_2}{2}
\end{array}
\,.
$$

To analyze the stability properties of the resulting methods, as in Section 5.4.2
we work with the polynomial

$$
\widetilde{p}(\eta, z) = (1 - \lambda z)^2 p(\eta, z),
$$

where $p(\eta, z)$ is the stability function. It can be verified that this polynomial
takes the form

$$
\widetilde{p}(\eta, z) = \left((1 - \lambda z)^2 \eta^3 - \widetilde{p}_2(z)\eta^2 + \widetilde{p}_1(z)\eta - \widetilde{p}_0(z) \right)\eta, \tag{5.5.3}
$$

with

$$
\widetilde{p}_2(z) = \widetilde{p}_{20} + \widetilde{p}_{21}z + \widetilde{p}_{22}z^2, \quad \widetilde{p}_1(z) = \widetilde{p}_{11}z + \widetilde{p}_{12}z^2, \quad \widetilde{p}_0(z) = \widetilde{p}_{02}z^2.
$$

Next we solve the equation $\widetilde{p}_{02} = 0$ with respect to a_{12}, and the system $\widetilde{p}_{11} = 0$ and $\widetilde{p}_{12} = 0$ with respect to v_1 and v_2. This leads to a one-parameter family of TSRK methods with coefficients given by

$$
\begin{array}{c|cc|cc}
0 & \lambda & & 0 & -\lambda \\
0 & \frac{2}{1+2\lambda} & \lambda & \frac{-1+2\lambda}{2} & \frac{-1+2\lambda-8\lambda^2}{2(1+2\lambda)} \\
\hline
0 & \frac{5-2\lambda+12\lambda^2+8\lambda^3}{4(1+2\lambda)} & \frac{1+4\lambda-4\lambda^2}{4} & \frac{5-2\lambda+12\lambda^2+8\lambda^3}{4(1+2\lambda)} & \frac{1+4\lambda-4\lambda^2}{4}
\end{array} \ .
$$

The stability polynomial of these formulas takes the form

$$
\widetilde{p}(\eta, z) = \left((1 - \lambda z)^2 \eta - 1 - (1 - 2\lambda)z - \frac{1}{2}(1 - 4\lambda + 2\lambda^2)z^2 \right)\eta^3.
$$

Observe that the only nonzero root of $\widetilde{p}(\eta, z)$ is equal to the stability function of SDIRK method of order $p = 2$. As a consequence, it follows from Section 2.7 that these methods are A-stable if and only if $\lambda \geq \frac{1}{4}$ and L-stable if and only if $\lambda = (2 \pm \sqrt{2})/2$. The coefficients of these A- and L-stable TSRK methods are given by

$$
\begin{array}{c|cc|cc}
0 & \frac{2\pm\sqrt{2}}{2} & & 0 & \frac{-2\mp\sqrt{2}}{2} \\
0 & \frac{2(3\mp\sqrt{2})}{7} & \frac{2\pm\sqrt{2}}{2} & \frac{1\pm\sqrt{2}}{2} & \frac{-19\mp10\sqrt{2}}{14} \\
\hline
0 & \frac{73\pm34\sqrt{2}}{28} & \frac{-1\mp2\sqrt{2}}{4} & \frac{-3\mp2\sqrt{2}}{4} & \frac{-17\mp6\sqrt{2}}{28}
\end{array} \ .
$$

Finally, consider type 4 methods, which correspond to $a_{21} = 0$. There is a three-parameter family of such methods with $\mathbf{c} = [0, 1]^T$, $\vartheta = 0$, and $\mathbf{u} = [0, 0]^T$ of order $p = 2$ and stage order $q = 2$ which depend on v_1, v_2, and λ. The stability polynomial of this family has the form (5.5.3). Solving the system $\widetilde{p}_{02} = 0$ and $\widetilde{p}_{12} = 0$ with respect to v_1 and v_2, we obtain a one-parameter family of TSRK methods of the form

$$
\begin{array}{c|cc|cc}
0 & \lambda & & 0 & -\lambda \\
0 & & \lambda & \frac{2\lambda-1}{2} & \frac{3-4\lambda}{2} \\
\hline
0 & \frac{\lambda(2\lambda-1)(2\lambda+1)}{4(\lambda-1)} & -\frac{\lambda(4\lambda^2-8\lambda+5)}{4(\lambda-1)} & \frac{(2\lambda-1)(2\lambda^2-3\lambda+2)}{4(\lambda-1)} & \frac{4\lambda^3-16\lambda^2+17\lambda-6}{4(\lambda-1)}
\end{array} \ ,
$$

$\lambda \neq 1$. The stability polynomial of this family of formulas assumes the form

$$
\widetilde{p}(\eta, z) = \left((1 - \lambda z)^2 \eta^2 - \widetilde{p}_2(z)\eta + \widetilde{p}_1(z) \right)\eta^2
$$

with

$$
\widetilde{p}_2(z) = 1 - \frac{3 - 8\lambda + 4\lambda^2}{2(\lambda - 1)}z + \frac{\lambda(3 - 6\lambda + 2\lambda^2)}{2(\lambda - 1)}z^2, \quad \widetilde{p}_1(z) = \frac{2\lambda - 1}{2(\lambda - 1)}z.
$$

It can be verified using the Schur criterion (Theorem 2.8.1) that these methods are A-stable if and only if $\frac{1}{2} \leq \lambda \leq \frac{3}{4}$ or $\lambda \geq \frac{3}{2}$. These methods are L-stable if and only if $\lambda = (3 \pm \sqrt{3})/2$, which are the roots of the equation $\widetilde{p}_{22} = 0$. The coefficients of these A- and L-stable methods are

$$
\begin{array}{c|ccc|c}
0 & \frac{3\pm\sqrt{3}}{2} & & & 0 & \frac{-3\mp\sqrt{3}}{2} \\
0 & & \frac{3\pm\sqrt{3}}{2} & & \frac{2\pm\sqrt{3}}{2} & \frac{-3\mp2\sqrt{3}}{2} \\
\hline
0 & \frac{18\pm11\sqrt{3}}{4} & \frac{-6\mp5\sqrt{3}}{4} & & \frac{-8\mp5\sqrt{3}}{4} & \mp\frac{\sqrt{3}}{4}
\end{array}
$$

5.5.3 Explicit TSRK methods: $s = 2$, $p = 4$ or 5

In this and the following sections we demonstrate that it is possible to construct methods of order $p = 4$ and $p = 5$ with only two stages. This section closely follows the presentation by Jackiewicz and Tracogna [183] and Tracogna [270]. It will usually be assumed that $\vartheta = 0$ and $u_1 = u_2 = 0$. We will look for methods of stage order q equal to at least 3, that is, methods which satisfy the stage order conditions

$$C_i = 0, \quad i = 1, 2, 3. \tag{5.5.4}$$

Solving in addition to (5.5.4) the order conditions

$$\widehat{C}_i = 0, \quad i = 1, 2, 3, 4, \tag{5.5.5}$$

we obtain two one-parameter families of methods depending on c_2 or c_1 of order $p = 4$ and stage order $q = 3$. The coefficients of the first family of methods that depend on c_2 are

$$c_1 = 0, \quad a_{21} = \frac{c_2(6 - c_2^2)}{6(1 - c_2)},$$

$$b_{11} = 0, \quad b_{12} = 0,$$

$$b_{21} = \frac{c_2(3 - c_2)}{6}, \quad b_{22} = \frac{c_2(3 + 2c_2)}{6(c_2 - 1)}, \tag{5.5.6}$$

$$v_1 = \frac{17 - 38c_2 + 18c_2^2}{12c_2(1 - c_2)}, \quad v_2 = \frac{17 - 10c_2}{12c_2(1 + c_2)},$$

$$w_1 = \frac{-7 + 14c_2 - 6c_2^2}{12c_2(1 + c_2)}, \quad w_2 = \frac{7 - 10c_2}{12c_2(1 - c_2)}.$$

The coefficients of the second family of methods that depend on the parameter c_1 are

$$c_2 = \frac{c_1^2 - 6}{3(c_1 - 2)}, \quad a_{21} = \frac{2(c_1^2 - 6)(c_1^2 - 3c_1 + 3)(2c_1^2 - 12c_1 + 15)}{27c_1(c_1 - 2)^2(2c_1 - 3)},$$

$$b_{11} = \frac{c_1^3}{4(3 - 3c_1 + c_1^2)}, \quad b_{12} = \frac{3c_1(2 - c_1)^2}{4(3 - 3c_1 + c_1^2)},$$

$$b_{21} = \frac{(6 - c_1^2)(72 - 162c_1 + 156c_1^2 - 63c_1^3 + 8c_1^4)}{108(2 - c_1)^2(3 - 3c_1 + c_1^2)},$$

$$b_{22} = \frac{(c_1^2 - 6)(180 - 486c_1 + 498c_1^2 - 225c_1^3 + 38c_1^4)}{36c_1(c_1 - 2)(2c_1 - 3)(3 - 3c_1 + c_1^2)},$$

$$v_1 = \frac{36 - 48c_1 - 31c_1^2 + 68c_1^3 - 30c_1^4 + 4c_1^5}{8c_1(3 - 2c_1)(3 - 3c_1 + c_1^2)},$$

$$v_2 = \frac{(2 - c_1)(42 - 135c_1 + 160c_1^2 - 78c_1^3 + 12c_1^4)}{8(3 - 3c_1 + c_1^2)(12 - 9c_1 + 2c_1^2)},$$

$$w_1 = \frac{12 - 24c_1 - c_1^2 + 44c_1^3 - 26c_1^4 + 4c_1^5}{8(3 - 3c_1 + c_1^2)(12 - 9c_1 + 2c_1^2)},$$

$$w_2 = \frac{(c_1 - 2)(18 - 39c_1 - 16c_1^2 + 42c_1^3 - 12c_1^4)}{8c_1(3 - 2c_1)(3 - 3c_1 + c_1^2)}.$$

$$(5.5.7)$$

The stability polynomial of these methods is given by

$$p(\eta, z) = \left(\eta^3 - \left(1 + \frac{24\mu - 1}{2}z + \frac{17 - 60\mu}{12}z^2\right)\eta^2 \right.$$

$$\left. + \left(\frac{3(8\mu - 1)}{2}z + \frac{96\mu - 7}{12}z^2\right)\eta - \mu z^2\right)\eta,$$

$$(5.5.8)$$

where

$$\mu = \frac{c_2(12 - 17c_2 - 2c_2^2)}{72(1 - c_2)(1 + c_2)}$$

for methods (5.5.6) and

$$\mu = \frac{(3 - 3c_1 + c_1^2)(126 - 315c_1 + 261c_1^2 - 75c_1^3 + 4c_1^4)}{54c_1(2 - c_1)(2c_1 - 3)(12 - 9c_1 + 2c_1^2)}$$

for methods (5.5.7). We have plotted in Fig. 5.5.3 the left-hand side of the interval of absolute stability of the polynomial $p(\eta, z)$ given by (5.5.8) versus μ for $\mu \in [0, 0.4]$. This polynomial has the maximal interval of absolute stability

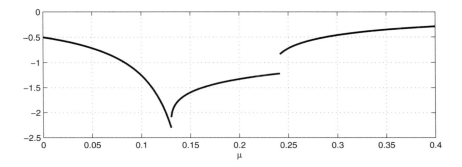

Figure 5.5.3 Interval of absolute stability of the polynomial $p(\eta, z)$ given by (5.5.8) versus the parameter μ for $\mu \in [0, 0.4]$

approximately equal to $(-2.307, 0)$ as μ approaches approximately 0.1305 from below. There is also a discontinuity of this stability interval at $\mu \approx 0.2414$. The abscissa vector c and the parameters of methods corresponding to $\mu = 0.13$ are

$$
\begin{array}{c|cc|cc}
0 & & & 0 & 0 \\
0 & 2.87785 & & -7.01513 & -1.0216 \\
\hline
0 & 1.8152 & 0.266403 & -0.927937 & -0.153665
\end{array}
$$

with $c = [0, -5.15889]^T$;

$$
\begin{array}{c|cc|cc}
0 & & & 2.12016 & 0.031664 \\
0 & 2.87785 & & -4.89497 & -0.989938 \\
\hline
0 & -0.304963 & 0.234739 & 1.19223 & -0.122001
\end{array}
$$

with $c = [2.15183, -3.00706]^T$; and

$$
\begin{array}{c|cc|cc}
0 & & & 3.10518 & 5.48132 \\
0 & 2.87785 & & -3.90995 & 4.45972 \\
\hline
0 & -1.28998 & -5.21492 & 2.17725 & 5.32766
\end{array}
$$

with $c = [8.5865, 3.42762]^T$. We have plotted in Fig.5.5.4 the stability regions of the polynomial $p(\eta, z)$ defined by (5.5.8) for selected values of the parameter μ.

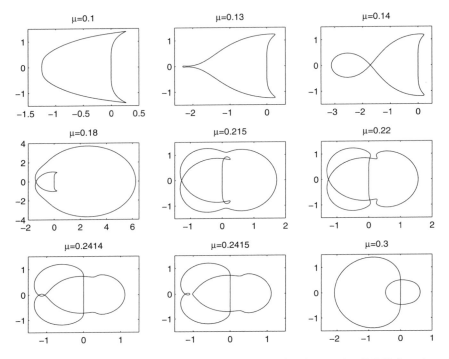

Figure 5.5.4   Stability regions of the polynomial $p(\eta, z)$ given by (5.5.8) for selected values of the parameter μ

Putting $a_{21} = 0$ in (5.5.6) or (5.5.7), we obtain methods of type 3. The coefficients of these methods corresponding to (5.5.6) are

$$
\begin{array}{c|cc|cc}
0 & & & 0 & 0 \\
0 & & & \frac{-2\pm\sqrt{6}}{2} & \frac{2\pm\sqrt{6}}{2} \\
\hline
0 & \frac{522\mp103\sqrt{6}}{360} & \frac{162\mp77\sqrt{6}}{360} & \frac{-342\pm127\sqrt{6}}{360} & \frac{18\pm53\sqrt{6}}{360}
\end{array}
$$

with $\mathbf{c} = [0, \pm\sqrt{6}]^T$. The coefficients of these methods corresponding to (5.5.7) are

$$
\begin{array}{c|cc|cc}
0 & & & \frac{16\mp\sqrt{6}}{8} & \frac{4\mp\sqrt{6}}{4} \\
0 & & & \frac{4\pm\sqrt{6}}{4} & \frac{16\pm\sqrt{6}}{8} \\
\hline
0 & \frac{-99\mp29\sqrt{6}}{180} & \frac{-99\pm29\sqrt{6}}{180} & \frac{189\pm4\sqrt{6}}{180} & \frac{189\mp41\sqrt{6}}{180}
\end{array}
$$

with $\mathbf{c} = [(6 \mp \sqrt{6})/2, 6 \pm \sqrt{6})/2]^T$; and

$$
\begin{array}{c|c|cc}
0 & & \frac{2\mp\sqrt{6}}{2} & \frac{-2\pm\sqrt{6}}{2} \\
0 & & 0 & 0 \\
\hline
0 & \frac{162\mp77\sqrt{6}}{360} \quad \frac{522\mp103\sqrt{6}}{360} & \frac{18\pm53\sqrt{6}}{360} & \frac{-342\mp127\sqrt{6}}{360}
\end{array}
$$

with $\mathbf{c} = [\pm\sqrt{6}, 0]^T$. The stability polynomial of all these methods corresponds to $\mu = 17/60$ and is given by

$$
p(\eta, z) = \left(\eta^3 - \left(1 + \frac{29}{10} z \right) \eta^2 + \left(\frac{19}{10} z + \frac{101}{60} z^2 \right) \eta - \frac{17}{60} z^2 \right) \eta,
$$

and the interval of absolute stability is

$$
\left(\frac{2\sqrt{411} - 72}{59}, 0 \right) \approx \left(-0.53, 0 \right).
$$

We conclude this section with the construction of TSRK methods of order $p = 5$ and stage order $q = 4$. Solving the system of stage order and order conditions $C_i = 0$, $i = 1, 2, 3, 4$, $\widehat{C}_i = 0$, $i = 1, 2, 3, 4, 5$, leads to families of such methods which depend on the parameter ϑ. We have tried to choose ϑ to obtain methods with large interval of absolute stability. The method with coefficients

$$
\begin{array}{cc|cc|cc}
-0.548071 & & & & 2.30708 & -0.0113337 \\
6.97276 & -3.2631 & & & 6.43125 & -0.829075 \\
\hline
-0.6 & -0.189159 & -0.0947644 & & 0.635485 & 0.0484391
\end{array}
$$

seems to be close to be optimal in this respect. For this method the abscissa vector $\mathbf{c} = [2.84382, -4.63369]^T$, and the interval of absolute stability is approximately equal to $(-1, 23, 0)$. We have also found TSRK methods of type 3 with $p = 5$ and $q = 4$. The coefficients of such a method are

$$
\begin{array}{cc|cc|cc}
-0.210299 & & & & 1.97944 & 0.0387917 \\
-0.0995138 & & & & 2.5617 & 2.3738 \\
\hline
-0.186912 & -0.812426 & -0.0761097 & & 1.45338 & 0.248242
\end{array}
$$

with $\mathbf{c} = [2.22853, 5.03502]^T$, and the interval of absolute stability is approximately equal to $(-0.486, 0)$. This method is unique in the class of type 3 zero-stable two-stage TSRK methods of order $p = 5$ and stage order $q = 4$.

5.5.4   Implicit TSRK methods: $s = 2$, $p = 4$ or 5

In this section we consider implicit TSRK methods with two stages specified by the abscissa vector $\mathbf{c} = [c_1, c_2]^T$ and the table of coefficients (5.5.2) with $\lambda > 0$. Similarly as in Section 5.4.2, we work with the polynomial $\widetilde{p}(\eta, z)$ defined by

$$\widetilde{p}(\eta, z) = (1 - \lambda z)^2 p(\eta, z),$$

where $p(\eta, z)$ is the stability function. Assuming that $u_1 = u_2 = 0$, and $\vartheta = 0$ and solving the system of order and stage order conditions (5.5.4) and (5.5.5) corresponding to (5.5.2), we obtain a family of TSRK methods of order $p = 4$ and stage order $q = 3$, depending on the parameters c_1 and λ. The coefficients of these methods are not listed here; they are reduced to (5.5.7) for $\lambda = 0$. The stability polynomial of these methods takes the form (5.5.3) with

$$\widetilde{p}_2(z) = 1 + \frac{24\mu - 1 + 16\lambda - 24\lambda^2}{2} z + \frac{17 - 60\mu - 96\lambda + 96\lambda^2}{12} z^2,$$

$$\widetilde{p}_1(z) = \frac{24\mu - 3 + 20\lambda - 24\lambda^2}{2} z + \frac{96\mu - 7 + 48\lambda - 60\lambda^2}{12} z^2,$$

and

$$\widetilde{p}_0(z) = \mu\, z^2,$$

where $\mu = \mu(c_1, \lambda)$ depends on the unknown coefficients of the method. This can also be verified directly from the exponential fitting condition

$$\widetilde{p}(e^z, z) = O(z^5)$$

(compare with Theorem 3.2.3).

We have performed an extensive computer search looking for TSRK methods of the form (5.5.2) with $p = 4$ and $q = 3$, which are A-stable. The results of this search in the parameter space (c_1, λ) are presented in Fig. 5.5.5. The results of the search in the parameter space (λ, μ) are given by Jackiewicz and Tracogna [183] and Tracogna [270]. It can be verified that inside the shaded regions in Fig. 5.5.5 the roots of the polynomial

$$\varphi(\eta) = \lim_{z \to \infty} \left(\widetilde{p}(\eta, z)/(1 - \lambda z)^2 \right)$$

have modulus less than 1. Choosing, for example $c_1 = 3/4$ and $\lambda = 3/4$, we have $c_2 = 9/4$, and the coefficients of the A-stable TSRK method are

$$
\begin{array}{cc|cc|cc}
0 & \frac{3}{4} & & & \frac{3}{16} & -\frac{3}{16} \\[6pt]
0 & \frac{21}{8} & \frac{3}{4} & & -\frac{3}{16} & -\frac{15}{16} \\[6pt]
\hline
0 & \frac{107}{720} & -\frac{43}{144} & & \frac{161}{144} & \frac{23}{720}
\end{array}.
$$

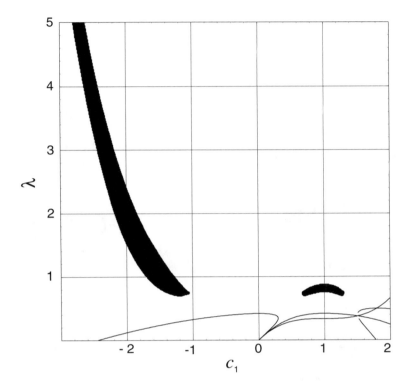

Figure 5.5.5 *A*-stable methods of the form (5.5.2) of type 2 of order $p = 4$ and stage order $q = 3$ with $u_1 = u_2 = \vartheta = 0$

For the method above the roots of the polynomial $\varphi(\eta)$ are $\eta_1 = 0$, $\eta_2 \approx 0.568756$, $\eta_3 \approx -0.414087$, and $\eta_4 \approx -0.331542$. Other examples of A-stable methods of order $p = 4$ and stage order $q = 3$ with $\lambda = 3/4$ and $\mu = 0$ are presented in [183, 270].

We demonstrate next that there are no A-stable TSRK methods of type 4 in the class of methods above with $p = 4$, $q = 3$, and $u_1 = u_2 = \vartheta = 0$, for the range of parameters displayed in Fig 5.5.5. Imposing the additional condition $a_{21} = 0$ leads to the relations

$$\lambda = \lambda_i(c_1), \quad i = 1, 2, 3, 4,$$

where

$$\lambda_1(c_1) = \frac{c_1(c_1^2 - 3c_1 + 3)}{3},$$

and $\lambda_i(c_1)$, $i = 2, 3, 4$, are solutions to the polynomial equation

$$216\lambda^3 - 18(5 + 24c_1 + 12c_1^2)\lambda^2 + 6c_1(30 + 24c_1 - 38c_1^2 + 9c_1^3)\lambda$$

$$-90c_1^2 + 72c_1^3 + 3c_1^4 - 12c_1^5 + 2c_1^6 = 0.$$

These solutions are shown in Fig. 5.5.5 by solid lines and they do not intersect the region of A-stability.

A somewhat different argument for the non-existence of type 4 methods in this class is presented in [183, 270].

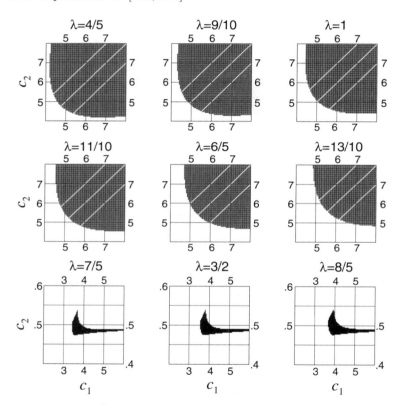

Figure 5.5.6 A-stable methods of the form (5.5.2) of type 2 of order $p = 4$ and stage order $q = 3$

We consider next TSRK methods (5.5.2) of order $p = 4$ and stage order $q = 3$, where we removed the restrictions on the parameters u_1, u_2, and ϑ. Solving the system of order and stage order conditions (5.5.4) and (5.5.5), we obtain a four-parameter family of methods depending on c_1, c_2, λ and ϑ. The polynomial that determines the stability properties of these methods takes the form

$$\widetilde{p}(\eta, z) = (1 - \lambda z)^2 \eta^4 - \widetilde{p}_3(z)\eta^3 + \widetilde{p}_2(z)\eta^2 - \widetilde{p}_1(z)\eta + \widetilde{p}_0(z),$$

with

$$\widetilde{p}_3(z) = p_{30} + p_{31}z + p_{32}z^2, \quad \widetilde{p}_2(z) = p_{20} + p_{21}z + p_{22}z^2,$$
$$\widetilde{p}_1(z) = p_{11}z + p_{12}z^2, \quad \text{and} \quad \widetilde{p}_0(z) = p_{02}z^2.$$

We have performed an extensive computer search looking for methods that are A-stable, where we have assumed that one root of the polynomial $\widetilde{p}(\eta, z)$ is $\eta = 0$ (i.e., that $p_{20} = 0$). This condition was used to eliminate the parameter ϑ, and the search was performed in the three-dimensional space (c_1, c_2, λ). The results of this search are presented in Fig. 5.5.6 for $\lambda = 4/5$, $9/10$, 1, $11/10$, $6/5$, $13/10$, and $4 \leq c_1, c_2 \leq 8$, and for $\lambda = 7/5$, $3/2$, $8/5$, and $2 \leq c_1 \leq 6$, $2/5 \leq c_2 \leq 3/5$. An example of the method with $\lambda = 7/5$, $c_1 = 7/2$, and $c_2 = 1/2$ is

$$
\begin{array}{cc|cc|cc}
\frac{476}{5} & & \frac{7}{5} & 0 & \frac{91}{120} & \frac{2317}{24} \\
-\frac{118}{5} & & 0 & \frac{7}{5} & -\frac{41}{120} & -\frac{2899}{120} \\
\hline
-\frac{61}{490} & -\frac{41}{11760} & \frac{1607}{1680} & & \frac{15}{784} & -\frac{379}{3920}
\end{array} .
$$

An example of the method with $\lambda = 1$, $c_1 = 5$ and $c_2 = 7$ is presented in [183, 270].

We have also investigated TSRK methods (5.5.2) of order $p = 5$ and stage order $q = 4$. Solving the corresponding system of order and stage order conditions leads in this case to the two-parameter family of methods depending on c_1 and c_2. The computer search that we performed for $-6 \leq c_1, c_2 \leq 6$ did not reveal A-stable methods in this class.

5.6   ANALYSIS OF TSRK METHODS WITH THREE STAGES

5.6.1   Explicit TSRK methods: $s = 3$, $p = 3$, $q = 2$ or 3

In this section we are concerned primarily with methods of high stage order. We will consider explicit methods given by the abscissa vector $\mathbf{c} = [c_1, c_2, c_3]^T$ and the table of coefficients

$$
\frac{\mathbf{u} \mid \mathbf{A} \mid \mathbf{B}}{\vartheta \mid \mathbf{v}^T \mid \mathbf{w}^T} =
\begin{array}{c|cc|ccc}
u_1 & & & b_{11} & b_{12} & b_{13} \\
u_2 & a_{21} & & b_{21} & b_{22} & b_{23} \\
u_3 & a_{31} & a_{32} & b_{31} & b_{32} & b_{33} \\
\hline
\vartheta & v_1 & v_2 \; v_3 & w_1 & w_2 & w_3
\end{array} .
\tag{5.6.1}
$$

Following Chollom and Jackiewicz [91], we assume that $\mathbf{c} = [0, \frac{1}{2}, 1]^T$ and $\vartheta = 0$. Solving stage order and order conditions (5.3.1) and (5.3.2) corresponding to $p = 3$ and $q = 2$, we obtain a twelve-parameter family of methods depending on u_1, u_2, u_3, a_{21}, a_{31}, a_{32}, b_{13}, b_{23}, b_{33}, v_1, v_2, and v_3. The coefficients of these methods are given by

$$b_{11} = b_{13}, \quad b_{12} = u_1 - 2b_{13},$$

$$b_{21} = \frac{4a_{21} + 4b_{23} - 3}{4}, \quad b_{22} = \frac{5 - 8a_{21} - 8b_{23} + 4u_2}{4},$$

$$b_{31} = a_{31} + 2a_{32} + b_{33} - 2, \quad b_{32} = 3 - 2a_{31} - 3a_{32} - 2b_{33} + u_3,$$

$$w_1 = \frac{7 - 6v_2 - 18v_3}{6}, \quad w_2 = \frac{9v_2 + 24v_3 - 10}{3}, \quad w_3 = \frac{19 - 6v_1 - 18v_2 - 36v_3}{6}.$$

We choose these free parameters of the method trying to maximize the area of the intersection of the region of absolute stability of TSRK method with the negative half-plane. As in Section 5.5.1, this area can be approximated using numerical integration in polar coordinates. The resulting objective function for the "negative area" is then minimized using the subroutine `fminsearch` from Matlab starting from some random initial guesses. This computer search leads to a method with coefficients

$$\mathbf{u} = \begin{bmatrix} 0.0213802 & 0.0991119 & 0.353437 \end{bmatrix}^{\mathbf{T}},$$

$$\mathbf{A} = \begin{bmatrix} 0 & 0 & 0 \\ 0.392328 & 0 & 0 \\ 0.582927 & 0.262283 & 0 \end{bmatrix},$$

$$\mathbf{B} = \begin{bmatrix} 0.264446 & -0.507512 & 0.264446 \\ 0.240292 & -0.631471 & 0.597963 \\ 0.0969341 & -0.578148 & 0.98944 \end{bmatrix},$$

$$\mathbf{v} = \begin{bmatrix} 0.0867149 & 0.621047 & 0.115883 \end{bmatrix}^{\mathbf{T}},$$

$$\mathbf{w} = \begin{bmatrix} 0.197972 & -0.543131 & 0.521515 \end{bmatrix}^{\mathbf{T}},$$

whose stability region is shown by a thin line in Fig. 5.6.1.

Solving in addition to (5.3.1) and (5.3.2) stage order condition (5.3.3), we obtain a nine-parameter family of TSRK methods depending on u_1, u_2, u_3,

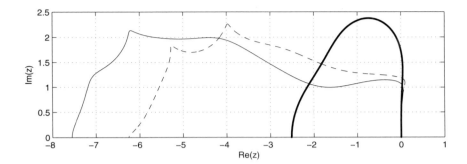

Figure 5.6.1   Stability region of TSRK methods of type 1 with $s = 3$ and $p = q + 1 = 3$ (thin line), TSRK method with $s = 3$ and $p = q = 3$ (thin dashed line), and RK method of order 3 (thick line)

a_{21}, a_{31}, a_{32}, v_1, v_2, and v_3. The coefficients of these methods are

$$b_{11} = \frac{u_1}{6}, \quad b_{12} = \frac{2u_1}{3}, \quad b_{13} = \frac{u_1}{6},$$

$$b_{21} = \frac{5 + 4u_2}{24}, \quad b_{22} = \frac{2(u_2 - 1)}{3}, \quad b_{23} = \frac{23 - 24a_{21} + 4u_2}{24},$$

$$b_{31} = \frac{7 - 6a_{32} + u_3}{6}, \quad b_{32} = \frac{9a_{32} + 2u_3 - 10}{3},$$

$$b_{33} = \frac{19 - 6a_{31} - 18a_{32} + u_3}{6}, \quad w_1 = \frac{7 - 6v_2 - 18v_3}{6},$$

$$w_2 = \frac{9v_2 + 24v_3 - 10}{3}, \quad w_3 = \frac{19 - 6v_1 - 18v_2 - 36v_3}{6}.$$

A computer search in this nine-parameter space leads to a method with coefficients

$$\mathbf{u} = \begin{bmatrix} 0.149831 & -0.0104229 & 0.0542442 \end{bmatrix}^{\mathbf{T}},$$

$$\mathbf{A} = \begin{bmatrix} 0 & 0 & 0 \\ 0.708342 & 0 & 0 \\ 1.03838 & 0.556873 & 0 \end{bmatrix},$$

$$\mathbf{B} = \begin{bmatrix} 0.0249718 & 0.0998873 & 0.0249718 \\ 0.206596 & -0.673615 & 0.248254 \\ 0.618834 & -1.62655 & 0.466708 \end{bmatrix},$$

$$\mathbf{v} = \begin{bmatrix} 1.40014 & 0.122363 & 0.141148 \end{bmatrix}^{\mathbf{T}},$$

$$\mathbf{w} = \left[\begin{array}{ccc} 0.620859 & -1.83706 & 0.552552 \end{array}\right]^{\mathbf{T}},$$

whose stability region is shown by a thin dashed line in Fig. 5.6.1.

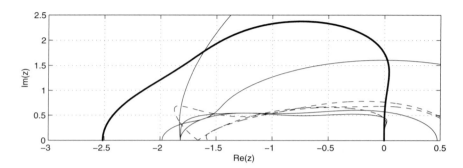

Figure 5.6.2   Stability region of TSRK methods of type 3 with $s = 3$ and $p = q + 1 = 3$ (thin line), TSRK method with $s = 3$ and $p = q = 3$ (thin dashed line), and RK method of order 3 (thick line)

We consider next type 3 methods with $p = 3$ and $q = 2$ (i.e., methods for which $a_{21} = a_{31} = a_{32} = 0$). Assuming as before that $\mathbf{c} = [0, \frac{1}{2}, 1]^T$ and $\vartheta = 0$, a computer search for methods with a large region of absolute stability leads to a formula of the form

0.774352				-5.25807	11.2905	-5.25807
2.89678				-7.12077	16.8883	-6.37077
6.18955				-9.88612	24.9618	-7.88612
0	0.407808	-0.700253	1.05397	-1.29498	2.99765	-1.46419

This search was performed in a parameter space u_1, u_2, u_3, b_{13}, b_{23}, b_{33}, v_1, v_2, and v_3. The stability region of this formula is shown by a thin line in Fig. 5.6.2.

Consider next type 3 methods with $p = q = 3$. Assuming again that $\mathbf{c} = [0, \frac{1}{2}, 1]^T$ and $\vartheta = 0$, a computer search in a parameter space u_1, u_2, u_3, v_1, v_2, and v_3 leads to a TSRK formula with coefficients given by

0.258133				0.0430222	0.17089	0.0430222
0.382867				0.272144	-0.411422	1.02214
0.417524				1.23625	-3.05498	3.23625
0	0.27882	2.10743	-3.70769	10.1823	-26.6726	18.8116

whose region of absolute stability is shown by a thin dashed line in Fig 5.6.2. Observe that these regions are much smaller than the corresponding regions of type 1 methods.

5.6.2 Implicit TSRK methods: $s = 3$, $p = 3$, $q = 2$ or 3

Implicit TSRK methods of type 2 with three stages are specified by the abscissa vector $\mathbf{c} = [c_1, c_2, c_3]^T$ and the table of coefficients

$$
\frac{\mathbf{u} \mid \mathbf{A} \mid \mathbf{B}}{\vartheta \mid \mathbf{v}^T \mid \mathbf{w}^T} =
\begin{array}{c|ccc|ccc}
u_1 & \lambda & & & b_{11} & b_{12} & b_{13} \\
u_2 & a_{21} & \lambda & & b_{21} & b_{22} & b_{23} \\
u_3 & a_{31} & a_{32} & \lambda & b_{31} & b_{32} & b_{33} \\
\hline
\vartheta & v_1 & v_2 & v_3 & w_1 & w_2 & w_3
\end{array}, \qquad (5.6.2)
$$

where u_1, u_2, u_3, ϑ, λ, a_{21}, a_{31}, a_{32}, b_{11}, b_{12}, b_{13}, b_{21}, b_{22}, b_{23}, b_{31}, b_{32}, b_{33}, v_1, v_2, v_3, w_1, w_2, and w_3 are real parameters. We assume throughout this section that $\mathbf{c} = [0, \frac{1}{2}, 1]^T$, $\vartheta = 0$, and $\mathbf{u} = [0, 0, 0]^T$. As in Section 5.6.1, solving the system of order conditions (5.3.1) and (5.3.2) corresponding to $p = 3$ and $q = 2$ with respect to w_1, w_2, w_3, b_{11}, b_{21}, b_{31}, b_{12}, b_{22}, and b_{32}, we obtain a ten-parameter family of formulas (5.6.2) of order $p = 3$ and stage order $q = 2$. The coefficients of these methods are not given here, and in what follows we restrict our attention to methods of stage order $q = 3$. Solving the stage order condition $C_3 = 0$ with respect to b_{13}, b_{23}, and b_{33}, we obtain a seven-parameter family of methods of order $p = 3$ and stage order $q = 3$ which depend on the parameters λ, a_{21}, a_{31}, a_{32}, v_1, v_2, and v_3. The coefficients of these methods are

$$
\begin{array}{c|ccc|ccc}
0 & \lambda & & & 0 & 0 & -\lambda \\
0 & a_{21} & \lambda & & \frac{5-24\lambda}{24} & \frac{-2+9\lambda}{3} & \frac{23-24a_{21}-72\lambda}{24} \\
0 & a_{31} & a_{32} & \lambda & \frac{7-6a_{32}-18\lambda}{6} & \frac{-10+9a_{32}+24\lambda}{3} & \frac{19-6a_{31}-18a_{32}-36\lambda}{6} \\
\hline
0 & v_1 & v_2 & v_3 & \frac{7-6v_2-18v_3}{6} & \frac{-10+9v_2+24v_3}{3} & \frac{19-6v_1-18v_2-36v_3}{6}
\end{array}.
$$

Similarly as in Sections 5.4.2 and 5.5.2, to analyze stability properties of the resulting methods we work with the polynomial

$$
\widetilde{p}(\eta, z) = (1 - \lambda z)^3 p(\eta, z),
$$

where $p(\eta, z)$ is the stability function. It can be verified that this polynomial takes the form

$$
\widetilde{p}(\eta, z) = \Big((1 - \lambda z)^3 \eta^4 - \widetilde{p}_3(z)\eta^3 + \widetilde{p}_2(z)\eta^2 - \widetilde{p}_1(z)\eta + \widetilde{p}_0(z)\Big)\eta, \qquad (5.6.3)
$$

with

$$\widetilde{p}_3(z) = \widetilde{p}_{30} + \widetilde{p}_{31}z + \widetilde{p}_{32}z^2 + \widetilde{p}_{33}z^3, \quad \widetilde{p}_2(z) = \widetilde{p}_{21}z + \widetilde{p}_{22}z^2 + \widetilde{p}_{23}z^3,$$

$$\widetilde{p}_1(z) = \widetilde{p}_{12}z^2 + \widetilde{p}_{13}z^3, \quad \widetilde{p}_0(z) = \widetilde{p}_{03}z^3.$$

Next we solve the system of polynomial equations

$$\widetilde{p}_{12} = 0, \quad \widetilde{p}_{13} = 0, \quad \widetilde{p}_{03} = 0 \tag{5.6.4}$$

with respect to v_1, v_2, and v_3. This leads to a four-parameter family of formulas for which the stability polynomial $\widetilde{p}(\eta, z)$ takes the form

$$\widetilde{p}(\eta, z) = \Big((1 - \lambda z)^3 \eta^2 - \widetilde{p}_3(z)\eta + \widetilde{p}_2(z) \Big)\eta^3. \tag{5.6.5}$$

In our search for highly stable methods, we impose the conditions

$$\widetilde{p}_{13} = 0, \quad \widetilde{p}_{03} = 0, \tag{5.6.6}$$

which are solved with respect to a_{31} and a_{32}. This results in a two-parameter family of TSRK formulas with respect to λ and a_{21}. Choosing $\lambda = 1$ and $a_{21} = 0$, we obtain TSRK methods with coefficients given by

0	1			0	0	-1
0	0	1		-0.791667	2.33333	-2.04167
0	-2.07846	1.34863	1	-3.18196	8.71255	-4.80076
0	3.17415	0.428068	-0.468128	2.14298	-5.79415	1.51708

and

0	1			0	0	-1
0	0	1		-0.791667	2.33333	-2.04167
0	1.30253	-2.40364	1	0.570305	-2.54425	3.07505
0	7.43253	-15.7509	3.5766	6.18771	-21.9731	21.5271

It can be verified using the Schur criterion (Theorem 2.8.1) that these methods are A-stable. Since they satisfy (5.6.6), they are also L-stable.

Finally, consider type 4 methods corresponding to $a_{21} = a_{31} = a_{32} = 0$. Again solving system (5.6.4) with respect to v_1, v_2, and v_3, we obtain a one-parameter family of TSRK formulas depending on λ, with coefficients given by

$$\mathbf{A} = \begin{bmatrix} \lambda & & \\ & \lambda & \\ & & \lambda \end{bmatrix}, \quad \mathbf{B} = \begin{bmatrix} 0 & 0 & -\lambda \\ \frac{5-24\lambda}{24} & \frac{-2+9\lambda}{3} & \frac{23-72\lambda}{24} \\ \frac{7-18\lambda}{6} & \frac{2(-5+12\lambda)}{3} & \frac{19-36\lambda}{6} \end{bmatrix},$$

$$\mathbf{v} = \left[\begin{array}{ccc} \dfrac{(1+3\lambda)(1-6\lambda+12\lambda^2)}{18(1-2\lambda)} & \dfrac{(1+6\lambda)(-1+6\lambda-12\lambda^2)}{9(1-2\lambda)} & \dfrac{-2+39\lambda-114\lambda^2+108\lambda^3}{18(1-2\lambda)} \end{array} \right]^{\mathbf{T}},$$

$$\mathbf{w} = \left[\begin{array}{ccc} \dfrac{29-159\lambda+294\lambda^2-180\lambda^3}{18(1-2\lambda)} & \dfrac{-41+216\lambda-384\lambda^2+216\lambda^3}{9(1-2\lambda)} & \dfrac{74-345\lambda+546\lambda^2-252\lambda^3}{18(1-2\lambda)} \end{array} \right]^{\mathbf{T}},$$

$\lambda \neq \frac{1}{2}$. The stability polynomial $\widetilde{p}(\eta, z)$ of these methods has the form (5.6.5), where $\widetilde{p}_3(z)$ and $\widetilde{p}_2(z)$ are cubic polynomials with respect to z. It can be verified using the Schur criterion (Theorem 2.8.1) that these methods are A-stable if $\lambda > 0.783491$. It can also be verified that these methods are not L-stable.

5.7   TWO-STEP COLLOCATION METHODS

We examine in this section a class of continuous two-step methods for the numerical solution of the initial value problem (2.1.1), which is based on the collocation approach proposed by D'Ambrosio et al. [104]. We consider a class of methods defined by

$$P(t_{n-1} + \theta h) = \varphi_1(\theta)y_{n-1} + \varphi_0(\theta)y_{n-2}$$

$$+ h\sum_{j=1}^{s} \Big(\psi_j(\theta)f\big(P(t_{n-1} + c_j h)\big) + \chi_j(\theta)f\big(P(t_{n-2} + c_j h)\big) \Big), \qquad (5.7.1)$$

$$y_n = P(t_n),$$

$n = 2, 3, \ldots, N$, $\theta \in (0, 1]$, $t_n = t_0 + nh$, $Nh = T - t_0$. Here $P(t)$ is a continuous approximation to the solution $y(t)$ of (2.1.1), $\mathbf{c} = [c_1, \ldots, c_s]^T$ is the abscissa vector, and $\varphi_0(\theta)$, $\varphi_1(\theta)$, $\chi_j(\theta)$, and $\psi_j(\theta)$, $j = 1, 2, \ldots, s$, are polynomials that define the method. Observe that this method requires a starting procedure to compute an approximation $P(t_0 + \theta h)$ to $y(t_0 + \theta h)$ for $\theta \in [0, 1]$, which corresponds to the initial interval $[t_0, t_1]$. For this purpose we generally use for this purpose starting procedures for TSRK methods, which are discussed in Section 6.2. These starting procedures are based on continuous RK methods constructed by Owren and Zenanro [231, 232, 233] (see also [230]).

Setting

$$Y_i^{[n]} = P(t_{n-1} + c_i h), \quad Y_i^{[n-1]} = P(t_{n-2} + c_i h), \quad i = 1, 2, \ldots, s,$$

method (5.7.1) corresponding to $\theta = c_i$, $i = 1, 2, \ldots, s$, can be written as a TSRK method of the form

$$Y_i^{[n]} = \widetilde{u}_i y_{n-1} + u_i y_{n-2} + h\sum_{j=1}^{s} \Big(a_{ij} f(Y_j^{[n]}) + b_{ij} f(Y_j^{[n-1]}) \Big),$$

$$y_n = \widetilde{\vartheta} y_{n-1} + \vartheta y_{n-2} + h\sum_{j=1}^{s} \Big(v_j f(Y_j^{[n]}) + w_j f(Y_j^{[n-1]}) \Big),$$

(5.7.2)

$i = 1, 2, \ldots, s$, $n = 2, 3, \ldots, N$, with

$$\widetilde{u}_i = \varphi_1(c_i), \quad u_i = \varphi_0(c_i), \quad a_{ij} = \psi_j(c_i), \quad b_{ij} = \chi_j(c_i),$$

$$\widetilde{\vartheta} = \varphi_1(1), \quad \vartheta = \varphi_0(1), \quad v_j = \psi_j(1), \quad w_j = \chi_j(1).$$

Putting

$$\mathbf{A} = \left[\, a_{ij} \,\right]_{i,j=1}^{s}, \qquad \mathbf{B} = \left[\, b_{ij} \,\right]_{i,j=1}^{s},$$

$$\mathbf{u} = \left[\, u_1 \; \cdots \; u_s \,\right]^{T}, \qquad \widetilde{\mathbf{u}} = \left[\, \widetilde{u}_1 \; \cdots \; \widetilde{u}_s \,\right]^{T},$$

$$\mathbf{v} = \left[\, v_1 \; \cdots \; v_s \,\right]^{T}, \qquad \mathbf{w} = \left[\, w_1 \; \cdots \; w_s \,\right]^{T},$$

method (5.7.2) can be written as a GLM with coefficients given by

$$\left[\begin{array}{c|ccc} \mathbf{A} & \widetilde{\mathbf{u}} & \mathbf{u} & \mathbf{B} \\ \hline \mathbf{v}^T & \widetilde{\vartheta} & \vartheta & \mathbf{w}^T \\ \mathbf{0} & 1 & 0 & \mathbf{0} \\ \mathbf{I} & 0 & 0 & \mathbf{0} \end{array}\right]$$

(compare Section 2.1).

Next we derive the continuous order conditions for method (5.7.1) assuming that $P(t_{n-1} + \theta h)$ is an approximation of uniform order p to $y(t_{n-1} + \theta h)$ for $\theta \in [0, 1]$. As a result, the stage values $P(t_{n-1} + c_i h)$, $i = 1, 2, \ldots, s$, have stage order $q = p$. To derive these order conditions we investigate the local discretization error $\mathrm{le}(t_{n-1} + \theta h)$ of method (5.7.1), which similarly as for the TSRK methods (5.1.1), is defined as the residuum obtained by replacing $P(t_{n-1} + \theta h)$ by $y(t_{n-1} + \theta h)$, $P(t_{n-1} + c_j h)$ by $y(t_{n-1} + c_j h)$, $P(t_{n-2} + c_j h)$ by $y(t_{n-2} + c_j h)$, y_{n-1} by $y(t_{n-1})$, and y_{n-2} by $y(t_{n-2})$, where $y(t)$ is the solution to (2.1.1). This leads to

$$\mathrm{le}(t_{n-1} + \theta h) = y(t_{n-1} + \theta h) - \varphi_1(\theta) y(t_{n-1}) - \varphi_0(\theta) y(t_{n-1} - h)$$

$$- h \sum_{j=1}^{s} \left(\psi_j(\theta) y'(t_{n-1} + c_j h) + \chi_j(\theta) y'\big(t_{n-1} + (c_j - 1)h\big) \right), \qquad (5.7.3)$$

$n = 2, 3, \ldots, N$, $\theta \in (0, 1]$. We have the following theorem.

Theorem 5.7.1 (D'Ambrosio et al. [104]) *Assume that the function $f(y)$ is sufficiently smooth. Then method (5.7.1) has uniform order p if the following conditions are satisfied:*

$$\varphi_1(\theta) + \varphi_0(\theta) = 1,$$

$$\frac{(-1)^k}{k!} \varphi_0(\theta) + \sum_{j=1}^{m} \left(\frac{c_j^{k-1}}{(k-1)!} \psi_j(\theta) + \frac{(c_j - 1)^{k-1}}{(k-1)!} \chi_j(\theta) \right) = \frac{\theta^k}{k!}, \qquad (5.7.4)$$

$\theta \in [0,1]$, $k = 1, 2, \ldots, p$. Moreover, the local discretization error (5.7.3) of method (5.7.1) of uniform order p takes the form

$$le(t_{n-1} + \theta h) = h^{p+1} C_p(\theta) y^{(p+1)}(t_{n-1}) + O(h^{p+2}), \qquad (5.7.5)$$

as $h \to 0$, where the error function $C_p(\theta)$ is defined by

$$C_p(\theta) = \frac{\theta^{p+1}}{(p+1)!} - \frac{(-1)^{p+1}}{(p+1)!} \varphi_0(\theta) - \sum_{j=1}^{m} \left(\frac{c_j^p}{p!} \psi_j(\theta) + \frac{(c_j - 1)^p}{p!} \chi_j(\theta) \right). \quad (5.7.6)$$

Proof: Expanding the expressions $y(t_{n-1} + \theta h)$, $y(t_{n-1} - h)$, $y'(t_{n-1} + c_j h)$, and $y'(t_{n-1} + (c_j - 1)h)$ into a Taylor series around the point t_{n-1} and collecting terms with the same powers of h, we obtain

$$
\begin{aligned}
le(t_{n-1} + \theta h) = & \left(1 - \varphi_1(\theta) - \varphi_0(\theta)\right) y(t_{n-1}) \\
& + \sum_{k=1}^{p+1} \left(\frac{\theta^k}{k!} - \frac{(-1)^k}{k!} \varphi_0(\theta) \right) h^k y^{(k)}(t_{n-1}) \\
& - \sum_{k=1}^{p+1} \sum_{j=1}^{s} \left(\frac{c_j^{k-1}}{(k-1)!} \psi_j(\theta) + \frac{(c_j-1)^{k-1}}{(k-1)!} \chi_j(\theta) \right) h^k y^{(k)}(t_{n-1}) \\
& + O(h^{p+2}).
\end{aligned}
$$

Equating to zero the terms of order k, $k = 0, 1, \ldots, p$, we obtain order conditions (5.7.4). Comparing the terms of order $p + 1$, we obtain (5.7.5) with error function $C_p(\theta)$ defined by (5.7.6). ∎

The condition

$$\varphi_1(\theta) + \varphi_0(\theta) = 1, \quad \theta \in [0, 1],$$

is the generalization of preconsistency conditions for TSRK methods (5.7.2). It follows from the Definition 2.2.1 that this condition takes the form

$$\mathbf{U} \mathbf{q}_0 = \mathbf{e}, \quad \mathbf{V} \mathbf{q}_0 = \mathbf{q}_0,$$

where

$$\mathbf{U} = \begin{bmatrix} \tilde{\mathbf{u}} & \mathbf{u} & B \end{bmatrix}, \quad \mathbf{V} = \begin{bmatrix} \tilde{\vartheta} & \vartheta & \mathbf{w}^T \\ 1 & 0 & 0 \\ 0 & 0 & 0 \end{bmatrix},$$

and

$$\mathbf{q}_0 = \begin{bmatrix} 1 & 1 & | & 0 & \cdots & 0 \end{bmatrix}^T \in \mathbb{R}^{s+2}$$

(compare Section 2.2). It is easy to verify that these conditions imply that

$$\tilde{u}_j + u_j = 1, \quad j = 1, 2, \ldots, s, \quad \tilde{\vartheta} + \vartheta = 1.$$

We are interested primarily in methods corresponding to $p = s + r$, where $r = 1, 2, \ldots, s+1$, and the next theorem examines the solvability of the linear systems of equations (5.7.4) corresponding to these orders.

Theorem 5.7.2 (D'Ambrosio et al. [104]) *Assume that $c_i \neq c_j$ and that $c_i \neq c_j - 1$ for $i \neq j$. Then the system of continuous order conditions (5.7.4) corresponding to $p = s + r$, where $r = 1, 2, \ldots, s$, has a unique solution $\varphi_1(\theta)$, $\psi_j(\theta)$, $j = 1, 2, \ldots, m$, and $\chi_j(\theta)$, $j = s - r + 1, s - r + 2, \ldots, s$, for any given polynomials $\varphi_0(\theta)$ and $\chi_j(\theta)$, $j = 1, 2, \ldots, s - r$. System (5.7.4) corresponding to $p = 2s + 1$ has a unique solution $\varphi_0(\theta)$, $\varphi_1(\theta)$, $\psi_j(\theta)$, and $\chi_j(\theta)$, $j = 1, 2, \ldots, s$, which are polynomials of degree $\leq 2s + 1$.*

Proof: Observe that the polynomial $\varphi_1(\theta)$ is determined uniquely from the first equation of (7.5.4). The proof of the first part of the theorem for $p = s + r$, $r = 1, 2, \ldots, s$, follows from the fact that the matrices of the systems (5.7.4) corresponding to $\chi_j(\theta)$, $j = s - r + 1, s - r + 2, \ldots, s$, are Vandermonde matrices. The second part of the theorem, corresponding to $p = 2s + 1$, is technically more complicated and the details are given by D'Ambrosio [103]. ∎

The next result shows that the polynomials $\varphi_1(\theta)$, $\varphi_0(\theta)$, $\psi_j(\theta)$, and $\chi_j(\theta)$, $j = 1, 2, \ldots, s$, corresponding to the methods of order $p = 2s + 1$ satisfy some interpolation and collocation conditions.

Theorem 5.7.3 (D'Ambrosio et al. [104]) *Assume that $\varphi_1(\theta)$, $\varphi_0(\theta)$, $\psi_j(\theta)$, and $\chi_j(\theta)$, $j = 1, 2, \ldots, s$, satisfy (5.7.4) for $p = 2s + 1$. Then these polynomials satisfy the interpolation conditions*

$$\begin{aligned}
\varphi_1(0) = 0, \quad & \varphi_0(0) = 1, \quad & \psi_j(0) = 0, \quad & \chi_j(0) = 0, \\
\varphi_1(-1) = 0, \quad & \varphi_0(-1) = 1, \quad & \psi_j(-1) = 0, \quad & \chi_j(-1) = 0,
\end{aligned} \tag{5.7.7}$$

and the collocation conditions

$$\begin{aligned}
\varphi_1'(c_i) = 0, \quad & \varphi_0'(c_i) = 0, \quad & \psi_j'(c_i) = \delta_{ij}, \quad & \chi_j'(c_i) = 0, \\
\varphi_1'(c_i - 1) = 0, \quad & \varphi_0'(c_i - 1) = 0, \quad & \psi_j'(c_i - 1) = 0, \quad & \chi_j'(c_i - 1) = \delta_{ij},
\end{aligned} \tag{5.7.8}$$

$i, j = 1, 2, \ldots, s$. *Here δ_{ij} is the Kronecker delta (i.e., $\delta_{ii} = 1$ and $\delta_{ij} = 0$ for $i \neq j$).*

Proof: Conditions (5.7.7) follow immediately by substituting $\theta = 0$ and $\theta = -1$ into (5.7.4) corresponding to $p = 2s + 1$. To show (5.7.8) we differentiate (5.7.4) to get

$$\varphi_1'(\theta) + \varphi_0'(\theta) = 0,$$

$$\frac{(-1)^k}{k!} \varphi_0'(\theta) + \sum_{j=1}^{s} \left(\frac{c_j^{k-1}}{(k-1)!} \psi_j'(\theta) + \frac{(c_j - 1)^{k-1}}{(k-1)!} \chi_j'(\theta) \right) = \frac{\theta^{k-1}}{(k-1)!}, \tag{5.7.9}$$

$k = 1, 2, \ldots, 2s + 1$. Substituting $\theta = c_i$ and $\theta = c_i - 1$, $i = 1, 2, \ldots, s$, into (5.7.9) we obtain (5.7.8). This completes the proof.   ■

It follows from (5.7.7) and (5.7.8) that the polynomial $P(t)$ defined by (5.7.1) satisfies the interpolation conditions

$$P(t_{n-1}) = y_{n-1}, \quad P(t_{n-2}) = y_{n-2}$$

and the collocation conditions

$$P'(t_{n-1} + c_i h) = f\big(P(t_{n-1} + c_i h)\big), \quad P'(t_{n-2} + c_i h) = f\big(P(t_{n-2} + c_i h)\big),$$

$i = 1, 2, \ldots, s$. It also follows from (5.7.7) that the error constant of the methods described in Theorem 5.7.3 satisfy the conditions $C_p(-1) = 0$ and $C_p(0) = 0$.

For methods of order $p = s + r$, $r = 1, 2, \ldots, s$, we choose $\varphi_0(\theta)$ and $\chi_j(\theta)$, $j = 1, 2, \ldots, s - r$, as polynomials of degree $\leq s + r$ which satisfy the interpolation conditions

$$\varphi_0(0) = 0, \quad \chi_j(0) = 0, \quad j = 1, 2, \ldots, s - r \qquad (5.7.10)$$

and the collocation conditions

$$\varphi_0'(c_i) = 0, \quad \chi_j'(c_i) = 0, \quad j = 1, 2, \ldots, s - r. \qquad (5.7.11)$$

This leads to the polynomials $\varphi_0(\theta)$ and $\chi_j(\theta)$, $j = 1, 2, \ldots, s - r$, of the form

$$\varphi_0(\theta) = \theta\big(q_0 + q_1\theta + \cdots + q_{s+r-1}\theta^{s+r-1}\big),$$

$$\chi_j(\theta) = \theta\big(r_{j,0} + r_{j,1}\theta + \cdots + r_{j,s+r-1}\theta^{s+r-1}\big),$$

$j = 1, 2, \ldots, s - r$, where

$$q_0 + 2q_1 c_i + \cdots + (s+r)q_{s+r-1}c_i^{s+r-1} = 0,$$

$$r_{j,0} + 2r_{j,1}c_i + \cdots + (s+r)r_{j,s+r-1}c_i^{s+r-1} = 0,$$

$j = 1, 2, \ldots, s - r$, $i = 1, 2, \ldots, s$. The methods obtained in this way satisfy some of the interpolation and collocation conditions (5.7.7) and (5.7.8). We have the following theorem.

Theorem 5.7.4 (D'Ambrosio et al. [104]) *Assume that $\varphi_0(\theta)$ and $\chi_j(\theta)$, $j = 1, 2, \ldots, s - r$, satisfy (5.7.10) and (5.7.11). Then the solution $\varphi_1(\theta)$, $\psi_j(\theta)$, $j = 1, 2, \ldots, s$, and $\chi_j(\theta)$, $j = s - r + 1, s - r + 2, \ldots, s$ satisfy the interpolation conditions*

$$\varphi_1(0) = 1, \quad \psi_j(0) = 0, \quad j = 1, 2, \ldots, s,$$

$$\chi_j(0) = 0, \quad j = s - r + 1, s - r + 2, \ldots, s, \qquad (5.7.12)$$

and the collocation conditions

$$\varphi_1'(c_i) = 0, \quad \psi_j'(c_i) = \delta_{ij}, \quad j = 1, 2, \ldots, s,$$

$$\chi_j'(c_i) = 0, \quad j = s - r + 1, s - r + 2, \ldots, s, \tag{5.7.13}$$

$i = 1, 2, \ldots, s.$

Proof: Substituting $\theta = 0$ into (5.7.4) corresponding to $p = s + r$, $r = 1, 2, \ldots, s$, and taking into account that the solution to (5.7.4) is unique, condition (5.7.12) follows. Differentiating (5.7.4) with respect to θ and substituting $s = c_i$, $i = 1, 2, \ldots, s$, into the resulting relations for $k = 1, 2, \ldots, s + r$, we obtain (5.7.13). This completes the proof. ∎

The formulas obtained by imposing conditions (5.7.10) and (5.7.11) will then be called almost two-step collocation methods. It follows from Theorem 5.7.4 that the polynomial $P(t)$ defined by method (5.7.1) of order $p = s + r$, $r = 1, 2, \ldots, s$, satisfies the interpolation condition

$$P(t_{n-1}) = y_{n-1}$$

and the collocation conditions at the points c_i; that is,

$$P'(t_{n-1} + c_i h) = f\big(P(t_{n-1} + c_i h)\big), \quad i = 1, 2, \ldots, s.$$

However, in general, these methods do not satisfy the interpolation condition

$$P(t_{n-2}) = y_{n-2}$$

and the collocation conditions

$$P'(t_{n-2} + c_i h) = f\big(P(t_{n-2} + c_i h)\big), \quad i = 1, 2, \ldots, s.$$

5.8   LINEAR STABILITY ANALYSIS OF TWO-STEP COLLOCATION METHODS

Applying method (5.7.1) to the standard test equation

$$y' = \xi y, \quad t \geq 0,$$

and computing the resulting expression at the points $\theta = c_i$, $i = 1, 2, \ldots, s$, and $\theta = 1$, we obtain

$$P(t_{n-1} + c_i h) = \varphi_1(c_i)y_{n-1} + \varphi_0(c_i)y_{n-2}$$

$$+ h\xi \sum_{j=1}^{s} \Big(\psi_j(c_i)P(t_{n-1} + c_j h) + \chi_j(c_i)P(t_{n-2} + c_j h)\Big),$$

$$y_n = \varphi_1(1)y_{n-1} + \varphi_0(1)y_{n-2} \tag{5.8.1}$$

$$+ h\xi \sum_{j=1}^{s} \Big(\psi_j(1)P(t_{n-1} + c_j h) + \chi_j(1)P(t_{n-2} + c_j h)\Big),$$

$i = 1, 2, \ldots, s$, $n = 2, 3, \ldots, N$. Introducing the notation $z = h\xi$,

$$
P(t+\mathbf{c}h) = \begin{bmatrix} P(t + c_1 h) \\ \vdots \\ P(t + c_s h) \end{bmatrix}, \quad \varphi_1(\mathbf{c}) = \begin{bmatrix} \varphi_1(c_1) \\ \vdots \\ \varphi_1(c_s) \end{bmatrix}, \quad \varphi_0(\mathbf{c}) = \begin{bmatrix} \varphi_0(c_1) \\ \vdots \\ \varphi_0(c_s) \end{bmatrix},
$$

$$
\mathbf{A} = \begin{bmatrix} \psi_j(c_i) \end{bmatrix}_{i,j=1}^{s}, \quad \mathbf{B} = \begin{bmatrix} \chi_j(c_i) \end{bmatrix}_{i,j=1}^{s},
$$

and

$$
\mathbf{v}^T = \begin{bmatrix} \psi_1(1) & \cdots & \psi_s(1) \end{bmatrix}^T, \quad \mathbf{w}^T = \begin{bmatrix} \chi_1(1) & \cdots & \chi_s(1) \end{bmatrix}^T,
$$

(compare Section 5.7 for the definition of $\mathbf{A}$, $\mathbf{B}$, $\mathbf{v}$, and $\mathbf{w}$), relation (5.8.1) can be written in the vector form

$$
\begin{aligned}
P(t_{n-1} + \mathbf{c}h) &= \varphi_1(\mathbf{c})y_{n-1} + \varphi_0(\mathbf{c})y_{n-2} \\
&\quad + z\big(\mathbf{A}P(t_{n-1} + \mathbf{c}h) + \mathbf{B}P(t_{n-2} + \mathbf{c}h)\big), \\
y_n &= \varphi_1(1)y_{n-1} + \varphi_0(1)y_{n-2} \\
&\quad + z\left(\mathbf{v}^T P(t_{n-1} + \mathbf{c}h) + \mathbf{w}^T P(t_{n-2} + \mathbf{c}h)\right),
\end{aligned} \tag{5.8.2}
$$

$n = 2, 3, \ldots, N$. Hence,

$$
P(t_{n-1} + \mathbf{c}h) = \big(\mathbf{I} - z\mathbf{A}\big)^{-1}\Big(\varphi_1(\mathbf{c})y_{n-1} + \varphi_0(\mathbf{c})y_{n-2} + z\mathbf{B}P(t_{n-2} + \mathbf{c}h)\Big), \tag{5.8.3}
$$

and substituting this relation into the equation for y_n leads to

$$
\begin{aligned}
y_n &= \big(\varphi_1(1) + z\mathbf{v}^T(\mathbf{I} - z\mathbf{A})^{-1}\varphi_1(\mathbf{c})\big)y_{n-1} \\
&\quad + \big(\varphi_0(1) + z\mathbf{v}^T(\mathbf{I} - z\mathbf{A})^{-1}\varphi_0(\mathbf{c})\big)y_{n-2} \\
&\quad + z\big(\mathbf{w}^T + z\mathbf{v}^T(\mathbf{I} - z\mathbf{A})^{-1}\mathbf{B}\big)P(t_{n-2} + \mathbf{c}h).
\end{aligned} \tag{5.8.4}
$$

Relations (5.8.3) and (5.8.4) are equivalent to

$$
\begin{bmatrix} y_n \\ y_{n-1} \\ P(t_{n-1} + \mathbf{c}h) \end{bmatrix} = \begin{bmatrix} M_{11}(z) & M_{12}(z) & M_{13}(z) \\ 1 & 0 & 0 \\ \mathbf{Q}\varphi_1(\mathbf{c}) & \mathbf{Q}\varphi_0(\mathbf{c}) & z\mathbf{QB} \end{bmatrix} \begin{bmatrix} y_{n-1} \\ y_{n-2} \\ P(t_{n-2} + \mathbf{c}h) \end{bmatrix}, \tag{5.8.5}
$$

where

$$
M_{11}(z) = \varphi_1(1) + z\mathbf{v}^T\mathbf{Q}\varphi_1(\mathbf{c}),
$$

$$
M_{12}(z) = \varphi_0(1) + z\mathbf{v}^T\mathbf{Q}\varphi_0(\mathbf{c}),
$$

$$
M_{13}(z) = z(\mathbf{w}^T + z\mathbf{v}^T\mathbf{QB}),
$$

and

$$\mathbf{Q} = (\mathbf{I} - z\mathbf{A})^{-1} \in \mathbb{C}^{s \times s}.$$

As in Section 2.6, the matrix appearing in (5.8.5) is called the stability matrix of method (5.7.1) and is denoted by $\mathbf{M}(z)$. We have $\mathbf{M}(z) \in \mathbb{C}^{(s+2) \times (s+2)}$. We also define the stability function of method (5.7.1) as

$$p(\eta, z) = \det\left(\eta \mathbf{I} - \mathbf{M}(z)\right). \tag{5.8.6}$$

We are interested mainly in methods that are A-stable. This means that all the roots $w_1, w_2, \ldots, w_{s+2}$ of the polynomial $p(\eta, z)$ defined by (5.8.6) are in the unit circle for all $z \in \mathbb{C}$ such that $\mathrm{Re}(z) \leq 0$. By the maximum principle this will be the case if the denominator of $p(\eta, z)$ does not have poles in the negative half-plane $\mathbb{C}_-$ and if the roots of $p(\eta, iy)$ are in the unit circle for all $y \in \mathbb{R}$. This last condition will be investigated using the Schur criterion (Theorem 2.8.1).

We are also interested in two-step collocation methods (5.7.1) that are L-stable, i.e., methods that are A-stable and all the roots of the stability function $p(\eta, z)$ given by (5.8.6) are equal to zero as $z \to -\infty$. Examples of such methods are given in Sections 5.9 and 5.10.

5.9   TWO-STEP COLLOCATION METHODS WITH ONE STAGE

In this section we analyze the two-step collocation methods (5.7.1) with $s = 1$. Consider first methods (5.7.1) of order $p = 2s + 1 = 3$. Solving the order conditions (5.7.4) corresponding to $s = 1$ and $p = 3$, we obtain a one-parameter family of two-step methods depending on the abscissa c. The coefficients of these methods are

$$\varphi_1(\theta) = \frac{(1+\theta)\left(6c^2 - 1 + (1 - 6c)\theta + 2\theta^2\right)}{1 - 6c^2},$$

$$\varphi_0(\theta) = \frac{\theta\left(6c(c-1) + 3(1 - 2c)\theta + 2\theta^2\right)}{1 - 6c^2},$$

$$\psi(\theta) = \frac{\theta(1+\theta)\left(1 - 4c + 3c^2 + (1 - 2c)\theta\right)}{1 - 6c^2},$$

$$\chi(\theta) = \frac{\theta(1+\theta)\left(2c + 3c^2 - (1 + 2c)\theta\right)}{1 - 6c^2},$$

and the error constant $C_3(1)$ is given by

$$C_3(1) = \frac{1 - 3c - 3c^2 + 12c^3 - 6c^4}{6(1 - 6c^2)},$$

$c \neq \pm\sqrt{6}/6$. To investigate the stability properties of (5.7.1), it is more convenient to work with the polynomial obtained by multiplying the stability function (5.8.6) by its denominator. The resulting polynomial, which is denoted by the same symbol $p(\eta, z)$, for this family of methods takes the form

$$p(\eta, z) = p_3(z)\eta^3 + p_2(z)\eta^2 + p_1(z)\eta + p_0(z), \qquad (5.9.1)$$

where the polynomials $p_i(z)$, $i = 0, 1, 2, 3$, assume the form

$$p_0(z) = -(c-1)^2 c^2 z,$$

$$p_1(z) = 5 - 12c + 6c^2 + (2 - 5c + 6c^2 - 6c^3 + 3c^4)z,$$

$$p_2(z) = -4 + 12c - 12c^2 + (4 - 8c - 3c^2 + 6c^3 - 3c^4)z,$$

and

$$p_3(z) = -1 + 6c^2 + (1 - 2c - 2c^2 + c^3)cz.$$

Next we investigate if there exist A-stable methods in this class of two-step formulas of order $p = 3$. Let

$$\widetilde{p}(\eta, y) := p(\eta, iy),$$

where $p(\eta, z)$ is the stability polynomial (5.9.1). Next we compute the constant polynomial with respect to η, which we denote by $\widetilde{p}_0(y)$, using the recursive procedure described in Section 2.8, leading to the Schur theorem (Theorem 2.8.1). This polynomial takes the form

$$\widetilde{p}_0(y) = \alpha(c)y^4 + \beta(c)y^6 + \gamma(c)y^8,$$

where $\alpha(c)$, $\beta(c)$, and $\gamma(c)$ are polynomials with respect to the abscissa c. It follows from the Schur criterion that the condition

$$\widetilde{p}_0(y) \geq 0 \quad \text{for all} \quad y \geq 0$$

is the necessary condition for A-stability. However, it can be verified that the polynomials $\alpha(c)$, $\beta(c)$, and $\gamma(c)$ are not simultaneously greater or equal to zero for any c. This proves that A-stable methods do not exist in this class of methods of order $p = 3$.

Consider next methods (5.7.1) of order $p = 2s = 2$. We choose the polynomial $\varphi_0(\theta)$ of degree less than or equal to 2 that satisfies the interpolation condition (5.7.10) and collocation condition (5.7.11):

$$\varphi_0(0) = 0 \quad \text{and} \quad \varphi_0'(c) = 0.$$

This leads to a polynomial $\varphi_0(\theta)$ of the form

$$\varphi_0(\theta) = q_0\theta\left(1 - \frac{1}{2c}\theta\right), \qquad (5.9.2)$$

where q_0 is a real parameter. Solving the order conditions (5.7.4) correspond-ing to $s = 1$ and $p = 2$, where $\varphi_0(\theta)$ is given by (5.9.2), we obtain a two-parameter family of two-step methods depending on q_0 and abscissa c. The coefficients of these formulas are given by

$$\varphi_1(\theta) = 1 - q_0\theta + \frac{q_0}{2c}\theta^2,$$

$$\psi(\theta) = \left(1 - c + \frac{q_0}{2} - q_0c\right)\theta + \left(\frac{1}{2} + \frac{q_0}{2} - \frac{q_0}{4c}\right)\theta^2,$$

$$\chi(\theta) = \left(c + \frac{q_0}{2} + cq_0\right)\theta - \left(\frac{1}{2} + \frac{q_0}{2} + \frac{q_0}{4c}\right)\theta^2,$$

and the error constant $C_2(1)$ takes the form

$$C_2(1) = \frac{10c - 24c^2 + 12c^3 + q_0 - 2q_0c - 6q_0c^2 + 12q_0c^3}{24c},$$

$c \neq 0$. The stability polynomial of this family of methods is

$$p(\eta, z) = \eta\big(p_2(z)\eta^2 + p_1(z)\eta + p_0(z)\big), \qquad (5.9.3)$$

where polynomials $p_0(z)$, $p_1(z)$, and $p_2(z)$ are now given by

$$p_0(z) = 2q_0 - 4q_0c + (2c - 4c^2 + 2c^3 + q_0 - 2q_0c - q_0c^2 + 2q_0c^3)z,$$

$$p_1(z) = -4c - 2q_0 + 4q_0c - (6c - 8c^2 + 4c^3 - q_0 + 2q_0c - 2q_0c^2 + 4q_0c^3)z,$$

and

$$p_2(z) = 4c - c^2(4 - 2c + q_0 - 2q_0c)z.$$

We have performed a computer search based on the Schur criterion using the polynomial $p(\eta, z)$ given by (5.9.3) with $p_0(z)$, $p_1(z)$, and $p_2(z)$ defined above. This search was performed in the parameter space (q_0, c) and the results are presented in Fig. 5.9.1 for $-3 \leq q_0 \leq 1$ and $0 \leq c \leq 2$, where the shaded region corresponds to the A-stable formulas. Choosing, for example, $q_0 = -1$ and $c = \frac{3}{4}$, we obtain the A-stable two-step method with coefficients given by

$$\varphi_1(\theta) = \frac{3 + 3\theta - 2\theta^2}{3}, \quad \varphi_0(\theta) = \frac{(2\theta - 3)\theta}{3},$$

$$\psi(\theta) = \frac{(2\theta + 3)\theta}{6}, \qquad \chi(\theta) = \frac{(2\theta - 3)\theta}{6}. \qquad (5.9.4)$$

For this method the stability polynomial $p(\eta, z)$ is given by

$$p(\eta, z) = \eta\left(\left(3 - \frac{27}{16}z\right)\eta^2 - \left(4 + \frac{5}{8}z\right)\eta + \left(1 + \frac{5}{16}z\right)\right),$$

and the error constant is $C_2(1) = -\frac{17}{144}$.

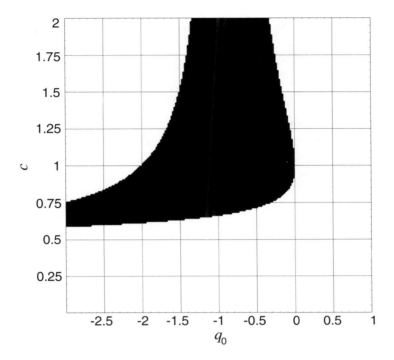

Figure 5.9.1   Region of A-stability in the (q_0, c)-plane for the two-step methods (5.7.1) with $m = 1$ and $p = 2$

We look next for L-stable methods, that is, methods for which all roots of the polynomial $p(\eta, z)/p_2(z))$, where $p(\eta, z)$ is given by (5.9.3), are equal to zero as $z \to -\infty$. Such methods correspond to solutions of the nonlinear system of equations

$$\lim_{z \to -\infty} \frac{p_0(z)}{p_2(z)} = 0, \qquad \lim_{z \to -\infty} \frac{p_1(z)}{p_2(z)} = 0.$$

It can be verified that for methods with stability polynomial (5.9.3), this system takes the form

$$(c - 1)(2c - 2c^2 + q_0 - q_0 c - 2q_0 c^2) = 0,$$

$$6c - 8c^2 + 4c^3 - q_0 + 2q_0 c - 2q_0 c^2 + 4q_0 c^3 = 0,$$

and has solutions

$$q_0 = -\frac{2}{3}, \quad c = 1 \qquad \text{and} \qquad q_0 = -\frac{4}{9}, \quad c = 2.$$

The coefficients of the method corresponding to the first set of the parameters above are

$$\varphi_1(\theta) = \frac{3 + 2\theta - \theta^2}{3}, \quad \varphi_0(\theta) = \frac{(s-2)\theta}{3},$$

$$\psi(\theta) = \frac{(\theta+1)\theta}{3}, \qquad \chi(\theta) = 0,$$

(5.9.5)

and the error constant is $C_2(1) = -\frac{2}{9}$. The coefficients of the method corresponding to the second set of the parameters q_0 and c are

$$\varphi_1(\theta) = \frac{9 + 4\theta - \theta^2}{9}, \quad \varphi_0(\theta) = \frac{\theta(\theta-4)}{9},$$

$$\psi(\theta) = \frac{(\theta-1)\theta}{9}, \qquad \chi(\theta) = \frac{2(\theta-4)\theta}{9},$$

(5.9.6)

and the error constant is, as before, $C_2(1) = -\frac{2}{9}$. It can be verified that for $\theta = 1$, methods (5.9.5) and (5.9.6) both reduce to the backward differentiation method of order $p = 2$ (compare [52, 194]).

5.10 TWO-STEP COLLOCATION METHODS WITH TWO STAGES

We consider first methods (5.7.1) of order $p = 2s + 1 = 5$. Solving order conditions (5.7.4) corresponding to $s = 2$ and $p = 5$, we obtain a family of methods depending on the components of the abscissa vector $\mathbf{c}$, c_1 and c_2. We have plotted in Fig. 5.10.1 the contour plots of error constant $C_5(1)$ of these formulas for $0 \le c_1 \le 1$ and $0 \le c_2 \le 1$. Choosing, for example, $c_1 = \frac{1}{2}$ and $c_2 = 1$, we obtain a two-step formula of uniform order $p = 5$ with coefficients given by

$$\varphi_1(\theta) = \frac{(1+\theta)(29 - 29\theta + 44\theta^2 - 54\theta^3 + 24\theta^4)}{29},$$

$$\varphi_0(\theta) = -\frac{(15 - 10\theta - 30\theta^2 + 24\theta^3)\theta^2}{29},$$

$$\psi_1(\theta) = \frac{\theta^2(1+\theta)(19 + 7\theta - 16\theta^2)}{29},$$

$$\psi_2(\theta) = -\frac{\theta^2(1+\theta)(7 - 2\theta - 12\theta^2)}{87},$$

$$\chi_1(\theta) = -\frac{\theta^2(1+\theta)(89 - 187\theta + 96\theta^2)}{87},$$

$$\chi_2(\theta) = \frac{\theta(1+\theta)(29 - 31\theta - 16\theta^2 + 20\theta^3)}{29}.$$

The error constant of this method is $C_5(1) = \frac{113}{83520}$.

The stability polynomial of this two-parameter family of methods takes the form

$$p(\eta, z) = p_4(z)\eta^4 + p_3(z)\eta^3 + p_2(z)\eta^2 + p_1(z)\eta + p_0(z),$$

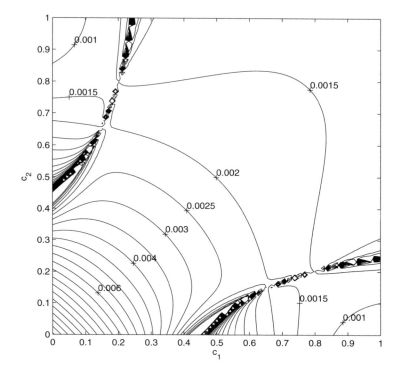

Figure 5.10.1   Contour plots of error constant $C_5(1)$ for $0 \leq c_1 \leq 1$ and $0 \leq c_2 \leq 1$

where $p_i(z)$, $i = 0, 1, 2, 3, 4$ are quadratic polynomials with respect to z. These polynomials also depend on c_1 and c_2. We have performed an extensive computer search based on the Schur criterion in the two-dimensional space (c_1, c_2) looking for methods with good stability properties, but so far we have not been able to find methods that are A-stable. We suspect that such methods do not exist in this class of formulas with $s = 2$ and $p = 5$.

We consider next methods of order $p = 2s = 4$. We choose the polynomial $\varphi_0(\theta)$, which satisfies the interpolation condition (5.7.10) and collocation conditions (5.7.11):

$$\varphi_0(0) = 0 \quad \text{and} \quad \varphi_0'(c_i) = 0, \quad i = 1, 2.$$

This leads to a polynomial of the form

$$\varphi_0(\theta) = \theta(q_0 + q_1\theta + q_2\theta^2 + q_3\theta^3),$$

where q_2 and q_3 are given by

$$q_2 = -\frac{7q_0 + 6q_1}{3}, \quad q_3 = \frac{3q_0 + 2q_1}{2}.$$

Choosing, for example, $c_1 = \frac{3}{4}$, $c_2 = 1$, $q_0 = q_1 = -1$, we obtain the method with coefficients given by

$$\varphi_1(\theta) = \frac{27 + 27\theta + 27\theta^2 - 79\theta^3 + 39\theta^4}{27},$$

$$\varphi_0(\theta) = -\frac{s(27 + 27\theta - 79\theta^2 + 39\theta^3)}{27},$$

$$\psi_1(\theta) = -\frac{2\theta(27 + 18\theta - 97\theta^2 + 57\theta^3)}{27},$$

$$\psi_2(\theta) = \frac{s(837 + 594\theta - 2881\theta^2 + 1857\theta^3)}{810}, \qquad (5.10.1)$$

$$\chi_1(\theta) = -\frac{2\theta(783 + 1026\theta - 2669\theta^2 + 1293\theta^3)}{405},$$

$$\chi_2(\theta) = \frac{\theta(783 + 756\theta - 2249\theta^2 + 1113\theta^3)}{162}.$$

The error constant of this method is $C_4(1) = \frac{1085}{248832}$.

The stability polynomial of the four-parameter family of methods of order $p = 4$ takes the form

$$p(\eta, z) = \eta\big(p_3(z)\eta^3 + p_2(z)\eta^2 + p_1(z)\eta + p_0(z)\big),$$

where $p_i(z)$, $i = 0, 1, 2, 3$ are quadratic polynomials with respect to z. These polynomials also depend on the parameters q_0, q_1, c_1, and c_2. We have performed an extensive computer search based on the Schur criterion in the four-dimensional space (q_0, q_1, c_1, c_2), but so far we were not able to find methods that are A-stable. We suspect again that such methods do not exist in this class of formulas with $s = 2$ and $p = 4$.

Finally, consider methods of order $p = 2s - 1 = 3$. We choose the polynomials $\varphi_0(\theta)$ and $\chi_1(\theta)$ of degree less than or equal to 3 that satisfy conditions (5.7.10) and (5.7.11):

$$\varphi_0(0) = 0, \quad \chi_1(0) = 0, \quad \varphi_0'(c_i) = 0, \quad \chi_1'(c_i) = 0, \quad i = 1, 2.$$

These polynomials take the form

$$\varphi_0(\theta) = \theta(q_0 + q_1\theta + q_2\theta^2), \quad \chi_1(\theta) = \theta(r_0 + r_1\theta + r_2\theta^2),$$

where

$$q_1 = r_1 = -\frac{(c_1 + c_2)q_0}{2c_1 c_2}, \quad q_2 = r_2 = \frac{q_0}{3c_1 c_2}.$$

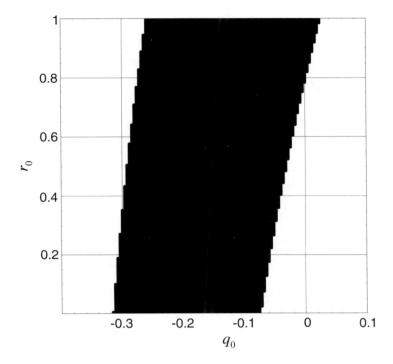

Figure 5.10.2   Region of A-stability in the (q_0, r_0)-plane for the two-step methods (5.7.1) with $s = 2$ and $p = 3$

Solving order conditions (5.7.4) corresponding to $s = 2$ and $p = 3$, we obtain a four parameter family of methods (5.7.1) depending on q_0, r_0, c_1, and c_2. The stability polynomial of this family of methods is given by

$$p(\eta, z) = \eta^2 \big(p_2(z)\eta^2 + p_1(z)\eta + p_0(z) \big),$$

where $p_i(z)$, $i = 0, 1, 2$, are polynomials of degree less than or equal to 2 with respect to z. These polynomials also depend on q_0, r_0, c_1, and c_2. We have performed again an extensive computer search looking for methods that are A-stable. We have found such methods only if both components of the abscissa vector are outside the interval $[0, 1]$. The results of this search for $c_1 = \frac{5}{2}$ and $c_2 = \frac{9}{2}$ are presented in Fig. 5.10.2 for $-0.4 \le q_0 \le 0.1$ and $0 \le r_0 \le 1$, where the shaded region corresponds to A-stable methods. The coefficients of the resulting methods with $s = 2$ and $p = 3$ and $\mathbf{c} = [\frac{5}{2}, \frac{9}{2}]^T$ are given by:

$$\varphi_1(\theta) = \frac{135 - 135q_0\theta + 42q_0\theta^2 - 4q_0\theta^3}{135},$$

$$\varphi_0(\theta) = \frac{q_0\theta(135 - 42\theta + 4\theta^2)}{135},$$

$$\psi_1(\theta) = \left(\frac{63}{8} + \frac{241}{24}q_0 - 3r_0\right)\theta - \left(2 + \frac{1687}{540}q_0 - \frac{14}{15}r_0\right)\theta^2$$
$$+ \left(\frac{1}{6} + \frac{241}{810}q_0 - \frac{4}{45}r_0\right)\theta^3,$$

$$\psi_2(\theta) = \left(\frac{35}{8} + \frac{145}{24}q_0 - r_0\right)\theta - \left(\frac{3}{2} + \frac{203}{108}q_0 - \frac{14}{45}r_0\right)\theta^2$$

$$+ \left(\frac{1}{6} + \frac{29}{162}q_0 - \frac{4}{135}r_0\right)\theta^3,$$

$$\chi_1(\theta) = \frac{r_0\theta(135 - 42\theta + 4\theta^2)}{135},$$

$$\chi_2(\theta) = -\frac{(135 + 181q_0 - 36r_0)(135 - 42\theta + 4\theta^2)\theta}{1620}.$$

$(5.10.2)$

The error constant $C_3(1)$ is

$$C_3(1) = \frac{4494825 + 6019723q_0 - 1229184r_0}{77760}.$$

We have also found methods in this class that are L-stable. Such methods correspond to solutions of the nonlinear system

$$\lim_{z \to -\infty} \frac{p_0(z)}{p_2(z)} = 0, \quad \lim_{z \to -\infty} \frac{p_1(z)}{p_2(z)} = 0.$$

One such solution is

$$q_0 \approx -\frac{21225899}{77647080} \approx -0.273364, \quad r_0 \approx \frac{113887980}{163068619} \approx 0.698405,$$

the error constant is $C_3(1) \approx 25.6016$, and the resulting method is A- and L-stable.

5.11 TSRK METHODS WITH QUADRATIC STABILITY FUNCTIONS

In this section, which follows the presentation by Conte et al. [92], we analyze a special class of implicit TSRK methods (5.1.1) for which the stability function has only two nonzero roots. In this analysis it will be convenient to use the representation of TSRK methods as GLMs, which was introduced in Section 2.1. For easy reference, this representation, corresponding to the problem (2.1.1) with $m = 1$, which is relevant in linear stability analysis, is reproduced below:

$$
\begin{bmatrix} Y^{[n]} \\ \hline y_n \\ y_{n-1} \\ hf(Y^{[n]}) \end{bmatrix} = \begin{bmatrix} \mathbf{A} & \mathbf{e} - \mathbf{u} & \mathbf{u} & \mathbf{B} \\ \hline \mathbf{v}^T & 1 - \vartheta & \vartheta & \mathbf{w}^T \\ \mathbf{0} & 1 & 0 & \mathbf{0} \\ \mathbf{I} & 0 & 0 & \mathbf{0} \end{bmatrix} \begin{bmatrix} hf(Y^{[n]}) \\ \hline y_{n-1} \\ y_{n-2} \\ hf(Y^{[n-1]}) \end{bmatrix}.
\tag{5.11.1}
$$

We also define the coefficients matrices $\mathbf{A}$, $\mathbf{U}$, $\mathbf{P}$, and $\mathbf{V}$ of the corresponding GLM according to the representation

$$
\begin{bmatrix} \mathbf{A} & \mathbf{U} \\ \hline \mathbf{P} & \mathbf{V} \end{bmatrix} = \begin{bmatrix} \mathbf{A} & \mathbf{e} - \mathbf{u} & \mathbf{u} & \mathbf{B} \\ \hline \mathbf{v}^T & 1 - \vartheta & \vartheta & \mathbf{w}^T \\ \mathbf{0} & 1 & 0 & \mathbf{0} \\ \mathbf{I} & 0 & 0 & \mathbf{0} \end{bmatrix}.
\tag{5.11.2}
$$

The stability properties of (5.11.1) with respect to the linear test equation $y' = \xi y$, $t \geq 0$, where $\xi \in \mathbb{C}$, are determined by the stability function

$$
\widetilde{p}(\eta, z) = \det\left(\eta \mathbf{I} - \mathbf{M}(z)\right),
\tag{5.11.3}
$$

where $\mathbf{M}(z)$ is the stability matrix defined by $\mathbf{M}(z) = \mathbf{V} + z\mathbf{P}(\mathbf{I} - z\mathbf{A})^{-1}\mathbf{U}$ (compare Section 2.6). Here $\mathbf{A}$, $\mathbf{U}$, $\mathbf{P}$, and $\mathbf{V}$ are the coefficient matrices of the GLM defined by (5.11.2). The function $\widetilde{p}(\eta, z)$ defined by (5.11.3) is a polynomial of degree $s + 2$ with respect to η whose coefficients are rational functions with respect to z. In this section we are only interested in methods for which the coefficient matrix $\mathbf{A}$ has a one-point spectrum of the form

$$
\sigma(\mathbf{A}) = \{\lambda\}, \quad \lambda > 0.
\tag{5.11.4}
$$

This feature would allow for efficient implementation of these methods, similarly as in the case of SIRK methods considered by Burrage [27], Butcher [39], and Burrage et al. [34] (see also [41, 52]).

To analyze the stability properties of such methods, it is more convenient to work with the function $p(\eta, z)$ defined by

$$
p(\eta, z) = (1 - \lambda z)^s \widetilde{p}(\eta, z),
\tag{5.11.5}
$$

since the coefficients of η^i, $i = 0, 1, \ldots, s+2$, are polynomials of degree s with respect to z instead of rational functions, as is the case for (5.11.3). We are interested in TSRK methods (5.11.1) of order $p = s$ and stage order $q = p$, for which the stability polynomial $p(\eta, z)$ takes a simple form,

$$p(\eta, z) = \eta^s \big((1 - \lambda z)^s \eta^2 - p_1(z)\eta + p_0(z)\big), \qquad (5.11.6)$$

where $p_1(z)$ and $p_0(z)$ are polynomials of degree s with respect to z. Methods for which this is the case are said to possess quadratic stability (QS). In what follows we present a characterization of such methods which was inspired by recent work of Butcher and Wright [79, 80, 81] and Wright [293, 294] on GLMs with inherent Runge-Kutta stability. This work of Butcher and Wright is also described in detail in Chapter 7.

We derive first the stage order and order conditions of TSRK methods (5.11.1) using the order theory of GLMs presented in Section 2.4. The GLM with coefficients given by (5.11.2) takes the form

$$
\begin{aligned}
Y_i^{[n]} &= h \sum_{j=1}^{s} a_{ij} f(Y_j^{[n]}) + \sum_{j=1}^{s+2} u_{ij} y_j^{[n-1]}, \quad i = 1, 2, \ldots, s, \\
y_i^{[n]} &= h \sum_{j=1}^{s} p_{ij} f(Y_j^{[n]}) + \sum_{j=1}^{s+2} v_{ij} y_j^{[n-1]}, \quad i = 1, 2, \ldots, s+2,
\end{aligned}
\qquad (5.11.7)
$$

$n = 1, 2, \ldots, N$, where $Y_i^{[n]}$ are defined as in (5.11.1) and $y^{[n]}$ and $y^{[n-1]}$ are given by

$$
y^{[n]} = \begin{bmatrix} y_n \\ y_{n-1} \\ hF(Y^{[n]}) \end{bmatrix}, \qquad
y^{[n-1]} = \begin{bmatrix} y_{n-1} \\ y_{n-2} \\ hF(Y^{[n-1]}) \end{bmatrix}.
$$

It follows from Theorem 2.4.1 that GLM (5.11.7) has order $p = s$ and stage order $q = p$ if and only if the relations (2.4.5) and (2.4.6) are satisfied (with $\mathbf{B}$ replaced by $\mathbf{P}$), where we recall $\mathbf{w}(z) = \sum_{k=0}^{p} \mathbf{q}_k z^k$ and $\mathbf{q}_k = [q_{1k}, \ldots, q_{rk}]^T$ with $r = s + 2$. Expanding $e^{\mathbf{c}z}$ and e^z in (2.4.5) and (2.4.6) into power series around $z = 0$ and comparing constant terms in the resulting expressions, we obtain the preconsistency conditions $\mathbf{U}\mathbf{q}_0 = \mathbf{e}$ and $\mathbf{V}\mathbf{q}_0 = \mathbf{q}_0$ (i.e., the conditions (2.2.1)). Comparing the terms of order z^k for $k = 1, 2, \ldots, s$, in the resulting expressions the stage order and order conditions (2.4.5) and (2.4.6) can be written in the form (2.4.8) and (2.4.9) or, equivalently,

$$\frac{\mathbf{c}^k}{k!} - \frac{\mathbf{A}\mathbf{c}^{k-1}}{(k-1)!} - \mathbf{U}\mathbf{q}_k = 0, \quad k = 1, 2, \ldots, s, \qquad (5.11.8)$$

and

$$\sum_{l=0}^{k} \frac{\mathbf{q}_{k-l}}{l!} - \frac{\mathbf{P}\mathbf{c}^{k-1}}{(k-1)!} - \mathbf{V}\mathbf{q}_k = 0, \quad k = 1, 2, \ldots, s. \qquad (5.11.9)$$

We determine now the vectors $\mathbf{q}_k$, $k = 1, 2, \ldots, s$, appearing in (5.11.8) and (5.11.9). We have

$$
y^{[n]} = \begin{bmatrix} y_n \\ y_{n-1} \\ hF(Y^{[n]}) \end{bmatrix} = \begin{bmatrix} y(t_n) \\ y(t_n - h) \\ hy'(t_n + (\mathbf{c} - \mathbf{e})h) \end{bmatrix} + O(h^{p+1}),
$$

and expanding $y(t_n - h)$ and $y'(t_n + (\mathbf{c} - \mathbf{e})h)$ into a Taylor series around the point t_n, we obtain

$$
y^{[n]} = \begin{bmatrix} y(t_n) \\ y(t_n) - hy'(t_n) + \dfrac{h^2}{2!}y''(t_n) + \cdots + (-1)^p \dfrac{h^p}{p!}y^{(p)}(t_n) \\ hy'(t_n)\mathbf{e} + h^2 \dfrac{(\mathbf{c} - \mathbf{e})}{1!}y''(t_n) + \cdots + h^p \dfrac{(\mathbf{c} - \mathbf{e})^{p-1}}{(p-1)!}y^{(p)}(t_n) \end{bmatrix} + O(h^{p+1}).
$$

Comparing this expression with (2.4.3) in Section 2.4 leads to the following formulas for the vector $\mathbf{q}_k$:

$$
\mathbf{q}_0 = \begin{bmatrix} 1 \\ 1 \\ \mathbf{0} \end{bmatrix}, \quad \mathbf{q}_k = \begin{bmatrix} 0 \\ \dfrac{(-1)^k}{k!} \\ \dfrac{(\mathbf{c} - \mathbf{e})^{k-1}}{(k-1)!} \end{bmatrix}, \quad k = 1, 2, \ldots, s,
$$

where $\mathbf{0}$ in $\mathbf{q}_0$ stands for a zero vector of dimension s. The vector $\mathbf{q}_0$ is called the preconsistency vector. It can be verified, using the representation (5.11.2), that the preconsistency conditions (2.2.1) are satisfied automatically for TSRK method (5.11.1). The vector $\mathbf{q}_1$ is called a consistency vector and conditions corresponding to $k = 1$ in (5.11.8) and (5.11.9) are called consistency and stage-consistency conditions. These conditions take the form

$$
\mathbf{Pe} + \mathbf{Vq}_1 = \mathbf{q}_0 + \mathbf{q}_1, \quad \mathbf{Ae} + \mathbf{Uq}_1 = \mathbf{c},
$$

which corresponds to (2.2.2) and (2.2.3).

It is convenient to express the stage order and order conditions (5.11.8) and (5.11.9) directly in terms of coefficients $\mathbf{c}$, ϑ, $\mathbf{u}$, $\mathbf{v}$, $\mathbf{w}$, $\mathbf{A}$, and $\mathbf{B}$ of the original TSRK method (5.1.2). We have the following theorem.

Theorem 5.11.1 (Conte et al. [92]) *Assume that the TSRK method (5.1.2) or (5.11.1) has order $p = s$ and stage order $q = p$. Then the stage order and order conditions take the form*

$$
C_k := \frac{\mathbf{c}^k}{k!} - \frac{(-1)^k}{k!}\mathbf{u} - \frac{\mathbf{Ac}^{k-1}}{(k-1)!} - \frac{\mathbf{B}(\mathbf{c} - \mathbf{e})^{k-1}}{(k-1)!} = 0, \tag{5.11.10}
$$

$k = 1, 2, \ldots, s,$ *and*

$$\widehat{C}_k := \frac{1}{k!} - \frac{(-1)^k}{k!}\vartheta - \frac{\mathbf{v}^T \mathbf{c}^{k-1}}{(k-1)!} - \frac{\mathbf{w}^T (\mathbf{c} - \mathbf{e})^{k-1}}{(k-1)!} = 0, \qquad (5.11.11)$$

$k = 1, 2, \ldots, s,$ *where* $\mathbf{c}^\nu := [c_1^\nu, \ldots, c_s^\nu]^T.$

Proof: To reformulate (5.11.8) and (5.11.9) in terms of the coefficients of TSRK method (5.1.2), we use the representation of matrices $\mathbf{A}$, $\mathbf{U}$, $\mathbf{P}$, and $\mathbf{V}$ in (5.11.2). The stage order conditions (5.11.10) follow directly from (5.11.8). To reformulate (5.11.9), observe that

$$\sum_{l=0}^{k} \frac{\mathbf{q}_{k-l}}{l!} = \begin{bmatrix} \dfrac{1}{k!} \\[2ex] \displaystyle\sum_{l=0}^{k} \dfrac{(-1)^l}{(k-l)!l!} \\[2ex] \displaystyle\sum_{l=0}^{k} \dfrac{(\mathbf{c} - \mathbf{e})^l}{(k-1-l)!l!} \end{bmatrix}.$$

Since

$$\sum_{l=0}^{k} \frac{(-1)^l}{(k-l)!l!} = \frac{1}{k!}\sum_{l=0}^{k} \binom{k}{l}(-1)^l = 0$$

and

$$\sum_{l=0}^{k-1} \frac{(\mathbf{c} - \mathbf{e})^l}{(k-1-l)!l!} = \frac{1}{(k-1)!}\sum_{l=0}^{k-1} \binom{k-1}{l}(\mathbf{c} - \mathbf{e})^l = \frac{\mathbf{c}^{k-1}}{(k-1)!},$$

it follows that the last $s + 1$ components of the left-hand side of (5.11.9) are automatically equal to zero, and comparing the first components of (5.11.9), we obtain order conditions (5.11.11). This completes the proof. ∎

The vectors C_k and constants $\widehat{C}_k$ were introduced in Section 5.2 as equations (5.2.3) and (5.2.4). For easy reference they are reproduced in the formulation of Theorem 5.11.1 in formulas (5.11.8) and (5.11.9).

Setting $k = 1$ in (5.11.10) and (5.11.11), the stage-consistency and consistency conditions take the form

$$(\mathbf{A} + \mathbf{B})\mathbf{e} - \mathbf{u} = \mathbf{c}, \quad (\mathbf{v}^T + \mathbf{w}^T)\mathbf{e} = 1 + \vartheta.$$

These relations were obtained in Section 2.2.

We now turn out attention to the main topic of this section: TSRK methods with quadratic stability functions. To investigate such methods it is convenient to introduce some equivalence relation between matrices of the same dimensions. We say that the two matrices $\mathbf{D}$ and $\mathbf{E}$ are equivalent, which we denote by $\mathbf{D} \equiv \mathbf{E}$ if they are equal except for the first two rows.

This relation has several useful properties that will aid in the derivation of TSRK methods with appropriate stability properties. It can be verified that if $\mathbf{F} \in R^{(\nu+2) \times (\nu+2)}$ is a matrix partitioned as

$$\mathbf{F} = \left[\begin{array}{c|c} \mathbf{F}_{11} & \mathbf{F}_{12} \\ \hline \mathbf{F}_{21} & \mathbf{F}_{22} \end{array} \right],$$

where $\mathbf{F}_{11} \in R^{2 \times 2}$, $\mathbf{F}_{12} \in R^{2 \times \nu}$, $\mathbf{F}_{21} \in R^{\nu \times 2}$, $\mathbf{F}_{22} \in R^{\nu \times \nu}$, and if $\mathbf{F}_{21} = \mathbf{0}$, then

$$\mathbf{D} \equiv \mathbf{E} \quad \text{implies that} \quad \mathbf{FD} \equiv \mathbf{FE}.$$

Moreover, for any matrix $\mathbf{F}$ we also have

$$\mathbf{D} \equiv \mathbf{E} \quad \text{implies that} \quad \mathbf{DF} \equiv \mathbf{EF}.$$

In general, it is a very complicated task to construct TSRK methods (5.11.1) that possess QS, especially for methods with a large number of stages s, since this requires the solution of large systems of polynomial equations of high degree for the unknown coefficients of the methods. However, if we restrict the class of methods, it is possible to find interrelations between the coefficient matrices $\mathbf{A}$, $\mathbf{U}$, $\mathbf{P}$, and $\mathbf{V}$ defined by (5.11.2), which ensure that this is the case (i.e., that TSRK method (5.11.1) possesses QS). For GLMs with RK stability such conditions were discovered recently by Butcher and Wright [80, 293] (see also Theorem 7.2.4). They take a similar form for TSRK methods with QS. This is formalized in the following definition.

Definition 5.11.2 *TSRK method (5.11.1) with coefficients $\mathbf{A}$, $\mathbf{U}$, $\mathbf{P}$, and $\mathbf{V}$ defined by (5.11.2) has inherent quadratic stability (IQS) if there exists a matrix $\mathbf{X} \in R^{(s+2) \times (s+2)}$ such that*

$$\mathbf{PA} \equiv \mathbf{XP} \tag{5.11.12}$$

and

$$\mathbf{PU} \equiv \mathbf{XV} - \mathbf{VX}. \tag{5.11.13}$$

The significance of this definition follows from the following theorem.

Theorem 5.11.3 (Conte et al. [92]) *Assume that TSRK method (5.11.1) has IQS. Then its stability function $\widetilde{p}(w, z)$ defined by (5.11.3) assumes the form*

$$\widetilde{p}(\eta, z) = \eta^s \big(\eta^2 - \widetilde{p}_1(z)\eta + \widetilde{p}_0(z) \big), \tag{5.11.14}$$

where $\widetilde{p}_1(z)$ and $\widetilde{p}_0(z)$ are rational functions with respect to z.

Proof: The proof of this theorem follows along the lines of the corresponding result for GLMs with IRKS [80, 293] (compare also Theorem 7.2.4). IQS relation (5.11.12) is equivalent to

$$\mathbf{P}(\mathbf{I} - z\mathbf{A}) \equiv (\mathbf{I} - z\mathbf{X})\mathbf{P},$$

and assuming that $\mathbf{I} - z\mathbf{A}$ is nonsingular, it follows that

$$\mathbf{P} \equiv (\mathbf{I} - z\mathbf{X})\mathbf{P}(\mathbf{I} - z\mathbf{A})^{-1}. \tag{5.11.15}$$

To investigate characteristic polynomial of the stability matrix $\mathbf{M}(z)$ it is more convenient to consider the matrix related to $\mathbf{M}(z)$ by similarity transformation. Using (5.11.13) and (5.11.15) and assuming that $\mathbf{I} - z\mathbf{X}$ is nonsingular, we have

$$(\mathbf{I} - z\mathbf{X})\mathbf{M}(z)(\mathbf{I} - z\mathbf{X})^{-1}$$
$$= (\mathbf{I} - z\mathbf{X})\big(\mathbf{V} + z\mathbf{P}(\mathbf{I} - z\mathbf{A})^{-1}\mathbf{U}\big)(\mathbf{I} - z\mathbf{X})^{-1}$$
$$= (\mathbf{V} - z\mathbf{X}\mathbf{V} + z(\mathbf{I} - z\mathbf{X})\mathbf{P}(\mathbf{I} - z\mathbf{A})^{-1}\mathbf{U})(\mathbf{I} - z\mathbf{X})^{-1}$$
$$\equiv (\mathbf{V} - z\mathbf{X}\mathbf{V} + z\mathbf{P}\mathbf{U})(\mathbf{I} - z\mathbf{X})^{-1}$$
$$\equiv \big(\mathbf{V} - z\mathbf{X}\mathbf{V} + z(\mathbf{X}\mathbf{V} - \mathbf{V}\mathbf{X})\big)(\mathbf{I} - z\mathbf{X})^{-1}$$
$$\equiv (\mathbf{V} - z\mathbf{V}\mathbf{X})(\mathbf{I} - z\mathbf{X})^{-1}.$$

Hence,

$$(\mathbf{I} - z\mathbf{X})\mathbf{M}(z)(\mathbf{I} - z\mathbf{X})^{-1} \equiv \mathbf{V}. \tag{5.11.16}$$

It follows from the structure of the matrix $\mathbf{V}$ and relation (5.11.16) that the matrix $(\mathbf{I} - z\mathbf{X})\mathbf{M}(z)(\mathbf{I} - z\mathbf{X})^{-1}$ can be partitioned as

$$(\mathbf{I} - z\mathbf{X})\mathbf{M}(z)(\mathbf{I} - z\mathbf{X})^{-1} = \left[\begin{array}{c|c} \widetilde{\mathbf{M}}_{11}(z) & \widetilde{\mathbf{M}}_{12}(z) \\ \hline \mathbf{0} & \mathbf{0} \end{array}\right], \tag{5.11.17}$$

where $\widetilde{\mathbf{M}}_{11}(z) \in R^{2\times 2}$, $\widetilde{\mathbf{M}}_{12}(z) \in R^{2\times s}$, and $\mathbf{0}$ stands for a zero matrix of dimension $s \times 2$ and $s \times s$, respectively. This relation implies that the characteristic polynomial $\widetilde{p}(\eta, z)$ of the matrix $(\mathbf{I} - z\mathbf{X})\mathbf{M}(z)(\mathbf{I} - z\mathbf{X})^{-1}$ and $\mathbf{M}(z)$ assumes the form (5.11.14). This concludes the proof. ∎

To express IQS conditions (5.11.12) and (5.11.13) in terms of the coefficients ϑ, $\mathbf{u}$, $\mathbf{v}$, $\mathbf{w}$, $\mathbf{A}$, and $\mathbf{B}$ of TSRK method (5.11.1), we partition matrix $\mathbf{X}$ as

$$\mathbf{X} = \left[\begin{array}{c|c} \mathbf{X}_{11} & \mathbf{X}_{12} \\ \hline \mathbf{X}_{21} & \mathbf{X}_{22} \end{array}\right], \tag{5.11.18}$$

where $\mathbf{X}_{11} \in R^{2\times 2}$, $\mathbf{X}_{12} \in R^{2\times s}$, $\mathbf{X}_{21} \in R^{s\times 2}$, $\mathbf{X}_{22} \in R^{s\times s}$. We also partition accordingly matrices $\mathbf{P}$, $\mathbf{U}$, and $\mathbf{V}$:

$$\mathbf{P} = \left[\begin{array}{c} \mathbf{P}_{11} \\ \hline \mathbf{I} \end{array}\right], \quad \mathbf{U} = \left[\begin{array}{c|c} \mathbf{U}_{11} & \mathbf{B} \end{array}\right], \quad \mathbf{V} = \left[\begin{array}{c|c} \mathbf{V}_{11} & \mathbf{V}_{12} \\ \hline \mathbf{0} & \mathbf{0} \end{array}\right],$$

where $\mathbf{P}_{11} \in R^{2 \times s}$, $\mathbf{U}_{11} \in R^{s \times 2}$, $\mathbf{V}_{11} \in R^{2 \times 2}$, $\mathbf{V}_{12} \in R^{2 \times s}$ are given by

$$\mathbf{P}_{11} = \begin{bmatrix} \mathbf{v}^T \\ \mathbf{0} \end{bmatrix}, \ \mathbf{U}_{11} = \begin{bmatrix} \mathbf{e} - \mathbf{u} & \mathbf{u} \end{bmatrix}, \ \mathbf{V}_{11} = \begin{bmatrix} 1 - \vartheta & \vartheta \\ 1 & 0 \end{bmatrix}, \ \mathbf{V}_{12} = \begin{bmatrix} \mathbf{w}^T \\ \mathbf{0} \end{bmatrix},$$

$\mathbf{I}$ is the identity matrix of dimension $s \times s$, and $\mathbf{0}$ in $\mathbf{V}$ stands for zero matrices of dimension $s \times 2$ and $s \times s$, respectively. Then it is easy to verify that IQS conditions (5.11.12) and (5.11.13) are equivalent to

$$\mathbf{A} = \mathbf{X}_{21}\mathbf{P}_{11} + \mathbf{X}_{22} \qquad (5.11.19)$$

and

$$\mathbf{U}_{11} = \mathbf{X}_{21}\mathbf{V}_{11}, \quad \mathbf{B} = \mathbf{X}_{21}\mathbf{V}_{12}. \qquad (5.11.20)$$

Set

$$\mathbf{X}_{21} = \begin{bmatrix} \alpha & \beta \end{bmatrix} \in R^{s \times 2},$$

where $\alpha, \beta \in R^s$. Since

$$\mathbf{X}_{21}\mathbf{P}_{11} = \alpha\mathbf{v}^T, \quad \mathbf{X}_{21}\mathbf{V}_{11} = \begin{bmatrix} (1 - \vartheta)\alpha + \beta & \vartheta\alpha \end{bmatrix}, \quad \mathbf{X}_{21}\mathbf{V}_{12} = \alpha\mathbf{w}^T,$$

conditions (5.11.19) and (5.11.20) take the form

$$\mathbf{A} = \alpha\mathbf{v}^T + \mathbf{X}^* \qquad (5.11.21)$$

and

$$\begin{bmatrix} \mathbf{e} - \mathbf{u} & \mathbf{u} \end{bmatrix} = \begin{bmatrix} (1 - \vartheta)\alpha + \beta & \vartheta\alpha \end{bmatrix}, \quad \mathbf{B} = \alpha\mathbf{w}^T,$$

where we have written $\mathbf{X}^*$ instead of $\mathbf{X}_{22}$. The conditions above can be simplified to

$$\mathbf{e} = \alpha + \beta, \quad \mathbf{u} = \vartheta\alpha, \quad \mathbf{B} = \alpha\mathbf{w}^T. \qquad (5.11.22)$$

Summing up the discussion above, TSRK method (5.11.1) has IQS if there exist vectors $\alpha, \beta \in R^s$ and a matrix $\mathbf{X}^* \in R^{s \times s}$ such that conditions (5.11.21) and (5.11.22) are satisfied. The construction of TSRK methods that satisfy these conditions is described in the next section.

5.12   CONSTRUCTION OF TSRK METHODS WITH INHERENT QUADRATIC STABILITY

In this section we describe an algorithm, proposed by Conte et al. [92], for the construction of TSRK methods with IQS and such that the coefficient matrix $\mathbf{A}$ has a one-point spectrum (5.11.4). In this algorithm we first compute the coefficient matrix $\mathbf{B}$ and the vector $\mathbf{w}$ from stage order and order conditions (5.11.10) and (5.11.11). Introducing the notation

$$\mathbf{C} = \begin{bmatrix} \mathbf{c} & \dfrac{\mathbf{c}^2}{2!} & \cdots & \dfrac{\mathbf{c}^s}{s!} \end{bmatrix}, \quad \widetilde{\mathbf{C}} = \begin{bmatrix} \mathbf{e} & \dfrac{\mathbf{c}}{1!} & \cdots & \dfrac{\mathbf{c}^{s-1}}{(s-1)!} \end{bmatrix},$$

$$\mathbf{d} = \left[\begin{array}{cccc} -1 & \dfrac{1}{2!} & \cdots & \dfrac{(-1)^s}{s!} \end{array} \right]^T, \quad \mathbf{g} = \left[\begin{array}{cccc} 1 & \dfrac{1}{2!} & \cdots & \dfrac{1}{s!} \end{array} \right]^T,$$

$$\mathbf{E} = \left[\begin{array}{cccc} \mathbf{e} & \dfrac{\mathbf{c} - \mathbf{e}}{1!} & \cdots & \dfrac{(\mathbf{c} - \mathbf{e})^{s-1}}{(s-1)!} \end{array} \right],$$

conditions (5.11.10) and (5.11.11) are equivalent to

$$\mathbf{BE} = \mathbf{C} - \mathbf{u}\mathbf{d}^T - \mathbf{A}\widetilde{\mathbf{C}}$$

and

$$\mathbf{w}^T \mathbf{E} = \mathbf{g}^T - \vartheta \mathbf{d}^T - \mathbf{v}^T \widetilde{\mathbf{C}}.$$

Hence, assuming that $\mathbf{E}$ is nonsingular, we obtain

$$\mathbf{B} = (\mathbf{C} - \mathbf{u}\mathbf{d}^T - \mathbf{A}\widetilde{\mathbf{C}})\mathbf{E}^{-1} \tag{5.12.1}$$

and

$$\mathbf{w}^T = (\mathbf{g}^T - \vartheta \mathbf{d}^T - \mathbf{v}^T \widetilde{\mathbf{C}})\mathbf{E}^{-1}. \tag{5.12.2}$$

To obtain TSRK methods with IQS, we compute the matrix $\mathbf{X}^*$ from the condition (5.11.21):

$$\mathbf{X}^* = \mathbf{A} - \alpha \mathbf{v}^T,$$

and the vectors β and $\mathbf{u}$ from the first two conditions of (5.11.22):

$$\beta = \mathbf{e} - \alpha, \quad \mathbf{u} = \vartheta \alpha.$$

Then we enforce the condition $\mathbf{B} = \alpha \mathbf{w}^T$ in (5.11.22) using the representations of $\mathbf{B}$ and $\mathbf{w}$ given by (5.12.1) and (5.12.2). This leads to

$$\mathbf{C} - \mathbf{u}\mathbf{d}^T - \mathbf{A}\widetilde{\mathbf{C}} = \alpha \mathbf{g}^T - \vartheta \alpha \mathbf{d}^T - \alpha \mathbf{v}^T \widetilde{\mathbf{C}}$$

and using the condition $\mathbf{u} = \vartheta \alpha$ and assuming that $\widetilde{\mathbf{C}}$ is nonsingular, we obtain

$$\mathbf{A} = \big(\mathbf{C} - \alpha(\mathbf{g}^T - \mathbf{v}^T \widetilde{\mathbf{C}}) \big) \widetilde{\mathbf{C}}^{-1}. \tag{5.12.3}$$

Computing matrix $\mathbf{A}$ from (5.12.3) and then matrix $\mathbf{B}$ from (5.12.1), where $\mathbf{u} = \vartheta \alpha$, and vector $\mathbf{w}$ from (5.12.2), we obtain a family of TSRK methods (5.11.1) of order $p = s$ and stage order $q = p$, which depends on the parameters ϑ, α, $\mathbf{c}$, and $\mathbf{v}$. By construction, these methods satisfy IQS conditions (5.11.21) and (5.11.22). Next we impose the condition (5.11.4): that matrix $\mathbf{A}$ has a one-point spectrum $\sigma(\mathbf{A}) = \{\lambda\}$, where λ will be chosen in such a way that the resulting method is A- and L-stable. The condition that $\mathbf{A}$ has a one-point spectrum (5.11.4) is equivalent to the requirement that the characteristic polynomial of $\mathbf{A}$ assumes the simple form

$$\det(\eta \mathbf{I} - \mathbf{A}) = (\eta - \lambda)^s.$$

Since

$$\det(\eta\mathbf{I} - \mathbf{A}) = \sum_{k=0}^{s} b_k \eta^{s-k},$$

where $b_0 = 1$, $b_k = b_k(\vartheta, \alpha, \mathbf{c}, \mathbf{v})$, $k = 1, 2, \ldots, s$, and

$$(\eta - \lambda)^s = \sum_{k=0}^{s} \binom{s}{k} (-1)^k \lambda^k \eta^{s-k},$$

this is equivalent to the system of equations

$$b_k(\vartheta, \alpha, \mathbf{c}, \mathbf{v}) = \binom{s}{k} (-1)^k \lambda^k, \quad k = 1, 2, \ldots, s. \tag{5.12.4}$$

Since it follows from (5.12.3) that

$$\mathbf{A} = (\mathbf{C} - \alpha\mathbf{g}^T)\widetilde{\mathbf{C}}^{-1} + \alpha\mathbf{v}^T,$$

system (5.12.4) is linear with respect to $\mathbf{v}$, and its solution leads to methods for which stability polynomial $p(\eta, z)$ takes the form (5.11.6):

$$p(\eta, z) = \eta^s \Big((1 - \lambda z)^s \eta^2 - p_1(z)\eta + p_0(z) \Big).$$

The polynomials $p_1(z)$ and $p_0(z)$ appearing in $p(\eta, z)$ take the form

$$p_1(z) = p_{10} + p_{11}z + \cdots + p_{1,s-1}z^{s-1} + p_{1s}z^s,$$

$$p_0(z) = p_{00} + p_{01}z + \cdots + p_{0,s-1}z^{s-1} + p_{0s}z^s.$$

We have

$$p(\eta, 0) = \det(\eta\mathbf{I} - \mathbf{V}) = \eta^s(\eta - 1)(\eta + \vartheta) = \eta^s\big(\eta^2 - p_1(0)\eta + p_0(0)\big)$$

and it follows that

$$p_1(0) = p_{10} = 1 - \vartheta, \quad p_0(0) = p_{00} = -\vartheta. \tag{5.12.5}$$

For a TSRK method of order $p = s$, the stability polynomial $p(\eta, z)$ for $\eta = e^z$ satisfies the condition

$$p(e^z, z) = O(z^{s+1}), \quad z \to 0. \tag{5.12.6}$$

Expanding (5.12.6) into a power series around $z = 0$ it follows from (5.12.5) that the constant term vanishes, and comparing to zero terms of order z^k, $k = 1, 2, \ldots, s$, we obtain a system of s linear equations for the $2s$ coefficients $p_{1j}, p_{0j}, j = 1, 2, \ldots, s$, of the polynomials $p_1(z)$ and $p_0(z)$. This system has a family of solutions depending on λ, ϑ, and s additional parameters which may be chosen from p_{1j} and p_{0j}. These free parameters will be chosen in such a way that the resulting stability polynomials $p(\eta, z)$ are A-stable and, if possible, also L-stable. Assuming A-stability, the latter requirement is equivalent to

$$\lim_{z \to \infty} \frac{p_1(z)}{(1 - \lambda z)^s} = 0, \quad \lim_{z \to \infty} \frac{p_0(z)}{(1 - \lambda z)^s} = 0,$$

which leads to the system of equations

$$p_{1s} = 0, \quad p_{0s} = 0. \tag{5.12.7}$$

To formulate the system of equations consisting of (5.12.7) and additional restrictions on the parameters p_{1j} and p_{0j}, we compute the stability polynomial $\widetilde{p}(\eta, z)$ of TSRK method (5.11.1) from the relation

$$\widetilde{p}(\eta, z) = \eta^s \det \left(\eta \mathbf{I} - \widetilde{\mathbf{M}}_{11}(z) \right), \tag{5.12.8}$$

where $\mathbf{I}$ is the identity matrix of dimension 2 and $\widetilde{\mathbf{M}}_{11}(z) \in R^{2 \times 2}$ is the matrix appearing in (5.11.17). Since IQS conditions (5.11.21) and (5.11.22) do not depend on the blocks $\mathbf{X}_{11}$ and $\mathbf{X}_{12}$ of the matrix $\mathbf{X}$ in (5.11.8), we can assume without loss of generality that $\mathbf{X}_{11} = \mathbf{0}$ and $\mathbf{X}_{12} = \mathbf{0}$. Partitioning the stability matrix $\mathbf{M}(z)$ as

$$\mathbf{M}(z) = \left[\begin{array}{c|c} \mathbf{M}_{11}(z) & \mathbf{M}_{12}(z) \\ \hline \mathbf{M}_{21}(z) & \mathbf{M}_{22}(z) \end{array} \right],$$

where $\mathbf{M}_{11}(z) \in R^{2 \times 2}$, $\mathbf{M}_{12}(z) \in R^{2 \times s}$, $\mathbf{M}_{21}(z) \in R^{s \times 2}$, and $\mathbf{M}_{22}(z) \in R^{s \times s}$, it follows from (5.11.17) that

$$\left[\begin{array}{c|c} \mathbf{I} & \mathbf{0} \\ \hline -z\mathbf{X}_{21} & \mathbf{I} - z\mathbf{X}_{22} \end{array} \right] \left[\begin{array}{c|c} \mathbf{M}_{11}(z) & \mathbf{M}_{12}(z) \\ \hline \mathbf{M}_{21}(z) & \mathbf{M}_{22}(z) \end{array} \right]$$

$$= \left[\begin{array}{c|c} \widetilde{\mathbf{M}}_{11}(z) & \widetilde{\mathbf{M}}_{12}(z) \\ \hline \mathbf{0} & \mathbf{0} \end{array} \right] \left[\begin{array}{c|c} \mathbf{I} & \mathbf{0} \\ \hline -z\mathbf{X}_{21} & \mathbf{I} - z\mathbf{X}_{22} \end{array} \right].$$

Hence,

$$\mathbf{M}_{11}(z) = \widetilde{\mathbf{M}}_{11}(z) - z\widetilde{\mathbf{M}}_{12}(z)\mathbf{X}_{21}, \quad \mathbf{M}_{12}(z) = \widetilde{\mathbf{M}}_{12}(z)(\mathbf{I} - z\mathbf{X}_{22}),$$

which, taking into account that $\mathbf{X}_{21} = [\alpha \ \beta]$ and $\mathbf{X}_{22} = \mathbf{X}^*$, leads to the following formula for the matrix $\widetilde{\mathbf{M}}_{11}(z)$:

$$\widetilde{\mathbf{M}}_{11}(z) = \mathbf{M}_{11}(z) + z\mathbf{M}_{12}(z)(\mathbf{I} - z\mathbf{X}^*)^{-1}[\ \alpha \ \beta \ ]. \tag{5.12.9}$$

System (5.12.7) and additional restrictions on parameters p_{1j} and p_{0j} can be satisfied for specific values of the parameter vector α. This process is illustrated in Section 5.13 for methods of order $p = s$ and stage order $q = p$ for $s = 1, 2, 3$, and 4.

The construction of highly stable TSRK methods (5.11.1) with IQS properties and coefficient matrix $\mathbf{A}$ with a one-point spectrum $\sigma(\mathbf{A}) = \{\lambda\}$ can be summarized in the following algorithm:

1. Choose abscissa vector $\mathbf{c}$ with distinct components and such that matrix $\mathbf{E}$ defined at the beginning of this section is nonsingular.

2. Choose parameters ϑ and $\lambda > 0$ and free parameters among p_{1j} and p_{0j} so that the stability polynomial $p(\eta, z)$ is A-stable and, if possible, also L-stable.

3. Compute coefficient matrix $\mathbf{A}$ from formula (5.12.3). This matrix depends on vectors α and $\mathbf{v}$.

4. Compute vectors β and $\mathbf{u}$ from the first two conditions of (5.11.22): $\beta = \mathbf{e} - \alpha$ and $\mathbf{u} = \vartheta\alpha$.

5. Compute coefficient matrix $\mathbf{B}$ from (5.12.1) and vector $\mathbf{w}$ from (5.12.2). They depend on α and $\mathbf{v}$.

6. Solve system (5.12.4) with respect to $\mathbf{v}$. This leads to a family of methods with IQS for which matrix $\mathbf{A}$ has a one-point spectrum $\sigma(\mathbf{A}) = \{\lambda\}$.

7. Compute matrix $\widetilde{\mathbf{M}}_{11}(z)$ from relation (5.12.9) and stability polynomial $\widetilde{p}(\eta, z)$ from (5.12.8) and formulate a system consisting of (5.12.7) and additional restrictions on p_{1j} and p_{0j} consistent with the choice made in point 2.

8. Solve the system obtained in point 7 with respect to parameter vector α. Then the stability polynomial of the resulting TSRK method (5.11.1) corresponds to polynomial $p(\eta, z)$ in point 2.

5.13 EXAMPLES OF HIGHLY STABLE QUADRATIC POLYNOMIALS AND TSRK METHODS

In this section we apply the algorithm described in Section 5.12 to derive quadratic polynomials that are A- and L-stable and the corresponding TSRK methods with IQS properties for $s = 1, 2, 3$, and 4.

For $s = 1$, the stability polynomial (5.11.6) takes the form

$$p(\eta, z) = \eta\big((1 - \lambda z)\eta^2 - p_1(z)\eta + p_0(z)\big),$$

with

$$p_1(z) = 1 - \vartheta + p_{11}z, \qquad p_0(z) = -\vartheta + p_{01}z.$$

The solution of the equation corresponding to (5.12.6) with $s = 1$ is

$$p_{11} = 1 - \lambda - p_{01} + \vartheta.$$

Assuming that $p_{01} = 0$, it can be verified using the Schur criterion [194, 253] in Theorem 2.8.1 that $p(\eta, z)$ is A-stable if and only if

$$2\lambda + \vartheta \geq 1, \quad 2\lambda - \vartheta \geq 1, \quad \text{and} \quad \lambda \leq 2.$$

Moreover, $p_{11} = 0$ leads to $\vartheta = \lambda - 1$, and the resulting polynomial is L-stable if and only if $\frac{2}{3} \leq \lambda \leq 2$. This is illustrated in Fig. 5.13.1, where the range of parameters (ϑ, λ) for which $p(\eta, z)$ is A-stable corresponds to the shaded region and the range of (ϑ, λ) for which $p(\eta, z)$ is L-stable is plotted by a thick line.

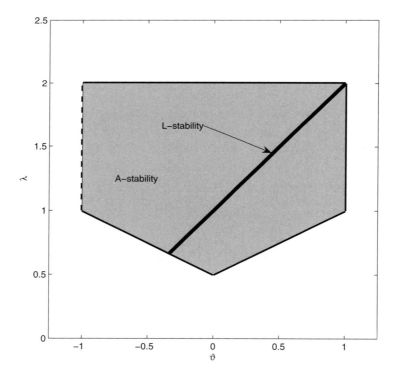

Figure 5.13.1   Regions of A- and L-stability in the (ϑ, λ)-plane for $p(\eta, z)$ with $p = s = 1$

The coefficients of the TSRK method corresponding to $\lambda = 1$, $\vartheta = 0$, and abscissa c are given by

$$\frac{\mathbf{u} \;\big|\; \mathbf{A} \;\big|\; \mathbf{B}}{\vartheta \;\big|\; \mathbf{v} \;\big|\; \mathbf{w}} = \frac{0 \;\big|\; 1 \;\big|\; c - 1}{0 \;\big|\; 2 - c \;\big|\; c - 1}.$$

The stability polynomial $p(\eta, z)$ of this family of methods is

$$p(\eta, z) = \eta\big((1 - z)\eta - 1\big)$$

for any c. In particular, for $c = 1$ this method is equivalent to the backward Euler method.

For $s = 2$ the stability polynomial (5.11.6) takes the form

$$p(\eta, z) = \eta^2\big((1 - \lambda z)^2\eta^2 - p_1(z)\eta + p_0(z)\big),$$

with

$$p_1(z) = 1 - \vartheta + p_{11}z + p_{12}z^2, \quad p_0(z) = -\vartheta + p_{01}z + p_{02}z^2.$$

The system of equations corresponding to (5.12.6) with $s = 2$ takes the form

$$p_{11} - p_{01} = 1 - 2\lambda + \vartheta, \quad 2p_{11} - 2p_{02} + 2p_{12} = 3 - 8\lambda + 2\lambda^2 + \vartheta,$$

and assuming that $p_{02} = 0$ and $p_{12} = 0$ for L-stability, the unique solution to this system is given by

$$p_{11} = \frac{3 - 8\lambda + 2\lambda^6 + \vartheta}{2}, \quad p_{01} = \frac{1 - 4\lambda + 2\lambda^2 - \vartheta}{2}.$$

The range of parameters (ϑ, λ) for which the $p(\eta, z)$ is A-stable, and since $p_{12} = 0$ and $p_{02} = 0$, also L-stable, is plotted in Fig. 5.13.2 by the shaded region.

The coefficients of the TSRK method corresponding to $\lambda = \frac{5}{4}$, $\vartheta = 0$, and abscissa vector $\mathbf{c} = [0, 1]^T$ are given by

$$\frac{\mathbf{u} \;\big|\; \mathbf{A} \;\big|\; \mathbf{B}}{\vartheta \;\big|\; \mathbf{v}^T \;\big|\; \mathbf{w}^T} = \begin{array}{c|cc|cc} 0 & \frac{75}{32} & -\frac{25}{32} & -\frac{25}{32} & -\frac{25}{32} \\ \hline 0 & \frac{49}{32} & \frac{5}{32} & -\frac{11}{32} & -\frac{11}{32} \\ \hline 0 & \frac{49}{32} & \frac{5}{32} & -\frac{11}{32} & -\frac{11}{32} \end{array}.$$

The stability polynomial $p(\eta, z)$ of this method is

$$p(\eta, z) = \eta^2\left(\left(1 - \frac{5}{4}z\right)^2\eta^2 - \left(1 - \frac{31}{16}z\right)\eta - \frac{7}{16}z\right).$$

For $s = 3$, stability polynomial (5.11.6) takes the form

$$p(\eta, z) = \eta^3\big((1 - \lambda z)^3\eta^2 - p_1(z)\eta + p_0(z)\big),$$

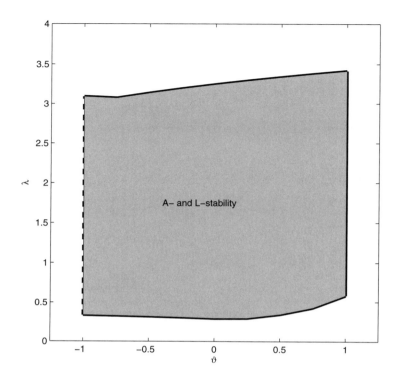

Figure 5.13.2   Regions of A- and L-stability in the (ϑ, λ)-plane for $p(\eta, z)$ with $p = s = 2$

with

$$p_1(z) = 1 - \vartheta + p_{11}z + p_{12}z^2 + p_{13}z^3, \quad p_0(z) = -\vartheta + p_{01}z + p_{02}z^2 + p_{03}z^3.$$

The system of equations corresponding to (5.12.6) with $s = 3$ takes the form

$$p_{11} - p_{01} = 1 - 3\lambda + \vartheta, \quad 2p_{11} - 2p_{02} + 2p_{12} = 3 - 12\lambda + 6\lambda^2 + \vartheta,$$

$$3p_{11} + 6p_{12} + 6p_{13} - 6p_{03} = 7 - 36\lambda + 36\lambda^2 - 6\lambda^3 + \vartheta,$$

and assuming that $p_{13} = 0$, $p_{03} = 0$ for L-stability, and in addition that $p_{02} = 0$, the unique solution to this system is given by

$$p_{11} = \frac{2(1 - 9\lambda^2 + 3\lambda^3 + \vartheta)}{3}, \quad p_{12} = \frac{5 - 36\lambda + 54\lambda^2 - 12\lambda^3 - \vartheta}{6},$$

$$p_{01} = \frac{1 - 9\lambda + 18\lambda^2 - 6\lambda^3 + \vartheta}{3}.$$

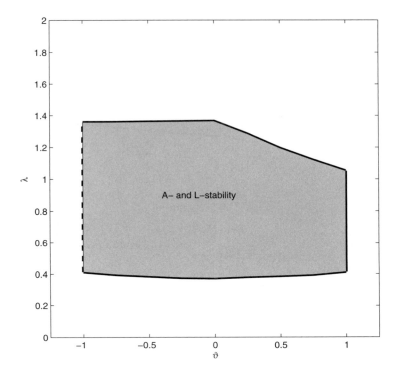

Figure 5.13.3   Regions of A- and L-stability in the (ϑ, λ)-plane for $p(\eta, z)$ with $p = s = 3$

The range of parameters (ϑ, λ) for which the $p(\eta, z)$ is A-stable, and since $p_{13} = 0$ and $p_{03} = 0$, also L-stable is shown by the shaded region in Fig. 5.13.3.

The coefficients $(\mathbf{u}, \mathbf{A}, \mathbf{B}, \vartheta, \mathbf{v}^T, \mathbf{w}^T)$ of the method corresponding to $\lambda = \frac{3}{4}$, $\vartheta = 0$, and abscissa vector $\mathbf{c} = [0, \frac{1}{2}, 1]^T$ are given by

$$
\begin{array}{c|ccc|ccc}
0 & \dfrac{3955778}{915873} & -\dfrac{573724}{492365} & \dfrac{253229}{1575340} & \dfrac{1371718}{2008359} & -\dfrac{1349029}{610487} & -\dfrac{598537}{334774} \\[2ex]
0 & \dfrac{4717083}{411104} & -\dfrac{3938351}{1455396} & \dfrac{307583}{814540} & \dfrac{1996151}{1120476} & -\dfrac{3899713}{676582} & -\dfrac{4599017}{986185} \\[2ex]
0 & \dfrac{6683188}{522061} & -\dfrac{3272705}{1193527} & \dfrac{472108}{741259} & \dfrac{2289675}{1145977} & -\dfrac{2640065}{408409} & -\dfrac{4106281}{785118} \\[2ex]
\hline
0 & \dfrac{6683188}{522061} & -\dfrac{3272705}{1193527} & \dfrac{472108}{741259} & \dfrac{2289675}{1145977} & -\dfrac{2640065}{408409} & -\dfrac{4106281}{785118}
\end{array} \;.
$$

The stability polynomial $p(\eta, z)$ of this method is

$$
p(\eta, z) = \eta^3 \left(\left(1 - \frac{3}{4} z \right)^3 \eta^2 - \left(1 - \frac{179}{96} z + \frac{53}{96} z^2 \right) \eta - \frac{59}{96} z \right).
$$

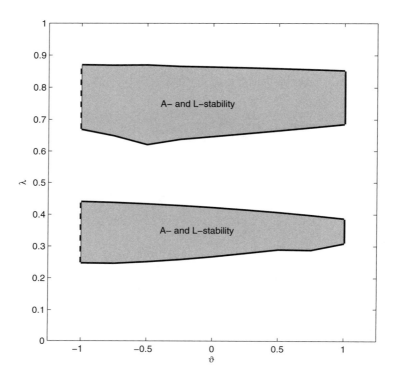

Figure 5.13.4   Regions of A- and L-stability in the (ϑ, λ)-plane for $p(\eta, z)$ with $p = s = 4$

Finally, consider the case $s = 4$. The stability polynomial (5.11.6) now takes the form

$$p(\eta, z) = \eta^4\big((1 - \lambda z)^4 \eta^2 - p_1(z)\eta + p_0(z)\big),$$

with

$$p_1(z) = 1 - \vartheta + p_{11} z + p_{12} z^2 + p_{13} z^3 + p_{14} z^4,$$

$$p_0(z) = -\vartheta + p_{01} z + p_{02} z^2 + p_{03} z^3 + p_{04} z^4.$$

The system of equations corresponding to (5.12.6) with $s = 4$ takes the form

$$p_{11} - p_{01} = 1 - 4\lambda + \vartheta, \quad 2p_{11} + 2p_{12} - 2p_{02} = 3 - 16\lambda + 12\lambda^2 + \vartheta,$$

$$3p_{11} + 6p_{12} + 6p_{13} - 6p_{03} = 7 - 48\lambda + 72\lambda^2 - 24\lambda^3 + \vartheta,$$

$$4p_{11} + 12p_{12} + 24p_{13} + 24p_{14} - 24p_{04} = 15 - 128\lambda + 288\lambda^2 - 192\lambda^3 + 24\lambda^4 + \vartheta,$$

and assuming that $p_{14} = 0$, $p_{04} = 0$ for L-stability, and in addition that $p_{13} = 0$ and $p_{03} = 0$, the unique solution to this system is given by

$$p_{11} = \frac{1 - 32\lambda + 144\lambda^2 - 144\lambda^3 + 24\lambda^4 - \vartheta}{2},$$

$$p_{12} = \frac{17 - 192\lambda + 576\lambda^2 - 480\lambda^3 + 72\lambda^4 - \vartheta}{12},$$

$$p_{01} = \frac{3 - 40\lambda + 144\lambda^2 - 144\lambda^3 + 24\lambda^4 + \vartheta}{2},$$

$$p_{02} = \frac{7 - 96\lambda + 360\lambda^2 - 384\lambda^3 + 72\lambda^4 + \vartheta}{12}.$$

The range of parameters (ϑ, λ) for which the $p(\eta, z)$ is A-stable and, since $p_{14} = 0$ and $p_{04} = 0$, also L-stable is shown by the shaded region in Fig. 5.13.4. The regions for $s = 2, 3$, and 4 were obtained by computer searches in the parameter space (ϑ, λ) using the Schur criterion [194, 253].

The coefficients of the method corresponding to $\lambda = \frac{1}{3}$, $\vartheta = 0$, and abscissa vector $\mathbf{c} = [0, \frac{1}{3}, \frac{2}{3}, 1]^T$ are given by

$$\mathbf{A} = \begin{bmatrix} \frac{1082275}{789096} & -\frac{47158}{1102905} & -\frac{20658}{230377} & \frac{16548}{733283} \\ \frac{2053468}{392523} & \frac{173881}{1660851} & -\frac{337517}{836884} & \frac{86197}{880374} \\ \frac{13765224}{1684843} & \frac{119918}{620675} & -\frac{387828}{932779} & \frac{214966}{1621163} \\ \frac{8694859}{954168} & \frac{68987}{727614} & -\frac{198815}{935168} & \frac{90358}{331129} \end{bmatrix},$$

$$\mathbf{B} = \begin{bmatrix} -\frac{73571}{418565} & \frac{316790}{450193} & -\frac{383309}{370547} & -\frac{1102057}{1459404} \\ -\frac{324116}{495273} & \frac{3108022}{1186313} & -\frac{2008351}{521461} & -\frac{1905671}{677809} \\ -\frac{813738}{787901} & \frac{4021146}{972541} & -\frac{6409321}{1054477} & -\frac{6349415}{1430988} \\ -\frac{426460}{370257} & \frac{4154204}{900915} & -\frac{12185608}{1797671} & -\frac{6621076}{1338039} \end{bmatrix},$$

$$\mathbf{v} = \begin{bmatrix} \frac{8694859}{954168} & \frac{68987}{727614} & -\frac{198815}{935168} & \frac{90358}{331129} \end{bmatrix}^T,$$

$$\mathbf{w} = \begin{bmatrix} -\frac{426460}{370257} & \frac{4154204}{900915} & -\frac{12185608}{1797671} & -\frac{6621076}{1338039} \end{bmatrix}^T.$$

The stability polynomial $p(\eta, z)$ of this method is

$$p(\eta, z) = \eta^4 \left(\left(1 - \frac{1}{3}z\right)^4 \eta^2 - p_1(z)\eta + p_0(z) \right)$$

with

$$p_1(z) = 1 - \frac{744347}{1148421}z + \frac{2965}{320219}z^2 \quad \text{and} \quad p_0(z) = -\frac{241021}{765596}z - \frac{198226}{1427227}z^2.$$

CHAPTER 6

IMPLEMENTATION OF TSRK METHODS

6.1 VARIABLE STEP SIZE FORMULATION OF TSRK METHODS

Much of this chapter follows the presentation by Bartoszewski and Jackiewicz
[17]. For the numerical integration of initial value problem (2.1.1), we consider
the class of variable step size TSRK methods defined on the nonuniform grid

$$t_0 < t_1 < \cdots < t_N, \quad t_N \geq T,$$

by the formulas:

$$
\begin{aligned}
Y_i^{[n]} &= (1 - u_i)y_{n-1} + u_i\overline{y}_{n-2} \\
&\quad + h_n \sum_{j=1}^{s} \left(a_{ij} f(Y_j^{[n]}) + b_{ij} f(\overline{Y}_j^{[n-1]}) \right), \\
y_n &= (1 - \vartheta)y_{n-1} + \vartheta\overline{y}_{n-2} \\
&\quad + h_n \sum_{j=1}^{s} \left(v_j f(Y_j^{[n]}) + w_j f(\overline{Y}_j^{[n-1]}) \right),
\end{aligned}
\tag{6.1.1}
$$

General Linear Methods for Ordinary Differential Equations. By Zdzisław Jackiewicz   **315**
Copyright © 2009 John Wiley & Sons, Inc.

$i = 1, 2, \ldots, s$, $n = 2, 3, \ldots, N$. Here $[t_0, T]$ is the interval of integration. In these formulas $h_n = t_n - t_{n-1}$, $n = 1, 2, \ldots, N$, y_n, y_{n-1}, and $\overline{y}_{n-2}$ are approximations to $y(t_n)$, $y(t_{n-1})$, and $y(t_{n-1} - h_n)$, and $Y_j^{[n]}$, $\overline{Y}_j^{[n-1]}$ are approximations to $y(t_{n-1} + c_j h_n)$, $y(t_{n-1} + (c_j - 1)h_n)$. Observe that, in general, $t_{n-1} - h_n$ does not coincide with grid point, and $t_{n-1} + (c_j - 1)h_n$, $j = 1, 2, \ldots, s$, do not coincide with points at which stage values are computed.

Similarly as in Sections 2.1 and 5.1 by introducing the standard notation

$$
Y^{[n]} = \begin{bmatrix} Y_1^{[n]} \\ \vdots \\ Y_s^{[n]} \end{bmatrix}, \quad F(Y^{[n]}) = \begin{bmatrix} f(Y_1^{[n]}) \\ \vdots \\ f(Y_s^{[n]}) \end{bmatrix},
$$

$$
\overline{Y}^{[n-1]} = \begin{bmatrix} \overline{Y}_1^{[n-1]} \\ \vdots \\ \overline{Y}_s^{[n-1]} \end{bmatrix}, \quad F(\overline{Y}^{[n-1]}) = \begin{bmatrix} f(\overline{Y}_1^{[n-1]}) \\ \vdots \\ f(\overline{Y}_s^{[n-1]}) \end{bmatrix},
$$

these methods can be written in the vector form

$$
\begin{aligned}
Y^{[n]} &= ((\mathbf{e} - \mathbf{u}) \otimes \mathbf{I})) y_{n-1} + (\mathbf{u} \otimes \mathbf{I}) \overline{y}_{n-2} \\
&\quad + h\Big((\mathbf{A} \otimes \mathbf{I}) F(Y^{[n]}) + (\mathbf{B} \otimes \mathbf{I}) F(\overline{Y}^{[n-1]})\Big), \\
y_n &= (1 - \vartheta) y_{n-1} + \vartheta \overline{y}_{n-2} \\
&\quad + h\Big((\mathbf{v} \otimes \mathbf{I}) F(Y^{[n]}) + (\mathbf{w} \otimes \mathbf{I}) F(\overline{Y}^{[n-1]})\Big),
\end{aligned}
\tag{6.1.2}
$$

$i = 1, 2, \ldots, s$, $n = 2, 3, \ldots, N$, where $\mathbf{u}$, $\mathbf{A}$, $\mathbf{B}$, $\mathbf{v}$, and $\mathbf{w}$ are defined as in (5.1.3) and $\mathbf{e} = [1, 1, \ldots, 1]^T \in \mathbb{R}^s$.

The definition of the algorithm for the numerical solution of (2.1.1) is not yet complete, and we have to specify in addition to (6.1.1) or (6.1.2) the starting procedure to advance the step from t_0 to t_1 with initial step size h_1 and the procedure for computing approximations $\overline{y}_{n-2}$ and $\overline{Y}_j^{[n-1]}$ to $y(t_{n-1} - h_n)$ and $y(t_{n-1} + (c_j - 1)h_n)$, $n = 2, 3, \ldots, N$, $j = 1, 2, \ldots, s$. These approximations are expressed in terms of approximation $\widetilde{z}(t_n, h_n)$ to the Nordsieck vector $z(t_n, h_n)$, defined by

$$
z(t, h) = \begin{bmatrix} y(t) \\ h y'(t) \\ \vdots \\ h^p y^{(p)}(t) \end{bmatrix}.
\tag{6.1.3}
$$

Observe that this is the same definition as that given in (3.11.1). The starting procedure based on continuous RK methods are described in Section 6.2. In

Section 6.3 we discuss error propagation for TSRK methods (6.1.2) which will lead to order conditions for variable step size TSRK methods and formulas for error constant and a vector of error constants of stage values. Computation of the approximation to the Nordsieck vector and to $h_n^{p+1} y^{(p+1)}(t_n)$ which is needed for local error estimation is then discussed in Section 6.4. The computation of $\overline{y}_{n-2}$ and $h_n F(\overline{Y}^{[n-1]})$ in discussed in Section 6.5. In Section 6.6 we describe the construction of TSRK methods with error constant given in advance and large regions of absolute stability. We also discuss the assessment of local error estimation for the resulting formulas. In Section 6.7 we describe construction of continuous extensions of TSRK methods. In Section 6.8 we present the results of numerical experiments with the variable step size implementation of the methods constructed in Section 6.6. In Section 6.9 we describe the local error estimation of two-step collocation methods introduced in Section 5.7. Finally, in Section 6.10 some recent work related to two-step collocation methods is described.

6.2 STARTING PROCEDURES FOR TSRK METHODS

TSRK methods of the form (6.1.1) or (6.1.2) require a starting procedure to compute $y_1 \approx y(t_1)$, $\overline{Y}_j^{[1]} \approx y(t_1 + (c_j - 1)h_2)$, $j = 1, 2, \ldots, s$, and $\overline{y}_0 \approx y(t_1 - h_2)$ using the given initial value y_0. It was observed by Hairer and Wanner [147] and reiterated recently by Verner [277, 278] that if the order p of a TSRK formula is at least greater by 2 than its stage order q, special starting values are necessary for the first step, which must be compatible with the TSRK method. This means that the terms of order up to $p - 1$ in the B-series corresponding to $\overline{Y}^{[1]}$ computed by this starting procedure must coincide with the corresponding terms in the TSRK formula (compare [147, 272]; for the B-series concept we refer the reader to the book by Hairer et al. [143]). The construction of starting procedures that satisfy these requirements was undertaken by Verner for TSRK of order $p = 6$ and stage order $q = 3$ [277] and for TSRK methods of stage order $p - 3$ [278]. Such methods were investigated by Jackiewicz and Verner [189].

It was pointed out by Hairer and Wanner [147] and Tracogna and Welfert [272] that the situation is much simpler for TSRK methods of order p and stage order $q = p$ or $q = p - 1$. In such a case it is possible to choose as a starting procedure any continuous RK method of uniform order p, or to use repeatedly a discrete RK method of order p with step sizes $h_1 + (c_j - 1)h_2$, $j = 1, 2, \ldots, s$, $h_2 \leq h_1$, to compute $\overline{Y}_j^{[1]}$, and with a step size $h_1 - h_2$ to compute $\overline{y}_0$.

In the reminder of this section we concentrate on the approach based on a continuous RK method. Following this approach, the first step from t_0 to t_1 with a step size h_1 will be computed by an explicit continuous RK method of uniform order p with abscissa vector $\tilde{\mathbf{c}} = [\tilde{c}_1, \ldots, \tilde{c}_s]^T$, coefficients $\tilde{a}_{ij}$, and continuous weights $\widetilde{b}_j(\theta)$. These methods are given by

$$Y_i = y_0 + h_1 \sum_{j=1}^{i-1} \widetilde{a}_{ij} f(Y_j), \quad i = 1, 2, \ldots, \widetilde{s},$$

$$y_h(t_0 + \theta h_1) = y_0 + h_1 \sum_{j=1}^{\widetilde{s}} \widetilde{b}_j(\theta) f(Y_j), \quad \theta \in [0, 1].$$

(6.2.1)

Here Y_i are approximations to $y(x_0 + \widetilde{c}_i h_1)$, $i = 1, 2, \ldots, \widetilde{s}$, and $y_h(t_0 + \theta h_1)$ is an approximation to $y(t_0 + \theta h_1)$, $\theta \in [0, 1]$. Such methods were investigated by Zennaro [295, 296], Owren [230], and Owren and Zennaro [231, 232, 233]. Following Shampine [257], we define initial step size h_1 by the formula

$$h_1 = \min \left\{ |T - t_0|, \mathrm{Tol}^{1/(p+1)} \left\| f(y_0) \right\|^{-1} \right\},$$

where p is the order of the method and Tol is a given accuracy tolerance.

Assuming that $h_2 \leq h_1$, we can compute $\overline{y}_0$ and $\overline{Y}_i^{[1]}$, which are needed to start TSRK method (6.1.1) from the formulas

$$\overline{y}_0 = y_h(t_0 + (1 - \delta_2)h_1),$$

$$\overline{Y}_i^{[1]} = y_h(t_0 + (1 + (c_i - 1)\delta_2)h_1), \quad i = 1, 2, \ldots, s,$$

(6.2.2)

where $\delta_2 = h_2/h_1$ and y_h is defined by (6.2.1). In particular, if $h_2 = h_1$, then $\overline{y}_0 = y_0$ and $\overline{Y}_i^{[1]} = y_h(t_0 + c_i h_1)$, $i = 1, 2, \ldots, s$.

Owren and Zennaro [232] constructed efficient continuous RK methods by minimizing the continuous coefficients of the local discretization error. The coefficients $\widetilde{\mathbf{c}}$, $\widetilde{\mathbf{A}}$, and $\widetilde{\mathbf{b}}(\theta)$ of the resulting methods of order $p = 3$, $p = 4$, and $p = 5$ with $\widetilde{s} = 4$, $\widetilde{s} = 6$, and $\widetilde{s} = 8$, respectively, are given below.

Example 1. Continuous RK method with $p = 3$ and $\widetilde{s} = 4$:

$$\frac{\widetilde{\mathbf{c}} \quad | \quad \widetilde{\mathbf{A}}}{\quad | \quad \widetilde{\mathbf{b}}^T(\theta)} =
\begin{array}{c|cccc}
0 & & & & \\
\frac{12}{23} & \frac{12}{23} & & & \\
\frac{4}{5} & -\frac{68}{375} & \frac{368}{375} & & \\
1 & \frac{31}{144} & \frac{529}{1152} & \frac{125}{384} & \\
\hline
& \widetilde{b}_1(\theta) & \widetilde{b}_2(\theta) & \widetilde{b}_3(\theta) & \widetilde{b}_4(\theta)
\end{array} ,$$

$$\widetilde{b}_1(\theta) = \tfrac{41}{72}\theta^3 - \tfrac{65}{48}\theta^2 + \theta, \quad \widetilde{b}_2(\theta) = -\tfrac{529}{576}\theta^3 + \tfrac{529}{384}\theta^2,$$

$$\widetilde{b}_3(\theta) = -\tfrac{125}{192}\theta^3 + \tfrac{125}{128}\theta^2, \quad \widetilde{b}_4(\theta) = \theta^3 - \theta^2.$$

Example 2. Continuous RK method with $p = 4$ and $\widetilde{s} = 6$:

$$
\frac{\widetilde{\mathbf{c}} \;\Big|\; \widetilde{\mathbf{A}}}{\Big|\; \widetilde{\mathbf{b}}^T(\theta)}
=
\begin{array}{c|cccccc}
0 & & & & & & \\
\frac{1}{6} & \frac{1}{6} & & & & & \\
\frac{11}{37} & \frac{44}{1369} & \frac{363}{1369} & & & & \\
\frac{11}{17} & \frac{3388}{4913} & -\frac{8349}{4913} & \frac{8140}{4913} & & & \\
\frac{13}{15} & -\frac{36764}{408375} & \frac{767}{1125} & -\frac{32708}{136125} & \frac{210392}{408375} & & \\
1 & \frac{1697}{18876} & 0 & \frac{50653}{116160} & \frac{299693}{1626240} & \frac{3375}{11648} & \\
\hline
& \widetilde{b}_1(\theta) & \widetilde{b}_2(\theta) & \widetilde{b}_3(\theta) & \widetilde{b}_4(\theta) & \widetilde{b}_5(\theta) & \widetilde{b}_6(\theta)
\end{array}
\quad,
$$

$$
\widetilde{b}_1(\theta) = -\frac{866577}{824252}\theta^4 + \frac{1806901}{618189}\theta^3 - \frac{104217}{37466}\theta^2 + \theta,
$$

$$
\widetilde{b}_2(\theta = 0,
$$

$$
\widetilde{b}_3(\theta) = \frac{12308679}{5072320}\theta^4 - \frac{2178079}{380424}\theta^3 + \frac{861101}{230560}\theta^2,
$$

$$
\widetilde{b}_4(\theta) = -\frac{7816583}{10144640}\theta^4 + \frac{6244423}{5325936}\theta^3 - \frac{63869}{293440}\theta^2,
$$

$$
\widetilde{b}_5(\theta) = -\frac{624375}{217984}\theta^4 + \frac{982125}{190736}\theta^3 - \frac{1522125}{762944}\theta^2,
$$

$$
\widetilde{b}_6(\theta) = \frac{296}{131}\theta^4 - \frac{461}{131}\theta^3 + \frac{165}{131}\theta^2.
$$

Example 3. Continuous RK method $(\widetilde{\mathbf{c}}, \widetilde{\mathbf{A}}, \widetilde{\mathbf{b}}^T(\theta))$ with $p = 5$ and $\widetilde{s} = 8$:

$$
\begin{array}{c|cccccccc}
0 & & & & & & & & \\
\frac{1}{6} & \frac{1}{6} & & & & & & & \\
\frac{1}{4} & \frac{1}{16} & \frac{3}{16} & & & & & & \\
\frac{1}{2} & \frac{1}{4} & -\frac{3}{4} & 1 & & & & & \\
\frac{1}{2} & -\frac{3}{4} & \frac{15}{4} & -3 & \frac{1}{2} & & & & \\
\frac{9}{14} & \frac{369}{1372} & -\frac{243}{343} & \frac{297}{343} & \frac{1485}{9604} & \frac{297}{4802} & & & \\
\frac{7}{8} & -\frac{133}{4512} & \frac{1113}{6016} & \frac{7945}{16544} & -\frac{12845}{24064} & -\frac{315}{24064} & \frac{156065}{198528} & & \\
1 & \frac{83}{945} & 0 & \frac{248}{825} & \frac{41}{180} & \frac{1}{36} & \frac{2401}{38610} & \frac{6016}{20475} & \\
\hline
& \widetilde{b}_1(\theta) & \widetilde{b}_2(\theta) & \widetilde{b}_3(\theta) & \widetilde{b}_4(\theta) & \widetilde{b}_5(\theta) & \widetilde{b}_6(\theta) & \widetilde{b}_7(\theta) & \widetilde{b}_8(\theta)
\end{array}
\quad,
$$

$$\widetilde{b}_1(\theta) = \tfrac{596}{315}\theta^5 - \tfrac{4969}{819}\theta^4 + \tfrac{17893}{2457}\theta^3 - \tfrac{3292}{819}\theta^2 + \theta,$$

$$\widetilde{b}_2(\theta) = 0,$$

$$\widetilde{b}_3(\theta) = -\tfrac{1984}{275}\theta^5 + \tfrac{1344}{65}\theta^4 - \tfrac{43568}{2145}\theta^3 + \tfrac{5112}{715}\theta^2,$$

$$\widetilde{b}_4(\theta) = \tfrac{118}{15}\theta^5 - \tfrac{1465}{78}\theta^4 + \tfrac{3161}{234}\theta^3 - \tfrac{123}{52}\theta^2,$$

$$\widetilde{b}_5(\theta) = 2\theta^5 - \tfrac{413}{78}\theta^4 + \tfrac{1061}{234}\theta^3 - \tfrac{63}{52}\theta^2,$$

$$\widetilde{b}_6(\theta) = -\tfrac{9604}{6435}\theta^5 + \tfrac{2401}{1521}\theta^4 + \tfrac{60025}{50193}\theta^3 - \tfrac{40817}{33462}\theta^2,$$

$$\widetilde{b}_7(\theta) = -\tfrac{48128}{6825}\theta^5 + \tfrac{96256}{5915}\theta^4 - \tfrac{637696}{53235}\theta^3 + \tfrac{18048}{5915}\theta^2,$$

$$\widetilde{b}_8(\theta) = 4\theta^5 - \tfrac{109}{13}\theta^4 + \tfrac{75}{13}\theta^3 - \tfrac{18}{13}\theta^2.$$

Owren and Zennaro [232] also constructed embedded discrete RK methods of order $p-1$ which can be used for error estimation and control over the first step. These methods take the form

$$\widehat{y}_1 = y_0 + h_1 \sum_{j=1}^{\widetilde{s}} \widehat{b}_j f(Y_j), \tag{6.2.3}$$

where Y_j, $j = 1, 2, \ldots, \widetilde{s}$, are given by (6.2.1) and the coefficient vector $\widehat{\mathbf{b}} = [\widehat{b}_1, \ldots, \widehat{b}_{\widetilde{s}}]^T$ is given by

$$\widehat{\mathbf{b}} = \begin{bmatrix} \tfrac{1}{24} & \tfrac{23}{24} & 0 & 0 \end{bmatrix}^T$$

for the method with $p = 3$ and $\widetilde{s} = 4$ in Example 1,

$$\widehat{\mathbf{b}} = \begin{bmatrix} \tfrac{101}{363} & 0 & -\tfrac{1369}{14520} & \tfrac{11849}{14520} & 0 & 0 \end{bmatrix}^T$$

for the method with $p = 4$ and $\widetilde{s} = 6$ in Example 2, and

$$\widehat{\mathbf{b}} = \begin{bmatrix} -\tfrac{1}{9} & 0 & \tfrac{40}{33} & -\tfrac{7}{4} & -\tfrac{1}{12} & \tfrac{343}{198} & 0 & 0 \end{bmatrix}^T$$

for the method with $p = 5$ and $\widetilde{s} = 8$ in Example 3.

Alternatively, error estimate after the first step $\mathrm{est}(t_1)$ can be computed by the Richardson extrapolation

$$\mathrm{est}(t_1) = \frac{2^p \big(y_h(t_1) - y_h^*(t_1)\big)}{2^p - 1}, \tag{6.2.4}$$

where $y_h^*(t_1)$ is an approximation to $y(t_1)$ computed by a continuous RK method (6.2.1) over two steps of size $h_1/2$. This estimate is usually very reliable, and although its cost is quite high, it is used only on the first step of the integration, so its contribution to the overall cost of the algorithm is not significant.

Local error estimation for subsequent steps taken with TSRK methods (6.1.1) is discussed in the next section.

6.3 ERROR PROPAGATION, ORDER CONDITIONS, AND ERROR CONSTANT

In this section we consider only TSRK methods of order p and stage order $q = p$. We demonstrate that for these variable step size methods (6.1.1) or (6.1.2), the stage order and order conditions take the form

$$C_\nu = 0, \quad \nu = 1, 2, \ldots, p, \tag{6.3.1}$$

and

$$\widehat{C}_\nu = 0, \quad \nu = 1, 2, \ldots, p, \tag{6.3.2}$$

where C_ν and $\widehat{C}_\nu$ are defined by (5.1.3) and (5.1.4), respectively. To justify (6.3.1) and (6.3.2), we investigate error propagation of TSRK methods defined by (6.1.1) or (6.1.2) on the interval $[t_{n-1}, t_n]$. Similarly as in the work by Butcher and Jackiewicz [69, 70, 71], we determine the error constant $E = E_p$ and the vector of error constants $\xi = [\xi_1, \ldots, \xi_s]^T$ of stage values such that the localizing assumptions

$$y_{n-1} = y(t_{n-1}),$$

$$\overline{y}_{n-2} = y(t_{n-1} - h_n) - (-1)^{p+1} E h_n^{p+1} y^{(p+1)}(t_{n-1}) + O(h_n^{p+2}), \tag{6.3.3}$$

$$\overline{Y}^{[n-1]} = y(t_{n-1} + (\mathbf{c} - \mathbf{e})h_n) + O(h_n^{p+1})$$

imply that

$$y_n = y(t_{n-1} + h_n) - E h_n^{p+1} y^{(p+1)}(t_{n-1}) + O(h_n^{p+2}),$$

$$Y^{[n]} = y(t_{n-1} + \mathbf{c} h_n) - (\xi \otimes \mathbf{I}) h_n^{p+1} y^{(p+1)}(t_{n-1}) + O(h_n^{p+2}). \tag{6.3.4}$$

We recall that in the formulas above $y(t + \mathbf{c}h)$ stands for

$$y(t + \mathbf{c}h) := \begin{bmatrix} y(t + c_1 h) \\ \vdots \\ y(t + c_s h) \end{bmatrix}.$$

This analysis will also lead to stage order and order conditions. We have the following theorem.

Theorem 6.3.1 (Bartoszewski and Jackiewicz [17]) *The method defined by (6.1.2) or (6.1.2) has order p and stage order $q = p$ if and only if (6.3.1) and (6.3.2) are satisfied. Moreover, the error constant $E = E_p$ and vector of error constants of stage values ξ are given by*

$$E = E_p = \frac{1}{(p+1)!} - \frac{\mathbf{v}^T \mathbf{c}^p + \mathbf{w}^T(\mathbf{c}-\mathbf{e})^p}{p!\left(1 - (-1)^{p+1}\vartheta\right)} \tag{6.3.5}$$

and

$$\xi = \frac{\mathbf{c}^{p+1}}{(p+1)!} - (-1)^{p+1}\left(\frac{1}{(p+1)!} - E\right)\mathbf{u} - \frac{\mathbf{A}\mathbf{c}^p}{p!} - \frac{\mathbf{B}(\mathbf{c}-\mathbf{e})^p}{p!}. \tag{6.3.6}$$

Proof: It follows from (6.3.3) and (6.3.4) that

$$h_n f(\overline{Y}^{[n-1]}) = h_n y'\big(t_{n-1} + (\mathbf{c}-\mathbf{e})h_n\big) + O(h_n^{p+2})$$

and

$$h_n f(Y^{[n]}) = h_n y'(t_{n-1} + \mathbf{c}h_n) + O(h_n^{p+2}).$$

Substituting (6.3.3), (6.3.4), and the relations above into (6.1.1), we obtain

$$
\begin{aligned}
y(t_{n-1} + \mathbf{c}h_n) &- (\xi \otimes \mathbf{I})h_n^{p+1} y^{(p+1)}(t_{n-1}) \\
&= (\mathbf{u} \otimes \mathbf{I})\Big(y(t_{n-1} - h_n) - (-1)^{p+1} E h_n^{p+1} y^{(p+1)}(t_{n-1})\Big) \\
&\quad + \big((\mathbf{e}-\mathbf{u}) \otimes \mathbf{I}\big) y(t_{n-1}) + h_n(\mathbf{A} \otimes \mathbf{I})y'(t_{n-1} + \mathbf{c}h_n) \\
&\quad + h_n(\mathbf{B} \otimes \mathbf{I})y'\big(t_{n-1} + (\mathbf{c}-\mathbf{e})h_n\big) + O(h_n^{p+2})
\end{aligned}
$$

and

$$
\begin{aligned}
y(t_{n-1} + h_n) &- E h_n^{p+1} y^{(p+1)}(t_{n-1}) \\
&= \vartheta\Big(y(t_{n-1} - h_n) - (-1)^{p+1} E h_n^{p+1} y^{(p+1)}(t_{n-1})\Big) + (1-\vartheta)y(t_{n-1}) \\
&\quad + h_n(\mathbf{v}^T \otimes \mathbf{I})y'(t_{n-1} + \mathbf{c}h_n) + h_n(\mathbf{w}^T \otimes \mathbf{I})y'\big(t_{n-1} + (\mathbf{c}-\mathbf{e})h_n\big) \\
&\quad + O(h_n^{p+2}).
\end{aligned}
$$

Expanding $y(t_{n-1} + \mathbf{c}h_n)$, $y(t_{n-1} - h_n)$, $y(t_{n-1} + h_n)$, $y'(t_{n-1} + (\mathbf{c} - \mathbf{e})h_n)$, and $y'(t_{n-1} + \mathbf{c}h_n)$ into a Taylor series around t_{n-1} leads to

$$\sum_{j=1}^{p+1} \left(\frac{\mathbf{c}^j}{j!} \otimes \mathbf{I} \right) h_n^j y^{(j)}(t_{n-1}) - (\xi \otimes \mathbf{I}) h_n^{p+1} y^{(p+1)}(t_{n-1})$$

$$= (\mathbf{u} \otimes \mathbf{I}) \left(\sum_{j=1}^{p+1} \frac{(-1)^j}{j!} h_n^j y^{(j)}(t_{n-1}) - (-1)^{p+1} E h_n^{p+1} y^{(p+1)}(t_{n-1}) \right)$$

$$+ (\mathbf{A} \otimes \mathbf{I}) \sum_{j=1}^{p+1} \left(\frac{\mathbf{c}^{j-1}}{(j-1)!} \otimes \mathbf{I} \right) h_n^j y^{(j)}(t_{n-1})$$

$$+ (\mathbf{B} \otimes \mathbf{I}) \sum_{j=1}^{p+1} \left(\frac{(\mathbf{c} - \mathbf{e})^{j-1}}{(j-1)!} \otimes \mathbf{I} \right) h_n^j y^{(j)}(t_{n-1}) + O(h_n^{p+2})$$

and

$$\sum_{j=1}^{p+1} \frac{1}{j!} h_n^j y^{(j)}(t_{n-1}) - E h_n^{p+1} y^{(p+1)}(t_{n-1})$$

$$= \vartheta \left(\sum_{j=1}^{p+1} \frac{(-1)^j}{j!} h_n^j y^{(j)}(t_{n-1}) - (-1)^{p+1} E h_n^{p+1} y^{(p+1)}(t_{n-1}) \right)$$

$$+ (\mathbf{v}^T \otimes \mathbf{I}) \sum_{j=1}^{p+1} \left(\frac{\mathbf{c}^{j-1}}{(j-1)!} \otimes \mathbf{I} \right) h_n^j y^{(j)}(t_{n-1})$$

$$+ (\mathbf{w}^T \otimes \mathbf{I}) \sum_{j=1}^{p+1} \left(\frac{(\mathbf{c} - \mathbf{e})^{j-1}}{(j-1)!} \otimes \mathbf{I} \right) h_n^j y^{(j)}(t_{n-1}) + O(h_n^{p+2}).$$

Comparing terms of order j, $j = 1, 2, \ldots, p$, in the relations above we obtain stage order and order conditions (6.3.1) and (6.3.2). Comparing terms of order $p + 1$, we obtain (6.3.5) and (6.3.6). This completes the proof.   ∎

It follows from (6.3.3) and (6.3.4) that the local discretization error $\mathrm{le}(t_n)$ of method (6.1.1) of order p and stage order $q = p$ at the point t_n has the form

$$\mathrm{le}(t_n) := y(t_n) - y_n = E h_n^{p+1} y^{(p+1)}(t_n) + O(h_n^{p+2}).$$

The computable estimate of the principal part of this error is be derived in the next section. For methods with $c_1 = 0$ and $c_s = 1$, high stage order has the additional advantage of reducing the number of evaluations of the function f since we can approximate $f(Y_1^{[n]})$ by $f(Y_s^{[n-1]})$ computed in the preceding step. This is similar to the idea of FSAL (first same as last) introduced by Dormand and Prince [117] in the context of RK methods (see also [116, 257, 260]). In the context of DIMSIMs this property was referred to as FASAL (first approximately same as last) by VanWieren [273, 274, 275]. However, such implementation may affect the properties of the resulting formulas, and the effects of this modification on stability properties of the overall numerical algorithm require further study.

6.4   COMPUTATION OF APPROXIMATIONS TO THE NORDSIECK VECTOR AND LOCAL ERROR ESTIMATION

To implement the method defined by (6.1.1) or (6.1.2) in a variable step size environment, we need to compute approximations

$$h_{n+1}F(\overline{Y}^{[n]}) \approx h_{n+1}y'\big(t_n + (\mathbf{c}-\mathbf{e})h_{n+1}\big) \quad \text{and} \quad \overline{y}_{n-1} \approx y(t_n - h_{n+1})$$

after the step from t_{n-1} to t_n is completed. Here h_{n+1} is a new step size from t_n to t_{n+1} chosen according to some step size changing strategy. These approximations are expressed in terms of approximation $\widetilde{z}(t_n, h_n)$ to the Nordsieck vector $z(t_n, h_n)$ defined by (6.1.3), which is derived in this section.

We look for the approximation $\widetilde{z}(t_n, h_n)$ to $z(t_n, h_n)$ of the form

$$\widetilde{z}(t_n, h_n) = (\alpha \otimes \mathbf{I})y_{n-1} + (\beta \otimes \mathbf{I})y_n + h_n(\Gamma \otimes \mathbf{I})F(Y^{[n]}), \qquad (6.4.1)$$

where

$$\alpha = \begin{bmatrix} \alpha_0 & \alpha_1 & \cdots & \alpha_p \end{bmatrix}^T,$$

$$\beta = \begin{bmatrix} \beta_0 & \beta_1 & \cdots & \beta_p \end{bmatrix}^T,$$

$$\Gamma = \begin{bmatrix} \gamma_{01} & \gamma_{02} & \cdots & \gamma_{0s} \\ \gamma_{11} & \gamma_{12} & \cdots & \gamma_{1s} \\ \vdots & \vdots & \ddots & \vdots \\ \gamma_{p1} & \gamma_{p2} & \cdots & \gamma_{ps} \end{bmatrix}.$$

The representation (6.4.1) is more convenient in a variable step size environment than the representation considered by Tracogna [271], which depends on stage values at two consecutive steps.

The coefficients α, β, and Γ are computed by requiring that

$$\widetilde{z}(t_n, h_n) = z(t_n, h_n) + O(h_n^{p+2}),$$

where it is assumed that y_{n-1}, y_n, and $Y^{[n]}$ satisfy (6.3.3) and (6.3.4). This leads to the following theorem.

Theorem 6.4.1 (Bartoszewski and Jackiewicz [17]) *Assume that function* y *is sufficiently smooth and that* y_n, y_{n+1}, *and* $Y^{[n]}$ *satisfy (6.3.3) and (6.3.4). Then*

$$z(t_n, h_n) = (\alpha \otimes \mathbf{I})y_{n-1} + (\beta \otimes \mathbf{I})y_n + h_n(\Gamma \otimes \mathbf{I})F(Y^{[n]}) + O(h_n^{p+2}), \quad (6.4.2)$$

where α, β, and Γ satisfy the system of equations

$$\alpha\,\mathbf{t}^T + \beta\,\mathbf{e}_1^T + \Gamma\,\tilde{\mathbf{C}} = \mathbf{I}_{p+1},$$

$$\frac{(-1)^{p+1}\alpha}{(p+1)!} - E\beta + \frac{\Gamma(\mathbf{c}-\mathbf{e})^p}{p!} = 0. \tag{6.4.3}$$

Here $\mathbf{e}_1 = [1, 0, \dots, 0]^T \in \mathbb{R}^{p+1}$, $\mathbf{I}_{p+1}$ is the identity matrix of dimension $p+1$, and

$$\mathbf{t} = \left[\begin{array}{ccccc} 1 & -1 & \dfrac{1}{2!} & \cdots & \dfrac{(-1)^p}{p!} \end{array}\right]^T,$$

$$\tilde{\mathbf{C}} = \left[\begin{array}{ccccc} \mathbf{0} & \mathbf{e} & \mathbf{c}-\mathbf{e} & \cdots & \dfrac{(\mathbf{c}-\mathbf{e})^{p-1}}{(p-1)!} \end{array}\right].$$

Proof: Since y_{n-1}, y_n, and $Y^{[n]}$ satisfy (6.3.3) and (6.3.4), relation (6.4.2) is equivalent to

$$
\begin{aligned}
z(t_n, h_n) &= (\alpha \otimes \mathbf{I})y(t_n - h_n) \\
&\quad + (\beta \otimes \mathbf{I})\Big(y(t_n) - Eh_n^{p+1}y^{(p+1)}(t_n)\Big) \\
&\quad + h_n(\Gamma \otimes \mathbf{I})y'\big(t_n + (\mathbf{c}-\mathbf{e})h_n\big) + O(h_n^{p+2}).
\end{aligned}
$$

Expanding $y(t_n - h_n)$ and $y'\big(t_n + (\mathbf{c}-\mathbf{e})h_n\big)$ into a Taylor series around t_n, we obtain

$$
\begin{aligned}
z(t_n, h_n) &= (\alpha \otimes \mathbf{I})\sum_{j=0}^{p+1} \frac{(-1)^j}{j!}h_n^j y^{(j)}(t_n) \\
&\quad + (\beta \otimes \mathbf{I})\Big(y(t_n) - Eh_n^{p+1}y^{(p+1)}(t_n)\Big) \\
&\quad + (\Gamma \otimes \mathbf{I})\sum_{j=1}^{p+1}\left(\frac{(\mathbf{c}-\mathbf{e})^{j-1}}{(j-1)!} \otimes I_m\right)h_n^j y^{(j)}(t_n) + O(h_n^{p+2}).
\end{aligned}
$$

This relation is equivalent to

$$
\begin{aligned}
z(t_n, h_n) &= \left(\big(\alpha\,\mathbf{t}^T + \beta\,\mathbf{e}_1^T + \Gamma\,\tilde{\mathbf{C}}\big) \otimes I_m\right)z(t_n, h_n) \\
&\quad + \left(\left(\frac{(-1)^{p+1}}{(p+1)!}\alpha - E\beta + \frac{\Gamma(\mathbf{c}-\mathbf{e})^p}{p!}\right) \otimes I\right)h_n^{p+1}y^{(p+1)}(t_n) \\
&\quad + O(h_n^{p+2})
\end{aligned}
$$

and implies (6.4.3). ∎

The quantity $h_n^{p+1} y^{(p+1)}(t_n)$ can be estimated in a similar way. We look for an estimate of the form

$$
\begin{aligned}
h_n^{p+1} y^{(p+1)}(t_n) &= \alpha_{p+1}\, y_{n-1} + \beta_{p+1}\, y_n \\
&\quad + (\gamma_{p+1} \otimes \mathbf{I})\, h_n F(Y^{[n]}) + O(h_n^{p+2}),
\end{aligned}
\tag{6.4.4}
$$

where $\alpha_{p+1}, \beta_{p+1} \in \mathbb{R}$ and

$$
\gamma_{p+1} = \begin{bmatrix} \gamma_{p+1,1} & \gamma_{p+1,2} & \cdots & \gamma_{p+1,s} \end{bmatrix} \in \mathbb{R}^s.
$$

We have the following theorem.

Theorem 6.4.2 (Bartoszewski and Jackiewicz [17]) *Assume that function y is sufficiently smooth and that y_{n-1}, y_n, and $Y^{[n]}$ satisfy (6.3.3) and (6.3.4). Then (6.4.4) holds where α_{p+1}, β_{p+1}, and γ_{p+1} satisfy the system of equations*

$$
\begin{aligned}
& \alpha_{p+1}\, \mathbf{t}^T + \beta_{p+1}\, \mathbf{e}_1^T + \gamma_{p+1}\, \widetilde{\mathbf{C}} = 0, \\
& \frac{(-1)^{p+1}}{(p+1)!}\, \alpha_{p+1} - E\, \beta_{p+1} + \gamma_{p+1}\, \frac{(\mathbf{c} - \mathbf{e})^p}{p!} = 1,
\end{aligned}
\tag{6.4.5}
$$

where $\mathbf{t}$, $\mathbf{e}_1$, and $\widetilde{\mathbf{C}}$ are defined as in Theorem 6.4.1.

Proof: It follows from (6.3.3) and (6.3.4) that (6.4.4) is equivalent to

$$
\begin{aligned}
h_n^{p+1} y^{(p+1)}(t_n) &= \alpha_{p+1}\, y(t_n - h_n) + \beta_{p+1} \left(y(t_n) - E\, h_n^{p+1} y^{(p+1)}(t_n) \right) \\
&\quad + (\gamma_{p+1} \otimes \mathbf{I})\, h_n y'\big(t_n + (\mathbf{c} - \mathbf{e}) h_n\big) + O(h_n^{p+2}).
\end{aligned}
$$

Expanding $y(t_n - h_n)$ and $y'\big(t_n + (\mathbf{c} - \mathbf{e}) h_n\big)$ into a Taylor series around the point t_n, we get

$$
\begin{aligned}
h_n^{p+1} y^{(p+1)}(t_n) &= \alpha_{p+1} \sum_{j=0}^{p+1} \frac{(-1)^j}{j!} h_n^j y^{(j)}(t_n) \\
&\quad + \beta_{p+1} \left(y(t_n) - E\, h_n^{p+1} y^{(p+1)}(t_n) \right) \\
&\quad + \gamma_{p+1} \sum_{j=1}^{p+1} \frac{(\mathbf{c} - \mathbf{e})^{j-1}}{(j-1)!} h_n^j y^{(j)}(t_n) + O(h_n^{p+2}).
\end{aligned}
$$

The last relation can be written in the form

$$
\begin{aligned}
h_n^{p+1} y^{(p+1)}(t_n) &= \left((\alpha_{p+1}\, \mathbf{t}^T + \beta_{p+1}\, \mathbf{e}_1^T + \gamma_{p+1}\, \widetilde{\mathbf{C}}) \otimes I_m \right) z(t_n, h_n) \\
&\quad + \left(\frac{(-1)^{p+1}}{(p+1)!}\, \alpha_{p+1} - E\, \beta_{p+1} + \gamma_{p+1}\, \frac{(\mathbf{c} - \mathbf{e})^p}{p!} \right) h_n^{p+1} y^{(p+1)}(t_n) \\
&\quad + O(h_n^{p+2}),
\end{aligned}
$$

and comparing the corresponding terms, we obtain (6.4.5).   ∎

A different approach to the computation of approximations to $z(t_n, h_n)$ and $h_n^{p+1} y^{(p+1)}(t_n)$ is presented by Tracogna [271], where the quantities $F(Y^{[n]})$ and $F(\overline{Y}^{[n-1]})$ are utilized.

It follows from (6.4.4) that the local discretization error of TSRK method defined by (6.1.1) or (6.1.2) can be estimated by the formula

$$\text{est}(t_n) = E\,\eta(t_n, h_n),$$

where E is the error constant given by (6.3.5) and $\eta(t_n, h_n)$ is defined by

$$\eta(t_n, h_n) = \alpha_{p+1}\, y_{n-1} + \beta_{p+1}\, y_n + (\gamma_{p+1} \otimes I)\, h_n F(Y^{[n]}). \qquad (6.4.6)$$

6.5   COMPUTATION OF APPROXIMATIONS TO THE SOLUTION AND STAGE VALUES BETWEEN GRID POINTS

In this section we describe the efficient computation of the quantities $\overline{y}_{n-1}$ and $h_{n+1} F(\overline{Y}^{[n]})$, which are needed to advance the step from t_n to t_{n+1} with the method (6.1.1) or (6.1.2) with a new step size h_{n+1}. To derive the formula for $\overline{y}_{n-1}$, observe that

$$\overline{y}_{n-1} = y(t_n - h_{n+1}) - E(h_n - h_{n+1})^{p+1} y^{(p+1)}(t_n) + O\big((h_n - h_{n+1})^{p+2}\big). \quad (6.5.1)$$

Note also that if $h_{n+1} = h_n$, then $\overline{y}_{n-1} = y(t_{n-1}) = y_{n-1}$, which is consistent with (6.3.3). On the other hand, if $h_{n+1} = 0$, then $\overline{y}_{n-1}$ approximates the quantity $y(t_n) - E\, h_n^{p+1} y^{(p+1)}(t_n)$ up to terms of order $O(h_n^{p+2})$.

Set $\delta_{n+1} = h_{n+1}/h_n$ and as in Section 4.1, denote by $\mathbf{D}(\delta)$ the rescaling matrix defined by

$$\mathbf{D}(\delta) = \text{diag}\big(1, \delta, \ldots, \delta^p\big).$$

In actual implementation of TSRK methods, the ratios δ will be restricted by the requirement

$$\delta_{\min} \leq \delta \leq \delta_{\max}, \qquad (6.5.2)$$

where $\delta_{\min}$ and $\delta_{\max}$ are the minimum and maximum ratios permitted in a code.

Expanding $y(t_n - h_{n+1})$ in (6.5.1) into a Taylor series around t_n, we obtain

$$\begin{aligned} \overline{y}_{n-1} = {} & \Big(\mathbf{t}^T \mathbf{D}(\delta_{n+1}) \otimes \mathbf{I}\Big) z(t_n, h_n) \\ & + \left(\frac{(-1)^{p+1}}{(p+1)!} \delta_{n+1}^{p+1} - E\big(1 - \delta_{n+1}\big)^{p+1}\right) h_n^{p+1} y^{(p+1)}(t_n) \\ & + O\big((h_n - h_{n+1})^{p+2}\big), \end{aligned}$$

where the vector $\mathbf{t}$ is as defined in Theorem 6.4.1. This suggests the following formula for $\overline{y}_{n-1}$:

$$
\begin{aligned}
\overline{y}_{n-1} &= \left(\mathbf{t}^T \mathbf{D}(\delta_{n+1}) \otimes \mathbf{I}\right)\widetilde{z}(t_n, h_n) \\
&+ \left(\frac{(-1)^{p+1}}{(p+1)!}\delta_{n+1}^{p+1} - E\left(1 - \delta_{n+1}\right)^{p+1}\right)\eta(t_n, h_n),
\end{aligned}
\tag{6.5.3}
$$

where $\widetilde{z}(t_n, h_n)$ is defined by (6.4.1) and $\eta(t_n, h_n)$ is defined by (6.4.6).

To arrive at the formula for $h_{n+1}F(\overline{Y}^{[n]})$, we start with the relation

$$
h_{n+1}f(\overline{Y}^{[n]}) = h_{n+1}y'\left(t_n + (\mathbf{c} - \mathbf{e})h_{n+1}\right) + O(h_{n+1}^{p+2}).
$$

Expanding $y'\left(t_n + (\mathbf{c} - \mathbf{e})h_{n+1}\right)$ into a Taylor series around t_n, we obtain

$$
\begin{aligned}
h_{n+1}f(\overline{Y}^{[n]}) &= \left(\widetilde{\mathbf{C}}\mathbf{D}(\delta_{n+1}) \otimes \mathbf{I}\right)z(t_n, h_n) \\
&+ \left(\frac{(\mathbf{c} - \mathbf{e})^p}{p!} \otimes \mathbf{I}\right)\delta_{n+1}^{p+1}h_n^{p+1}y^{(p+1)}(t_n) + O(h_{n+1}^{p+2}),
\end{aligned}
$$

where the matrix $\widetilde{\mathbf{C}}$ is defined as in Theorem 6.4.1. Similarly as before, this suggests the following formula for $h_{n+1}F(\overline{Y}^{[n]})$:

$$
\begin{aligned}
h_{n+1}f(\overline{Y}^{[n]}) &= \left(\widetilde{\mathbf{C}}\mathbf{D}(\delta_{n+1}) \otimes \mathbf{I}\right)\widetilde{z}(t_n, h_n) \\
&+ \left(\frac{(\mathbf{c} - \mathbf{e})^p}{p!} \otimes \mathbf{I}\right)\delta_{n+1}^{p+1}\eta(t_n, h_n).
\end{aligned}
\tag{6.5.4}
$$

Observe that (6.5.3) and (6.5.4) can be computed without any extra evaluations of the right-hand side of equation (2.1.1).

6.6 CONSTRUCTION OF TSRK METHODS WITH A GIVEN ERROR CONSTANT AND ASSESSMENT OF LOCAL ERROR ESTIMATION

In this section we describe the construction of TSRK formulas given by (6.1.1) or (6.1.2) with error constants E given in advance and with large regions of absolute stability. We illustrate our approach in the case of methods with $p = q = s = 3$. Assuming that $\vartheta = 0$ and $\mathbf{c} = [0, 1/2, 1]^T$ and solving stage order and order conditions (6.3.1) and (6.3.2) with respect to $\mathbf{B}$, $\mathbf{w}$, and v_3, we obtain an eight-parameter family of methods depending on u_1, u_2, u_3, a_{21}, a_{31}, a_{32}, v_1, and v_2. It follows from Jackiewicz and Tracogna [183] that the local discretization error of these methods has the form

$$
\mathrm{le}(t_n) = E\,h_n^4 y^{(4)}(t_{n-1})
$$

$$
+ \left(F\,y^{(5)}(t_{n-1}) + (\mathbf{v}+\mathbf{w})^T \mathbf{G} \frac{\partial f}{\partial y}\big(y(t_{n-1})\big) y^{(4)}(t_{n-1}) \right) h_n^5 \qquad (6.6.1)
$$

$$
+ O(h_n^6),
$$

where $E = E_4$ and $F = E_5$ are defined by (6.3.5) with $p = 3$ and $p = 4$, respectively, and $\mathbf{G}$ is given by

$$
\mathbf{G} = \frac{\mathbf{c}^4}{24} - \frac{\mathbf{u}}{24} - \frac{A\mathbf{c}^3}{6} - \frac{B(\mathbf{c}-\mathbf{e})^3}{6}. \qquad (6.6.2)
$$

Striving for reliable estimation of the local discretization error, we then enforce the condition

$$
(\mathbf{v}+\mathbf{w})^T \mathbf{G} = 0, \qquad (6.6.3)
$$

which is solved with respect to v_2. Although this is desirable, it is not necessary and there are implementations of Adams methods in PECE mode [143] and general linear methods [67] where this condition is not met. This leads to methods with $\mathbf{B}$, v_2, v_3, and $\mathbf{w}$ given by

$$
b_{11} = \frac{u_1}{6}, \qquad b_{12} = \frac{2u_1}{3}, \qquad b_{13} = \frac{u_1}{6},
$$

$$
b_{21} = \frac{5+4u_2}{24}, \qquad b_{22} = \frac{2(u_2-1)}{3}, \qquad b_{23} = \frac{23 - 24a_{21} + 4u_2}{24},
$$

$$
b_{31} = \frac{7 - 6a_{32} + u_3}{6}, \qquad b_{32} = \frac{9a_{32} - 10 + 2u_3}{3}, \qquad b_{33} = \frac{19 - 6a_{31} - 18a_{32} + u_3}{6},
$$

$$
v_2 = \frac{36a_{32} - 45 - 8(31 - 30a_{32})E + 8(4 - 3a_{32})v_1}{42a_{32} - 47},
$$

$$
v_3 = \frac{60a_{32} - 53 + 144(13 - 12a_{32})E - 24(4 - 3a_{32})v_1}{12(42a_{32} - 47)},
$$

$$
w_1 = \frac{41 - 24a_{32} - 48(55 - 48a_{32})E - 24(4 - 3a_{32})v_1}{12(42a_{32} - 47)},
$$

$$
w_2 = \frac{24a_{32} - 41 + 216(7 - 6a_{32})E + 24(4 - 3a_{32})v_1}{3(42a_{32} - 47)},
$$

$$
w_3 = \frac{38 - 15a_{32} - 9(64 - 47a_{32})E - 3(1 + 6a_{32})v_1}{3(42a_{32} - 47)},
$$

and local discretization error (6.6.1) now takes the form

$$
\mathrm{le}(t_n) = E\,h_n^4 y^{(4)}(t_{n-1}) + F\,h_n^5 y^{(5)}(t_{n-1}) + O(h_n^6), \qquad (6.6.4)
$$

E	$\frac{1}{12}$	$\frac{1}{24}$	$\frac{1}{48}$	$\frac{1}{120}$
u_1	0.147239	−0.363883	−1.353015	0.0736696
u_2	−0.0128864	−0.228023	−0.128392	−0.0204487
u_3	0.0896426	0.224976	−0.565685	0.544967
a_{21}	0.825400	0.921151	1.511248	0.985434
a_{31}	1.571173	1.602293	0.990546	1.766083
a_{32}	0.475788	0.564620	0.882220	0.467017
v_1	1.759708	1.139034	0.694921	1.489838

Table 6.6.1   Coefficients of TSRK formulas with $p = q = s = 3$

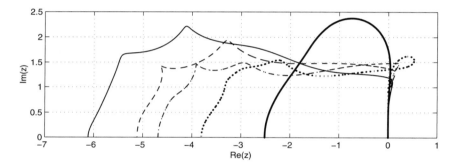

Figure 6.6.1   Stability regions of RK method of order 3 (thick solid line) and TSRK methods with $E = 1/12$ (solid line), $E = 1/24$ (dashed line), $E = 1/48$ (dashed-dotted line), and $E = 1/120$ (dotted line)

where $F = (120\,E - 1)/480$. The remaining free parameters u_1, u_2, u_3, a_{21}, a_{31}, a_{32}, and v_1 are then computed trying to maximize the area of the intersection of the region of absolute stability with the negative half-plane using the procedure described in Sections 5.5.1 and 5.6.1. For $E = 1/12$, $1/24$, $1/48$, and $1/120$, this leads to methods whose coefficients are listed in Table 6.6.1. Coefficients of these methods up to about sixteen significant digits are given by Bartoszewski and Jackiewicz [17]. The regions of absolute stability of the resulting methods are plotted in Fig. 6.6.1. We can see that there is an apparent trade-off between accuracy and stability — the region of absolute stability is becoming smaller as the accuracy of the method increases or the error constant E decreases.

To better assess the quality of local error estimation, we also investigate terms of order 6 in (6.6.4). It follows from Jackiewicz and Tracogna [183] that the principal part of $O(h_n^6)$ is a linear combination of elementary differentials

$$y^{(6)}(t_{n-1}), \qquad \frac{\partial f}{\partial y}\big(y(t_{n-1})\big)y^{(5)}(t_{n-1}),$$

$$\left(\frac{\partial f}{\partial y}\right)^2\big(y(t_{n-1})\big)y^{(4)}(t_{n-1}), \quad \left(\frac{\partial f}{\partial y}\right)'\big(y(t_{n-1})\big)y^{(4)}(t_{n-1}),$$

with weights $P_1 = E_5$,

$$P_2 = (\mathbf{v}+\mathbf{w})^T\mathbf{Q}-\mathbf{v}^T\mathbf{G}, \quad P_3 = (\mathbf{v}+\mathbf{w})^T(\mathbf{A}+\mathbf{B})\mathbf{G}, \quad P_4 = (\mathbf{v}+\mathbf{w})^T\Gamma_\mathbf{c}\mathbf{G}-\mathbf{v}^T\mathbf{G},$$

where E_5 is defined by (6.3.5) with $p = 5$, $\mathbf{G}$ is defined by (6.6.2), $\mathbf{Q}$ is defined by

$$\mathbf{Q} = \frac{\mathbf{c}^5}{120} + \frac{\mathbf{u}}{120} - \frac{\mathbf{A}\mathbf{c}^4}{24} - \frac{\mathbf{B}(\mathbf{c}-\mathbf{e})^4}{24}, \tag{6.6.5}$$

and $\Gamma_\mathbf{c} = \mathrm{diag}(c_1, c_2, c_3)$. We then define

$$T = |P_1| + |P_2| + |P_3| + |P_4|$$

and compute the ratios F/E and T/E, which in addition to the size of the error constant E, should give some indication of the quality of error estimation. These ratios are given in Table 6.6.2 for the methods in Table 6.6.1.

E	$\frac{1}{12}$	$\frac{1}{24}$	$\frac{1}{48}$	$\frac{1}{120}$
F/E	0.23	0.20	0.15	0
T/E	0.91	1.65	3.83	12.19

Table 6.6.2 Ratios F/E and T/E for the TSRK formulas in Table 6.6.1

We can see that as the error constant E gets smaller (or the TSRK method becomes more accurate) the ratio F/E also decreases slowly which should have a positive effect on the quality of error estimation. However, higher order terms also play a role, and the ratio T/E shows an opposite trend (they increase rather fast). The overall effect of smaller E, smaller F/E, and much larger T/E is a deterioration of the quality of the local error estimation for methods with smaller error constants. This is confirmed by the numerical experiments presented in Section 6.8.

6.7 CONTINUOUS EXTENSIONS OF TSRK METHODS

Continuous extensions of TSRK methods have been investigated by Bartoszewski and Jackiewicz [18], Jackiewicz and Tracogna [184], and Tracogna

[271]. We consider variable step size continuous TSRK methods of the form

$$Y_i^{[n]} = (1 - u_i)\, y_{n-1} + u_i\, \overline{y}_{n-2}$$

$$+ h_n \sum_{j=1}^{s} \left(a_{ij} f(Y_j^{[n]}) + b_{ij} f(\overline{Y}_j^{[n-1]}) \right),$$

$$y_h(t_{n-1} + \theta h_n) = (1 - \vartheta)\, y_{n-1} + \vartheta\, \overline{y}_{n-2}$$

$$+ h_n \sum_{j=1}^{s} \left(v_j(\theta) f(Y_j^{[n]}) + w_j(\theta) f(\overline{Y}_j^{[n-1]}) \right),$$

(6.7.1)

$i = 1, 2, \ldots, s$, $n = 2, 3, \ldots, N$, $\theta \in [0, 1]$, where $\overline{y}_{n-2}$ and $\overline{Y}_j^{[n-1]}$ have the same meaning as in Section 6.1. In (6.7.1), $y_h(t_{n-1} + \theta h_n)$ is an approximation to the solution $y(t_{n-1} + \theta h_n)$ defined on the subinterval $[t_{n-1}, t_n]$, and

$$\mathbf{v}(\theta) = \left[\begin{array}{ccc} v_1(\theta) & \cdots & v_s(\theta) \end{array} \right]^T, \quad \mathbf{w}(\theta) = \left[\begin{array}{ccc} w_1(\theta) & \cdots & w_s(\theta) \end{array} \right]^T$$

are continuous weights. It will always be assumed that $\mathbf{v}(0) = \mathbf{0}$ and $\mathbf{w}(0) = \mathbf{0}$ to assure the continuity of the interpolant $y_h(t_{n-1} + \theta h_n)$ and that $\mathbf{v}(1) = \mathbf{v}$ and $\mathbf{w}(1) = \mathbf{w}$, where $\mathbf{v}$ and $\mathbf{w}$ are coefficients of the underlying discrete TSRK method corresponding to $\theta = 1$, so that $y_h(t_{n-1} + h_n) = y_n$.
    Set

$$\widehat{C}_\nu(\theta) = \frac{\theta^\nu}{\nu!} - \frac{(-1)^\nu}{\nu!} \vartheta - \frac{\mathbf{v}(\theta)^T \mathbf{c}^{\nu-1}}{(\nu-1)!} - \frac{\mathbf{w}(\theta)(\mathbf{c} - \mathbf{e})^{\nu-1}}{(\nu-1)!}, \quad (6.7.2)$$

$\nu = 1, 2, \ldots$. To derive continuous order conditions for (6.7.1) we follow an approach similar to that adopted in Section 6.3. We determine continuous error constant $E(\theta) = E_p(\theta)$ and vector of error constants ξ of stage values so that the localizing assumptions (6.3.3) with $E = E(1)$, imposed on y_{n-1}, $\overline{y}_{n-2}$ and $\overline{Y}^{[n-1]}$, imply that

$$y_h(t_{n-1} + \theta h_n) = y(t_{n-1} + \theta h_n)$$

$$- E(\theta) h_n^{p+1} y^{(p+1)}(t_{n-1}) + O(h_n^{p+2}), \quad (6.7.3)$$

$$Y^{[n]} = y(t_{n-1} + \mathbf{c} h_n) - (\xi \otimes \mathbf{I}) h_n^{p+1} y^{(p+1)}(t_{n-1}) + O(h_n^{p+2}).$$

We have the following theorem.

Theorem 6.7.1 *Continuous TSRK method (6.7.1) has uniform order p and stage order $q = p$ if (6.3.1) is satisfied and*

$$\widehat{C}_\nu(\theta) = 0, \quad \nu = 1, 2, \ldots, p, \quad (6.7.4)$$

where $\widehat{C}_\nu(\theta)$ *is defined by (6.7.2). Moreover, the continuous error constant* $E(\theta) = E_p(\theta)$ *is given by*

$$E(\theta) = E_p(\theta) = \frac{\theta^{p+1}}{(p+1)!} - \frac{\mathbf{v}^T(\theta)\mathbf{c}^p + \mathbf{w}^T(\theta)(\mathbf{c} - \mathbf{e})^p}{p!\big(1 - (-1)^{p+1}\vartheta\big)}, \qquad (6.7.5)$$

and the vector of error constants ξ *of stage values is given by (6.3.6) with* $E = E(1)$.

Proof: The proof of this theorem follows along the lines of the proof of Theorem 6.3.1 and is therefore omitted. ∎

Somewhat different representation and proof of continuous order conditions is given by Tracogna [271] (compare also [184]).

In what follows we apply Theorem 6.7.1 to the construction of continuous extensions of TSRK methods with $p = q = s = 3$ with a given error constant, which were considered in Section 6.6. We look for methods (6.7.1) with continuous weights of the form

$$\mathbf{v}(\theta) = \begin{bmatrix} v_{11}\theta + v_{12}\theta^2 + v_{13}\theta^3 \\ v_{21}\theta + v_{22}\theta^2 + v_{23}\theta^3 \\ v_{31}\theta + v_{32}\theta^2 + v_{33}\theta^3 \end{bmatrix}, \quad \mathbf{w}(\theta) = \begin{bmatrix} w_{11}\theta + w_{12}\theta^2 + w_{13}\theta^3 \\ w_{21}\theta + w_{22}\theta^2 + w_{23}\theta^3 \\ w_{31}\theta + w_{32}\theta^2 + w_{33}\theta^3 \end{bmatrix},$$

for which $\mathbf{v}(0) = \mathbf{0}$ and $\mathbf{w}(0) = \mathbf{0}$. Solving the continuous order conditions (6.7.4) corresponding to $p = 3$ with respect to w_{ij}, $i,j = 1,2,3$, and then the conditions $\mathbf{v}(1) = \mathbf{v}$, $\mathbf{w}(1) = \mathbf{w}$ with respect to $v_{i,1}$, $i = 1,2,3$, we obtain a six-parameter family of continuous TSRK methods (6.7.1) depending on v_{ij}, $i = 1,2,3$, $j = 1,2$. We then determine the parameters v_{12} and v_{13} to enforce the continuous analog of the condition (6.6.3), that is,

$$\big(\mathbf{v}(\theta) + \mathbf{w}(\theta)\big)^T \mathbf{G} = 0, \qquad (6.7.6)$$

where $\mathbf{G}$ is defined by (6.6.2). As a result, the continuous local discretization error $\mathrm{le}(t_{n-1} + \theta h_n)$ takes the form

$$\mathrm{le}(t_{n-1} + \theta h_n) = E(\theta)\, h_n^4 y^{(4)}(t_{n-1}) + F(\theta)\, h_n^5 y^{(5)}(t_{n-1}) + O(h_n^6), \quad (6.7.7)$$

$\theta \in [0,1]$, where $E(\theta) = E_4(\theta)$ and $F(\theta) = E_5(\theta)$, are defined by (6.7.5) for $p = 3$ and $p = 4$. The remaining parameters v_{22}, v_{23}, v_{32}, and v_{33} are then determined so that the continuous error constant $E_4(\theta)$ is monotonic for $\theta \in [0,1]$, or deviate from the θ-axis as little as possible, finally reaching error constant E for $\theta = 1$. Coefficients $\mathbf{v}(\theta)$ and $\mathbf{w}(\theta)$ of continuous extensions of TSRK methods with coefficients listed in Table 6.6.1 and error constants $E = \frac{1}{12}, \frac{1}{24}, \frac{1}{48}$, and $\frac{1}{120}$ are listed below.

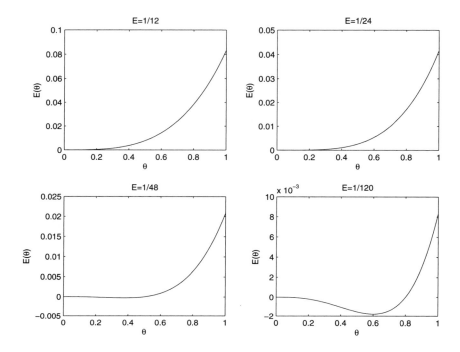

Figure 6.7.1 Continuous error constants $E(\theta)$ corresponding to TSRK methods with coefficients listed in Table 6.6.1 with error constants $E = \frac{1}{12}, \frac{1}{24}, \frac{1}{48}$, and $\frac{1}{120}$

Continuous extension of TSRK method with $E = \frac{1}{12}$ and $E(\theta) = \frac{\theta^4}{24} + \frac{\theta^2}{24}$:

$$
\mathbf{v}(\theta) = \begin{bmatrix} 1.73881\,\theta + 1.06271\,\theta^2 - 1.0418\,\theta^3 \\ -0.562812\,\theta + 0.666667\,\theta^3 \\ 0.140703\,\theta \end{bmatrix},
$$

$$
\mathbf{w}(\theta) = \begin{bmatrix} 0.140703\,\theta + 0.5\,\theta^2 \\ -0.562812\,\theta - 2\,\theta^2 + 0.666667\,\theta^3 \\ 0.105412\,\theta + 0.437295\,\theta^2 - 0.29153\,\theta^3 \end{bmatrix}.
$$

Continuous extension of TSRK method with $E = \frac{1}{24}$ and $E(\theta) = \frac{\theta^4}{24}$:

$$\mathbf{v}(\theta) = \begin{bmatrix} 1.80976\,\theta + 0.337391\,\theta^2 - 1.00811\,\theta^3 \\ -0.641559\,\theta + 0.333333\,\theta^2 + 0.666667\,\theta^3 \\ 0.16039\,\theta \end{bmatrix},$$

$$\mathbf{w}(\theta) = \begin{bmatrix} 0.16039\,\theta + 0.166667\,\theta^2 \\ -0.641559\,\theta - \theta^2 + 0.666667\,\theta^3 \\ 0.152581\,\theta + 0.162609\,\theta^2 - 0.325219\,\theta^3 \end{bmatrix}.$$

Continuous extension of TSRK method with $E = \frac{1}{48}$ and $E(\theta) = \frac{\theta^4}{24} - \frac{\theta^3}{48}$:

$$\mathbf{v}(\theta) = \begin{bmatrix} 1.47407\,\theta + 0.222908\,\theta^2 - 1.00206\,\theta^3 \\ -0.516009\,\theta + 0.333333\,\theta^2 + 0.833333\,\theta^3 \\ 0.129002\,\theta \end{bmatrix},$$

$$\mathbf{w}(\theta) = \begin{bmatrix} 0.129002\,\theta + 0.166667\,\theta^2 - 0.166667\,\theta^3 \\ -0.516009\,\theta - \theta^2 + 1.66667\,\theta^3 \\ 0.299944\,\theta + 0.277092\,\theta^2 - 0.831277\,\theta^3 \end{bmatrix}.$$

Continuous extension of TSRK method with $E = \frac{1}{120}$ and $E(\theta) = \frac{\theta^4}{24} - \frac{\theta^3}{30}$:

$$\mathbf{v}(\theta) = \begin{bmatrix} 2.74802\,\theta + 0.355711\,\theta^2 - 1.61389\,\theta^3 \\ -1.13714\,\theta + 0.333333\,\theta^2 + 0.933333\,\theta^3 \\ 0.331785\,\theta \end{bmatrix},$$

$$\mathbf{w}(\theta) = \begin{bmatrix} 0.331785\,\theta + 0.166667\,\theta^2 - 0.26667\,\theta^3 \\ -1.32714\,\theta - \theta^2 + 1.146667\,\theta^3 \\ 0.242691\,\theta + 0.144289\,\theta^2 - 0.519441\,\theta^3 \end{bmatrix}.$$

The conditions given in Theorem 6.7.1 are sufficient but not necessary for uniform convergence of order p of continuous TSRK methods (6.7.1). It has been demonstrated [18, 184], that these conditions can be relaxed to

$$C_\nu = 0, \quad \nu = 1, 2, \ldots, p - 1, \tag{6.7.8}$$

and

$$\widehat{C}_\nu(\theta) = 0, \quad \nu = 1, 2, \ldots, p-1, \quad \widehat{C}_p(1) = 0. \tag{6.7.9}$$

We have the following theorem.

Theorem 6.7.2 *Assume that* $y_0 = O(h^p)$, $y_1 = O(h^p)$, $\overline{y}_0 = O(h^p)$, *and* $\overline{Y}_j^{[1]} = O(h^p)$, $j = 1, 2, \ldots, s$. *Assume also that conditions (6.7.8) and (6.7.9) are satisfied. Then TSRK method (6.7.1) is uniformly convergent with order* p *(i.e.,* $\|y_h - y\|_{[t_0,T]} = O(h^p)$ *as* $h = \max\{h_n\} \to 0$*). Here* $\|y\|_{[t_0,T]} = \sup\{\|y(s)\| : s \in [t_0, T]\}$.

Proof: This theorem follows from Jackiewicz and Tracogna [184, Theorem 3] (compare also [18]). The discrete version of this theorem follows from Theorem 2.4.3. ∎

Assuming in addition to (6.7.8) that $C_p = 0$ leads to methods for which the local discretization error takes the convenient form

$$\mathrm{le}(t_n) = E h_n^{p+1} y^{(p+1)}(t_{n-1}) + O(h_n^{p+2}),$$

with $E = E_p$; that is, the principal part of the error depends on $y^{(p+1)}(t_{n-1})$ but is independent of $\frac{\partial f}{\partial y}\big(y(t_{n-1})\big) y^{(p)}(t_{n-1})$. This is the case for many methods discussed in Sections 5.4, 5.5, 5.6, and 6.6.

The construction of continuous extensions of TSRK methods that satisfy conditions (6.7.8) and (6.7.9) and such that $C_p = 0$ is described by Bartoszewski and Jackiewicz [18] for TSRK formulas with $p = q = s = 3$ considered in Section 6.6.

6.8 NUMERICAL EXPERIMENTS

The methods constructed in Section 6.6 were implemented in a variable step size environment with the standard step size changing strategy described for DIMSIMs in Section 4.5 (compare also [143]). This strategy is based on the estimate $\mathrm{est}(t_n) = E\eta(t_n, h_n)$ of the local discretization error, with error constant E defined by (6.3.5) and $\eta(t_n, h_n)$ defined by (6.4.6). This estimate was made to satisfy

$$\big|\mathrm{est}_i(t_n)\big| \leq \mathrm{Rtol} \cdot \max\big\{|y_{n-1,i}|, |y_{n,i}|\big\} + \mathrm{Atol}, \tag{6.8.1}$$

where Rtol and Atol are given relative and absolute error tolerances and $y_{n-1,i}$ and $y_{n,i}$ stand for the ith component of the vectors y_{n-2} and y_{n-1} and $\mathrm{est}_i(t_n)$ stands for the ith component of $\mathrm{est}(t_n)$. In our numerical experiments we have used Rtol = Atol = Tol. The required starting values $\overline{y}_0$ and $\overline{Y}_i^{[1]}$, $i = 1, 2, 3, 4$, were computed by formula (6.2.2) using continuous RK method (6.2.1) presented in Example 1 in Section 6.2.

It is known (compare [156, 255, 256]) that the maximum global error resulting from the step size selection based on (6.8.1) is proportional to $\mathrm{Tol}^{p/(p+1)}$,

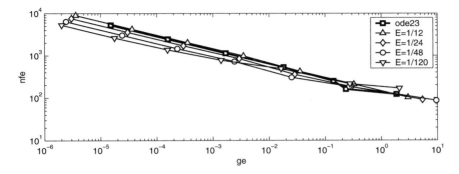

Figure 6.8.1 Number of functions evaluations nfe versus global error ge at the end of the interval of integration for BUBBLE (1.2.1)

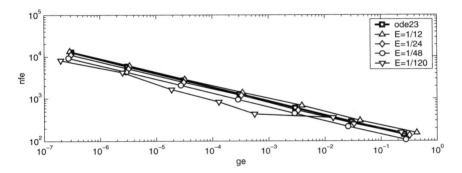

Figure 6.8.2 Number of functions evaluations nfe versus global error ge at the end of the interval of integration for EULR (1.2.5)

where p is the order of a numerical scheme. To obtain proportionality between Tol and global error, we follow Shampine [256] and inside the code use a smaller tolerance Tol$'$ defined by

$$\text{Tol}' = C \cdot \text{Tol}^{(p+1)/p}, \tag{6.8.2}$$

where C is a constant that is problem dependent. This constant was chosen so that global errors obtained by `ode23` are comparable to the global errors obtained by codes based on TSRK methods defined by (6.1.1) or (6.1.2), and we have found that a choice $C = 7$ works quite well for the test problems we experimented with. These are BUBBLE defined by (1.2.2), EULR defined by (1.2.5), and ROPE defined by (1.2.7) or (1.2.8). The selection of numerical results is displayed in Figs. 6.8.1, 6.8.2, and 6.8.3 (see also Tables 6.8.1, 6.8.2, and 6.8.3). We can observe that this choice of a constant C leads to similar

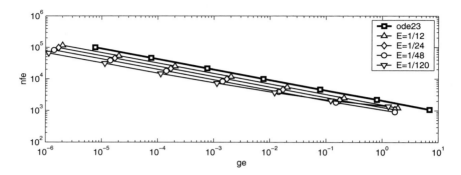

Figure 6.8.3 Number of functions evaluations nfe versus global error ge at the end of the interval of integration for ROPE defined by (1.2.7) or (1.2.8)

accuracy for the BUBBLE and EULR problems and to results that are more accurate by about half to one decimal digit than the results obtained by **ode23** for the ROPE problem. As observed in Section 1.2, the BUBBLE problem places a great demand on the precision and step size control and leads to a considerable range in step sizes.

code	$E = 1/12$	$E = 1/24$	$E = 1/48$	$E = 1/120$	ode23
ns	201	169	142	231	82
nr	1	1	1	7	1
nfe	421	357	303	490	250
ge	$3.6 \cdot 10^{-2}$	$2.9 \cdot 10^{-2}$	$2.5 \cdot 10^{-2}$	$1.6 \cdot 10^{-2}$	$1.4 \cdot 10^{-1}$

Table 6.8.1 Cost statistics and global error for the BUBBLE problem (1.2.1) for Tol $= 10^{-4}$

To illustrate the potential of the new TSRK formulas, we compared the variable step size implementation of these methods with state-of-the-art **ode23** code from the Matlab ODE suite [263]. This code is based on an embedded pair of RK formulas of order 3 and 2 constructed by Bogacki and Shampine [22] and uses local extrapolation. In Figs. 6.8.1, 6.8.2, and 6.8.3, we have plotted the number of function calls versus global error at the end of the interval of integration for the BUBBLE, EULR, and ROPE problems corresponding to Tol $= 10^{-k}$, $k = 2, 3, \ldots, 8$, with the tolerance Tol$'$ used inside the code defined by (6.8.2). Observe that all these codes exhibit quite good tolerance proportionality for small enough tolerances. The more precise cost and accuracy statistics are given in Tables 6.8.1, 6.8.2, and 6.8.3 for Tol $= 10^{-4}$,

code	$E = 1/12$	$E = 1/24$	$E = 1/48$	$E = 1/120$	ode23
ns	327	254	214	208	199
nr	4	0	0	0	0
nfe	676	522	442	430	598
ge	$3.9 \cdot 10^{-3}$	$3.4 \cdot 10^{-3}$	$2.9 \cdot 10^{-3}$	$5.6 \cdot 10^{-4}$	$3.0 \cdot 10^{-3}$

Table 6.8.2 Cost statistics and global error for the EULR problem (1.2.5) for $\text{Tol} = 10^{-4}$

where ns is the number of steps, nr is the number of rejected steps, nfe is the number of functions calls, and ge is the global error.

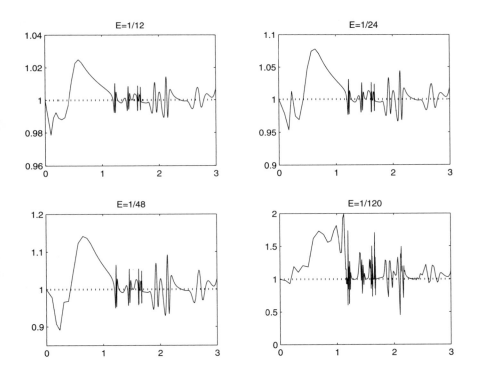

Figure 6.8.4 Ratios between $\|\text{le}(t_n)\|$ and $\|\text{est}(t_n)\|$ versus t for methods listed in Table 6.6.1 applied to the PLEI problem (1.2.6) with $\text{Tol}' = 10^{-4}$ Note the different scales on the vertical axes

code	$E = 1/12$	$E = 1/24$	$E = 1/48$	$E = 1/120$	ode23
ns	2718	2289	1927	1819	1534
nr	6	5	5	12	11
nfe	5475	4612	3888	3686	4636
ge	$2.0 \cdot 10^{-2}$	$1.7 \cdot 10^{-2}$	$1.4 \cdot 10^{-2}$	$1.2 \cdot 10^{-2}$	$7.8 \cdot 10^{-2}$

Table 6.8.3 Cost statistics and global error for the ROPE problem (1.2.7) or (1.2.8) for Tol $= 10^{-4}$

We can observe that the codes based on TSRK formulas with $E = 1/24$, $E = 1/48$, and $E = 1/120$ are more efficient (except for BUBBLE) and in most cases also more accurate than the ode23. We can also see that compared with other formulas, the quality of the method with the smallest error constant $(E = 1/120)$ deteriorates for large tolerances. In general, ode23 is taking fewer steps, but its formula involves more function evaluations per step than the new TSRK formulas (except for $E = 1/12$). We expect even bigger gains in accuracy and efficiency for TSRK methods of higher order, as indicated by our preliminary experiments with TSRK methods of order $p = 5$. However, the construction of such methods leads to very large and difficult optimization problems and requires very time consuming and sophisticated computer searches. This work will be described elsewhere.

To test numerically the quality of error estimation proposed here we have applied these methods to various test problems discussed in [143] and in Section 1.2. We have observed that there is a trade-off between the accuracy of the methods and the quality of error estimation and that this quality tends to be higher for TSRK methods with larger error constants E. This is illustrated in Fig. 6.8.4, where we have plotted the ratio between the norm of the local discretization error $\|\mathrm{le}(t_n)\|$ and the norm of error estimate $\|\mathrm{est}(t_n)\|$ versus t for the methods in Table 6.6.1 with error constants $E = 1/12$, $1/24$, $1/48$, and $1/120$, applied to PLEI problem (1.2.6) with $\mathrm{Tol}' = 10^{-4}$. Note that there are different scales on the vertical axes. In practice, we would recommend using methods in the middle of the range of error constant E considered in Table 6.6.1: for example, formulas with $E = 1/24$ or $E = 1/48$.

6.9 LOCAL ERROR ESTIMATION OF TWO-STEP COLLOCATION METHODS

It was demonstrated in Section 5.7 that the local discretization error $\mathrm{le}(t_n)$ of two-step collocation method (5.7.1) takes the form

$$\mathrm{le}(t_n) = h^{p+1} C_p(1) y^{(p+1)}(t_{n-1}) + O(h^{p+2}), \tag{6.9.1}$$

which corresponds to (5.7.5) for $\theta = 1$. Here $y(t)$ is the solution to (2.1.1), and $C_p(1)$ is the error constant of the method of order p. This error constant is defined by (5.7.6) corresponding to $\theta = 1$. Observe that we are using a different notation for the error constant than in Section 6.3, where the error constant was denoted by E. We also consider a different variant of local error $\widetilde{\text{le}}(t_n)$ defined by

$$\widetilde{\text{le}}(t_n) = C_p(1)h^{p+1}\widetilde{y}^{(p+1)}(t_{n-1}) + O(h^{p+2}), \tag{6.9.2}$$

where $\widetilde{y}(t)$ is the local solution, that is, the solution to the initial value problem

$$\begin{aligned}
\widetilde{y}'(t) &= f\big(\widetilde{y}(t)\big), \quad t \in [t_{n-1}, t_n], \\
\widetilde{y}(t_{n-1}) &= y_{n-1}.
\end{aligned} \tag{6.9.3}$$

We make the standard assumption that the function $f(y)$ appearing in (2.1.1) and (6.9.3) satisfies a Lipschitz condition of the form

$$\big\|f(y) - f(z)\big\| \le L\|y - z\|,$$

with a constant $L \ge 0$. Subtracting the integral forms of (2.1.1) and (6.9.3), we obtain

$$\big\|y(t) - \widetilde{y}(t)\big\| \le \big\|y(t_{n-1}) - y_{n-1}\big\| + L\int_{t_{n-1}}^{t} \big\|y(s) - \widetilde{y}(s)\big\|ds,$$

$t \in [t_{n-1}, t_n]$. Using Gronwall's lemma 1.4.1 (compare also [137, 257]) yields

$$\big\|y(t) - \widetilde{y}(t)\big\| \le \big\|y(t_{n-1}) - y_{n-1}\big\|e^{L(t - t_{n-1})}.$$

Hence,

$$\big\|y(t) - \widetilde{y}(t)\big\| = O(h^p), \quad t \in [t_{n-1}, t_n].$$

Assuming that the function $f(y)$ is sufficiently smooth, we have a similar relation for the derivatives of $y(t)$ and $\widetilde{y}(t)$:

$$\big\|y^{(i)}(t) - \widetilde{y}^{(i)}(t)\big\| = O(h^p), \quad t \in [t_{n-1}, t_n] \quad i = 1, 2, \ldots,$$

(compare [201, 262]). Therefore, we can conclude that the principal parts (i.e., terms of order $p + 1$) of the local discretization error $\text{le}(t_n)$ (6.9.1) and the local error $\widetilde{\text{le}}(t_n)$ (6.9.2) are the same.

In the remainder of this section we look for estimates of $h^{p+1}\widetilde{y}^{(p+1)}(t_{n-1})$ of the form

$$\begin{aligned}
h^{p+1}\widetilde{y}^{(p+1)}(t_{n-1}) &= \alpha_1 y_{n-1} + \alpha_0 y_{n-2} \\
&+ h\sum_{j=1}^{s}\Big(\beta_j f\big(P(t_{n-1} + c_j h)\big) + \gamma_j f\big(P(t_{n-2} + c_j h)\big)\Big),
\end{aligned} \tag{6.9.4}$$

where α_1, α_0, β_j, γ_j, $j = 1, 2, \ldots, s$, are some constants. We have the following theorem.

Theorem 6.9.1 (D'Ambrosio et al. [104]) *Assume that the solution $\widetilde{y}(t)$ to (6.9.3) is sufficiently smooth. Then the constants α_1, α_0, β_j, and γ_j, $j = 1, 2, \ldots, s$, appearing in (6.9.4) satisfy the system of equations*

$$\alpha_1 + \alpha_0 = 0,$$

$$\frac{(-1)^k}{k!}\alpha_0 + \sum_{j=1}^{s}\left(\beta_j \frac{c_j^{k-1}}{(k-1)!} + \gamma_j \frac{(c_j-1)^{k-1}}{(k-1)!}\right) = 0,$$

$$k = 1, 2, \ldots, p,$$

$$\left(\frac{(-1)^{p+1}}{(p+1)!} - C_p(-1)\right)\alpha_0 + \sum_{j=1}^{s}\left(\beta_j \frac{c_j^p}{p!} + \gamma_j \frac{(c_j-1)^p}{p!}\right) = 1.$$

(6.9.5)

Proof: Since method (5.7.1) is of order p, it is locally of order $p+1$ and we have

$$y_{n-2} = \widetilde{y}(t_{n-2}) - C_p(-1)h^{p+1}\widetilde{y}^{(p+1)}(t_{n-1}) + O(h^{p+2}).$$

We also have

$$P(t_{n-1} + \theta h) = \widetilde{y}(t_{n-1} + \theta h) + O(h^{p+1}), \quad \theta \in [-1, 1].$$

Substituting these relations and $y_{n-1} = \widetilde{y}(t_{n-1})$ into (6.9.4), we obtain

$$h^{p+1}\widetilde{y}(t_{n-1}) = \alpha_1\widetilde{y}(t_{n-1}) + \alpha_0\left(\widetilde{y}(t_{n-1} - h) - C_p(-1)h^{p+1}\widetilde{y}^{(p+1)}(t_{n-1})\right)$$

$$+ h\sum_{j=1}^{s}\left(\beta_j\widetilde{y}'(t_{n-1} + c_j h) + \gamma_j\widetilde{y}'(t_{n-1} + (c_j - 1)h)\right).$$

Expanding $\widetilde{y}(t_{n-1} - h)$, $\widetilde{y}'(t_{n-1} + c_j h)$, and $\widetilde{y}'(t_{n-1} + (c_j - 1)h)$ into a Taylor series around the point t_{n-1} and comparing the terms of order $O(h^k)$ for $k = 0, 1, \ldots, p+1$ leads to system (6.9.5). ∎

Observe that (6.9.5) constitutes a system of $p+2$ equations with respect to $2s+2$ unknown coefficients α_1, α_0, α_2, β_j, and γ_j, $j = 1, 2, \ldots, s$. We have the following theorem.

Theorem 6.9.2 (D'Ambrosio et al. [104]) *Assume that $c_i \neq c_j$ and $c_i \neq c_j - 1$ for $i \neq j$. Then system (6.9.5) corresponding to $p = s + r$, where $r = 1, 2, \ldots, s$, has a family of solutions depending on $s - r$ free parameters which may be chosen as, for example, $\beta_{r+1}, \beta_{r+2}, \ldots, \beta_s$ or $\gamma_{r+1}, \gamma_{r+2}, \ldots, \gamma_s$. In particular, if $r = s$, the solution to the system (6.9.5) is unique. This system does not have solutions if $r = s + 1$.*

Proof: The proof is similar to that of Theorem 5.7.2 and is therefore omitted. The interested reader can find complete details in D'Ambrosio's thesis [103]. ∎

Choices of free parameters other than those indicated in Theorem 6.9.2 are also possible. For example, if $r = s - 1 \geq 1$, there is one free parameter, which may be chosen as α_1; if $r = s - 2 \geq 1$, there are two free parameters, which may be chosen as α_1 and α_0; if $r = s - 3 \geq 1$, there are three free parameters, which may be chosen as α_1, α_0, and β_1 or γ_1; and if $r = s - k \geq 1$, $k > 3$, there are k free parameters, which may be chosen as α_1, α_0, and β_j or γ_j, $j = 1, 2, \ldots, k - 2$.

For method (5.7.1) with $s = 1$, $p = 2s = 2$, and the polynomials $\varphi_1(\theta)$, $\varphi_0(\theta)$, $\psi(\theta)$, and $\chi(\theta)$ defined by (5.9.4), the constants α_1, α_0, $\beta = \beta_1$, and $\gamma = \gamma_1$ appearing in the estimator of $h^3 \widetilde{y}^{(3)}(t_{n-1})$ are unique. They correspond to the solution of system (6.9.5) with $s = 1$ and $p = 2$ and are given by

$$\alpha_1 = \frac{288}{95}, \quad \alpha_0 = -\frac{288}{95}, \quad \beta = \frac{72}{95}, \quad \gamma = -\frac{72}{19}.$$

Similarly, for the method with coefficients $\varphi_1(\theta)$, $\varphi_0(\theta)$, $\psi(\theta)$, and $\chi(\theta)$ given by (5.9.5), the constants α_1, α_0, $\beta = \beta_1$, and $\gamma = \gamma_1$ assume the form

$$\alpha_1 = \frac{12}{5}, \quad \alpha_0 = -\frac{12}{5}, \quad \beta = \frac{6}{5}, \quad \gamma = -\frac{18}{5},$$

and for the method with coefficients $\varphi_1(\theta)$, $\varphi_0(\theta)$, $\psi(\theta)$, and $\chi(\theta)$ given by (5.9.6), the constants α_1, α_0, $\beta = \beta_1$, and $\gamma = \gamma_1$ have the form

$$\alpha_1 = \frac{108}{115}, \quad \alpha_0 = -\frac{108}{115}, \quad \beta = \frac{162}{115}, \quad \gamma = -\frac{54}{23}.$$

Consider next the methods (5.7.1) with $s = 2$ and $p = 2s = 4$. For the methods with coefficients $\varphi_1(\theta)$, $\varphi_0(\theta)$, $\psi_1(\theta)$, $\psi_2(\theta)$, $\chi_1(\theta)$, and $\chi_2(\theta)$ given by (5.10.1), the unique solution α_1, α_0, β_1, β_2, γ_1, and γ_2 is given by

$$\alpha_1 = -\frac{1244160}{21127}, \quad \alpha_0 = \frac{1244160}{21127},$$

$$\beta_1 = \frac{2488320}{21127}, \quad \beta_2 = -\frac{1285632}{21127},$$

and

$$\gamma_1 = \frac{4810752}{21127}, \quad \gamma_2 = -\frac{4769280}{21127}.$$

For a family of methods with $s = 2$ and $p = 2s - 1 = 3$ with coefficients $\varphi_1(\theta)$, $\varphi_0(\theta)$, $\psi_1(\theta)$, $\psi_2(\theta)$, $\chi_1(\theta)$, and $\chi_2(\theta)$ given by (5.10.2) there is a one-parameter family of solutions to system (6.9.5) corresponding to $s = 2$ and $p = 3$. Assuming that $\beta_1 = \beta_2$, the solution α_1, α_0, β_1, β_2, γ_1, and γ_2 is given by

$$\alpha_1 = -\alpha_0 = -\frac{77760}{9019485 + 11232679 q_0 - 2293632 r_0},$$

$$\beta_1 = \beta_2 = \frac{314280}{9019485 + 11232679q_0 - 2293632r_0},$$

$$\gamma_1 = \frac{155520}{9019485 + 11232679q_0 - 2293632r_0},$$

and

$$\gamma_2 = -\frac{706320}{9019485 + 11232679q_0 - 2293632r_0}.$$

Here q_0 and r_0 are free parameters that appear in (5.10.2).

6.10 RECENT WORK ON TWO-STEP COLLOCATION METHODS

Work is in progress on the construction and implementation of a family of two-step collocation methods of order p and stage order $q = p = s$, where s is the number of stages, for $1 \leq s \leq 8$, with desirable accuracy and stability properties. We are aiming at methods that are A- and L-stable with small error constants and favorable error propagation, including terms of order $p+2$. Preliminary numerical experiments with methods of order up to 4 indicate that local error estimation is very accurate and reliable for small and large step sizes for stiff systems of differential equations. We are now working on implementation of these methods of order $1 \leq p \leq 8$ in a variable step size variable order environment in a Matlab code intended for stiff differential systems [103, 105, 106].

We are also working on the construction and implementation of TSRK methods of high order with quadratic stability functions. This work, which is a continuation of work by Conte et al. [92], is reported by Conte et al. [93] and D'Ambrosio [103].

CHAPTER 7

GENERAL LINEAR METHODS WITH INHERENT RUNGE-KUTTA STABILITY

7.1 REPRESENTATION OF METHODS AND ORDER CONDITIONS

In this chapter we investigate GLMs that possess inherent Runge-Kutta stability (IRKS). These are methods which have the property that the stability matrix has only one nonzero eigenvalue, which is an approximation of order p to exponential function $\exp(z)$. As a consequence, the stability behavior of such methods is similar to that of RK methods of the same order, and it is said that they possess RK stability. This and IRKS will be made more precise later in the chapter. In general, it is a very nontrivial task to find GLMs that satisfy this property. In the case of the DIMSIMs discussed in Chapters 3 and 4, and the TSRK methods discussed in Chapters 5 and 6, after satisfying the appropriate order and stage order conditions, the RK stability requirement leads to large systems of nonlinear equations of high degree which are very difficult to solve. In view of this, it is quite remarkable that choosing the appropriate values of the parameters p, q, r, and s (i.e., the order, the stage order, the number of external approximations, and the number of internal approximations), and assuming that the vector $y^{[n]}$ of external approximations approximates the Nordsieck vector $z(t_n, h_n)$, it is possible to characterize all

General Linear Methods for Ordinary Differential Equations. By Zdzisław Jackiewicz **345**
Copyright © 2009 John Wiley & Sons, Inc.

explicit and diagonally-implicit GLMs with RK stability. This characterization was discovered recently by Butcher [50, 53], Butcher and Wright [79, 80], and Wright [293, 294]. Practical algorithms for the construction of both explicit and implicit methods of this type which use only linear operations were also reported [80, 293, 294]. A special variant of the algorithm presented by Butcher and Wright [80] was then used by Butcher and Jackiewicz to construct GLMs with RK stability which achieve good balance between accuracy and stability [72], and to construct GLMs of this type which are unconditionally stable for any step size pattern [73].

Following Butcher and Wright [80] and Wright [293], it is assumed throughout this chapter that $p = q$ and $r = s = p + 1$. As observed in [80], methods characterized in this way seem to have considerable potential as the basis of good solvers for initial value problems. We consider GLMs of the form

$$Y^{[n]} = h(\mathbf{A} \otimes \mathbf{I})F(Y^{[n]}) + (\mathbf{U} \otimes \mathbf{I})\mathbf{y}^{[n-1]},$$

$$\mathbf{y}^{[n]} = h(\mathbf{B} \otimes \mathbf{I})F(Y^{[n]}) + (\mathbf{V} \otimes \mathbf{I})\mathbf{y}^{[n-1]},$$

(7.1.1)

$n = 1, 2, \ldots, N$, where all the coefficient matrices have the same dimension (i.e., $\mathbf{A}, \mathbf{U}, \mathbf{B}, \mathbf{V} \in \mathbb{R}^{s \times s}$). We also assume that $\mathbf{y}^{[n-1]}$ and $\mathbf{y}^{[n]}$ are approximations to the Nordsieck vectors $z(t_{n-1}, h)$ and $z(t_n, h)$, where $z(t, h)$ is defined by (3.11.1). For easy reference this definition is repeated here:

$$z(t, h) = \begin{bmatrix} y(t) \\ hy'(t) \\ \vdots \\ h^p y^{(p)}(t) \end{bmatrix}.$$

(7.1.2)

Observe that the notation for (7.1.1) differs from that adopted for (3.11.8) in Section 3.11, where the coefficient matrices were denoted by $\mathbf{A}$, $\mathbf{P}$, $\mathbf{G}$, and $\mathbf{Q}$, and they are not of the same dimension.

As in the proof of Theorem 3.12.1 and the comments following it, it can be demonstrated that the order conditions for (7.1.1) take the form

$$e^{cz} = z\mathbf{A}e^{cz} + \mathbf{U}\mathbf{Z} + O(z^{p+1}),$$

(7.1.3)

$$e^z\mathbf{Z} = z\mathbf{B}e^{cz} + \mathbf{V}\mathbf{Z} + O(z^{p+1}),$$

(7.1.4)

where the vector $\mathbf{Z}$ is defined by

$$\mathbf{Z} = \begin{bmatrix} 1 & z & \cdots & z^p \end{bmatrix}^T.$$

(7.1.5)

We consider only type 1 or 2 methods: formulas for which the coefficient matrix $\mathbf{A}$ has the form (2.7.1) with $\lambda = 0$ or $\lambda > 0$, respectively. We also assume that GLM (7.1.1) is zero-stable (i.e., the coefficient matrix $\mathbf{V}$ satisfies (2.2.5) and the conditions given in Theorem 2.2.5).

Let $\mathbf{J}$ and $\mathbf{K}$ be shifting matrices of dimension $p + 1$ defined by

$$
\mathbf{J} = \begin{bmatrix} 0 & 0 & \cdots & 0 & 0 \\ 1 & 0 & \cdots & 0 & 0 \\ 0 & 1 & \vdots & 0 & 0 \\ \vdots & \vdots & \ddots & \vdots & \vdots \\ 0 & 0 & \cdots & 1 & 0 \end{bmatrix}, \quad
\mathbf{K} = \begin{bmatrix} 0 & 1 & 0 & \cdots & 0 \\ 0 & 0 & 1 & \cdots & 0 \\ \vdots & \vdots & \vdots & \ddots & \vdots \\ 0 & 0 & 0 & \cdots & 1 \\ 0 & 0 & 0 & \cdots & 0 \end{bmatrix}.
$$

Observe that the matrix $\mathbf{K}$ was already defined in Section 3.12. Let $\mathbf{Q} \in \mathbb{R}^{(p+1) \times (p+1)}$ be a matrix that is partitioned into column and row vectors as follows:

$$
\mathbf{Q} = \begin{bmatrix} \mathbf{c}_1 & \mathbf{c}_2 & \cdots & \mathbf{c}_{p+1} \end{bmatrix} = \begin{bmatrix} \mathbf{r}_1 \\ \mathbf{r}_2 \\ \vdots \\ \mathbf{r}_{p+1} \end{bmatrix}.
$$

Then it is easy to verify that the effects of multiplying this matrix by $\mathbf{J}$ and $\mathbf{K}$ from the right and the left are

$$
\begin{bmatrix} \mathbf{c}_1 & \mathbf{c}_2 & \cdots & \mathbf{c}_{p+1} \end{bmatrix} \mathbf{J} = \begin{bmatrix} \mathbf{c}_2 & \cdots & \mathbf{c}_{p+1} & \mathbf{0} \end{bmatrix},
$$

$$
\begin{bmatrix} \mathbf{c}_1 & \cdots & \mathbf{c}_p & \mathbf{c}_{p+1} \end{bmatrix} \mathbf{K} = \begin{bmatrix} \mathbf{0} & \mathbf{c}_1 & \cdots & \mathbf{c}_p \end{bmatrix},
$$

and

$$
\mathbf{J} \begin{bmatrix} \mathbf{r}_1 \\ \vdots \\ \mathbf{r}_p \\ \mathbf{r}_{p+1} \end{bmatrix} = \begin{bmatrix} \mathbf{0} \\ \mathbf{r}_1 \\ \vdots \\ \mathbf{r}_p \end{bmatrix}, \quad
\mathbf{K} \begin{bmatrix} \mathbf{r}_1 \\ \mathbf{r}_2 \\ \vdots \\ \mathbf{r}_{p+1} \end{bmatrix} = \begin{bmatrix} \mathbf{r}_2 \\ \vdots \\ \mathbf{r}_{p+1} \\ \mathbf{0} \end{bmatrix}.
$$

We have demonstrated in Corollary 3.12.2 that in the case of DIMSIMs with $p = q = s$ and $r = s + 1$, the coefficient matrices $\mathbf{P}$ and $\mathbf{Q}$ in (3.11.8) are determined completely by the abscissa vector $\mathbf{c}$ and the coefficient matrices $\mathbf{A}$ and $\mathbf{G}$. The same is true in the case of GLMs (7.1.1) with $p = q$ and $r = s = p + 1$ (i.e., the matrices $\mathbf{U}$ and $\mathbf{V}$ are completely determined by $\mathbf{c}$, $\mathbf{A}$, and $\mathbf{B}$). We justify this below using arguments by Butcher and Wright [80, 293] which differ slightly from that used in the proof of Corollary 3.12.2. Observe first that

$$
e^{\mathbf{c}z} = \mathbf{C} \mathbf{Z} + O(z^{p+1}),
$$

$$e^z \, \mathbf{Z} = \mathbf{E}\,\mathbf{Z} + O(z^{p+1}),$$

where the scaled Vandermonde matrix $\mathbf{C}$ is given by

$$\mathbf{C} = \left[\begin{array}{cccc} \mathbf{e} & \mathbf{c} & \dfrac{\mathbf{c}^2}{2!} & \cdots & \dfrac{\mathbf{c}^p}{p!} \end{array} \right] \in \mathbb{R}^{(p+1) \times (p+1)},$$

and the special Toeplitz matrix $\mathbf{E}$ is given by

$$\mathbf{E} = \exp(\mathbf{K}) = \begin{bmatrix} 1 & 1 & \frac{1}{2!} & \cdots & \frac{1}{p!} \\ 0 & 1 & 1 & \cdots & \frac{1}{(p-1)!} \\ 0 & 0 & 1 & \cdots & \frac{1}{(p-2)!} \\ \vdots & \vdots & \vdots & \ddots & \vdots \\ 0 & 0 & 0 & \cdots & 1 \end{bmatrix} \in \mathbb{R}^{(p+1) \times (p+1)}.$$

This matrix was already defined in Section 3.12. Using the relations above for $e^{\mathbf{c}z}$ and $e^z\mathbf{Z}$, the order and stage order conditions (7.1.3) and (7.1.4) can be reformulated as

$$\mathbf{C}\,\mathbf{Z} = z\,\mathbf{A}\,\mathbf{C}\,\mathbf{Z} + \mathbf{U}\,\mathbf{Z} + O(z^{p+1}),$$

$$\mathbf{E}\,\mathbf{Z} = z\,\mathbf{B}\,\mathbf{C}\,\mathbf{Z} + \mathbf{V}\,\mathbf{Z} + O(z^{p+1}).$$

Since $z\mathbf{Z} = \mathbf{K}\mathbf{Z} + O(z^{p+1})$, it follows that

$$\mathbf{C}\,\mathbf{Z} = \mathbf{A}\,\mathbf{C}\,\mathbf{K}\,\mathbf{Z} + \mathbf{U}\,\mathbf{Z} + O(z^{p+1}),$$

$$\mathbf{E}\,\mathbf{Z} = \mathbf{B}\,\mathbf{C}\,\mathbf{K}\,\mathbf{Z} + \mathbf{V}\,\mathbf{Z} + O(z^{p+1}),$$

and comparing the coefficients of $z^0, z^1, \ldots, z^p$, we obtain representation formulas for $\mathbf{U}$ and $\mathbf{V}$ of the form

$$\mathbf{U} = \mathbf{C} - \mathbf{A}\,\mathbf{C}\,\mathbf{K} \tag{7.1.6}$$

and

$$\mathbf{V} = \mathbf{E} - \mathbf{B}\,\mathbf{C}\,\mathbf{K}. \tag{7.1.7}$$

Formula (7.1.6) was first derived by Butcher and Jackiewicz [69] in the context of Nordsieck representation of DIMSIMs.

7.2   INHERENT RUNGE-KUTTA STABILITY

We now turn our attention to stability. As discussed in Section 2.6, linear stability properties of method (7.1.1) are determined by the stability matrix $\mathbf{M}(z)$, defined by (2.6.4): that is,

$$\mathbf{M}(z) = \mathbf{V} + z\,\mathbf{B}(\mathbf{I} - z\,\mathbf{A})^{-1}\mathbf{U}. \tag{7.2.1}$$

It turns out that the principal eigenvalue and the corresponding eigenvector of this matrix have special forms given in the following lemma.

Lemma 7.2.1 (Butcher and Wright [80, 293]) *Stability matrix* $\mathbf{M}(z)$ *defined by (7.2.1) of the GLM in Nordsieck form (7.1.1) with $p = q$ has principal eigenvalue* $\exp(z) + O(z^{p+1})$ *and corresponding eigenvector* $\mathbf{Z} + O(z^{p+1})$, *where* $\mathbf{Z}$ *is defined by (7.1.5).*

Proof: It follows from (7.1.3) that

$$e^{\mathbf{c}z} = (\mathbf{I} - z\,\mathbf{A})^{-1}\mathbf{U}\mathbf{Z} + O(z^{p+1}),$$

where $\mathbf{I} \in \mathbb{R}^{(p+1)\times(p+1)}$ is the identity matrix. Substituting this relation into (7.1.4), we obtain

$$e^{z}\mathbf{Z} = \big(\mathbf{V} + z\,\mathbf{B}(\mathbf{I} - z\,\mathbf{A})^{-1}\mathbf{U}\big)\mathbf{Z} + O(z^{p+1}),$$

so that

$$\mathbf{M}(z)\,\mathbf{Z} = e^{z}\mathbf{Z} + z^{p+1}\Phi(z) \tag{7.2.2}$$

for a vector-valued power series $\Phi(z)$. Since $\mathbf{M}(z) \to \mathbf{V}$ and $\mathbf{Z} \to \mathbf{e}_1$ as $z \to 0$, we have

$$\mathbf{V}\,\mathbf{e}_1 = \mathbf{e}_1.$$

This is one of the preconsistency conditions given in (2.2.1) with preconsistency vector equal to $\mathbf{e}_1$. Denote by $\lambda(z)$ the principal eigenvalue of $\mathbf{M}(z)$: the one that tends to 1 as $z \to 0$, and let the corresponding eigenvector be $\mathbf{v}(z)$. Since $\mathbf{M}(0)\mathbf{e}_1 = \mathbf{e}_1$, we can assume that the first component of $\mathbf{v}(z)$ is equal to 1. Suppose that

$$e^{z} = \lambda(z) + \theta(z)z^{m},$$

where $\theta(0) \neq 0$, and that

$$\mathbf{Z} = \mathbf{v}(z) + \Psi(z)z^{n},$$

where $\Psi(z) \neq 0$ and $\Psi_1(z) = 0$. Here $\Psi_1(z)$ stands for the first component of $\Psi(z)$. We have to show that $m, n \geq p + 1$. This will be done in two steps by proving that (i) if $m < p+1$, then $n \leq m$, and (ii) it is impossible that $n \leq m$ and that $n < p + 1$.

To prove (i), observe that

$$z^{n}\big(\mathbf{M}(z) - e^{z}\mathbf{I}\big)\Psi(z) - z^{m}\theta(z)\mathbf{v}(z)$$

$$= \big(\mathbf{M}(z) - e^{z}\mathbf{I}\big)\big(\mathbf{Z} - \mathbf{v}(z)\big) + \big(\lambda(z) - e^{z}\big)\mathbf{v}(z)$$

$$= \big(\mathbf{M}(z) - e^{z}\mathbf{I}\big)\mathbf{Z} - \big(\mathbf{M}(z) - \lambda(z)\mathbf{I}\big)\mathbf{v}(z),$$

and since $(\mathbf{M}(z) - \lambda(z)\mathbf{I})\mathbf{v}(z) = 0$ using (7.2.2), we obtain

$$z^{n}\big(\mathbf{M}(z) - e^{z}\mathbf{I}\big)\Psi(z) - z^{m}\theta(z)\mathbf{v}(z) = z^{p+1}\Phi(z). \tag{7.2.3}$$

Hence, if $m < p + 1$, then necessarily $n \leq m$, which proves (i).

To show (ii), divide equation (7.2.3) by z^n and then compute the limit as $z \to 0$. Then

$$(\mathbf{V} - \mathbf{I})\Psi(0) = 0 \tag{7.2.4}$$

if $m > n$, or

$$(\mathbf{V} - \mathbf{I})\Psi(0) = \theta(0)\mathbf{v}(0) \tag{7.2.5}$$

if $m = n$. Since $\mathbf{V}\mathbf{e}_1 = \mathbf{e}_1$ and $\Psi_1(0) = 0$, the matrix $\mathbf{V}$ and the vector $\Psi(0)$ can be partitioned as follows:

$$\mathbf{V} = \left[\begin{array}{c|c} 1 & v^T \\ \hline 0 & V \end{array}\right], \quad \Psi(0) = \left[\begin{array}{c} 0 \\ \hline \psi(0) \end{array}\right].$$

Hence, it follows from (7.2.4) or (7.2.5) and the fact that $\mathbf{v}(0) = \mathbf{e}_1$ that

$$(V - I)\psi(0) = 0.$$

Here I is the identity matrix of dimension p. It follows from zero-stability of GLM (7.1.1) that $V - I$ is a nonsingular matrix which implies that $\psi(0) = 0$. However, this is impossible since $\Psi(0) \neq 0$ and $\Psi_1(0) = 0$. This proves (ii). ∎

Lemma 7.2.1 characterizes the principal eigenvalue and the corresponding eigenvector of the stability matrix $\mathbf{M}(z)$ given by (7.2.1). If this eigenvalue is the only nonzero eigenvalue of $\mathbf{M}(z)$, the stability properties of the corresponding GLMs (7.1.1) are very similar to that of RK methods. This observation motivates the following definition.

Definition 7.2.2 *If the characteristic polynomial $p(w, z)$ of the stability matrix $\mathbf{M}(z)$ given by (7.2.1) has the special form*

$$p(w, z) = \det\left(w\mathbf{I} - \mathbf{M}(z)\right) = w^p\left(w - R(z)\right), \tag{7.2.6}$$

the GLM (7.1.1) is said to possess RK stability.

If GLM possesses RK stability, then the rational function $R(z)$ in (7.2.6) plays the same role as the stability function of the RK method of the same order. Since RK methods have stability properties which are far superior, for example, to that of linear multistep methods, it is desirable to find GLMs with RK stability. The characterization of such methods is the goal of this chapter.

We introduce next some equivalence relation between matrices and column vectors, denoted by "$\equiv$", which will be useful to define the property of inherent Runge-Kutta stability (IRKS). We say that two matrices $\mathbf{A}$ and $\mathbf{B}$ are equivalent, written as $\mathbf{A} \equiv \mathbf{B}$, if and only if they are identical except possibly their first rows.

This relation has several useful properties which will aid in the stability analysis of GLMs. For example, if $\mathbf{F}\mathbf{e}_1 = \lambda\mathbf{e}_1$, then

$$\mathbf{D} \equiv \mathbf{E} \quad \text{implies that} \quad \mathbf{F}\mathbf{D} \equiv \mathbf{F}\mathbf{E}.$$

Moreover, if can be easily verified that for any matrix $\mathbf{G}$ we have

$$\mathbf{D} \equiv \mathbf{E} \quad \text{implies that} \quad \mathbf{D}\mathbf{G} \equiv \mathbf{E}\mathbf{G}.$$

We also have

$$z\mathbf{J}\mathbf{Z} \equiv \mathbf{Z}, \tag{7.2.7}$$

where $\mathbf{J}$ is the shifting matrix defined in Section 7.1 and $\mathbf{Z}$ is the column vector given by (7.1.5).

It is a complicated task to find conditions which ensure that GLMs (7.1.1) possess RK stability. However, it is possible to find interrelations between the coefficients matrices which ensure that this is the case. Such conditions were discovered by Butcher and Wright [80, 293] and formulated as IRKS. This is formalized in the following definition.

Definition 7.2.3 *GLM (7.1.1) satisfying preconsistency condition* $\mathbf{V}\mathbf{e}_1 = \mathbf{e}_1$ *has an IRKS property if*

$$\mathbf{B}\mathbf{A} \equiv \mathbf{X}\mathbf{B}, \tag{7.2.8}$$

$$\mathbf{B}\mathbf{U} \equiv \mathbf{X}\mathbf{V} - \mathbf{V}\mathbf{X}, \tag{7.2.9}$$

for some matrix $\mathbf{X}$, *and*

$$\det(w\mathbf{I} - \mathbf{V}) = w^p(w - 1). \tag{7.2.10}$$

The importance of this property follows from the following theorem.

Theorem 7.2.4 (Butcher and Wright [80, 293]) *Assume that GLM defined by (7.1.1) has IRKS. Then its stability function* $p(w, z) = \det(w\mathbf{I}-\mathbf{M}(z))$ *takes the form*

$$p(w, z) = w^p\big(w - R(z)\big).$$

Moreover, $R(z)$ *has the form*

$$R(z) = \mathbf{e}_1^T(\mathbf{I} - z\mathbf{X})\mathbf{M}(z)(\mathbf{I} - z\mathbf{X})^{-1}\mathbf{e}_1. \tag{7.2.11}$$

Proof: Relation (7.2.8) is equivalent to

$$\mathbf{B}(\mathbf{I} - z\mathbf{A}) \equiv (\mathbf{I} - z\mathbf{X})\mathbf{B},$$

and it follows that

$$\mathbf{B} \equiv (\mathbf{I} - z\mathbf{X})\mathbf{B}(\mathbf{I} - z\mathbf{A})^{-1}. \tag{7.2.12}$$

To investigate a characteristic polynomial of the stability matrix $\mathbf{M}(z)$ it is more convenient to consider the matrix related to $\mathbf{M}(z)$ by similarity transformation. Using (7.2.9) and (7.2.12), it follows that

$$(\mathbf{I} - z\mathbf{X})\mathbf{M}(z)(\mathbf{I} - z\mathbf{X})^{-1}$$

$$= \big(\mathbf{V} - z\mathbf{X}\mathbf{V} + z(\mathbf{I} - z\mathbf{X})\mathbf{B}(\mathbf{I} - z\mathbf{A})^{-1}\mathbf{U}\big)(\mathbf{I} - z\mathbf{X})^{-1}$$

$$\equiv \big(\mathbf{V} - z\mathbf{X}\mathbf{V} + z\mathbf{B}\mathbf{U}\big)(\mathbf{I} - z\mathbf{X})^{-1}$$

$$\equiv \big(\mathbf{V} - z\mathbf{X} + z(\mathbf{X}\mathbf{V} - \mathbf{V}\mathbf{X})\big)(\mathbf{I} - z\mathbf{X})^{-1}$$

$$= (\mathbf{V} - z\mathbf{V}\mathbf{X})(\mathbf{I} - z\mathbf{X})^{-1}.$$

Hence,

$$(\mathbf{I} - z\mathbf{X})\mathbf{M}(z)(\mathbf{I} - z\mathbf{X})^{-1} \equiv \mathbf{V}. \qquad (7.2.13)$$

It follows from the condition $\mathbf{V}\mathbf{e}_1 = \mathbf{e}_1$ and the relation (7.2.13) that the matrix $(\mathbf{I} - z\mathbf{X})\mathbf{M}(z)(\mathbf{I} - z\mathbf{X})^{-1}$ can be partitioned as follows

$$(\mathbf{I} - z\mathbf{X})\mathbf{M}(z)(\mathbf{I} - z\mathbf{X})^{-1} = \left[\begin{array}{c|c} R(z) & r^T \\ \hline 0 & V \end{array} \right], \qquad (7.2.14)$$

where V is the $p \times p$ matrix obtained from $\mathbf{V}$ by deleting the first row and column and r is some vector. Since condition (7.2.10) and $\mathbf{V}\mathbf{e}_1 = \mathbf{e}_1$ ensures that V has only zero eigenvalues, it follows that $(\mathbf{I} - z\mathbf{X})\mathbf{M}(z)(\mathbf{I} - z\mathbf{X})^{-1}$ and $\mathbf{M}(z)$ has only one nonzero eigenvalue $R(z)$, which is the $(1,1)$ element in (7.2.14). This leads to formula (7.2.11). Observe also that $R(z) \to 1$, the eigenvalue of $\mathbf{V}$, as $z \to 0$.   ∎

Definition 7.2.3 and the proof of Theorem 7.2.4 make no reference to any special form of the matrix $\mathbf{X}$. However, it turns out that this matrix must have a very special form for GLMs (7.1.1) of order $p = q$. This is formulated in the next theorem.

Theorem 7.2.5 (Butcher and Wright [80, 293]) *For GLM (7.1.1) with $p = q$, the most general form of the matrix $\mathbf{X}$ appearing in conditions (7.2.8)* *and (7.2.9) of Definition 7.2.3 is of doubly companion form:*

$$
\mathbf{X} =
\begin{bmatrix}
-\alpha_1 & -\alpha_2 & -\alpha_3 & \cdots & -\alpha_{p-1} & -\alpha_p & -\alpha_{p+1} - \beta_{p+1} \\
1 & 0 & 0 & \cdots & 0 & 0 & -\beta_p \\
0 & 1 & 0 & \cdots & 0 & 0 & -\beta_{p-1} \\
\vdots & \vdots & \vdots & \ddots & \vdots & \vdots & \vdots \\
0 & 0 & 0 & \cdots & 0 & 0 & -\beta_3 \\
0 & 0 & 0 & \cdots & 1 & 0 & -\beta_2 \\
0 & 0 & 0 & \cdots & 0 & 1 & -\beta_1
\end{bmatrix} . \tag{7.2.15}
$$

Proof: Multiplying (7.2.8) by $z\mathbf{B}$ and (7.2.9) by $\mathbf{I} - z\mathbf{X}$ and then adding the resulting relations, we obtain

$$
e^z(\mathbf{I} - z\mathbf{X})\mathbf{Z} = z^2(\mathbf{BA} - \mathbf{XB})e^{cz} + z(\mathbf{BU} - \mathbf{XV})\mathbf{Z} + \mathbf{VZ} + O(z^{p+1})
$$

$$
= z^2(\mathbf{BA} - \mathbf{XB})e^{cz} + z(\mathbf{BU} - \mathbf{XV} + \mathbf{VX})\mathbf{Z} + \mathbf{V}(\mathbf{I} - z\mathbf{X})\mathbf{Z} + O(z^{p+1}).
$$

Using (7.2.8) and (7.2.9), the equation above leads to the equivalence relation

$$
e^z(\mathbf{I} - z\mathbf{X})\mathbf{Z} \equiv \mathbf{V}(\mathbf{I} - z\mathbf{X})\mathbf{Z} + O(z^{p+1})
$$

and it follows that

$$
(e^z\mathbf{I} - \mathbf{V})(\mathbf{I} - z\mathbf{X})\mathbf{Z} \equiv O(z^{p+1}). \tag{7.2.16}
$$

Set $\widetilde{\mathbf{Z}} = (\mathbf{I} - z\mathbf{X})\mathbf{Z}$ and partition the matrix $e^z\mathbf{I} - \mathbf{V}$ and the vector $\widetilde{\mathbf{Z}}$ as follows:

$$
e^z\mathbf{I} - \mathbf{V} =
\left[
\begin{array}{c|c}
e^z - 1 & v^T \\
\hline
0 & e^z I - V
\end{array}
\right], \quad
\widetilde{\mathbf{Z}} =
\left[
\begin{array}{c}
\widetilde{Z}_1 \\
\hline
\widetilde{Z}
\end{array}
\right],
$$

where $\widetilde{Z}_1$ stands for the first component of the vector $\widetilde{\mathbf{Z}}$. Then (7.2.16) takes the form

$$
\left[
\begin{array}{c|c}
e^z - 1 & v^T \\
\hline
0 & e^z I - V
\end{array}
\right]
\left[
\begin{array}{c}
\widetilde{Z}_1 \\
\hline
\widetilde{Z}
\end{array}
\right]
=
\left[
\begin{array}{c}
(e^z - 1)\widetilde{Z}_1 + v^T\widetilde{Z} \\
\hline
(e^z I - V)\widetilde{Z}
\end{array}
\right]
\equiv O(z^{p+1}). \tag{7.2.17}
$$

Since the matrix $I - V$ is nonsingular, there exists $\epsilon > 0$ such that $e^z I - V$ is also nonsingular for $|z| < \epsilon$. Multiplying (7.2.17) by the matrix

$$
\mathbf{F} =
\left[
\begin{array}{c|c}
1 & 0 \\
\hline
0 & (e^z I - V)^{-1}
\end{array}
\right]
$$

and taking into account that $\mathbf{F}\mathbf{e}_1 = \mathbf{e}_1$, we obtain

$$\left[\frac{(e^z - 1)\widetilde{Z}_1 + v^T \widetilde{Z}}{\widetilde{Z}} \right] \equiv O(z^{p+1}),$$

which is equivalent to

$$(\mathbf{I} - z\mathbf{X})\mathbf{Z} \equiv O(z^{p+1}). \tag{7.2.18}$$

Using (7.2.7), this relation simplifies to

$$(\mathbf{J} - \mathbf{X})\mathbf{Z} \equiv O(z^p).$$

It is easy to verify that this relation implies that all elements of $\mathbf{J} - \mathbf{X}$ must be zero except for the first row and the last column. Therefore, $\mathbf{X}$ must be of the form (7.2.15). This concludes the proof. ∎

The doubly companion matrix (7.2.15) is the most general matrix satisfying condition (7.2.18), and the conclusion of the theorem follows also directly from this condition. This can be verified by comparing the coefficients of $z, z^2, \ldots, z^p$ on the last p rows of the relation

$$z\mathbf{X}\mathbf{Z} \equiv \mathbf{Z} + O(z^{p+1}),$$

which is equivalent to relation (7.2.18).

To provide additional motivation for conditions (7.2.8) and (7.2.9) in Definition 7.2.3 and the role of the doubly companion matrix $\mathbf{X}$ given by (7.2.15), we reformulate the stage order and order conditions (7.1.3) and (7.1.4) assuming that relation (7.2.18), which defines such a matrix $\mathbf{X}$, is satisfied. Substituting (7.1.3) into (7.1.4), it follows that

$$e^z\mathbf{Z} = z^2\mathbf{B}\mathbf{A}e^{cz} + (z\mathbf{B}\mathbf{U} + \mathbf{V})\mathbf{Z} + O(z^{p+1}),$$

and using (7.2.18) we obtain

$$e^z\mathbf{Z} \equiv z^2\mathbf{B}\mathbf{A}e^{cz} + z(\mathbf{B}\mathbf{U} + \mathbf{V}\mathbf{X})\mathbf{Z} + O(z^{p+1}). \tag{7.2.19}$$

Multiplying (7.1.4) on the left by $z\mathbf{X}$ leads to

$$ze^z\mathbf{X}\mathbf{Z} = z^2\mathbf{X}\mathbf{B}e^{cz} + z\mathbf{X}\mathbf{V}\mathbf{Z} + O(z^{p+1})$$

and again using (7.2.18) yields

$$e^z\mathbf{Z} \equiv z^2\mathbf{X}\mathbf{B}e^{cz} + z\mathbf{X}\mathbf{V}\mathbf{Z} + O(z^{p+1}). \tag{7.2.20}$$

We can now observe that (7.2.19) and (7.2.20) are equivalent if we assume that conditions (7.2.8) and (7.2.9) are satisfied.

Theorem 7.2.5 and the discussion above give some indication that doubly companion matrices (7.2.15) will play an important role in the analysis of GLMs with IRKS. Such matrices were introduced in the study of effective order singly-implicit Runge-Kutta (ESIRK) methods by Butcher and Chartier [55, 56]. In the next section we review some of the properties of doubly companion matrices following the presentation by Butcher and Wright [80] and Wright [293].

7.3 DOUBLY COMPANION MATRICES

Consider the set Π of polynomials of degree less than or equal to $p+1$ which map 0 to 1 (i.e., if $\varphi \in \Pi$, then $\varphi(0) = 1$). Examples of such polynomials related to the doubly companion matrix $\mathbf{X}$ defined by (7.2.15) are

$$\alpha(w) = 1 + \alpha_1 w + \cdots + \alpha_p w^p + \alpha_{p+1} w^{p+1}, \qquad (7.3.1)$$

$$\beta(w) = 1 + \beta_1 w + \cdots + \beta_p w^p + \beta_{p+1} w^{p+1}. \qquad (7.3.2)$$

The set Π becomes a group if we define the product $\gamma = \alpha\beta$ of $\alpha, \beta \in \Pi$ to be a polynomial $\gamma \in \Pi$ such that

$$\gamma(w) = \alpha(w)\beta(w) + O(w^{p+2}).$$

The identity of this group is the constant polynomial $e(w) = 1$, and the inverse element $\alpha^{-1} \in \Pi$ of $\alpha \in \Pi$ is defined so that

$$\alpha^{-1}(w)\alpha(w) = 1 + O(w^{p+2}).$$

We denote by $\mathbf{X}(\alpha, \beta)$ the doubly companion matrix given by (7.2.15), where the elements in the first row and last column of this matrix are the coefficients of the polynomials $\alpha, \beta \in \Pi$ given by (7.3.1) and (7.3.2). When there is no possibility of confusion, we write $\mathbf{X}$ instead of $\mathbf{X}(\alpha, \beta)$.

We first consider the characteristic polynomial $\psi(w) = \det(w\mathbf{I} - \mathbf{X}(\alpha, \beta))$ of the matrix $\mathbf{X} = \mathbf{X}(\alpha, \beta)$ given by (7.2.15). It is known that in the special cases $\beta_1 = \beta_2 = \cdots = \beta_{p+1} = 0$ or $\alpha_1 = \alpha_2 = \cdots = \alpha_{p+1} = 0$, the characteristic polynomials are given by

$$\det\left(w\mathbf{I} - \mathbf{X}(\alpha, 1)\right) = w^{p+1} + \alpha_1 w^p + \cdots + \alpha_p w + \alpha_{p+1},$$

$$\det\left(w\mathbf{I} - \mathbf{X}(1, \beta)\right) = w^{p+1} + \beta_1 w^p + \cdots + \beta_p w + \beta_{p+1},$$

respectively (compare [160, 197, 228]). Here 1 stands for the identity in the group Π.

To determine the characteristic polynomial of $\mathbf{X}(\alpha, \beta)$ in the general case, let λ be an eigenvalue of this matrix and $\mathbf{x}(\lambda)$ the corresponding eigenvector. It is easy to verify that the last component of $\mathbf{x}(\lambda)$ cannot be zero. Hence, we can assume without loss of generality that this eigenvector has the form

$$\mathbf{x}(\lambda) = \left[\begin{array}{cccc} x_p(\lambda) & x_{p-1}(\lambda) & \cdots & x_1(\lambda) & 1 \end{array} \right]^T.$$

We have the following lemma.

Lemma 7.3.1 (Butcher and Chartier[56], see also [293]) *The characteristic polynomial $\psi(w)$ of the matrix $\mathbf{X}(\alpha, \beta)$ given by (7.2.15) assumes the form*

$$\psi(w) = x_{p+1}(w) + \sum_{k=1}^{p+1} \alpha_k x_{p+1-k}(w), \qquad (7.3.3)$$

where $x_{p+1}(w) = wx_p(w) + \beta_{p+1}$, and $x_k(w)$, $k = p, p-1, \ldots, 1$, are the first p components of the vector $\mathbf{x}(w)$ obtained from the eigenvector $\mathbf{x}(\lambda)$ by setting $\lambda = w$ and $x_0(\lambda) = 1$.

Proof: Comparing components $p+1, p, \ldots, 2$, of the equation

$$\mathbf{X}(\alpha, \beta)\mathbf{x}(\lambda) = \lambda\mathbf{x}(\lambda) \tag{7.3.4}$$

which define eigenvalue λ and the corresponding eigenvector $\mathbf{x}(\lambda)$, we obtain

$$x_1(\lambda) = \lambda + \beta_1,$$

$$x_2(\lambda) = \lambda x_1(\lambda) + \beta_2 = \lambda^2 + \beta_1\lambda + \beta_2,$$

$$\vdots$$

$$x_{p-1}(\lambda) = \lambda x_{p-2}(\lambda) + \beta_{p-1} = \lambda^{p-1} + \beta_1\lambda^{p-2} + \cdots + \beta_{p-2}\lambda + \beta_{p-1},$$

$$x_p(\lambda) = \lambda x_{p-1} + \beta_p = \lambda^p + \beta_1\lambda^{p-1} + \cdots + \beta_{p-1}\lambda + \beta_p.$$

This can be written more compactly as

$$x_k(\lambda) = \lambda x_{k-1}(\lambda) + \beta_k, \quad k = 1, 2, \ldots, p,$$

where $x_0(\lambda) = 1$, or in expanded form as

$$x_k(\lambda) = \lambda^k + \sum_{j=1}^{k} \beta_j\lambda^{k-j}, \quad k = 0, 1, \ldots, p.$$

We also define $x_{p+1}(\lambda)$ by the same formulas; that is,

$$x_{p+1}(\lambda) = \lambda x_p(\lambda) + \beta_{p+1} = \lambda^{p+1} + \sum_{j=1}^{p+1} \beta_j\lambda^{p+1-j}.$$

Comparing the first components of equation (7.3.4) and using the relation $x_{p+1}(\lambda) = \lambda x_p(\lambda) + \beta_{p+1}$, we obtain

$$x_{p+1}(\lambda) + \sum_{k=1}^{p+1} \alpha_k x_{p+1-k}(\lambda) = 0. \tag{7.3.5}$$

Consider the polynomial $\psi(w)$ defined by (7.3.3). Then $\deg(\psi) = p + 1$, its leading coefficient is equal to 1, and it follows from (7.3.5) that $\psi(\lambda) = 0$ if λ is an eigenvalue of $\mathbf{X}(\alpha, \beta)$. Hence, $\psi(w)$ has to be equal to the characteristic polynomial of $\mathbf{X}(\alpha, \beta)$, that is,

$$\psi(w) = \det\left(w\mathbf{I} - \mathbf{X}(\alpha, \beta)\right),$$

which concludes the proof.   ∎

The next lemma provides a more convenient way to compute $\psi(w)$.

Lemma 7.3.2 (Butcher and Wright [80, 293]) *The characteristic polynomial $\psi(w)$ of doubly companion matrix $\mathbf{X}(\alpha,\beta)$ given by (7.2.15) consists of terms with nonnegative degree in the expansion of the product*

$$w^{-(p+1)}\left(w^{p+1}+\alpha_1 w^p+\cdots+\alpha_{p+1}\right)\left(w^{p+1}+\beta_1 w^p+\cdots+\beta_{p+1}\right).$$

Proof: We have

$$w^{-(p+1)}\left(w^{p+1}+\sum_{k=1}^{p+1}\alpha_k w^{p+1-k}\right)\left(w^{p+1}+\sum_{j=1}^{p+1}\beta_j w^{p+1-j}\right)$$

$$=\left(1+\sum_{k=1}^{p+1}\alpha_k w^{-k}\right)\left(w^{p+1}+\sum_{j=1}^{p+1}\beta_j w^{p+1-j}\right)$$

$$=x_{p+1}(w)+\sum_{k=1}^{p+1}\alpha_k\left(w^{p+1-k}+\sum_{j=1}^{p+1}\beta_j w^{p+1-k-j}\right)$$

$$=x_{p+1}(w)+\sum_{k=1}^{p+1}\alpha_k\left(w^{p+1-k}+\sum_{j=1}^{p+1-k}\beta_j w^{p+1-k-j}\right)+O(w^{-1})$$

$$=x_{p+1}(w)+\sum_{k=1}^{p+1}\alpha_k x_{p+1-k}(w)+O(w^{-1}).$$

In the formulas above $O(w^{-1})$ stands for terms with negative degrees in w. Using (7.3.5), the last relation is equal to $\psi(w)+O(w^{-1})$ and the conclusion of the lemma follows.   ∎

Observe that the statement of the lemma can be reformulated as

$$\left(w^{p+1}+\alpha_1 w^p+\cdots+\alpha_{p+1}\right)\left(w^{p+1}+\beta_1 w^p+\cdots+\beta_{p+1}\right)=w^{p+1}\psi(w)+O(w^p),$$

where $O(w^p)$ stands for the terms of degree less than or equal to p. This observation can be used to compute an alternative form of the characteristic polynomial of $\mathbf{X}$ defined by

$$\widetilde{\psi}(w)=\det(\mathbf{I}-w\mathbf{X}).$$

We have

$$\widetilde{\psi}(w)=w^{p+1}\det(\widetilde{w}\mathbf{I}-\mathbf{X})=w^{p+1}\psi(\widetilde{w}),$$

where $\widetilde{w}=1/w$. This polynomial can be computed from the relation

$$\widetilde{\psi}(w)=(\alpha\beta)(w),$$

where α and β are polynomials defined by (7.3.1) and (7.3.2) and $\alpha\beta$ is the product in the group Π. This follows from the relations

$$
\begin{aligned}
\alpha(w)\beta(w) &= \left(1 + \alpha_1 w + \cdots + \alpha_{p+1} w^{p+1}\right)\left(1 + \beta_1 w + \cdots + \beta_{p+1} w^{p+1}\right) \\
&= w^{2(p+1)}\left(\widetilde{w}^{p+1} + \alpha_1\widetilde{w}^p + \cdots + \alpha_{p+1}\right)\left(\widetilde{w}^{p+1} + \beta_1\widetilde{w}^p + \cdots + \beta_{p+1}\right) \\
&= w^{2(p+1)}\left(\widetilde{w}^{p+1}\psi(\widetilde{w}) + O(\widetilde{w}^p)\right) \\
&= w^{p+1}\psi(\widetilde{w}) + O(\widetilde{w}^{-(p+2)}) \\
&= \widetilde{\psi}(w) + O(w^{p+2}).
\end{aligned}
$$

In applications of doubly companion matrices to the construction of GLMs which are of interest here, the eigenvalues of the matrix $\mathbf{X}(\alpha, \beta)$ and the β_k, $k = 1, 2, \ldots, p + 1$, are free parameters to obtain specific properties of the method. Once they are determined, the parameters α_k, $k = 1, 2, \ldots, p + 1$, can be computed from the relation

$$
\alpha(w) = \widetilde{\psi}(w)\beta^{-1}(w) + O(w^{p+2}) = (\widetilde{\psi}\beta^{-1})(w), \tag{7.3.6}
$$

where α and β are given by (7.3.1) and (7.3.2) and β^{-1} is the inverse of β in group Π.

In deriving GLMs (7.1.1) with IRKS it will be necessary to find decompositions of $\mathbf{X}(\alpha, \beta)$. We are interested in type 1 and 2 methods for which the coefficient matrix $\mathbf{A}$ has the form (2.7.1) with $\lambda = 0$ or $\lambda > 0$, respectively. In particular, $\mathbf{A}$ has a one-point spectrum, $\sigma(\mathbf{A}) = \{\lambda\}$, and it will be demonstrated later that this is equivalent to the matrix $\mathbf{X}(\alpha, \beta)$ having a one-point spectrum, $\sigma(\mathbf{X}(\alpha, \beta)) = \{\lambda\}$, where λ is the diagonal element of $\mathbf{A}$. In this case the decomposition of $\mathbf{X}(\alpha, \beta)$ reduces to Jordan canonical form. To describe this decomposition, we also need a left eigenvector $\mathbf{y}(\lambda)$ corresponding to the eigenvalue λ of $\mathbf{X}(\alpha, \beta)$, which is defined by

$$
\mathbf{y}^T(\lambda)\mathbf{X}(\alpha, \beta) = \lambda\mathbf{y}^T(\lambda). \tag{7.3.7}
$$

It is easy to verify that the first component of $\mathbf{y}(\lambda)$ is nonzero and we can assume without loss of generality that $\mathbf{y}(\lambda)$ has the form

$$
\mathbf{y}(\lambda) = \begin{bmatrix} 1 & y_1(\lambda) & \cdots & y_{p-1}(\lambda) & y_p(\lambda) \end{bmatrix}^T.
$$

Similarly as in the case of the (right) eigenvector $\mathbf{x}(\lambda)$, the components of $\mathbf{y}(\lambda)$ can be found from the recurrence relations

$$
y_k(\lambda) = \lambda y_{k-1}(\lambda) + \alpha_k, \quad k = 1, 2, \ldots, p,
$$

or, in expanded form, from

$$
y_k(\lambda) = \lambda^k + \sum_{j=1}^{k} \alpha_j \lambda^{k-j}, \quad k = 0, 1, \ldots, p,
$$

where $y_0(\lambda) = 1$. Assuming that $\mathbf{X}(\alpha, \beta)$ has a one-point spectrum $\{\lambda\}$, the factorization of this matrix is described by the following lemma.

Lemma 7.3.3 (see [56, 80, 293]) *Given the (right) eigenvector $\mathbf{x}(\lambda)$ and left eigenvector $\mathbf{y}(\lambda)$, the doubly companion matrix $\mathbf{X}$ given by (7.2.15) has a characteristic polynomial of the form*

$$\psi(w) = (w - \lambda)^{p+1}$$

if $\mathbf{X}$ can be factorized as

$$\mathbf{X} = \Psi(\mathbf{J} + \lambda\mathbf{I})\Psi^{-1}, \tag{7.3.8}$$

where the unit upper triangular matrices Ψ and Ψ^{-1} are given by

$$\Psi = \left[\begin{array}{ccccc} \dfrac{1}{p!}\mathbf{x}^{(p)}(\lambda) & \dfrac{1}{(p-1)!}\mathbf{x}^{(p-1)}(\lambda) & \cdots & \mathbf{x}'(\lambda) & \mathbf{x}(\lambda) \end{array} \right]$$

and

$$\Psi^{-1} = \left[\begin{array}{c} \mathbf{y}(\lambda)^T \\ \mathbf{y}'(\lambda)^T \\ \vdots \\ \dfrac{1}{(p-1)!}\mathbf{y}^{(p-1)}(\lambda)^T \\ \dfrac{1}{p!}\mathbf{y}^{(p)}(\lambda)^T \end{array} \right].$$

Proof: Since λ is an eigenvalue of $\mathbf{X}$ of multiplicity $p + 1$, relation (7.3.4) can be differentiated up to p times and we obtain

$$\mathbf{X}\frac{\mathbf{x}^{(k)}(\lambda)}{k!} = \lambda\frac{\mathbf{x}^{(k)}(\lambda)}{k!} + \frac{\mathbf{x}^{(k-1)}(\lambda)}{(k-1)!},$$

$k = 0, 1, \ldots, p$, where $\mathbf{x}^{(-1)}(\lambda) = \mathbf{0}$, $\mathbf{x}^{(0)}(\lambda) = \mathbf{x}(\lambda)$. Hence, the sequence $\mathbf{x}^{(k)}/k!$, $k = 0, 1, \ldots, p$, is a Jordan chain of length $p + 1$ associated with λ, and it is linearly independent (compare [197]). Writing the relation above in vector form, it follows that

$$\mathbf{X}\Psi = \lambda\Psi + \left[\begin{array}{ccccc} \dfrac{1}{(p-1)!}\mathbf{x}^{(p-1)}(\lambda) & \dfrac{1}{(p-2)!}\mathbf{x}^{(p-2)}(\lambda) & \cdots & \mathbf{x}(\lambda) & \mathbf{0} \end{array} \right] = \lambda\Psi + \Psi\mathbf{J},$$

which is equivalent to (7.3.8). The representation for the matrix Ψ^{-1} can be proved in a similar way by differentiating up to p times relation (7.3.7). ∎

There are two slight variants of the nilpotent matrix $\mathbf{K}$ that will be used in the representation of the matrices Ψ and Ψ^{-1} appearing in (7.3.8), when the matrix $\mathbf{X}$ given by (7.2.15) has a one-point spectrum $\sigma(\mathbf{X}) = \{\lambda\}$. These are denoted by $\mathbf{K}^+$ and $\mathbf{K}^-$ and are defined by

$$\mathbf{K}^+_{ij} = i\delta_{i+1,j}, \quad \mathbf{K}^-_{ij} = (p+1-i)\delta_{i+1,j}, \quad i,j = 1, 2, \ldots, p+1,$$

where δ_{ij} is the Kronecker delta. That is,

$$\mathbf{K}^+ = \begin{bmatrix} 0 & 1 & 0 & \cdots & 0 & 0 & 0 \\ 0 & 0 & 2 & \cdots & 0 & 0 & 0 \\ 0 & 0 & 0 & \cdots & 0 & 0 & 0 \\ \vdots & \vdots & \vdots & \ddots & \vdots & \vdots & \vdots \\ 0 & 0 & 0 & \cdots & 0 & p-1 & 0 \\ 0 & 0 & 0 & \cdots & 0 & 0 & p \\ 0 & 0 & 0 & \cdots & 0 & 0 & 0 \end{bmatrix},$$

$$\mathbf{K}^- = \begin{bmatrix} 0 & p & 0 & \cdots & 0 & 0 & 0 \\ 0 & 0 & p-1 & \cdots & 0 & 0 & 0 \\ 0 & 0 & 0 & \cdots & 0 & 0 & 0 \\ \vdots & \vdots & \vdots & \ddots & \vdots & \vdots & \vdots \\ 0 & 0 & 0 & \cdots & 0 & 2 & 0 \\ 0 & 0 & 0 & \cdots & 0 & 0 & 1 \\ 0 & 0 & 0 & \cdots & 0 & 0 & 0 \end{bmatrix}.$$

The columns of Ψ are generalized (right) eigenvectors, and the columns of $(\Psi^{-1})^T$ are generalized left eigenvectors. Matrices Ψ and Ψ^{-1} can be factorized into the product of two matrices, as shown in the following corollary.

Corollary 7.3.4 (Butcher and Wright [80, 293]) *Matrices Ψ and Ψ^{-1} can be factorized in the form*

$$\Psi = \beta(\mathbf{K})\exp(\lambda\mathbf{K}^-), \quad \Psi^{-1} = \exp(\lambda\mathbf{K}^+)\alpha(\mathbf{K}), \tag{7.3.9}$$

where $\alpha(\mathbf{K})$ and $\beta(\mathbf{K})$ are matrix polynomials with α and β defined by (7.3.1) and (7.3.2).

Proof: Consider first the formula for Ψ. Define the matrix $\Phi(w)$ by

$$\Phi(w) = \begin{bmatrix} \dfrac{1}{p!}\mathbf{x}^{(p)}(w) & \dfrac{1}{(p-1)!}\mathbf{x}^{(p-1)}(w) & \cdots & \mathbf{x}'(w) & \mathbf{x}(w) \end{bmatrix}$$

(i.e., by substituting $\lambda = w$ in Ψ). It can be verified by direct computations that

$$\Psi(0) = \begin{bmatrix} 1 & \beta_1 & \cdots & \cdots & \beta_{p-2} & \beta_{p-1} & \beta_p \\ 0 & 1 & \beta_1 & \cdots & \cdots & \beta_{p-2} & \beta_{p-1} \\ \vdots & 0 & 1 & \beta_1 & \cdots & \cdots & \beta_{p-2} \\ \vdots & \vdots & \ddots & \ddots & \ddots & & \vdots \\ 0 & \vdots & & \ddots & \ddots & \ddots & \vdots \\ 0 & 0 & \cdots & \cdots & 0 & 1 & \beta_1 \\ 0 & 0 & 0 & \cdots & \cdots & 0 & 1 \end{bmatrix}$$

$$= \mathbf{I} + \beta_1 \mathbf{K} + \beta_2 \mathbf{K}^2 + \cdots + \beta_p \mathbf{K}^p,$$

and since $\mathbf{K}^{p+1} = \mathbf{0}$, we obtain

$$\Phi(0) = \beta(\mathbf{K}).$$

Observe that the coefficient β_{p+1} does not affect the value of $\beta(\mathbf{K})$. We also have

$$\Phi'(w) = \begin{bmatrix} \dfrac{1}{p!}\mathbf{x}^{(p+1)}(w) & \dfrac{1}{(p-1)!}\mathbf{x}^{(p)}(w) & \cdots & \mathbf{x}''(w) & \mathbf{x}'(w) \end{bmatrix}$$

$$= \begin{bmatrix} \mathbf{0} & \dfrac{1}{(p-1)!}\mathbf{x}^{(p)}(w) & \cdots & \mathbf{x}''(w) & \mathbf{x}'(w) \end{bmatrix}$$

$$= \begin{bmatrix} \dfrac{1}{p!}\mathbf{x}^{(p)}(w) & \dfrac{1}{(p-1)!}\mathbf{x}^{(p-1)}(w) & \cdots & \mathbf{x}'(w) & \mathbf{x}(w) \end{bmatrix} \mathbf{K}^-$$

$$= \Phi(w)\mathbf{K}^-.$$

Hence,

$$\Phi'(w)\exp(-w\mathbf{K}^-) - \Phi(w)\mathbf{K}^- \exp(-w\mathbf{K}^-) = \mathbf{0}$$

or

$$\frac{d}{dw}\Big(\Phi(w)\exp(-w\mathbf{K}^-)\Big) = \mathbf{0}.$$

Integrating this relation from 0 to λ and taking into account that $\Phi(0) = \beta(\mathbf{K})$ we obtain $\Psi = \Phi(\lambda) = \beta(\mathbf{K})\exp(\lambda \mathbf{K}^-)$, which is the formula required for Ψ. The formula for Ψ^{-1} can be proved in a similar way. ∎

It may be more convenient to compute Ψ^{-1} from the relation

$$\Psi^{-1} = \exp(\lambda \mathbf{K}^-)^{-1}\beta(\mathbf{K})^{-1} = \exp(-\lambda \mathbf{K}^-)\beta(\mathbf{K})^{-1} \tag{7.3.10}$$

since this avoids the computation of matrix polynomial $\alpha(\mathbf{K})$, leaving everything in terms of $\beta(\mathbf{K})$ only. Observe that for any β_k, $k = 1, 2, \ldots, p$, and the given value of λ, we can always choose the coefficients α_k, $k = 1, 2, \ldots, p+1$, so that the doubly companion matrix $\mathbf{X}$ has a one-point spectrum $\sigma(\mathbf{X}) = \{\lambda\}$. This can be done using relation (7.3.6) with $\widetilde{\psi}(w)$ given by

$$\widetilde{\psi}(w) = w^{p+1}\psi\left(\frac{1}{w}\right) = (1 - \lambda w)^{p+1}.$$

We can then compute Ψ and Ψ^{-1} using (7.3.9) or the first equation of (7.3.9) and (7.3.10) and matrix $\mathbf{X}$ from (7.3.8). Observe also that the parameter β_{p+1} is redundant since it does not affect the values of $\beta(\mathbf{X})$. This parameter enters the definition of the matrix $\mathbf{X}$ only through the combination $-\alpha_{p+1} - \beta_{p+1}$, and without loss of generality we can assume that $\beta_{p+1} = 0$.

The next corollary corresponds to the special case when $\lambda = 0$, which is relevant, for example, to (explicit) type 1 methods.

Corollary 7.3.5 (Butcher and Wright [80, 293]) *Assume that the polynomials $\alpha, \beta \in \Pi$ are such that $(\alpha\beta)(w) = 1$. Then the corresponding doubly companion matrix $\mathbf{X}(\alpha, \beta)$ has a one-point spectrum $\sigma(\mathbf{X}(\alpha, \beta)) = \{0\}$ and the following factorization holds:*

$$\mathbf{X}(\alpha, \beta) = \beta(\mathbf{K})\,\mathbf{J}\,\alpha(\mathbf{K}) = \beta(\mathbf{K})\,\mathbf{J}\,\beta(\mathbf{K})^{-1}. \tag{7.3.11}$$

Proof: It follows from the relation (7.3.6) that $\widetilde{\psi}(w) = (\alpha\beta)(w) = 1$. Hence, $\psi(w) = w^{p+1}\widetilde{\psi}(w) = w^{p+1}$ and, as a consequence, $\mathbf{X}(\alpha, \beta)$ has a one-point spectrum $\sigma(\mathbf{X}(\alpha, \beta)) = \{0\}$. The factorization (7.3.11) is now a direct consequence of Lemma 7.3.3, Corollary 7.3.4, and formula (7.3.10) with $\lambda = 0$. ∎

It is useful to extend this result so that the polynomials $\alpha(w)$ or $\beta(w)$ could be multiplied by a further polynomial $\gamma(w)$. We have the following result.

Corollary 7.3.6 (Butcher and Wright [80, 293]) *Assume that $\alpha, \beta, \gamma \in \Pi$ and $(\alpha\beta\gamma)(w) = 1$. Then the doubly companion matrices defined by $\mathbf{X}(\alpha\gamma, \beta)$ and $\mathbf{X}(\alpha, \gamma\beta)$ have a one-point spectrum $\{0\}$. Moreover,*

$$\mathbf{X}(\alpha\gamma, \beta) = \mathbf{X}(\alpha, \beta)\,\gamma(\mathbf{K})$$

and

$$\mathbf{X}(\alpha, \gamma\beta) = \gamma(\mathbf{K})\,\mathbf{X}(\alpha, \beta).$$

Proof: Since $(\alpha\gamma)(w)\beta(w) = 1$ and $\alpha(w)(\gamma\beta)(w) = 1$, the conclusions follow directly from Corollary 7.3.5. ∎

The result from Corollary 7.3.5 can be extended so that it is applicable for $\lambda > 0$ by choosing $\alpha(w)$ and $\beta(w)$ so that $\alpha\beta = \widetilde{\psi}$, where $\widetilde{\psi}(w) = (1 - \lambda w)^{p+1}$. We have the following lemma.

Lemma 7.3.7 (Butcher and Wright [80, 293]) *Assume that $\alpha(w)\beta(w) = \widetilde{\psi}(w) + O(w^{p+2})$, where the characteristic function of $\mathbf{X}(\alpha, \beta)$ takes the form*

$\widetilde{\psi}(w) = (1 - \lambda w)^{p+1}$. *Then* $\mathbf{X}(\alpha, \beta)$ *has a one-point spectrum* $\sigma(\mathbf{X}(\alpha, \beta)) = \{\lambda\}$ *and the following factorization holds:*

$$\mathbf{X}(\alpha, \beta) = \beta(\mathbf{K}) \, \mathbf{X}(\alpha\beta, 1) \, \beta^{-1}(\mathbf{K}). \qquad (7.3.12)$$

Here $\mathbf{X}(\alpha\beta, 1)$ *is the companion matrix corresponding to* $\alpha\beta$.

Proof: We have

$$\psi(w) = w^{p+1} \widetilde{\psi}\left(\frac{1}{w}\right) = (w - \lambda)^{p+1}$$

and it follows that $\sigma(\mathbf{X}(\alpha, \beta)) = \{\lambda\}$. Let $\gamma = \alpha\beta$; that is,

$$\gamma(w) = 1 + \sum_{i=1}^{p+1} \gamma_i w^i = (1 - \lambda w)^{p+1} + O(w^{p+2}).$$

For convenience, we also introduce the notation $\overline{\beta} = \beta^{-1}$, where $\overline{\beta}$ has coefficients $\overline{\beta}_1, \overline{\beta}_2, \ldots, \overline{\beta}_{p+1}$. Using the relations

$$\mathbf{K}^i \mathbf{e}_1 = \mathbf{0}, \quad i > 0, \quad \mathbf{e}_j^T \mathbf{K}^k = \mathbf{e}_{j+k}^T,$$

where by convention $\mathbf{e}_{j+k} = \mathbf{0}$ if $j + k > p + 1$, we also have

$$\beta(\mathbf{K}) \, \mathbf{X}(\gamma, 1) \, \beta^{-1}(\mathbf{K}) = \left(\mathbf{I} + \sum_{i=1}^{p} \beta_i \mathbf{K}^i\right)\left(\mathbf{J} - \sum_{j=1}^{p+1} \gamma_j \mathbf{e}_1 \mathbf{e}_j^T\right)\left(\mathbf{I} + \sum_{k=1}^{p} \overline{\beta}_k \mathbf{K}^k\right)$$

$$= \mathbf{J} + \sum_{k=1}^{p} \overline{\beta}_k \mathbf{J} \mathbf{K}^k - \sum_{j=1}^{p+1} \gamma_j \mathbf{e}_1 \mathbf{e}_j^T - \sum_{j=1}^{p+1}\sum_{k=1}^{p} \gamma_j \overline{\beta}_k \mathbf{e}_1 \mathbf{e}_{j+k}^T$$

$$+ \sum_{i=1}^{p} \beta_i \mathbf{K}^i \mathbf{J} + \sum_{i=1}^{p}\sum_{k=1}^{p} \beta_i \overline{\beta}_k \mathbf{K}^i \mathbf{J} \mathbf{K}^k.$$

It can be verified that

$$\mathbf{K}^i \mathbf{J} \mathbf{K}^k = \begin{cases} \mathbf{J}, & i = k = 0, \\ \mathbf{K}^{k-1} - \mathbf{e}_1 \mathbf{e}_k^T, & i = 0, \ k > 0, \\ \mathbf{K}^{i-1} - \mathbf{e}_{p+2-i} \mathbf{e}_{p+1}^T, & i > 0, \ k = 0, \\ \mathbf{K}^{i+k-1}, & i, j > 0. \end{cases}$$

Using these relations and observing that $\mathbf{K}^p = \mathbf{e}_1 \mathbf{e}_{p+1}^T$ and $\mathbf{K}^j = \mathbf{0}$ for $j > p$, we obtain

$$\beta(\mathbf{K}) \, \mathbf{X}(\gamma, 1) \, \beta^{-1}(\mathbf{K}) = \mathbf{J} - \sum_{k=1}^{p+1} \overline{\beta}_k \mathbf{e}_1 \mathbf{e}_k^T - \sum_{j=1}^{p+1} \gamma_j \mathbf{e}_1 \mathbf{e}_j^T - \sum_{j=1}^{p+1}\sum_{k=1}^{p+1} \gamma_j \overline{\beta}_k \mathbf{e}_1 \mathbf{e}_{j+k}^T$$

$$- \sum_{i=1}^{p+1} \beta_i \mathbf{e}_{p+2-i} \mathbf{e}_{p+1}^T + \sum_{i=1}^{p+1} \beta_i \mathbf{K}^{i-1} + \sum_{k=1}^{p+1} \overline{\beta}_k \mathbf{K}^{k-1} + \sum_{i=1}^{p+1}\sum_{k=1}^{p+1} \beta_i \overline{\beta}_k \mathbf{K}^{i+k-1}.$$

Since $\alpha = \gamma\overline{\beta}$, that is,

$$(1 + \gamma_1 w + \cdots + \gamma_p w^p + \gamma_{p+1} w^{p+1})(1 + \overline{\beta}_1 w + \cdots + \overline{\beta}_p w^p + \overline{\beta}_{p+1} w^{p+1})$$
$$= 1 + \alpha_1 w + \cdots + \alpha_p w^p + \alpha_{p+1} w^{p+1} + O(w^{p+2}),$$

it follows that

$$\sum_{j=1}^{p+1} \gamma_j \mathbf{e}_1 \mathbf{e}_j^T + \sum_{k=1}^{p+1} \overline{\beta}_k \mathbf{e}_1 \mathbf{e}_k^T + \sum_{j=1}^{p+1}\sum_{k=1}^{p+1} \gamma_j \overline{\beta}_k \mathbf{e}_1 \mathbf{e}_{j+k}^T = \sum_{i=1}^{p+1} \alpha_i \mathbf{e}_1 \mathbf{e}_i^T.$$

Similarly, since $\beta(w)\overline{\beta}(w) = 1 + O(w^{p+2})$, all coefficients of $\beta\overline{\beta}$ are equal to zero and we have

$$\sum_{i=1}^{p+1} \beta_i \mathbf{K}^{i-1} + \sum_{k=1}^{p+1} \overline{\beta}_k \mathbf{K}^{k-1} + \sum_{i=1}^{p+1}\sum_{k=1}^{p+1} \beta_i \overline{\beta}_k \mathbf{K}^{i+k-1} = \sum_{k=1}^{p+1} (\beta\overline{\beta})_k \mathbf{K}^{k-1} = \mathbf{0}.$$

Hence,

$$\beta(\mathbf{K})\,\mathbf{X}(\gamma, 1)\,\beta^{-1}(\mathbf{K}) = \mathbf{J} - \sum_{i=1}^{p+1} \alpha_i \mathbf{e}_1 \mathbf{e}_i - \sum_{i=1}^{p+1} \beta_i \mathbf{e}_{p+2-i} \mathbf{e}_{p+1}^T = \mathbf{X}(\alpha, \beta),$$

which concludes the proof.   ∎

7.4   TRANSFORMATIONS BETWEEN METHOD ARRAYS

In this section we describe transformations between the coefficient matrices of methods that will enable us to find GLMs with IRKS property. The transformed methods and the transformed IRKS conditions will be derived. We assume that the coefficient matrix $\mathbf{A}$ has the form (2.7.1) with $\lambda = 0$ or $\lambda > 0$, respectively, which corresponds to GLMs (7.1.1) of type 1 or 2.

We first present two results that will aid in the process of deriving GLMs with IRKS. The first considers a special connection between lower triangular matrices.

Lemma 7.4.1 (Butcher and Wright [80, 293]) *Assume that* $\mathbf{L} \in \mathbb{R}^{n \times n}$ *is a strictly lower triangular matrix, with* $n = p + 1$, *satisfying*

$$\mathbf{ML} \equiv \mathbf{JM}.$$

Then $\mathbf{M}$ *is lower triangular.*

Proof: We prove this result by induction. The lemma is clearly true for $n = 1$. Assume that it is true for $p = n - 1$ and let $\mathbf{M}, \mathbf{J}, \mathbf{L} \in \mathbb{R}^{n \times n}$. Since $\mathbf{JM} \equiv \mathbf{ML}$ implies that $\mathbf{JMe}_{p+1} \equiv \mathbf{MLe}_{p+1}$ and we have $\mathbf{Le}_{p+1} = \mathbf{0}$, it follows that the first p components of $\mathbf{M}\mathbf{e}_{p+1}$ are equal to zero, so that the last column of $\mathbf{M}$ is a multiple of $\mathbf{e}_{p+1}$. Let us partition $\mathbf{M}$, $\mathbf{J}$, and $\mathbf{L}$ as follows:

$$
\mathbf{M} = \left[\begin{array}{c|c} M & 0 \\ \hline m^T & \widehat{m} \end{array}\right], \quad
\mathbf{J} = \left[\begin{array}{c|c} J & 0 \\ \hline \mathbf{e}_p^T & 0 \end{array}\right], \quad
\mathbf{L} = \left[\begin{array}{c|c} L & 0 \\ \hline l^T & 0 \end{array}\right],
$$

where $M, J, L \in \mathbb{R}^{(n-1)\times(n-1)}$ and $m, l \in \mathbb{R}^{n-1}$. Then $\mathbf{J}\mathbf{M} \equiv \mathbf{M}\mathbf{L}$ implies that $ML \equiv JM$, which, by the induction hypothesis, implies that M, and therefore $\mathbf{M}$, is lower triangular. ■

The second result shows that the first of the IRKS conditions (7.2.8) can be strengthened to equality between matrices. We have the following lemma.

Lemma 7.4.2 (Butcher and Wright [80, 293]) *Given the coefficients* $\beta_1, \beta_2, \ldots, \beta_p$ *and the real parameter* λ, *choose coefficients* $\alpha_1, \alpha_2, \ldots, \alpha_{p+1}$, *so that the characteristic polynomial* $\psi(w) = \det(w\mathbf{I} - \mathbf{X}(\alpha, \beta))$ *is of the form* $\psi(w) = (w - \lambda)^{p+1}$. *Then the condition* $\mathbf{B}\mathbf{A} \equiv \mathbf{X}\mathbf{B}$ *implies that*

$$\mathbf{B}\mathbf{A} = \mathbf{X}\mathbf{B}. \tag{7.4.1}$$

Proof: It follows from Lemma 7.3.3 that $\mathbf{X} = \Psi(\mathbf{J} + \lambda\mathbf{I})\Psi^{-1}$, so that the condition $\mathbf{B}\mathbf{A} \equiv \mathbf{X}\mathbf{B}$ takes the form

$$\mathbf{B}\mathbf{A} \equiv \Psi(\mathbf{J} + \lambda\mathbf{I})\Psi^{-1}\mathbf{B}.$$

Since Ψ^{-1} is unit upper triangular, we have $\Psi^{-1}\mathbf{e}_1 = \mathbf{e}_1$, and it follows that

$$\Psi^{-1}\mathbf{B}\mathbf{A} \equiv (\mathbf{J} + \lambda\mathbf{I})\Psi^{-1}\mathbf{B}.$$

This can be rearranged in the form

$$\Psi^{-1}\mathbf{B}(\mathbf{A} - \lambda\mathbf{I}) \equiv \mathbf{J}\Psi^{-1}\mathbf{B}. \tag{7.4.2}$$

Since $\mathbf{A} - \lambda\mathbf{I}$ is strictly lower triangular, it follows from Lemma 7.4.1 that $\Psi^{-1}\mathbf{B}$ is lower triangular, which implies further that the left- and right-hand sides of (7.4.2) are equal. This is equivalent to (7.4.1), which completes the proof. ■

The results of this section are also applicable in a somewhat more general case, where the coefficient matrix $\mathbf{A}$ is not necessarily of the form (2.7.1), but it is only assumed to have a one-point spectrum $\sigma(\mathbf{A}) = \{\lambda\}$. Then it follows from real Schur decomposition [136] that this matrix can be reduced to a matrix $\overline{\mathbf{A}}$ of diagonally implicit form (2.7.1) by an orthogonal similarity transformation

$$\overline{\mathbf{A}} = \mathbf{W}^T\mathbf{A}\mathbf{W}.$$

If $\mathbf{W} = \mathbf{I}$, then $\mathbf{A} = \overline{\mathbf{A}}$ and the method reduces to GLM of type 1 or 2. However, by allowing $\mathbf{W} \neq \mathbf{I}$, more general methods, such as, for example, ESIRK considered by Butcher et al. [55, 56, 59] or diagonally extended singly-implicit Runge-Kutta effective order (DESIRE) methods considered by Butcher and Diamantakis [61], can be derived. This is discussed in detail by Butcher and Wright [81, 293], where new ESIRK and DESIRE methods are also constructed.

The significance of Lemma 7.4.2 consists of the fact that relation (7.4.1) can be used to find $\mathbf{A}$ given that we know $\mathbf{B}$. Clearly,

$$\mathbf{A} = \mathbf{B}^{-1}\mathbf{X}\mathbf{B}$$

if $\mathbf{B}$ is nonsingular. However, the relation (7.4.1) can be used to find $\mathbf{A}$ even if $\mathbf{B}$ is singular, although in some cases $\mathbf{A}$ may not be unique [293]. Once $\mathbf{A}$ and $\mathbf{B}$ are known, the remaining coefficient matrices $\mathbf{U}$ and $\mathbf{V}$ can be determined from relations (7.1.6) and (7.1.7), which follow from stage order and order conditions. Therefore, the method is determined once we know $\mathbf{B}$, and it is crucial to find a convenient way to represent this matrix. The transformation is used to represent $\mathbf{B}$ in such a way that the remaining conditions (7.2.9) and (7.2.10) for IRKS are satisfied automatically.

Denote by $\mathbf{M}$ the original method and by $\widetilde{\mathbf{M}}$ the transformed method,

$$\mathbf{M} = \left[\begin{array}{c|c} \mathbf{A} & \mathbf{U} \\ \hline \mathbf{B} & \mathbf{V} \end{array}\right], \quad \widetilde{\mathbf{M}} = \left[\begin{array}{c|c} \widetilde{\mathbf{A}} & \widetilde{\mathbf{U}} \\ \hline \widetilde{\mathbf{B}} & \widetilde{\mathbf{V}} \end{array}\right],$$

and assume that both methods have the same abscissa vector $\mathbf{c} = \widetilde{\mathbf{c}}$. To construct $\mathbf{M}$, we first construct $\widetilde{\mathbf{M}}$ with some appropriate properties, and then back transform to find the original method $\mathbf{M}$. This approach was proposed by Butcher and Wright [79] in a somewhat more restricted setting, where the definition of IRKS property was less general and where the matrix $\mathbf{J}$ instead of $\mathbf{X}$ was used (compare [79, Definition 2.2]). We now use Lemma 7.3.3 to transform method $\mathbf{M}$ and the IRKS conditions (7.4.1) and (7.2.9). Substituting (7.3.8) into (7.4.1) and rearranging terms, we obtain

$$\Psi^{-1}\mathbf{B}(\mathbf{A} - \lambda\mathbf{I}) = \mathbf{J}\Psi^{-1}\mathbf{B}.$$

This can be written in the form

$$\widetilde{\mathbf{B}}\widetilde{\mathbf{A}} = \mathbf{J}\widetilde{\mathbf{B}}, \tag{7.4.3}$$

where

$$\widetilde{\mathbf{A}} = \mathbf{A} - \lambda\mathbf{I}, \quad \widetilde{\mathbf{B}} = \Psi^{-1}\mathbf{B}.$$

Since $\widetilde{\mathbf{A}}$ is strictly lower triangular, it follows from Lemma 7.4.1 that $\widetilde{\mathbf{B}}$ is lower triangular. Substituting (7.3.8) into (7.2.9) and simplifying, we obtain

$$\mathbf{B}\mathbf{U} \equiv \Psi\mathbf{J}\Psi^{-1}\mathbf{V} - \mathbf{V}\Psi\mathbf{J}\Psi^{-1}.$$

Using the properties of the equivalence relation $\equiv$ listed in Section 7.1, this can be rewritten in the form

$$\widetilde{\mathbf{B}}\widetilde{\mathbf{U}} \equiv \mathbf{J}\widetilde{\mathbf{V}} - \widetilde{\mathbf{V}}\mathbf{J}, \qquad (7.4.4)$$

where

$$\widetilde{\mathbf{U}} = \mathbf{U}\Psi, \quad \widetilde{\mathbf{V}} = \Psi^{-1}\mathbf{V}\Psi.$$

Consider next IRKS condition (7.2.10) in Definition 7.2.3. Since Ψ^{-1} and Ψ are unit upper triangular matrices and $\mathbf{V}\mathbf{e}_1 = \mathbf{e}_1$, we have

$$\widetilde{\mathbf{V}}\mathbf{e}_1 = \Psi^{-1}\mathbf{V}\Psi\mathbf{e}_1 = \Psi^{-1}\mathbf{V}\mathbf{e}_1 = \Psi^{-1}\mathbf{e}_1 = \mathbf{e}_1,$$

which is the preconsistency condition for the transformed method $\widetilde{\mathbf{M}}$. We also have

$$\det(w\mathbf{I} - \widetilde{\mathbf{V}}) = \det(w\mathbf{I} - \Psi^{-1}\mathbf{V}\Psi) = \det(w\mathbf{I} - \mathbf{V}),$$

so that the transformed condition (7.2.10) takes the form

$$\det(w\mathbf{I} - \widetilde{\mathbf{V}}) = w^p(w - 1). \qquad (7.4.5)$$

Next we express conditions (7.1.6) and (7.1.7) in terms of the transformed method $\widetilde{\mathbf{M}}$. Postmultiplying (7.1.6) by Ψ and rearranging terms, we obtain

$$\mathbf{U}\Psi = \mathbf{C}\Psi - \mathbf{A}\mathbf{C}\mathbf{K}\Psi = \mathbf{C}\Psi - \lambda\mathbf{C}\mathbf{K}\Psi - \widetilde{\mathbf{A}}\mathbf{C}\mathbf{K}\Psi.$$

Hence,

$$\widetilde{\mathbf{U}} = \mathbf{C}(\mathbf{I} - \lambda\mathbf{K})\Psi - \widetilde{\mathbf{A}}\mathbf{C}\mathbf{K}\Psi. \qquad (7.4.6)$$

Similarly, premultiplying (7.1.7) by Ψ^{-1} and postmultiplying by Ψ, we get

$$\Psi^{-1}\mathbf{V}\Psi = \Psi^{-1}\mathbf{E}\Psi - \widetilde{\mathbf{B}}\mathbf{C}\mathbf{K}\Psi.$$

Using the formula for ψ in (7.3.9) and the formula for Ψ^{-1} in (7.3.10), this can be written in the form

$$\widetilde{\mathbf{V}} = \exp(-\lambda\mathbf{K}^-)\beta^{-1}(\mathbf{K})\mathbf{E}\beta(\mathbf{K})\exp(\lambda\mathbf{K}^{-1}) - \widetilde{\mathbf{B}}\mathbf{C}\mathbf{K}\Psi.$$

It is easy to verify that the upper triangular Toeplitz matrices $\mathbf{E}$ and $\beta(\mathbf{K})$ commute and it follows that

$$\widetilde{\mathbf{V}} = \mathbf{F} - \widetilde{\mathbf{B}}\mathbf{C}\mathbf{K}\Psi, \qquad (7.4.7)$$

where

$$\mathbf{F} = \exp(-\lambda\mathbf{K}^-)\mathbf{E}\exp(\lambda\mathbf{K}^-).$$

The structure of this matrix is described by the following lemma.

Lemma 7.4.3 (Butcher and Wright [80, 293]) *Let λ be a real number and $\mathbf{K}^{-1}$, $\mathbf{E}$, and $\mathbf{K}$ be the matrices defined in Sections 7.1 and 7.3. Then*

$$\exp(-\lambda \mathbf{K}^-)\mathbf{E}\exp(\lambda \mathbf{K}^-) = \exp\left(\mathbf{K}(\mathbf{I}+\lambda \mathbf{K})^{-1}\right). \tag{7.4.8}$$

Proof: Observe first that it follows from the Leibnitz rule for the i-fold derivative applied to the product of x and $x^{-1}f(x)$, where $f(x)$ is a sufficiently smooth function, that

$$\left(x\frac{f(x)}{x}\right)^{(i)} = \sum_{j=0}^{i}\binom{i}{j}\left(\frac{f(x)}{x}\right)^{(i-j)}x^{(j)} = x\left(\frac{f(x)}{x}\right)^{(i)} + i\left(\frac{f(x)}{x}\right)^{(i-1)}$$

or

$$\frac{1}{x}\left(f(x)\right)^{(i)} = \left(\frac{f(x)}{x}\right)^{(i)} + \frac{i}{x}\left(\frac{f(x)}{x}\right)^{(i-1)}. \tag{7.4.9}$$

This formula will be used when $f(x)$ is a polynomial of degree p:

$$f(x) = x^p + a_1 x^{p-1} + \cdots + a_{p-1}x + a_p = \mathbf{a}^T\xi,$$

where

$$\mathbf{a} = \begin{bmatrix} 1 & a_1 & \cdots & a_{p-1} & a_p \end{bmatrix}^T, \quad \xi = \begin{bmatrix} x^p & x^{p-1} & \cdots & x & 1 \end{bmatrix}^T,$$

and we will be interested in terms with nonnegative powers of x. It can be verified by induction with respect to i that

$$\frac{1}{x}\left(f(x)\right)^{(i)} = \mathbf{a}^T(\mathbf{K}^-)^i\mathbf{K}\xi + O(x^{-1}),$$

$$\left(\frac{f(x)}{x}\right)^{(i)} = \mathbf{a}^T\mathbf{K}(\mathbf{K}^-)^i\xi + O(x^{-1}), \tag{7.4.10}$$

$$\frac{i}{x}\left(\frac{f(x)}{x}\right)^{(i-1)} = i\mathbf{a}^T\mathbf{K}(\mathbf{K}^-)\mathbf{K}\xi + O(x^{-1}),$$

where $O(x^{-1})$ stands for terms with negative powers of x. We illustrate the proof of the first of the formulas above, which is rewritten as

$$\left(f(x)\right)^{(i)} = x\mathbf{a}^T(\mathbf{K}^-)^i\mathbf{K}\xi + O(1),$$

where $O(1)$ includes constant terms and terms with a negative power of x. Since $\mathbf{a}^T\xi$ and $x\mathbf{a}^T\mathbf{K}\xi$ differ by a constant, this formula is satisfied for $i = 0$.

Assuming that it is true for i, we have

$$
\begin{aligned}
\left(f(x)\right)^{(i+1)} &= \mathbf{a}^T(\mathbf{K}^-)^i\mathbf{K}(\xi + x\xi') + O(1) \\
&= \mathbf{a}^T(\mathbf{K}^-)^i\mathbf{K}\left[\ (p+1)x^p \quad px^{p-1} \quad \cdots \quad 2x \quad 1\ \right]^T + O(1) \\
&= \mathbf{a}^T(\mathbf{K}^-)^i\left[\ px^{p-1} \quad (p-1)x^{p-2} \quad \cdots \quad 2x \quad 1 \quad 0\ \right]^T + O(1) \\
&= \mathbf{a}^T(\mathbf{K}^-)^i\mathbf{K}^-\left[\ x^p \quad x^{p-1} \quad \cdots \quad x \quad 1\ \right]^T + O(1) \\
&= \mathbf{a}^T(\mathbf{K}^-)^{i+1}\mathbf{K}\left[\ x^{p+1} \quad x^p \quad \cdots \quad x^2 \quad x\ \right]^T + O(1) \\
&= x\mathbf{a}^T(\mathbf{K}^-)^{i+1}\mathbf{K}\xi + O(1),
\end{aligned}
$$

which is the same formula as that corresponding to $i+1$.

Substituting (7.4.10) into (7.4.9) and comparing terms with nonnegative powers of x, we obtain

$$\mathbf{a}^T\left((\mathbf{K}^-)^i\mathbf{K} - \mathbf{K}(\mathbf{K}^-)^i - i\mathbf{K}(\mathbf{K}^-)^{i-1}\mathbf{K}\right)\xi = 0.$$

Since this relation holds for any value of x and any choice of the components of $\mathbf{a}$, it follows that

$$(\mathbf{K}^-)^i\mathbf{K} - \mathbf{K}(\mathbf{K}^-)^i - i\mathbf{K}(\mathbf{K}^-)^{i-1}\mathbf{K} = 0.$$

Multiplying this equation by $\lambda^i/i!$ and summing for $i = 0, 1, \ldots$, we obtain

$$\exp(\lambda\mathbf{K}^-)\mathbf{K} = \mathbf{K}\exp(\lambda\mathbf{K}^-)(\mathbf{I} + \lambda\mathbf{K}),$$

or, what is equivalent,

$$\exp(-\lambda\mathbf{K}^-)\mathbf{K}\exp(\lambda\mathbf{K}^-) = \mathbf{K}(\mathbf{I} - \lambda\mathbf{K})^{-1}.$$

This relation implies that

$$\exp(-\lambda\mathbf{K}^-)\mathbf{K}^j\exp(\lambda\mathbf{K}^-) = \left(\mathbf{K}(\mathbf{I} - \lambda\mathbf{K})^{-1}\right)^j,$$

$j = 0, 1, \ldots$, and multiplying this by $1/j!$ and summing over j, we obtain equation (7.4.8). This completes the proof. ∎

The next results analyze the structure of the matrix $\exp(\mathbf{S}(\mathbf{I} + \lambda\mathbf{S})^{-1})$, where $\mathbf{S}$ satisfies the condition $\rho(\lambda\mathbf{S}) < 1$. Here $\rho(\mathbf{A})$ stands for the spectral radius of the matrix $\mathbf{A}$. This condition is clearly satisfied for any λ if $\mathbf{S} = \mathbf{K}$. Although the structure of this matrix could be concluded from Lemma 3.5.1 with $\mathbf{S}$ playing the role of $\widehat{z}$, in what follows we describe it in a somewhat different way following Wright [293]. The reason for this is that it is more convenient to use somewhat different notation in the context of GLMs with

IRKS than that corresponding to the case of DIMSIMs of types 1 and 2 discussed in Sections 3.4 and 3.5. We first need the following lemma.

Lemma 7.4.4 (Wright [293]) *For any integer $l \geq 1$ and a complex number w such that $|w| < 1$, we have*

$$\sum_{k=0}^{\infty} \binom{k+l-1}{k} w^k = \left(\sum_{k=0}^{\infty} w^k\right)^l = \frac{1}{(1-w)^l}. \tag{7.4.11}$$

Proof: The result is clearly true for $l = 1$. Assuming that it is true for $l > 1$, it follows that

$$\left(\sum_{k=0}^{\infty} w^k\right)^{l+1} = \left(\sum_{m=0}^{\infty} w^m\right)^l \sum_{n=0}^{\infty} w^n = \sum_{m=0}^{\infty} \binom{m+l-1}{m} w^m \sum_{n=0}^{\infty} w^n.$$

Setting $k = m + n$ and rearranging the double sum, we obtain

$$\left(\sum_{k=0}^{\infty} w^k\right)^{l+1} = \sum_{k=0}^{\infty} \sum_{m=0}^{k} \binom{m+l-1}{m} w^k.$$

It can be verified by induction with respect to k that

$$\sum_{m=0}^{k} \binom{m+l-1}{m} = \binom{k+l}{k}.$$

Hence,

$$\left(\sum_{k=0}^{\infty} w^k\right)^{l+1} = \sum_{k=0}^{\infty} \binom{k+l}{k} w^k,$$

which corresponds to (7.4.11) with l replaced by $l+1$. This completes the proof. ∎

We are now ready to describe the structure of $\exp(\mathbf{S}(\mathbf{I} + \lambda \mathbf{S})^{-1})$. We have the following lemma.

Lemma 7.4.5 (Butcher and Wright [80, 293]) *For any real number λ and the matrix $\mathbf{S}$ such that $\rho(\lambda \mathbf{S}) < 1$, we have the following representation:*

$$\exp\left(\mathbf{S}(\mathbf{I} + \lambda \mathbf{S})^{-1}\right) = \mathbf{I} + \sum_{i=1}^{\infty} N_i(\lambda)\mathbf{S}^i, \tag{7.4.12}$$

where

$$N_i(\lambda) = \sum_{k=0}^{i-1} \binom{i-1}{k} \frac{(-\lambda)^k}{(i-k)!}. \tag{7.4.13}$$

Proof: Expanding $\exp(\mathbf{S}(\mathbf{I} + \lambda\mathbf{S})^{-1})$ into a Taylor series, we obtain

$$\exp\left(\mathbf{S}(\mathbf{I} + \lambda\mathbf{S})^{-1}\right) = \mathbf{I} + \sum_{l=1}^{\infty} \frac{1}{l!}\mathbf{S}^l(\mathbf{I} + \lambda\mathbf{S})^{-l}.$$

Since $\rho(\lambda\mathbf{S}) < 1$, it follows from Lemma 7.4.4 that the negative binomial expansion of $(\mathbf{I} + \lambda\mathbf{S})^{-l}$ is given by

$$(\mathbf{I} + \lambda\mathbf{S})^{-l} = \sum_{k=0}^{\infty} \binom{k+l-1}{k}(-\lambda\mathbf{S})^k.$$

Substituting this relation into the expansion of $\exp(\mathbf{S}(\mathbf{I} + \lambda\mathbf{S})^{-1})$, we get

$$\exp\left(\mathbf{S}(\mathbf{I} + \lambda\mathbf{S})^{-1}\right) = \mathbf{I} + \sum_{l=1}^{\infty} \frac{1}{l!}\mathbf{S}^l \sum_{k=0}^{\infty} \binom{k+l-1}{k}(-\lambda\mathbf{S})^k.$$

Setting $i = l + k$ and rearranging the double sum, it follows that

$$\exp\left(\mathbf{S}(\mathbf{I} + \lambda\mathbf{S})^{-1}\right) = \mathbf{I} + \sum_{i=1}^{\infty}\sum_{k=0}^{i-1} \binom{i-1}{k}\frac{(-\lambda)^k}{(i-k)!}\mathbf{S}^i.$$

This is equivalent to (7.4.12) with $N_i(\lambda)$ defined by (7.4.13), and the proof is complete. ∎

The polynomials $N_i(\lambda)$ are listed in Table 7.4.1 for $i = 1, 2, \ldots, 6$.

i	$N_i(\lambda)$
1	1
2	$\frac{1}{2} - \lambda$
3	$\frac{1}{6} - \lambda + \lambda^2$
4	$\frac{1}{24} - \frac{1}{2}\lambda + \frac{3}{2}\lambda^2 - \lambda^3$
5	$\frac{1}{120} - \frac{1}{6}\lambda + \lambda^2 - 2\lambda^3 + \lambda^4$
6	$\frac{1}{720} - \frac{1}{24}\lambda + \frac{5}{12}\lambda^2 - \frac{5}{3}\lambda^3 + \frac{5}{2}\lambda^4 - \lambda^5$

Table 7.4.1   Polynomials $N_i(\lambda)$ for $i = 1, 2, \ldots, 6$

Next we analyze further IRKS condition (7.4.4), which will aid in the derivation of the coefficient matrix $\widetilde{\mathbf{B}}$ of the transformed method $\widetilde{\mathbf{M}}$. Substituting (7.4.6) and (7.4.7) into (7.4.4), we obtain

$$\widetilde{\mathbf{B}}\mathbf{C}(\mathbf{I} - \lambda\mathbf{K})\Psi - \widetilde{\mathbf{B}}\widetilde{\mathbf{A}}\mathbf{C}\mathbf{K}\Psi \equiv \mathbf{J}(\mathbf{F} - \widetilde{\mathbf{B}}\mathbf{C}\mathbf{K}\Psi) - (\mathbf{F} - \widetilde{\mathbf{B}}\mathbf{C}\mathbf{K}\Psi)\mathbf{J},$$

and using IRKS condition (7.4.3) it follows that

$$\widetilde{\mathbf{B}}\mathbf{C}\big((\mathbf{I} - \lambda\mathbf{K})\Psi - \mathbf{K}\Psi\mathbf{J}\big) \equiv \mathbf{J}\mathbf{F} - \mathbf{F}\mathbf{J}.$$

Since

$$(\mathbf{I} - \lambda\mathbf{K})\Psi - \mathbf{K}\Psi\mathbf{J} = \Psi - \mathbf{K}\Psi(\mathbf{J} + \lambda\mathbf{I})\Psi^{-1}\Psi = (\mathbf{I} - \mathbf{K}\mathbf{X})\Psi,$$

using (7.3.8) we obtain

$$\widetilde{\mathbf{B}}\mathbf{C}(\mathbf{I} - \mathbf{K}\mathbf{X})\Psi \equiv \mathbf{J}\mathbf{F} - \mathbf{F}\mathbf{J}. \tag{7.4.14}$$

Since Ψ is unit upper triangular (compare Lemma 7.3.3) it can be verified by direct computations that the matrix $(\mathbf{I} - \mathbf{K}\mathbf{X})\Psi$ appearing on the left-hand side of (7.4.14) takes the simple form

$$(\mathbf{I} - \mathbf{K}\mathbf{X})\Psi = \beta(\mathbf{K})\mathbf{e}_{p+1}\mathbf{e}_{p+1}^T, \tag{7.4.15}$$

which has only the nonzero last column equal to

$$\beta(\mathbf{K})\mathbf{e}_{p+1} = \begin{bmatrix} \beta_p & \beta_{p-1} & \cdots & \beta_1 & 1 \end{bmatrix}^T.$$

Here β is defined by (7.3.2). The next result is useful to describe the structure of the matrix $\mathbf{J}\mathbf{F} - \mathbf{F}\mathbf{J}$ appearing on the right-hand side of (7.4.14).

Lemma 7.4.6 (Butcher and Wright [80, 293]) *Given the polynomial*

$$f(w) = 1 + a_1 w + \cdots + a_{p-1}w^{p-1} + a_p w^p,$$

it follows that $f(\mathbf{K})$ is an upper triangular Toeplitz matrix that satisfies the relation

$$\mathbf{J}f(\mathbf{K}) - f(\mathbf{K})\mathbf{J} = \begin{bmatrix} -a_1 & -a_2 & \cdots & -a_p & 0 \\ 0 & 0 & \cdots & 0 & a_p \\ \vdots & \vdots & \ddots & \vdots & \vdots \\ 0 & 0 & \cdots & 0 & a_2 \\ 0 & 0 & \cdots & 0 & a_1 \end{bmatrix}.$$

Proof: Partitioning $\mathbf{J}$ into columns and $\mathbf{K}^i$ into rows and performing block multiplication, we obtain

$$\mathbf{J}\mathbf{K}^i = \sum_{j=1}^{p-i+1} \mathbf{e}_{j+1}\mathbf{e}_{j+i}^T, \quad i = 1, 2, \ldots, p.$$

Similarly, partitioning $\mathbf{K}^i$ into columns and $\mathbf{J}$ into rows, we get

$$\mathbf{K}^i\mathbf{J} = \sum_{j=0}^{p-i} \mathbf{e}_{j+1}\mathbf{e}_{j+i}^T, \quad i = 1, 2, \ldots, p.$$

Hence, subtracting these relations it follows that

$$\mathbf{J}\mathbf{K}^i - \mathbf{K}^i\mathbf{J} = \mathbf{e}_{p-i+2}\mathbf{e}_{p+1}^T - \mathbf{e}_1\mathbf{e}_i^T, \quad i = 1, 2, \dots, p.$$

Multiplying the equation above by a_i and summing for $i = 1, 2, \dots, p$, leads to

$$\mathbf{J}(f(\mathbf{K}) - \mathbf{I}) - (f(\mathbf{K}) - \mathbf{I})\mathbf{J} = \sum_{i=1}^{p} a_i\mathbf{e}_{p-i+2}\mathbf{e}_{p+1}^T - \sum_{i=1}^{p} a_i\mathbf{e}_1\mathbf{e}_i^T,$$

which is equivalent to the statement of the lemma.   ∎

Since $\rho(\lambda\mathbf{K}) = 0$ for any λ, it follows from Lemmas 7.4.3 and 7.4.5 and the fact that $\mathbf{K}^i = \mathbf{0}$ for $i \geq p+1$ that

$$\mathbf{F} = \exp\left(\mathbf{K}(\mathbf{I} + \lambda\mathbf{K})^{-1}\right) = \mathbf{I} + \sum_{i=1}^{p} N_i(\lambda)\mathbf{K}^i.$$

Hence, applying Lemma 7.4.6 to the function $f(w)$ defined by

$$f(w) = 1 + \sum_{i=1}^{p} N_i(\lambda)w^i,$$

we obtain

$$\mathbf{J}\mathbf{F} - \mathbf{F}\mathbf{J} = \begin{bmatrix} -N_1(\lambda) & -N_2(\lambda) & \cdots & -N_p(\lambda) & 0 \\ 0 & 0 & \cdots & 0 & N_p(\lambda) \\ \vdots & \vdots & \ddots & \vdots & \vdots \\ 0 & 0 & \cdots & 0 & N_2(\lambda) \\ 0 & 0 & \cdots & 0 & N_1(\lambda) \end{bmatrix}. \tag{7.4.16}$$

Taking (7.4.15) into account, relation (7.4.14) can be written in the form

$$\widetilde{\mathbf{B}}C\beta(\mathbf{K})\mathbf{e}_{p+1}\mathbf{e}_{p+1}^T \equiv \mathbf{J}\mathbf{F} - \mathbf{F}\mathbf{J}, \tag{7.4.17}$$

where $\mathbf{J}\mathbf{F} - \mathbf{F}\mathbf{J}$ satisfies (7.4.16). Observe that the matrix on the left-hand side of (7.4.17) is zero except for the last column, and the matrix on the right-hand side is zero except for the first row and the last column. Hence, the equivalence relation (7.4.17) is automatically satisfied except for the last column, and it can be written in the form

$$\widetilde{\mathbf{B}}C\beta(\mathbf{K})\mathbf{e}_{p+1} \equiv (\mathbf{J}\mathbf{F} - \mathbf{F}\mathbf{J})\mathbf{e}_{p+1}. \tag{7.4.18}$$

This relation leads only to p additional conditions for the $(p+1)(p+2)/2$ coefficients of the lower triangular matrix $\widetilde{\mathbf{B}}$. The determination of this matrix is discussed further in Section 7.7.

7.5   TRANSFORMATIONS BETWEEN STABILITY FUNCTIONS

In this section we investigate the relationship between the stability matrices of the original and the transformed method to determine the connection between their stability functions. The stability matrix $\mathbf{M}(z)$ of the original method is given by formula (2.6.4), and we have

$$
\begin{aligned}
\mathbf{M}(z) &= \mathbf{V} + z\mathbf{B}(\mathbf{I} - z\mathbf{A})^{-1}\mathbf{U} \\
&= \mathbf{V} + \frac{z}{1 - \lambda z}\mathbf{B}\left(\frac{1}{1 - \lambda z}\mathbf{I} - \frac{z}{1 - \lambda z}\mathbf{A}\right)^{-1}\mathbf{U} \\
&= \mathbf{V} + \frac{z}{1 - \lambda z}\mathbf{B}\left(\mathbf{I} - \frac{z}{1 - \lambda z}(\mathbf{A} - \lambda\mathbf{I})\right)^{-1}\mathbf{U}.
\end{aligned}
$$

Setting

$$
\widetilde{z} = \frac{z}{1 - \lambda z} \qquad \text{or} \qquad z = \frac{\widetilde{z}}{1 + \lambda\widetilde{z}},
$$

it follows that

$$
\mathbf{M}\left(\frac{\widetilde{z}}{1 + \lambda\widetilde{z}}\right) = \mathbf{V} + \widetilde{z}\mathbf{B}(\mathbf{I} - \widetilde{z}\widetilde{\mathbf{A}})^{-1}\mathbf{U},
$$

where we recall that $\widetilde{\mathbf{A}} = \mathbf{A} - \lambda\mathbf{I}$. Denote by $\widetilde{\mathbf{M}}(z)$ the stability matrix of the transformed method; that is,

$$
\widetilde{\mathbf{M}}(z) = \widetilde{\mathbf{V}} + z\widetilde{\mathbf{B}}(\mathbf{I} - z\widetilde{\mathbf{A}})^{-1}\widetilde{\mathbf{U}}.
$$

It turns out that there is a direct connection between stability matrices of the original and transformed methods. This connection is given by a similarity transformation between matrices $\mathbf{M}(\widetilde{z}/(1 - \lambda\widetilde{z}))$ and $\widetilde{\mathbf{M}}(\widetilde{z})$; that is,

$$
\begin{aligned}
\Psi^{-1}\mathbf{M}\left(\frac{\widetilde{z}}{1 + \lambda\widetilde{z}}\right)\Psi &= \Psi^{-1}\mathbf{V}\Psi + \widetilde{z}\Psi^{-1}\mathbf{B}(\mathbf{I} - \widetilde{z}\widetilde{\mathbf{A}})^{-1}\mathbf{U}\Psi \\
&= \widetilde{\mathbf{V}} + \widetilde{z}\widetilde{\mathbf{B}}(\mathbf{I} - \widetilde{z}\widetilde{\mathbf{A}})^{-1}\widetilde{\mathbf{U}}
\end{aligned}
$$

or

$$
\Psi^{-1}\mathbf{M}\left(\frac{\widetilde{z}}{1 + \lambda\widetilde{z}}\right)\Psi = \widetilde{\mathbf{M}}(\widetilde{z}). \tag{7.5.1}
$$

We have demonstrated in Section 7.4 that the transformed method $\widetilde{\mathbf{M}}$ satisfies the preconsistency condition $\widetilde{\mathbf{V}}\mathbf{e}_1 = \mathbf{e}_1$ and the transformed IRKS conditions (7.4.3), (7.4.4), and (7.4.5), which correspond to IRKS conditions (7.2.8), (7.2.9), and (7.2.10) of the original method $\mathbf{M}$. This means that Definition 7.2.3 holds for the transformed method, with $\mathbf{A}$, $\mathbf{B}$, $\mathbf{U}$, $\mathbf{V}$, and $\mathbf{X}$ replaced by $\widetilde{\mathbf{A}}$, $\widetilde{\mathbf{B}}$, $\widetilde{\mathbf{U}}$, $\widetilde{\mathbf{V}}$, and $\mathbf{J}$. Using the same arguments as in the proof of Theorem 7.2.4, we can establish a similarity equivalence relation

$$
(\mathbf{I} - \widetilde{z}\mathbf{J})\widetilde{\mathbf{M}}(\widetilde{z})(\mathbf{I} - \widetilde{z}\mathbf{J})^{-1} \equiv \widetilde{\mathbf{V}}. \tag{7.5.2}
$$

This relation implies that $\widetilde{\mathbf{M}}(\widetilde{z})$ has a single nonzero eigenvalue which will be denoted by $\widetilde{R}(\widetilde{z})$ and referred to as the stability function of the transformed method $\widetilde{\mathbf{M}}$. It follows from (7.5.1) that the stability function $\widetilde{R}(\widetilde{z})$ of the transformed method is related to the stability function $R(z)$ of the original method through the relation

$$\widetilde{R}(\widetilde{z}) = R\left(\frac{\widetilde{z}}{1 + \lambda \widetilde{z}}\right). \tag{7.5.3}$$

To express $\widetilde{R}(\widetilde{z})$ in terms of the coefficient matrices of the transformed method, observe that similarly as for the original method, it follows from the preconsistency condition $\widetilde{\mathbf{V}}\mathbf{e}_1 = \mathbf{e}_1$ and the equivalence similarity relation (7.5.2) that the matrix $(\mathbf{I} - \widetilde{z}\mathbf{J})\widetilde{\mathbf{M}}(\widetilde{z})(\mathbf{I} - \widetilde{z}\mathbf{J})^{-1}$ can be partitioned as

$$(\mathbf{I} - \widetilde{z}\mathbf{J})\widetilde{\mathbf{M}}(\widetilde{z})(\mathbf{I} - \widetilde{z}\mathbf{J})^{-1} = \left[\begin{array}{c|c} \widetilde{R}(\widetilde{z}) & \widetilde{r}^T \\ \hline 0 & \widetilde{V} \end{array}\right], \tag{7.5.4}$$

where $\widetilde{V}$ is the $p \times p$ matrix obtained from $\widetilde{\mathbf{V}}$ by deleting the first row and column, and $\widetilde{r}$ is a vector. Hence, the stability function $\widetilde{R}(\widetilde{z})$ is the $(1,1)$ element of the matrix above; that is,

$$\widetilde{R}(\widetilde{z}) = \mathbf{e}_1^T (\mathbf{I} - \widetilde{z}\mathbf{J})\widetilde{\mathbf{M}}(\widetilde{z})(\mathbf{I} - \widetilde{z}\mathbf{J})^{-1}\mathbf{e}_1.$$

To simplify the expression above observe that $\mathbf{e}_1^T(\mathbf{I} - \widetilde{z}\mathbf{J}) = \mathbf{e}_1^T$ and denote by $\widetilde{\mathbf{Z}}$ the vector

$$\widetilde{\mathbf{Z}} = \left[\begin{array}{cccc} \widetilde{z}_1 & \widetilde{z}_2 & \cdots & \widetilde{z}_p & \widetilde{z}_{p+1} \end{array}\right]^T = (\mathbf{I} - \widetilde{z}\mathbf{J})^{-1}\mathbf{e}_1.$$

Then

$$(\mathbf{I} - \widetilde{z}\mathbf{J})\widetilde{\mathbf{Z}} = \mathbf{e}_1,$$

or, componentwise,

$$\widetilde{z}_1 = 1, \quad -\widetilde{z}\widetilde{z}_1 + \widetilde{z}_2 = 0, \quad \ldots, \quad -\widetilde{z}\widetilde{z}_p + \widetilde{z}_{p+1} = 0,$$

which implies that

$$\widetilde{z}_i = \widetilde{z}^{i-1}, \quad i = 1, 2, \ldots, p+1.$$

Hence,

$$\widetilde{R}(\widetilde{z}) = \mathbf{e}_1^T \left(\widetilde{\mathbf{V}} - \widetilde{z}\widetilde{\mathbf{B}}(\mathbf{I} - \widetilde{z}\widetilde{\mathbf{A}})^{-1}\widetilde{\mathbf{U}}\right)\widetilde{\mathbf{Z}},$$

where $\widetilde{\mathbf{Z}}$ takes the form

$$\widetilde{\mathbf{Z}} = \left[\begin{array}{ccccc} 1 & \widetilde{z} & \cdots & \widetilde{z}^{p-1} & \widetilde{z}^p \end{array}\right]^T.$$

To further simplify this expression, observe that $\widetilde{\mathbf{A}}$ is strictly lower triangular and it follows that $(\mathbf{I} - \widetilde{z}\widetilde{\mathbf{A}})^{-1}$ is unit lower triangular. Moreover, since $\widetilde{\mathbf{B}}$ is lower triangular, it can be verified that $\mathbf{e}_1^T\widetilde{\mathbf{B}}(\mathbf{I} - \widetilde{z}\widetilde{\mathbf{A}})^{-1} = \mathbf{e}_1^T\widetilde{\mathbf{B}}$. Hence, the stability function $\widetilde{R}(\widetilde{z})$ of the transformed method $\widetilde{\mathbf{M}}$ takes a simple form,

$$\widetilde{R}(\widetilde{z}) = \mathbf{e}_1^T(\widetilde{\mathbf{V}} + \widetilde{z}\widetilde{\mathbf{B}}\widetilde{\mathbf{U}})\widetilde{\mathbf{Z}}. \tag{7.5.5}$$

If the original method is explicit, the remaining free parameters in the stability function can be used to control the error constant and the size of the region of absolute stability. This is discussed in Section 7.8. If the original method is implicit, the parameter λ will be chosen to guarantee A-stability, and the free parameters in the numerator of the stability function can be used to guarantee L-stability. To achieve these goals it is convenient to reformulate various conditions on the original method as corresponding conditions on the transformed method. Consider first the stability function $R(z)$ of the original method as the rational approximation of order p to the exponential function $\exp(z)$ with error constant denoted by E. Since $\mathbf{A}$ is lower triangular with λ on the diagonal, this function takes the form

$$R(z) = \frac{P(z)}{(1 - \lambda z)^{p+1}} = \exp(z) - Ez^{p+1} + O(z^{p+2}), \tag{7.5.6}$$

where $P(z)$ is a polynomial of degree $p + 1$. Then it follows from (7.5.3) that the stability function $\widetilde{R}(\widetilde{z})$ satisfies the relation

$$\widetilde{R}(\widetilde{z}) = \widetilde{P}(\widetilde{z}) = \exp\left(\frac{\widetilde{z}}{1 + \lambda\widetilde{z}}\right) - E\left(\frac{\widetilde{z}}{1 + \lambda\widetilde{z}}\right)^{p+1} + O\left(\left(\frac{\widetilde{z}}{1 + \lambda\widetilde{z}}\right)^{p+2}\right),$$

where $\widetilde{P}(\widetilde{z})$ is a polynomial of degree $p + 1$. Since

$$\left(\frac{\widetilde{z}}{1 + \lambda\widetilde{z}}\right)^{p+1} = \widetilde{z}^{p+1} + O(\widetilde{z}^{p+2}) \quad \text{and} \quad O\left(\left(\frac{\widetilde{z}}{1 + \lambda\widetilde{z}}\right)^{p+2}\right) = O(\widetilde{z}^{p+2}),$$

$\widetilde{R}(\widetilde{z})$ can be written as

$$\widetilde{R}(\widetilde{z}) = \widetilde{P}(\widetilde{z}) = \exp\left(\frac{\widetilde{z}}{1 + \lambda\widetilde{z}}\right) - E\,\widetilde{z}^{p+1} + O(\widetilde{z}^{p+2}). \tag{7.5.7}$$

To derive explicit expressions for $P(z)$ and $\widetilde{P}(\widetilde{z})$, we have to introduce some definitions and notation. For integers m and n, $m \geq n$, define the polynomials $M_{n,m}(x)$ as

$$M_{n,m}(x) = \sum_{i=0}^{n} \binom{m}{i} \frac{(-x)^i}{(n - i)!}. \tag{7.5.8}$$

These polynomials can be expressed in terms of the generalized Laguerre polynomials $L_{n,\alpha}(x)$, [1, 107]. These are polynomials orthogonal with respect to the weight function $w(x) = x^\alpha \exp(x)$, $\alpha > -1$, on the interval $[0, \infty)$, with an explicit expression of the form

$$L_{n,\alpha}(x) = \sum_{i=0}^{n} \binom{n+\alpha}{n-i} \frac{(-x)^i}{i!}.$$

For integers $m \geq 0$, they satisfy the relations

$$L_{n,0}(x) = L_n(x), \quad L_{n,m}(x) = (-1)^m L_{n+m}^{(m)}(x) \tag{7.5.9}$$

(compare [1]), where $L_n(x)$ are Laguerre polynomials defined by (2.7.6) in Section 2.7 and $L_n^{(m)}(x)$ stands for an m-fold derivative of $L_n(x)$. Observe that we do not use the notation $L_n^{(\alpha)}(x)$ for generalized Laguerre polynomials which is employed in [1, 107], since for integer and nonnegative m, the notation $L_n^{(m)}(x)$ is reserved for an m-fold derivative of $L_n(x)$.

For $\lambda = 0$ and any m, we have

$$M_{n,m}(0) = \frac{1}{n!},$$

and for $\lambda \neq 0$ and $m \geq n$, we have

$$\begin{aligned}
M_{n,m}(\lambda) &= \sum_{i=0}^{n} \binom{m}{i} \frac{(-\lambda)^i}{(n-i)!} \\
&= (-\lambda)^n \sum_{i=0}^{n} \binom{m}{i} \frac{1}{(n-i)!} \left(\frac{1}{\lambda}\right)^{n-i} \\
&= (-\lambda)^n \sum_{i=0}^{n} \binom{m}{n-i} \frac{1}{i!} \left(\frac{1}{\lambda}\right)^i.
\end{aligned}$$

Hence, it follows that

$$M_{n,m}(\lambda) = (-\lambda)^n L_{n,m-n}\left(\frac{1}{\lambda}\right), \quad \lambda \neq 0. \tag{7.5.10}$$

To determine exact expression for the polynomial $P(z)$, we use the relation

$$P(z) = (1 - \lambda z)^{p+1} \exp(z) - E z^{p+1} + O(z^{p+2}), \tag{7.5.11}$$

which, since $z^{p+1}(1-\lambda z)^{p+1} = z^{p+1} + O(z^{p+2})$, is equivalent to (7.5.6). Denote by $\langle z^n, f(z) \rangle$ the coefficient of z^n in a Taylor series expansion of the function $f(z)$ about $z = 0$. We have the following result.

Lemma 7.5.1 (Butcher and Wright [80, 293]) *For integers m and n, $m \geq n$, the following relation holds:*

$$\langle z^n, (1 - \lambda z)^m \exp(z) \rangle = M_{n,m}(\lambda), \tag{7.5.12}$$

$n = 0, 1, \ldots, m.$

Proof: We have

$$(1 - \lambda z)^m \exp(z) = \sum_{i=0}^{m} \binom{m}{i} (-\lambda)^i z^i \sum_{j=0}^{\infty} \frac{z^j}{j!} = \sum_{i=0}^{m} \sum_{j=0}^{\infty} \binom{m}{i} \frac{(-\lambda)^i}{j!} z^{i+j}.$$

Putting $n = i + j$ and then rearranging the double sum and using (7.5.8), we obtain

$$(1 - \lambda z)^m \exp(z) = \sum_{n=0}^{m} \sum_{i=0}^{n} \binom{m}{i} \frac{(-\lambda)^i}{(n-i)!} z^n + O(z^{m+1})$$

$$= \sum_{n=0}^{m} M_{n,m}(\lambda) z^n + O(z^{m+1}),$$

which implies (7.5.12).   ∎

It follows from this lemma with $m = p + 1$ that

$$(1 - \lambda z)^{p+1} \exp(z) = \sum_{n=0}^{p+1} M_{n,p+1}(\lambda) z^n + O(z^{p+2}),$$

and comparing this with (7.5.11) and observing that $P(z)$ is a polynomial of degree $p + 1$, we obtain

$$P(z) = \sum_{n=0}^{p} M_{n,p+1}(\lambda) z^n + \epsilon \, z^{p+1}, \tag{7.5.13}$$

where

$$\epsilon = M_{p+1,p+1}(\lambda) - E. \tag{7.5.14}$$

The error constant E is presented in Table 7.5.1 for orders p up to 6.

Exact expression for $\widetilde{P}(\widetilde{z})$ in (7.5.7) can be derived using Lemma 7.4.5 with **S** replaced by $\widetilde{z}$ in formula (7.4.12). It follows that

$$\exp\left(\frac{\widetilde{z}}{1 + \lambda \widetilde{z}}\right) = 1 + \sum_{i=1}^{\infty} N_i(\lambda) \widetilde{z}^i, \tag{7.5.15}$$

where the polynomials $N_i(\lambda)$ are defined by (7.4.13). Substituting this relation into (7.5.7) and taking into account that $\widetilde{P}(\widetilde{z})$ is a polynomial of degree $p + 1$, we obtain

$$\widetilde{P}(\widetilde{z}) = 1 + \sum_{i=1}^{p} N_i(\lambda) \widetilde{z}^i + \widetilde{\epsilon} \, \widetilde{z}^{p+1}, \tag{7.5.16}$$

where

$$\widetilde{\epsilon} = N_{p+1}(\lambda) - E. \tag{7.5.17}$$

p	E
1	$\frac{1}{2} - 2\lambda + \lambda^2 - \epsilon$
2	$\frac{1}{6} - \frac{3}{2}\lambda + 3\lambda^2 - \lambda^3 - \epsilon$
3	$\frac{1}{24} - \frac{2}{3}\lambda + 3\lambda^2 - 4\lambda^3 + \lambda^4 - \epsilon$
4	$\frac{1}{120} - \frac{5}{24}\lambda + \frac{5}{3}\lambda^2 - 5\lambda^3 + 5\lambda^4 - \lambda^5 - \epsilon$
5	$\frac{1}{720} - \frac{1}{20}\lambda + \frac{5}{8}\lambda^2 - \frac{10}{3}\lambda^3 + \frac{15}{2}\lambda^4 - 6\lambda^5 + \lambda^6 - \epsilon$
6	$\frac{1}{5040} - \frac{7}{120}\lambda + \frac{7}{40}\lambda^2 - \frac{35}{24}\lambda^3 + \frac{35}{6}\lambda^4 - \frac{21}{2}\lambda^5 + 7\lambda^6 - \lambda^7 - \epsilon$

Table 7.5.1   Error constant $E = M_{p+1,p+1}(\lambda) - \epsilon$ for $p = 1, 2, \ldots, 6$

Alternatively, we can express $\widetilde{P}(\widetilde{z})$ in terms of the polynomials $M_{n,m}(\lambda)$ defined by (7.5.8). Denote by $\langle \widetilde{z}^n, \widetilde{f}(\widetilde{z}) \rangle$ the coefficient of $\widetilde{z}^n$ in a Taylor series expansion of $\widetilde{f}(\widetilde{z})$ about $\widetilde{z} = 0$. We have the following lemma.

Lemma 7.5.2 (Butcher and Wright [80, 293]) *For integers m and n such that $m \geq n$, the following relation holds:*

$$\left\langle \widetilde{z}^n, \exp\left(\frac{\widetilde{z}}{1 + \lambda\widetilde{z}}\right) \right\rangle = M_{n,n}(\lambda) + \lambda M_{n-1,n-1}(\lambda), \qquad (7.5.18)$$

$n = 0, 1, \ldots, m$, *with the convention that* $M_{-1,-1}(\lambda) = 0$.

Proof: Substituting $\widetilde{z} = z/(1 - \lambda z)$ into (7.5.15), we obtain

$$\exp(z) = 1 + \sum_{i=1}^{\infty} N_i(\lambda) \left(\frac{z}{1 - \lambda z}\right)^i.$$

Multiplying both sides of this equation by $(1 - \lambda z)^{n-1}$ yields

$$(1 - \lambda z)^{n-1} \exp(z) = (1 - \lambda z)^{n-1} + \sum_{i=1}^{n-1} N_i(\lambda) \left(\frac{z}{1 - \lambda z}\right)^i (1 - \lambda z)^{n-1}$$
$$+ N_n(\lambda)\frac{z^n}{1 - \lambda z} + O(z^{n+1}).$$

Since the terms with

$$\left(\frac{z}{1 - \lambda z}\right)^i (1 - \lambda z)^{n-1}$$

are polynomials of degree less than n for $i = 0, 1, \ldots, n - 1$, we obtain

$$
\begin{aligned}
N_n(\lambda) &= \left\langle \widetilde{z}^n, \exp\left(\frac{\widetilde{z}}{1 + \lambda\widetilde{z}}\right) \right\rangle = \left\langle z^n, (1 - \lambda z)^{n-1} \exp(z) \right\rangle \\
&= \left\langle z^n, \left((1 - \lambda z)^n + \lambda z (1 - \lambda z)^{n-1}\right) \exp(z) \right\rangle \\
&= \left\langle z^n, (1 - \lambda z)^n \exp(z) \right\rangle + \lambda \left\langle z^n, z(1 - \lambda z)^{n-1} \exp(z) \right\rangle \\
&= \left\langle z^n, (1 - \lambda z)^n \exp(z) \right\rangle + \lambda \left\langle z^{n-1}, (1 - \lambda z)^{n-1} \exp(z) \right\rangle \\
&= M_{n,n}(\lambda) + \lambda M_{n-1,n-1}(\lambda),
\end{aligned}
$$

where the last equality follows from Lemma 7.5.1. This completes the proof of (7.5.18).   ∎

This lemma implies that

$$
N_n(\lambda) = M_{n,n}(\lambda) + \lambda M_{n-1,n-1}(\lambda), \tag{7.5.19}
$$

where $N_n(\lambda)$ is defined by (7.4.13) and $M_{n,m}(\lambda)$ by (7.5.8). This equation can also be verified by direct computations using the relations

$$
\binom{n}{i} - \binom{n-1}{i-1} = \binom{n-1}{i}
$$

if $i < n$, and

$$
\binom{n}{i} - \binom{n-1}{i-1} = 0
$$

if $i = n$.

Using (7.5.19), it follows that the relation between ϵ defined by (7.5.14) and $\widetilde{\epsilon}$ defined by (7.5.17) is

$$
\widetilde{\epsilon} = \epsilon + \lambda M_{p,p}(\lambda). \tag{7.5.20}
$$

The constants $\widetilde{\epsilon}$ and ϵ can also be expressed in terms of the $(1, 1)$ element $\widetilde{b}_{11}$ of the transformed matrix $\widetilde{\mathbf{B}}$, the parameter λ, and the parameters $\beta_1, \beta_2, \ldots, \beta_p$ of the doubly companion matrix $\mathbf{X}$. It follows from (7.5.16), (7.5.5), and the definition of the vector $\widetilde{\mathbf{Z}}$ that

$$
\widetilde{\epsilon} = \left\langle \widetilde{z}^{p+1}, \widetilde{R}(\widetilde{z}) \right\rangle = \left\langle \widetilde{z}^{p+1}, \mathbf{e}_1^T (\widetilde{\mathbf{V}} + \widetilde{z}\widetilde{\mathbf{B}}\widetilde{\mathbf{U}})\widetilde{\mathbf{Z}} \right\rangle = \mathbf{e}_1^T \widetilde{\mathbf{B}}\widetilde{\mathbf{U}}\mathbf{e}_{p+1}.
$$

Substituting $\widetilde{\mathbf{U}} = \mathbf{U}\Psi$ with $\mathbf{U}$ given by (7.1.6) and Ψ given by (7.3.9), we obtain

$$
\widetilde{\epsilon} = \mathbf{e}_1^T \widetilde{\mathbf{B}}(\mathbf{C} - \mathbf{ACK})\beta(\mathbf{K}) \exp(\lambda\mathbf{K}^-)\mathbf{e}_{p+1}.
$$

Since $\widetilde{\mathbf{B}}$ and $\mathbf{A}$ are lower triangular, we have

$$\mathbf{e}_1^T \widetilde{\mathbf{B}} = \widetilde{b}_{11} \mathbf{e}_1^T, \quad \mathbf{e}_1^T \widetilde{\mathbf{B}} \mathbf{A} = \widetilde{b}_{11} \lambda \mathbf{e}_1^T,$$

which leads to

$$\widetilde{\epsilon} = \widetilde{b}_{11} \mathbf{e}_1^T \mathbf{C} (\mathbf{I} - \lambda \mathbf{K}) \beta(\mathbf{K}) \exp(\lambda \mathbf{K}^-) \mathbf{e}_{p+1}.$$

It can be verified using the expansion

$$\exp(\lambda \mathbf{K}^-) = \sum_{i=0}^{\infty} \frac{\lambda^i}{i!} (\mathbf{K}^-)^i = \sum_{i=0}^{p} \frac{\lambda^i}{i!} (\mathbf{K}^-)^i$$

that $\exp(\lambda \mathbf{K}^-) \mathbf{e}_{p+1} = \Lambda$, where the vector Λ is defined by

$$\Lambda = \begin{bmatrix} \lambda^p & \lambda^{p-1} & \cdots & \lambda & 1 \end{bmatrix}^T.$$

Also taking into account that the matrices $\mathbf{K}$ and $\beta(\mathbf{K})$ commute, the equation for $\widetilde{\epsilon}$ can be written in the form

$$\widetilde{\epsilon} = \widetilde{b}_{11} \mathbf{e}_1^T \mathbf{C} \beta(\mathbf{K}) (\mathbf{I} - \lambda \mathbf{K}) \Lambda.$$

This can be simplified further observing that $(\mathbf{I} - \lambda \mathbf{K}) \Lambda = \mathbf{e}_{p+1}$. Hence,

$$\widetilde{\epsilon} = \widetilde{b}_{11} \mathbf{e}_1^T \mathbf{C} \beta(\mathbf{K}) \mathbf{e}_{p+1}.$$

This equation and (7.5.20) lead to the condition

$$\widetilde{b}_{11} \mathbf{e}_1^T \mathbf{C} \beta(\mathbf{K}) \mathbf{e}_{p+1} = \epsilon + \lambda M_{p,p}(\lambda). \tag{7.5.21}$$

Observe that the vector $\beta(\mathbf{K}) \mathbf{e}_{p+1}$ has the simple form

$$\beta(\mathbf{K}) \mathbf{e}_{p+1} = \begin{bmatrix} \beta_p & \beta_{p-1} & \cdots & \beta_1 & 1 \end{bmatrix}^T.$$

7.6 LOWER TRIANGULAR MATRICES AND CHARACTERIZATION OF MATRICES WITH ZERO SPECTRAL RADIUS

IRKS condition (7.2.10) for the original method $\mathbf{M}$ and the corresponding condition (7.4.5) for the transformed method $\widetilde{\mathbf{M}}$ require that the characteristic polynomials of $\mathbf{V}$ and $\widetilde{\mathbf{V}}$ have the form

$$\det(w\mathbf{I} - \mathbf{V}) = \det(w\mathbf{I} - \widetilde{\mathbf{V}}) = w^p(w - 1).$$

Since $\mathbf{V}\mathbf{e}_1 = \mathbf{e}_1$ and $\widetilde{\mathbf{V}}\mathbf{e}_1 = \mathbf{e}_1$, these matrices can be partitioned as

$$\mathbf{V} = \left[\begin{array}{c|c} 1 & v^T \\ \hline 0 & V \end{array} \right], \quad \widetilde{\mathbf{V}} = \left[\begin{array}{c|c} 1 & \widetilde{v}^T \\ \hline 0 & \widetilde{V} \end{array} \right],$$

and it follows that conditions (7.2.10) and (7.4.5) are equivalent to the requirement that matrices V and $\widetilde{V}$ have zero spectral radii. In this section we discuss the characterization of such matrices which was discovered by Butcher and Wright [80, 293].

Denote by $\mathcal{L}_n$ the set of $n \times n$ lower triangular matrices and by $\mathcal{U}_n$ the corresponding set of upper triangular matrices. For a square matrix Ω we denote by $\Delta(\Omega)$ the lower triangular part of the matrix. We also use the notation $\mathcal{L}(\Omega)$ and $\mathcal{U}(\Omega)$ for unit lower triangular matrix and upper triangular matrix, respectively, such that

$$\Omega = \mathcal{L}(\Omega)\mathcal{U}(\Omega),$$

assuming that this LU decomposition exists.

The next result shows the connection between two upper triangular matrices and the lower triangular part of a matrix.

Lemma 7.6.1 (Butcher and Wright [80, 293]) *([80], [293]). Given matrices* $\mathbf{U}_1, \mathbf{U}_2 \in \mathcal{U}_n$, *and a matrix* $\mathbf{H} \in \mathbb{R}^{n \times n}$, *the following relation holds:*

$$\Delta\big(\mathbf{U}_1\Delta(\mathbf{H})\mathbf{U}_2\big) = \Delta(\mathbf{U}_1\mathbf{H}\mathbf{U}_2).$$

Proof: The lemma is clearly true for $n = 1$. Assume that it is true for $n - 1$ and partition the matrices $\mathbf{U}_1, \mathbf{U}_2 \in \mathcal{U}_n$, $\mathbf{H} \in \mathbb{R}^{n \times n}$, $\Delta(\mathbf{H}) \in \mathcal{L}_n$ as follows:

$$\mathbf{U}_1 = \left[\begin{array}{c|c} u_1 & \mathbf{u}_1^T \\ \hline \mathbf{0} & U_1 \end{array}\right], \quad \mathbf{U}_2 = \left[\begin{array}{c|c} u_2 & \mathbf{u}_2^T \\ \hline \mathbf{0} & U_2 \end{array}\right],$$

$$\mathbf{H} = \left[\begin{array}{c|c} h & \mathbf{r}^T \\ \hline \mathbf{c} & H \end{array}\right], \quad \Delta(\mathbf{H}) = \left[\begin{array}{c|c} h & \mathbf{0}^T \\ \hline \mathbf{c} & \Delta(H) \end{array}\right],$$

where $u_1, u_2, h \in \mathbb{R}$, $\mathbf{u}_1, \mathbf{u}_2, \mathbf{r}, \mathbf{c} \in \mathbb{R}^{n-1}$, $U_1, U_2 \in \mathcal{U}_{n-1}$, $H \in \mathbb{R}^{(n-1) \times (n-1)}$, and $\Delta(H) \in \mathcal{L}_{n-1}$. It can be verified by direct computations that

$$\Delta\big(\mathbf{U}_1\Delta(\mathbf{H})\mathbf{U}_2\big) = \left[\begin{array}{c|c} u_1 h u_2 + u_2 \mathbf{u}_1^T \mathbf{c} & \mathbf{0}^T \\ \hline u_2 U_1 \mathbf{c} & \Delta(U_1 \mathbf{c} \mathbf{u}_2^T) + \Delta\big(U_1\Delta(H)U_2\big) \end{array}\right]$$

and

$$\Delta(\mathbf{U}_1\mathbf{H}\mathbf{U}_2) = \left[\begin{array}{c|c} u_1 h u_2 + u_2 \mathbf{u}_1^T \mathbf{c} & \mathbf{0}^T \\ \hline u_2 U_1 \mathbf{c} & \Delta(U_1 \mathbf{c} \mathbf{u}_2^T) + \Delta(U_1 H U_2) \end{array}\right].$$

The lemma follows since by induction assumption we have $\Delta(U_1\Delta(H)U_2) = \Delta(U_1 H U_2)$. ∎

The next lemma investigates the inverse of some mapping between the lower triangular matrices. This result is essential in the derivation of GLMs with IRKS.

Lemma 7.6.2 (Butcher and Wright [80, 293]) *Assume that* $\mathbf{H} \in \mathcal{L}_n$*, and* $\mathbf{G}_1, \mathbf{G}_2 \in \mathbb{R}^{n \times n}$*, and define the mapping* $f : \mathcal{L}_n \to \mathcal{L}_n$ *by the formula*

$$f(\mathbf{H}) = \Delta(\mathbf{G}_1 \mathbf{H} \mathbf{G}_2).$$

Then the inverse mapping $f^{-1} : \mathcal{L}_n \to \mathcal{L}_n$*, if it exists, is defined by*

$$f^{-1}(\mathbf{F}) = \mathcal{L}(\mathbf{G}_1^{-1})\Delta\big(\mathcal{U}(\mathbf{G}_1^{-1})\mathbf{F}\mathcal{U}(\mathbf{G}_2)^{-1}\big)\mathcal{L}(\mathbf{G}_2)^{-1}.$$

Proof: Given that

$$\mathbf{G}_1^{-1} = \mathcal{L}(\mathbf{G}_1^{-1})\mathcal{U}(\mathbf{G}_1^{-1}), \quad \mathbf{G}_2 = \mathcal{L}(\mathbf{G}_2)\mathcal{U}(\mathbf{G}_2),$$

we have

$$f(\mathbf{H}) = \Delta\big(\mathcal{U}(\mathbf{G}_1^{-1})^{-1}\mathcal{L}(\mathbf{G}_1^{-1})^{-1}\mathbf{H}\mathcal{L}(\mathbf{G}_2)\mathcal{U}(\mathbf{G}_2)\big).$$

Multiplying this equation on the left by $\mathcal{U}(\mathbf{G}_1^{-1})$ and on the right by $\mathcal{U}(\mathbf{G}_2)^{-1}$, we obtain

$$\mathcal{U}(\mathbf{G}_1^{-1})f(\mathbf{H})\mathcal{U}(\mathbf{G}_2)^{-1}$$

$$= \mathcal{U}(\mathbf{G}_1^{-1})\Delta\big(\mathcal{U}(\mathbf{G}_1^{-1})^{-1}\mathcal{L}(\mathbf{G}_1^{-1})^{-1}\mathbf{H}\mathcal{L}(\mathbf{G}_2)\mathcal{U}(\mathbf{G}_2)\big)\mathcal{U}(\mathbf{G}_2)^{-1}.$$

Hence, taking lower triangular parts of both sides and applying Lemma 7.6.1 to the resulting right-hand side we get

$$\Delta\big(\mathcal{U}(\mathbf{G}_1^{-1})f(\mathbf{H})\mathcal{U}(\mathbf{G}_2)^{-1}\big) = \Delta\big(\mathcal{L}(\mathbf{G}_1^{-1})^{-1}\mathbf{H}\mathcal{L}(\mathbf{G}_2)\big).$$

Since the matrix $\mathcal{L}(\mathbf{G}_1^{-1})^{-1}\mathbf{H}\mathcal{L}(\mathbf{G}_2)$ is already lower triangular, we have

$$\mathcal{L}(\mathbf{G}_1^{-1})^{-1}\mathbf{H}\mathcal{L}(\mathbf{G}_2) = \Delta\big(\mathcal{U}(\mathbf{G}_1^{-1})f(\mathbf{H})\mathcal{U}(\mathbf{G}_2)^{-1}\big),$$

and it follows that

$$\mathbf{H} = \mathcal{L}(\mathbf{G}_1^{-1})\Delta\big(\mathcal{U}(\mathbf{G}_1^{-1})f(\mathbf{H})\mathcal{U}(\mathbf{G}_2)^{-1}\big)\mathcal{L}(\mathbf{G}_2)^{-1}.$$

Substituting $\mathbf{F}$ for $f(\mathbf{H})$ yields the required result. ∎

It follows from the proof of this lemma that the inverse of the mapping $f : \mathcal{L}_n \to \mathcal{L}_n$ exists if $\mathbf{G}_1$ and $\mathbf{G}_2$ are nonsingular and the matrices $\mathbf{G}_1^{-1}$ and $\mathbf{G}_2$ admit LU factorizations. It is demonstrated in the following sections that these conditions will often be satisfied in applications of this lemma to the construction of GLMs with IRKS.

The next lemma describes the characterization of matrices with zero spectral radius.

Lemma 7.6.3 (Butcher and Wright [80, 293]) *A matrix* $\Omega \in \mathbb{R}^{n \times n}$ *has spectral radius equal to zero if and only if there exists a permutation matrix* $\mathbf{P}$ *and a unit lower triangular nonsingular matrix* $\mathbf{L}$ *such that*

$$\Delta(\mathbf{L}^{-1}\mathbf{P}^T\Omega\mathbf{P}\mathbf{L}) = \mathbf{0}.$$

Proof: Assume first that $\Delta(\mathbf{L}^{-1}\mathbf{P}^T\Omega\mathbf{P}\mathbf{L}) = \mathbf{0}$. Then Ω is similar to a strictly upper triangular matrix and it follows that its spectral radius is zero. Assume next that the spectral radius of Ω is equal to zero, and denote by $\mathbf{T}$ a nonsingular matrix that transforms Ω to the Jordan canonical form

$$\mathbf{T}^{-1}\Omega\mathbf{T} = \mathbf{S}.$$

Since $\mathbf{T}$ is nonsingular, there exists LU decomposition with partial pivoting of this matrix,

$$\mathbf{T} = \mathbf{P}\mathbf{L}\mathbf{R},$$

where $\mathbf{P}$ is a permutation matrix, $\mathbf{L}$ is unit lower triangular, and $\mathbf{R}$ is upper triangular. Hence,

$$\mathbf{R}^{-1}\mathbf{L}^{-1}\mathbf{P}^T\Omega\mathbf{P}\mathbf{L}\mathbf{R} = \mathbf{S}$$

or

$$\mathbf{L}^{-1}\mathbf{P}^T\Omega\mathbf{P}\mathbf{L} = \mathbf{R}\mathbf{S}\mathbf{R}^{-1}.$$

The result now follows since $\mathbf{R}\mathbf{S}\mathbf{R}^{-1}$ is strictly upper triangular. ∎

7.7   CANONICAL FORMS OF METHODS

In this section we describe an approach to generating GLMs with IRKS, which will lead to the constructive algorithm for the derivation of such methods of any order using only linear operations. First, for easy reference, we collect all assumptions on the class of methods we are interested in, and recall some of the formulas that were derived in previous sections. We are interested in GLMs of the form (7.1.1) with the abscissa vector $\mathbf{c}$ and coefficient matrices $\mathbf{A}$, $\mathbf{U}$, $\mathbf{B}$, and $\mathbf{V}$, with the following properties:

- The stage order and order of the method are each equal to p.

- The number of external stages r and the number of internal stages s are equal to $r = s = p + 1$.

- The method is nonconfluent (i.e., the abscissa vector $\mathbf{c} = [c_1, \ldots, c_s]^T$ satisfies $c_i \neq c_j$ for $i \neq j$).

- $\mathbf{A}$ is a lower triangular matrix with diagonal elements each equal to $\lambda \geq 0$.

- The method has RK stability with stability function given by

$$R(z) = \frac{P(z)}{(1 - \lambda z)^{p+1}},$$

where

$$P(z) = \exp(z)(1 - \lambda z)^{p+1} - E z^{p+1} + O(z^{p+2}) = \sum_{n=0}^{p} M_{n,p+1}(\lambda) z^n + \epsilon z^{n+1}$$

and $\epsilon = M_{p+1,p+1}(\lambda) - E$ (compare (7.5.11), (7.5.13), and (7.5.14)).

- The coefficient matrix $\mathbf{V}$ satisfies the preconsistency condition $\mathbf{V} e_1 = e_1$.

- The method has IRKS. This means that the following conditions are satisfied:

$$\mathbf{BA} = \mathbf{XB},$$

$$\mathbf{BU} \equiv \mathbf{XV} - \mathbf{VX},$$

$$\det(w\mathbf{I} - \mathbf{V}) = w^p(w - 1)$$

(compare (7.2.8), (7.2.9), (7.2.10), and Lemma 7.4.2). Here $\mathbf{X} = \mathbf{X}(\alpha, \beta)$ is a doubly companion matrix such that $\sigma(\mathbf{X}) = \{\lambda\}$ and λ is the diagonal element of $\mathbf{A}$. The first row of this matrix, which corresponds to α, is defined once the coefficient λ and the last p elements of the last column, which corresponds to β, are chosen.

To determine GLMs with IRKS, we start with the construction of the coefficient matrix $\widetilde{\mathbf{B}}$ of the transformed method $\widetilde{\mathbf{M}}$. It was demonstrated in Section 7.4 that the transformed IRKS condition (7.4.4) led to the equivalence relation (7.4.18). It was also demonstrated in Section 7.5 that the stability function $\widetilde{R}(\widetilde{z})$ of the transformed method $\widetilde{\mathbf{M}}$ takes the form (7.5.16), which leads to the condition (7.5.21) with ϵ defined by (7.5.14). We can combine (7.4.18) and (7.5.21) into one condition. Let $\widetilde{\mathbf{b}} = [\widetilde{b}_1, \ldots, \widetilde{b}_{p+1}]^T$ be the vector defined by

$$\widetilde{\mathbf{b}} = \widetilde{\mathbf{B}} C \beta(\mathbf{K}) e_{p+1}.$$

Then it follows from (7.4.16) that equation (7.4.18) can be written in the form

$$\left[\begin{array}{ccc} \widetilde{b}_2 & \cdots & \widetilde{b}_{p+1} \end{array} \right]^T = \left[\begin{array}{ccc} N_p(\lambda) & \cdots & N_1(\lambda) \end{array} \right]^T,$$

and equation (7.5.21) takes the form

$$\widetilde{b}_1 = \epsilon + \lambda M_{p,p}(\lambda).$$

Hence, these two equations can be combined into one condition,

$$\widetilde{\mathbf{B}} C \beta(\mathbf{K}) e_{p+1} = \delta, \tag{7.7.1}$$

where the vector δ is defined by

$$\delta = \begin{bmatrix} \epsilon + \lambda M_{p,p}(\lambda) & N_p(\lambda) & \cdots & N_1(\lambda) \end{bmatrix}^T.$$

We examine next the transformed IRKS condition (7.4.5), where the matrix $\widetilde{V}$ is defined by (7.4.7). To analyze this condition it is convenient to introduce some matrices that will be used to remove the first row or the first column of a matrix or to insert an additional first row or first column of zeros to a given matrix. These $p \times (p+1)$ and $(p+1) \times p$ matrices are defined by

$$\mathbf{I}_c = \begin{bmatrix} \mathbf{0} & | & \mathbf{I} \end{bmatrix} = \begin{bmatrix} 0 & 1 & 0 & \cdots & 0 & 0 \\ 0 & 0 & 1 & \cdots & 0 & 0 \\ \vdots & \vdots & \vdots & \ddots & \vdots & \vdots \\ 0 & 0 & 0 & \cdots & 1 & 0 \\ 0 & 0 & 0 & \cdots & 0 & 1 \end{bmatrix},$$

$$\mathbf{I}_r = \begin{bmatrix} \mathbf{0} \\ \hline \mathbf{I} \end{bmatrix} = \begin{bmatrix} 0 & 0 & \cdots & 0 & 0 \\ 1 & 0 & \cdots & 0 & 0 \\ 0 & 1 & \cdots & 0 & 0 \\ \vdots & \vdots & \ddots & \vdots & \vdots \\ 0 & 0 & \cdots & 1 & 0 \\ 0 & 0 & \cdots & 0 & 1 \end{bmatrix},$$

where $\mathbf{I}$ is the identity matrix of dimension $p \times p$. Then, for example, if $\mathbf{A}$ is a $(p+1) \times (p+1)$ matrix, it follows that $\mathbf{I}_c \mathbf{A} \mathbf{I}_r$ is a $p \times p$ matrix obtained from $\mathbf{A}$ by removing the first row and the first column. Similarly, if $\mathbf{A}$ is a $p \times p$ matrix, then $\mathbf{I}_r \mathbf{A} \mathbf{I}_c$ is a $(p+1) \times (p+1)$ matrix obtained from $\mathbf{A}$ by adding an additional first row and first column of zeros.

Condition (7.4.5) with the matrix $\widetilde{V}$ defined by (7.4.7) can be written in the form

$$\rho(\mathbf{I}_c \widetilde{\mathbf{V}} \mathbf{I}_r) = \rho\big(\mathbf{I}_c (\mathbf{F} - \widetilde{\mathbf{B}} \mathbf{C} \mathbf{K} \Psi) \mathbf{I}_r\big) = 0.$$

It follows from Lemma 7.6.3 that this condition is equivalent to

$$\Delta\big(T^{-1} \mathbf{I}_c (\mathbf{F} - \widetilde{\mathbf{B}} \mathbf{C} \mathbf{K} \Psi) \mathbf{I}_r T\big) = \mathbf{0}, \tag{7.7.2}$$

where $T \in \mathbb{R}^{p \times p}$ is the product of a permutation matrix P and a unit lower triangular matrix L. Assuming that the permutation matrix is equal to identity (i.e., $T = L$), conditions (7.7.1) and (7.7.2) can be combined into one condition. Let $\mathbf{L}$ be the matrix defined by

$$\mathbf{L} = \begin{bmatrix} 1 & | & 0 \\ \hline 0 & | & L \end{bmatrix} \in \mathbb{R}^{(p+1) \times (p+1)}.$$

We can rewrite equation (7.7.1) in matrix form,

$$\widetilde{\mathbf{B}}\mathbf{C}\beta(\mathbf{K})\mathbf{e}_{p+1}\mathbf{e}_1^T = \delta\mathbf{e}_1^T,$$

where the last p columns on both sides of this relation are equal to zero. This relation is equivalent to

$$\mathbf{L}^{-1}\widetilde{\mathbf{B}}\mathbf{C}\beta(\mathbf{K})\mathbf{e}_{p+1}\mathbf{e}_1^T\mathbf{L} = \delta\mathbf{L}^{-1}\mathbf{e}_1^T\mathbf{L},$$

and since $\mathbf{L}$ is lower triangular, the last p columns of the matrices on both sides of this equation are again equal to zero. Hence, this relation is equivalent to

$$\Delta\big(\mathbf{L}^{-1}(\widetilde{\mathbf{B}}\mathbf{C}\beta(\mathbf{K})\mathbf{e}_{p+1}\mathbf{e}_1^T - \delta\mathbf{e}_1^T)\mathbf{L}\big) = \mathbf{0}. \tag{7.7.3}$$

We can also rewrite equation (7.7.2) in the equivalent form

$$\Delta\big(\mathbf{L}^{-1}\mathbf{I}_r\mathbf{I}_c(\mathbf{F} - \widetilde{\mathbf{B}}\mathbf{C}\mathbf{K}\Psi)\mathbf{I}_r\mathbf{I}_c\mathbf{L}\big) = \mathbf{0}, \tag{7.7.4}$$

since the multiplication by $\mathbf{I}_r$ from the left and by $\mathbf{I}_c$ from the right introduce only the additional first row and first column of zeros. Equations (7.7.3) and (7.7.4) can be combined into one relation,

$$\Delta\big(\mathbf{L}^{-1}(\widetilde{\mathbf{B}}\mathbf{C}\beta(\mathbf{K})\mathbf{e}_{p+1}\mathbf{e}_1^T + \mathbf{I}_r\mathbf{I}_c\widetilde{\mathbf{B}}\mathbf{C}\mathbf{K}\Psi\mathbf{I}_r\mathbf{I}_c)\mathbf{L}\big) = \Delta\big(\mathbf{L}^{-1}(\mathbf{I}_r\mathbf{I}_c\mathbf{F}\mathbf{I}_r\mathbf{I}_c + \delta\mathbf{e}_1)\mathbf{L}\big).$$

This can be simplified further observing that

$$\mathbf{K}\Psi\mathbf{I}_r\mathbf{I}_c = \mathbf{K}\Psi$$

and that the $(1,1)$ element of the matrix $\mathbf{I}_r\mathbf{I}_c\widetilde{\mathbf{B}}$ is equal to zero. This leads to the relation

$$\Delta(\mathbf{L}^{-1}\widetilde{\mathbf{B}}\Omega\mathbf{L}) = \Delta(\mathbf{L}^{-1}\Gamma\mathbf{L}), \tag{7.7.5}$$

where the matrices Ω and Γ are defined by

$$\Omega = \mathbf{C}\big(\beta(\mathbf{K})\mathbf{e}_{p+1}\mathbf{e}_1^T + \mathbf{K}\Psi\big)$$

and

$$\Gamma = \mathbf{I}_r\mathbf{I}_c\mathbf{F}\mathbf{I}_r\mathbf{I}_c + \delta\mathbf{e}_1.$$

It follows from Lemma 7.6.2 that equation (7.7.5) has a solution if $\Omega\mathbf{L}$ is nonsingular and admits LU decomposition $\Omega\mathbf{L} = \mathcal{L}(\Omega\mathbf{L})\mathcal{U}(\Omega\mathbf{L})$, where $\mathcal{L}(\Omega\mathbf{L})$ is unit lower triangular and $\mathcal{U}(\Omega\mathbf{L})$ is upper triangular. Moreover, if this solution exists, it is given by

$$\widetilde{\mathbf{B}} = \mathbf{L}\Delta\big(\Delta(\mathbf{L}^{-1}\Gamma\mathbf{L})\mathcal{U}(\Omega\mathbf{L})^{-1}\big)\mathcal{L}(\Omega\mathbf{L})^{-1}.$$

Using Lemma 7.6.1 the equation above reduces to

$$\widetilde{\mathbf{B}} = \mathbf{L}\Delta\big(\mathbf{L}^{-1}\Gamma\mathbf{L}\mathcal{U}(\Omega\mathbf{L})^{-1}\big)\mathcal{L}(\Omega\mathbf{L})^{-1}. \tag{7.7.6}$$

We now summarize the derivation of GLMs with IRKS by collecting all necessary information that is required in this process. This leads to the following practical algorithm, which we describe step by step.

1. Choose the order p and the vector $\mathbf{c}$ of distinct abscissas $c_1, c_2, \ldots, c_s$, where $s = p + 1$. It is usually assumed that $0 \le c_i \le 1$, $i = 1, 2, \ldots, s$. The typical choice are abscissas uniformly distributed in the interval $[0, 1]$ (i.e., $c_i = (i - 1)/(s - 1)$, $i = 1, 2, \ldots, s$).

2. Choose the diagonal element $\lambda \ge 0$ of the coefficient matrix $\mathbf{A}$. If $\lambda > 0$, which corresponds to implicit GLMs, λ is usually chosen to achieve A-stability of the resulting method.

3. Choose the z^{p+1} coefficient ϵ of the numerator $P(z)$ given by (7.5.13) of the stability function $R(z)$ (7.5.6). This is usually done to achieve some balance between accuracy and stability for explicit methods and to ensure L-stability for implicit methods.

4. Choose the parameters $\beta_1, \beta_2, \ldots, \beta_p$ appearing in the doubly companion matrix $\mathbf{X} = \mathbf{X}(\alpha, \beta)$ defined by (7.2.15). The appropriate choice of these parameters is discussed in Sections 7.9 and 7.11.

5. Compute the matrices

$$\Psi = \beta(\mathbf{K}) \exp(\lambda \mathbf{K}^-),$$

$$\mathbf{X} = \Psi(\mathbf{J} + \lambda \mathbf{I})\Psi^{-1},$$

$$\mathbf{F} = \exp(-\lambda \mathbf{K}^-)\mathbf{E} \exp(\lambda \mathbf{K}^-),$$

$$\Omega = \mathbf{C}\big(\beta(\mathbf{K})\mathbf{e}_{p+1}\mathbf{e}_1^T + \mathbf{K}\Psi\big),$$

$$\Gamma = \mathbf{I}_r \mathbf{I}_c \mathbf{F} \mathbf{I}_r \mathbf{I}_c + \delta \mathbf{e}_1^T.$$

The alternative computation of the matrix Γ is discussed in Section 7.10.

6. Choose a unit lower triangular matrix $\mathbf{L}$ and compute the LU decomposition of the matrix $\Omega \mathbf{L}$:

$$\Omega \mathbf{L} = \mathcal{L}(\Omega \mathbf{L})\, \mathcal{U}(\Omega \mathbf{L}).$$

7. Compute the coefficient matrix $\widetilde{\mathbf{B}}$ of the transformed method from formula (7.7.6):

$$\widetilde{\mathbf{B}} = \mathbf{L}\Delta\big(\mathbf{L}^{-1}\Gamma \mathbf{L}\mathcal{U}(\Omega \mathbf{L})^{-1}\big)\mathcal{L}(\Omega \mathbf{L})^{-1}.$$

8. Compute the coefficient matrices $\mathbf{B}$, $\mathbf{A}$, $\mathbf{U}$, and $\mathbf{V}$ of the original method from the formulas

$$\mathbf{B} = \Psi\widetilde{\mathbf{B}},$$

$$\mathbf{A} = \mathbf{BXB},$$

$$\mathbf{U} = \mathbf{C} - \mathbf{ACK},$$

$$\mathbf{V} = \mathbf{E} - \mathbf{BCK}.$$

These formulas for $\mathbf{B}$ and $\mathbf{A}$ were derived in Section 7.4, and the formulas for $\mathbf{U}$ and $\mathbf{V}$ were derived in Section 7.1.

This algorithm will fail if $\Omega\mathbf{L}$ is singular or does not admit LU decomposition, or if the matrix $\widetilde{\mathbf{B}}$, and as a result the matrix $\mathbf{B}$, is singular. If $\mathbf{B}$ is singular, the matrix $\mathbf{A}$ can still be determined on a case-by-case basis from the relation

$$\mathbf{BA} = \mathbf{XB},$$

although, in general, $\mathbf{A}$ is not unique in such cases. However, it will be demonstrated in Sections 7.9 and 7.11, that the overall algorithm can be carried out successfully for many choices of the abscissa vector $\mathbf{c}$, the parameter λ, the coefficient ϵ, parameters $\beta_1, \beta_2, \ldots, \beta_s$, and the unit lower triangular matrix $\mathbf{L}$, for both explicit and implicit GLMs.

7.8   CONSTRUCTION OF EXPLICIT METHODS WITH IRKS AND GOOD BALANCE BETWEEN ACCURACY AND STABILITY

This and the following section follow the presentation by Butcher and Jackiewicz [72]. In this section we describe the construction of explicit GLMs with IRKS, which achieve a good balance between accuracy, measured by the size of the error constant E of the method, and stability, measured by the size of region or interval of absolute stability. Since for explicit methods, $\lambda = 0$, it follows from (7.5.6) that the stability polynomial of the method of order p satisfies the relation

$$R(z) = \exp(z) - Ez^{p+1} + O(z^{p+2}), \tag{7.8.1}$$

where E is the error constant. It also follows from (7.5.13) and the relation $M_{n,m}(0) = 1/n!$, where $M_{n,m}(\lambda)$ is defined by (7.5.8), that this polynomial takes the form

$$R(z) = R_p(z; \eta) := 1 + z + \frac{z^2}{2!} + \cdots + \frac{z^p}{p!} + \eta \frac{z^{p+1}}{(p+1)!}, \tag{7.8.2}$$

where, for convenience, we introduced the scaled constant $\eta = (p+1)!\epsilon$. We will write $R_p(z)$ instead of $R_p(z; 0)$. Observe that $R_p(z; 1) = R_{p+1}(z)$. It follows from (7.5.14) with $\lambda = 0$ that the relationship between E and ϵ is

$$E = \frac{1}{(p+1)!} - \epsilon,$$

or, in terms of the scaled constant η,

$$E = \frac{1 - \eta}{(p+1)!}. \tag{7.8.3}$$

Denote by

$$\mathcal{A} = \left\{ z : |R_p(z; \eta)| < 1 \right\}$$

the region of absolute stability of $R_p(z; \eta)$ (see also Definition 2.6.2). The next two theorems are concerned with the stability properties of $R_p(z; \eta)$. The first theorem examines the intersection of the region of absolute stability with the real axis.

Theorem 7.8.1 (Butcher and Jackiewicz [72]) *The boundary $\partial \mathcal{A}$ of the region of absolute stability $\mathcal{A}$ of $R_p(z; \eta)$ intersects itself at $z = \overline{x}$ for $\eta = \overline{\eta}$, where $\overline{x}$ and $\overline{\eta}$ satisfy the system of equations*

$$\frac{dR_p}{dz}(\overline{x}; \overline{\eta}) = 0,$$

$$R_p(\overline{x}; \overline{\eta}) = (-1)^p. \tag{7.8.4}$$

Moreover, this value of $\overline{\eta}$ corresponds to the maximal interval of absolute stability $(\widetilde{x}, 0) \subset \mathcal{A}$, where $\widetilde{x}$ is a negative real root of the equation

$$R_p(\widetilde{x}; \overline{\eta}) = (-1)^{p+1}. \tag{7.8.5}$$

Proof: We show first that for any $p \geq 0$,

$$R_{2p}(z) > 0. \tag{7.8.6}$$

This inequality is clearly satisfied for $p = 0$. Assume that (7.8.6) holds for p. Since

$$\frac{dR_{2p+1}}{dz}(z) = R_{2p}(z) > 0,$$

the function $R_{2p+1}(z)$ is increasing for all real values of z. We also have

$$\lim_{z \to -\infty} R_{2p+1}(z) = -\infty \quad \text{and} \quad R_{2p+1}(0) = 1.$$

Hence, there exists exactly one point $\xi < 0$ such that $R_{2p+1}(\xi) = 0$. Since

$$R_{2(p+1)}(z) = 1 + \int_0^z R_{2p+1}(y)dy,$$

it follows that the polynomial $R_{2(p+1)}(z)$ is decreasing for $z \in (-\infty, \xi)$, increasing for $z \in (\xi, \infty)$, and attains its minimum value for $z = \xi$. Hence,

$$R_{2(p+1)}(z) \geq R_{2(p+1)}(\xi) = 1 + \int_0^\xi R_{2p+1}(y)dy$$

$$= R_{2p+1}(\xi) + \frac{\xi^{2p+2}}{(2p+2)!} = \frac{\xi^{2p+2}}{(2p+2)!} > 0,$$

which completes the proof of (7.8.6). Consider next the function $y = R_p(z; \eta)$

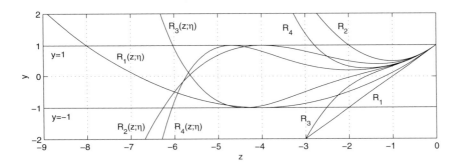

Figure 7.8.1   The functions $R_1(z)$, $R_2(z)$, $R_3(z)$, $R_4(z)$, and $R_1(z; \eta)$, $R_2(z; \eta)$, $R_3(z; \eta)$, $R_4(z; \eta)$ for $\eta = \overline{\eta}$ versus z

for $\eta > 0$. Since $R_p(z; \eta) \to \infty$ as $z \to -\infty$ if p is odd and $R_p(z; \eta) \to -\infty$ as $z \to -\infty$ if p is even, there exists $\overline{\eta} < 0$ such that the graph of this function is tangent to the line $y = 1$ or $y = -1$ at the point $z = \overline{x}$. This is illustrated in Fig. 7.8.1. These values of $\overline{x}$ and $\overline{\eta}$ are characterized by system (7.8.4). At the point $z = \overline{x}$ the boundary ∂A of the region of absolute stability intersects itself and the maximal interval of absolute stability $(\widetilde{x}, 0) \subset A$ can be determined from equation (7.8.5) (see also Fig. 7.8.1). This completes the proof. ∎

It is easy to verify that the solution $\overline{x}$ to (7.8.4) satisfies the polynomial equation

$$2(p+1) + pz + \frac{p-1}{2!}z^2 + \frac{p-2}{3!}z^3 + \cdots + \frac{1}{p!}z^p = 0$$

if p is odd, and

$$p + \frac{p-1}{2!}z + \frac{p-2}{3!}z^2 + \cdots + \frac{1}{p!}z^{p-1} = 0$$

if p is even. The solution $\overline{\eta}$ to (7.8.4) is then given by

$$\overline{\eta} = -p!\,\overline{x}^{-p} R_{p-1}(\overline{x})$$

for any $p \geq 1$.

The next theorem examines the intersection of the region of absolute stability with the imaginary axis.

Theorem 7.8.2 (Butcher and Jackiewicz [72]) *Assume that $p = 2r - 1$. Then the maximal $\bar{y}$ at which the boundary of the region of absolute stability of $R_p(z; \eta)$ intersects the positive part of the imaginary axis is the positive real root of the equation*

$$y - \frac{y^3}{3!} + \cdots + (-1)^{r-1} \frac{y^{2r-1}}{(2r-1)!} = (-1)^{r-1}, \tag{7.8.7}$$

and the corresponding $\bar{\eta}$ can be determined from the relation

$$1 - \frac{y^2}{2!} + \cdots + (-1)^{r-1} \frac{y^{2r-2}}{(2r-2)!} + \eta(-1)^r \frac{y^{2r}}{(2r)!} = 0. \tag{7.8.8}$$

Similarly, if $p = 2r$, the maximal $\bar{y}$ at which the boundary of the region of absolute stability of $R_p(z; \eta)$ intersects the positive part of the imaginary axis is the positive real root of the equation

$$1 - \frac{y^2}{2!} + \cdots + (-1)^r \frac{y^{2r}}{(2r)!} = (-1)^r, \tag{7.8.9}$$

and the corresponding $\bar{\eta}$ can be determined from the relation

$$y - \frac{y^3}{3!} + \cdots + (-1)^{r-1} \frac{y^{2r-1}}{(2r-1)!} + \eta(-1)^r \frac{y^{2r+1}}{(2r+1)!} = 0. \tag{7.8.10}$$

Proof: Assume first that $p = 2r - 1$. Put $z = iy$ and consider the equation

$$
\begin{aligned}
|R_p(iy; \eta)|^2 &= \left(1 - \frac{y^2}{2!} + \cdots + (-1)^{r-1} \frac{y^{2r-2}}{(2r-2)!} + \eta(-1)^r \frac{y^{2r}}{(2r)!} \right)^2 \\
&\quad + \left(y - \frac{y^3}{3!} + \cdots + (-1)^{r-1} \frac{y^{2r-1}}{(2r-1)!} \right)^2 = 1.
\end{aligned}
\tag{7.8.11}
$$

This relation defines implicitly the function $y = y(\eta)$ and we are interested in the values of η for which y attains its maximum value (i.e., $y'(\eta) = 0$). Differentiating equation (7.8.11) with respect to η and assuming that $y'(\eta) = 0$ leads to relation (7.8.8). The maximal value of y can then be determined from the equation $|R_p(iy; \eta)|^2 = 1$, which taking into account (7.8.8) and assuming that y is positive leads to (7.8.7). The proof of (7.8.9) and (7.8.10) corresponding to $p = 2r$ is similar and is therefore omitted. ∎

After finding values of $\bar{y}$ and $\bar{\eta}$ satisfying (7.8.7) and (7.8.8) or (7.8.9) and (7.8.10), we have to verify if

$$|R_p(iy; \bar{\eta})| < 1,$$

for all $y \in (0, \bar{y})$. This is discussed further for specific examples in Section 7.9.

7.9   EXAMPLES OF EXPLICIT METHODS WITH IRKS

In this section we apply the algorithm described at the end of Section 7.7 to construct specific examples of explicit GLMs with IRKS of order $p = 1, 2, 3$, and 4, with good accuracy and stability properties. The components of the abscissa vector $\mathbf{c}$ are chosen as

$$c_i = \frac{i-1}{s-1}, \quad i = 1, 2, \ldots, s, \tag{7.9.1}$$

$s = p + 1$. The parameters $\beta_1, \beta_2, \ldots, \beta_p$, are chosen as follows:

$$\beta_1 = \beta_2 = \cdots = \beta_p = E, \tag{7.9.2}$$

where E is the error constant of the method, which is related to the parameter η by equation (7.8.3). This choice of the parameters β_i, $i = 1, 2, \ldots, p$, is motivated by a result discovered by Wright [293]: that under some conditions

$$z_i^{[n]} = z_i(t_n, h) - \beta_{p+2-i} h^{p+1} y^{(p+1)}(t_n) + O(h^{p+2}), \tag{7.9.3}$$

$i = 2, 3, \ldots, p+1$, where $z(t, h)$ is the Nordsieck vector defined by (7.1.2). We also have

$$z_1^{[n]} = y(t_n) - h^{p+1} E y^{(p+1)}(t_n) + O(h^{p+2}). \tag{7.9.4}$$

This is discussed further in Chapter 8. This result means that the parameters β_{p+2-i} correspond to the errors of the external stages of the method $z_i^{[n]}$, $i = 2, 3, \ldots, p+1$. Therefore, it seems to be reasonable to choose them according to (7.9.2) so that they have the same order of magnitude as the error constant E, which corresponds to the error in the first component of $z^{[n]}$. The matrix $\mathbf{L}$ appearing in the formula for $\widetilde{\mathbf{B}}$ is chosen as $\mathbf{L} = \mathbf{I}$, so that the formula for $\widetilde{\mathbf{B}}$ simplifies to

$$\widetilde{\mathbf{B}} = \Delta\big(\Gamma\mathcal{U}(\Omega)^{-1}\big)\mathcal{L}(\Omega)^{-1}. \tag{7.9.5}$$

7.9.1   Methods with $p = q = 1$ and $s = 2$

To guide selection of the appropriate parameter η, we have plotted in Fig. 7.9.1 the left end of the interval of absolute stability $(a, 0) \subset \mathcal{A}$, corresponding to

$$R_1(z; \eta) = 1 + z + \eta\frac{z^2}{2},$$

versus η for $0 \leq \eta \leq 1$. The boundary of the stability region of $R_1(z; \eta)$ intersects itself at $z = -4$ (the root of $4+z$), which corresponds to $\eta = \frac{1}{4}$. This region corresponds to the stabilized RK method with $R_1(z; \frac{1}{4}) = T_2(1 + \frac{z}{4})$, where $T_2(x) = 2x^2 - 1$ is the Chebyshev polynomial of degree 2 (compare [146]). The boundary of this region is shown by a thick solid line in Fig. 7.9.2,

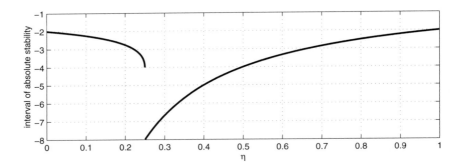

Figure 7.9.1   Left end of the interval of absolute stability versus η for the polynomial $R_1(z; \eta) = 1 + z + \eta z^2/2$

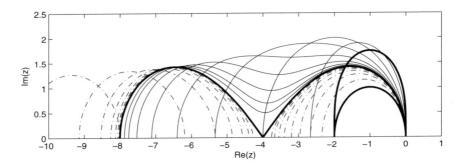

Figure 7.9.2   Regions of absolute stability of $R_1(z; \eta) = 1 + z + \eta z^2/2$ for selected values of the parameter η

together with   regions of absolute stability corresponding to $\eta = \frac{1}{4} - \frac{1}{2^k}$, $k = 3, 4, \ldots, 10$ (thin dashed-dotted lines), and $\eta = \frac{1}{4} + \frac{1}{2^k}$, $k = 1, 2, \ldots, 8$ (thin solid lines). We also show by thick solid lines the stability regions of the RK method of order 1 which corresponds to $\eta = 0$, and of order 2, which corresponds to $\eta = 1$.

There is a trade-off between the size of the error constant $E = (1 - \eta)/2$ and the size of the region of absolute stability. For example, $\eta = \frac{1}{4}$ corresponds to the maximal interval of absolute stability $(-8, 0)$ and quite large error constant $E = \frac{3}{8}$. As η increases from $\frac{1}{4}$ to 1, the interval of absolute stability decreases from $(-8, 0)$ to $(-2, 0)$, and the error constant decreases from $E = \frac{3}{8}$ to $E = 0$, which corresponds to the method of order 2. Choosing $\mathbf{c} = [0, 1]^T$, $\eta = \frac{1}{2}$, and $\beta_1 = \frac{1}{4}$ leads to the method with $E = \frac{1}{4}$ and stability interval

$(-4, 0)$. The coefficients of this method are given by

$$
\left[\begin{array}{c|c}
\mathbf{A} & \mathbf{U} \\
\hline
\mathbf{B} & \mathbf{V}
\end{array}\right]
=
\left[\begin{array}{cc|cc}
0 & 0 & 1 & 0 \\
\frac{4}{3} & 0 & 1 & -\frac{1}{3} \\
\hline
\frac{17}{16} & \frac{3}{16} & 1 & -\frac{1}{4} \\
\frac{1}{4} & \frac{3}{4} & 0 & 0
\end{array}\right].
$$

7.9.2 Methods with $p = q = 2$ and $s = 3$

The stability function of methods with $p = q = 2$ and $s = 3$ with IRKS takes the form $p(w, z) = w^2(w - R_2(z; \eta))$, where

$$
R_2(z; \eta) = 1 + z + \frac{z^2}{2} + \eta \frac{z^3}{6},
$$

$\eta \in R$, and the error constant is

$$
E = \frac{1 - \eta}{6}.
$$

We have plotted in Fig. 7.9.3 the left end of the interval of absolute sta-

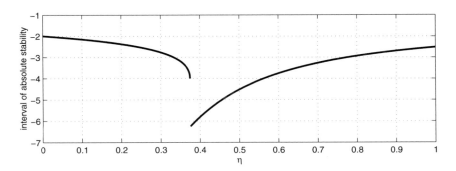

Figure 7.9.3 Left end of the interval of absolute stability versus η for the polynomial $R_2(z; \eta) = 1 + z + z^2/2 + \eta\, z^3/6$

bility $(a, 0) \subset \mathcal{A}$ versus η for $0 \leq \eta \leq 1$. The boundary of the stability region intersects itself at $z = -4$ (the root of $2 + \frac{1}{2}z$), which corresponds to $\eta = \frac{3}{8}$ and the interval of absolute stability is $(-6.2608, 0)$. This region is shown by a thick solid line in Fig. 7.9.4, together with stability regions corresponding to $\eta = \frac{3}{8} - \frac{6}{2^k}$, $k = 5, \ldots, 12$ (thin dashed-dotted lines) and $\eta = \frac{3}{8} + \frac{6}{2^k}$, $k = 4, \ldots, 11$ (thin solid lines). We also show by thick solid lines stability regions of the RK method of order 2 which corresponds to $\eta = 0$,

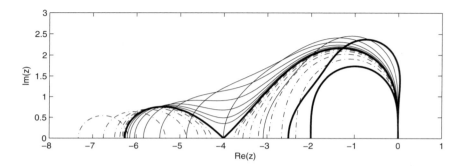

Figure 7.9.4   Regions of absolute stability of $R_2(z; \eta) = 1 + z + z^2/2 + \eta\, z^3/6$ for selected values of the parameter η

and the RK method of order 3 which corresponds to $\eta = 1$. The method with $\eta = \frac{9}{16}$ achieves a good balance between accuracy (the error constant is $E = \frac{7}{96}$) and stability (the interval of absolute stability is $(-4, 0)$), and choosing $\mathbf{c} = [0, \frac{1}{2}, 1]^T$ and $\beta_1 = \beta_2 = E = \frac{7}{96}$, the coefficients of this method are

$$
\left[\begin{array}{c|c} \mathbf{A} & \mathbf{U} \\ \hline \mathbf{B} & \mathbf{V} \end{array}\right] =
\left[\begin{array}{ccc|ccc}
0 & 0 & 0 & 1 & 0 & 0 \\
\frac{279}{574} & 0 & 0 & 1 & \frac{4}{287} & \frac{1}{8} \\
\frac{81}{14105} & \frac{1968}{2015} & 0 & 1 & \frac{8}{455} & \frac{47}{4030} \\
\hline
\frac{608663}{499968} & -\frac{2009}{35712} & \frac{455}{2304} & 1 & -\frac{241}{672} & \frac{41}{124} \\
-\frac{113815}{71424} & \frac{85567}{35712} & \frac{455}{2304} & 0 & 0 & -\frac{1177}{2976} \\
\frac{17}{24} & -\frac{41}{12} & \frac{65}{24} & 0 & 0 & 0
\end{array}\right].
$$

7.9.3   Methods with $p = q = 3$ and $s = 4$

The stability function of methods with $p = q = 3$ and $s = 4$ with IRKS takes the form $p(w, z) = w^3(w - R_3(z; \eta))$, where

$$
R_3(z; \eta) = 1 + z + \frac{z^2}{2} + \frac{z^3}{6} + \eta\, \frac{z^4}{24},
$$

$\eta \in \mathbb{R}$, and the error constant is

$$
E = \frac{1 - \eta}{24}.
$$

We have plotted in Fig. 7.9.5 the left end of the interval of absolute stability versus η for $0 \le \eta \le 1$. The boundary of the stability region intersects itself at

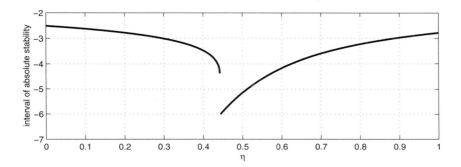

Figure 7.9.5   Left end of the interval of absolute stability versus η for the polynomial $R_3(z;\eta) = 1 + z + z^2/2 + z^3/6 + \eta\, z^4/24$

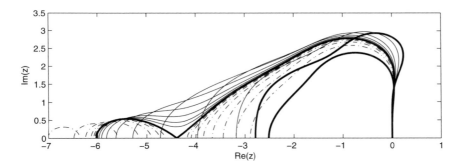

Figure 7.9.6   Regions of absolute stability of $R_3(z;\eta) = 1 + z + z^2/2 + z^3/6 + \eta\, z^4/24$ for selected values of the parameter η

$z = -4.39035$, which corresponds to $\overline{\eta} = 0.442937$, and the maximal interval of absolute stability is $(-6.02726, 0)$. This region is shown in Fig. 7.9.6 by a thick solid line together with stability regions corresponding to $\eta = \overline{\eta} - \frac{24}{2^k}$, $k = 7, 8, \ldots, 14$ (thin dashed-dotted lines) and stability regions corresponding to $\eta = \overline{\eta} + \frac{24}{2^k}$, $k = 6, 7, \ldots, 13$ (thin solid lines). We also show by thick solid lines in Fig. 7.9.6 stability regions corresponding to $\eta = 0$, which corresponds to the RK method of order 3, and $\eta = 1$, which corresponds to the RK method of order 4. The method with $\eta = \frac{3}{5}$ achieves a good balance between accuracy (the error constant is $E = \frac{1}{60}$) and stability properties (the interval of absolute stability is $(-4.1709, 0)$) and choosing $\mathbf{c} = [0, \frac{1}{3}, \frac{2}{3}, 1]^T$ and $\beta_1 = \beta_2 = \beta_3 =$

$E = \frac{1}{60}$, the coefficients of this method are

$$A = \begin{bmatrix} 0 & 0 & 0 & 0 \\ \frac{41}{324} & 0 & 0 & 0 \\ -\frac{1147432}{1657179} & \frac{20412}{20459} & 0 & 0 \\ -\frac{884267971}{603213156} & \frac{195810716}{150803289} & \frac{4990}{7371} & 0 \end{bmatrix},$$

$$U = \begin{bmatrix} 1 & 0 & 0 & 0 \\ 1 & \frac{67}{324} & \frac{1}{18} & \frac{1}{162} \\ 1 & \frac{14606}{40419} & -\frac{20318}{184131} & -\frac{10018}{1657179} \\ 1 & \frac{343643}{700596} & -\frac{5517035}{14362218} & -\frac{151762489}{2714459202} \end{bmatrix},$$

$$B = \begin{bmatrix} \frac{1368599}{1033200} & \frac{443273}{1033200} & -\frac{13343}{25200} & \frac{117}{400} \\ -\frac{11208301}{1033200} & \frac{12486353}{1033200} & -\frac{13343}{25200} & \frac{117}{400} \\ \frac{73403}{8400} & -\frac{169549}{8400} & \frac{93689}{8400} & \frac{117}{400} \\ -\frac{171}{20} & \frac{693}{20} & -\frac{873}{20} & \frac{351}{20} \end{bmatrix},$$

$$V = \begin{bmatrix} 1 & -\frac{31}{60} & \frac{1027}{2460} & \frac{53117}{464940} \\ 0 & 0 & -\frac{7301}{2460} & -\frac{4649}{23247} \\ 0 & 0 & 0 & -\frac{1903}{3780} \\ 0 & 0 & 0 & 0 \end{bmatrix}.$$

7.9.4   Methods with $p = q = 4$ and $s = 5$

The stability function of methods with $p = q = 4$ and $s = 5$ with IRKS takes the form $p(w, z) = w^4(w - R_4(z; \eta))$, where

$$R_4(z; \eta) = 1 + z + \frac{z^2}{2} + \frac{z^3}{6} + \frac{z^4}{24} + \eta \frac{z^5}{120},$$

$\eta \in R$, and the error constant is

$$E = \frac{1 - \eta}{120}.$$

We show in Fig. 7.9.7 the left end of the interval of absolute stability versus η for $0 \le \eta \le 1$. The boundary of the stability region intersects itself at $z = -4.68878$, which corresponds to $\bar{\eta} = 0.490435$, and the maximal interval

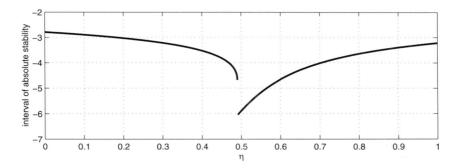

Figure 7.9.7   Left end of the interval of absolute stability versus η for the polynomial $R_4(z;\eta) = 1 + z + z^2/2 + z^3/6 + z^4/24 + \eta\,z^5/120$

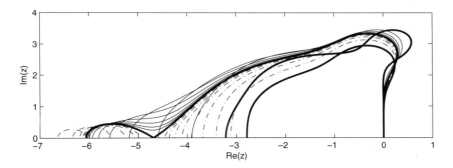

Figure 7.9.8   Regions of absolute stability of $R_4(z;\eta) = 1 + z + z^2/2 + z^3/6 + z^4/24 + \eta\,z^5/120$ for selected values of the parameter η

of absolute stability is $(-6.06060, 0)$. This region is shown in Fig. 7.9.8 by a thick solid line together with stability regions corresponding to $\eta = \overline{\eta} - \frac{120}{2^k}$, $k = 9, 10, \ldots, 16$ (thin dashed-dotted lines), and stability regions corresponding to $\eta = \overline{\eta} + \frac{120}{2^k}$, $k = 9, 10, \ldots, 16$ (thin solid lines). We also show by thick solid lines in Fig. 7.9.8 stability regions corresponding to $\eta = 0$, which corresponds to the RK method of order 4, and $\eta = 1$, which corresponds to $R_4(z; 1) = R_5(z)$. The method with $\eta = \frac{3}{5}$ achieves a good balance between accuracy (the error constant is $E = \frac{1}{300}$) and stability properties (the interval of absolute stability is $(-4.6568, 0)$), and choosing $\mathbf{c} = [0, \frac{1}{4}, \frac{1}{2}, \frac{3}{4}, 1]^T$ and

$\beta_1 = \beta_2 = \beta_3 = \beta_4 = E = \frac{1}{300}$, the coefficient matrix $\mathbf{A}$ is

$$
\begin{bmatrix}
0 & 0 & 0 & 0 & 0 \\
\frac{511}{11776} & 0 & 0 & 0 & 0 \\
-\frac{3101781017}{7028858144} & \frac{5166720}{9550079} & 0 & 0 & 0 \\
-\frac{83353109372729587}{145067197462387200} & -\frac{1001725806316}{12318885654075} & \frac{4242403}{5159700} & 0 & 0 \\
-\frac{41921552692643159297}{191921635567777915800} & -\frac{6767984667820805501}{4172209468864737300} & \frac{696338101127}{436876958700} & \frac{165375}{338684} & 0
\end{bmatrix},
$$

the coefficient matrix $\mathbf{U}$ is

$$
\begin{bmatrix}
1 & 0 & 0 & 0 & 0 \\
1 & \frac{2433}{11776} & \frac{1}{32} & \frac{1}{384} & \frac{1}{6144} \\
1 & \frac{5505879}{13755104} & -\frac{783361}{76400632} & \frac{1799999}{458403792} & \frac{4383359}{3667230336} \\
1 & \frac{6294407001}{10784001536} & -\frac{73807449337}{673853574240} & -\frac{47184261087409}{1576817363721600} & -\frac{94210576904551}{25229077819545600} \\
1 & \frac{47665961355}{62850508952} & -\frac{8094188961413}{31418422898940} & -\frac{17528291232939607}{147038219167039200} & -\frac{5780449858046290723}{267021406007343187200}
\end{bmatrix},
$$

and the coefficient matrices $\mathbf{B}$ and $\mathbf{V}$ are given by

$$
\mathbf{B} =
\begin{bmatrix}
\frac{172958424193}{101787367500} & -\frac{28218803078}{25446841875} & \frac{202394939}{99596250} & -\frac{2106322}{1276875} & \frac{2984}{5625} \\
-\frac{1692541208191}{50893683750} & \frac{848483163322}{25446841875} & \frac{202394939}{99596250} & -\frac{2106322}{1276875} & \frac{2984}{5625} \\
\frac{996053384}{16599375} & -\frac{689424962}{5533125} & \frac{1090797904}{16599375} & -\frac{2106322}{1276875} & \frac{2984}{5625} \\
-\frac{78390032}{1276875} & \frac{254922728}{1276875} & -\frac{91332664}{425625} & \frac{96787928}{1276875} & \frac{2984}{5625} \\
\frac{7136}{75} & -\frac{33344}{75} & \frac{19072}{25} & -\frac{42944}{75} & \frac{11936}{75}
\end{bmatrix},
$$

$$
\mathbf{V} =
\begin{bmatrix}
1 & -\frac{151}{300} & \frac{71723}{153300} & \frac{52375073}{358722000} & \frac{2425799989}{81429894000} \\
0 & 0 & -\frac{1172011}{153300} & -\frac{214263247}{358722000} & \frac{5298687019}{81429894000} \\
0 & 0 & 0 & -\frac{9131}{2925} & -\frac{2196853}{4249440} \\
0 & 0 & 0 & 0 & -\frac{63679}{136200} \\
0 & 0 & 0 & 0 & 0
\end{bmatrix}.
$$

7.9.5 Methods with large intervals of absolute stability on imaginary axis

In what follows we determine the coefficients of methods of order $p = 1$, 2, 3, and 4 with maximal intervals of absolute stability on the imaginary axis. Consider first the method of order 1. It follows from (7.8.7) and (7.8.8) with $r = 1$ ($p = 1$) that the maximal $\overline{y}$ at which the boundary of the stability region of $R_1(z; \eta)$ intersects the positive part of the imaginary axis is at $\overline{y} = 1$, which corresponds to $\overline{\eta} = 2$ (compare [146, p. 37], where optimal stability for hyperbolic problems is discussed). We can also verify that $|R_1(iy; \overline{\eta})| \leq 1$ for all $y \in [-1, 1]$. The error constant of this method is $E = -\frac{1}{2}$, and choosing $\mathbf{c} = [0, 1]^T$ and $\beta_1 = E = -\frac{1}{2}$, the coefficients of this method are

$$
\left[\begin{array}{c|c} \mathbf{A} & \mathbf{U} \\ \hline \mathbf{B} & \mathbf{V} \end{array} \right] = \left[\begin{array}{cc|cc} 0 & 0 & 1 & 0 \\ -\frac{4}{3} & 0 & 1 & \frac{7}{3} \\ \hline -\frac{7}{4} & -\frac{3}{4} & 1 & \frac{7}{2} \\ -\frac{1}{2} & \frac{3}{2} & 0 & 0 \end{array} \right].
$$

The region of absolute stability of the method above is plotted in Fig. 7.9.9.

Consider next the method of order 2. It follows from (7.8.9) and (7.8.10) with $r = 1$ ($p = 2$) that the maximal $\overline{y}$ at which the boundary of the stability region of $R_2(z; \eta)$ intersects the positive part of the imaginary axis is at $\overline{y} = 2$, which corresponds to $\overline{\eta} = \frac{3}{2}$. We can also verify that $|R_2(iy; \overline{\eta})| \leq 1$ for all $y \in [-2, 2]$. The error constant of this method is $E = -\frac{1}{12}$, and choosing $\mathbf{c} = [0, \frac{1}{2}, 1]^T$ and $\beta_1 = \beta_2 = E = -\frac{1}{12}$, the coefficients of this method are

$$
\left[\begin{array}{c|c} \mathbf{A} & \mathbf{U} \\ \hline \mathbf{B} & \mathbf{V} \end{array} \right] = \left[\begin{array}{ccc|ccc} 0 & 0 & 0 & 1 & 0 & 0 \\ -\frac{3}{7} & 0 & 0 & 1 & \frac{13}{14} & \frac{1}{8} \\ -\frac{12}{5} & \frac{21}{10} & 0 & 1 & \frac{13}{10} & -\frac{11}{20} \\ \hline -\frac{47}{18} & -\frac{7}{36} & -\frac{5}{18} & 1 & \frac{49}{12} & \frac{7}{8} \\ -\frac{55}{9} & \frac{133}{18} & -\frac{5}{18} & 0 & 0 & -\frac{29}{12} \\ \frac{4}{3} & -\frac{14}{3} & \frac{10}{3} & 0 & 0 & 0 \end{array} \right].
$$

The region of absolute stability of this method is plotted in Fig. 7.9.9.

Consider now the method of order 3. It follows from (7.8.7) and (7.8.8) with $r = 2$ ($p = 3$) that the maximal $\overline{y}$ at which the boundary of the stability region of $R_3(z; \eta)$ intersects the positive part of the imaginary axis is at $\overline{y} = 2.84732$, which corresponds to $\overline{\eta} = 1.11501$. We can also verify that $|R_3(iy; \overline{\eta})| \leq 1$ for $y \in [-\overline{y}, -0.74008] \cup [0.74008, \overline{y}]$, but $|R_3(iy; \overline{\eta})|$ is slightly greater than 1 for $y \in [-0.74008, 0.74008]$. In fact, $|R_3(iy; \overline{\eta})|$ attains its maximum value 1.0002 for $y = \pm 0.60187$. The error constant of this method is $E = -0.00479208$, and choosing $\mathbf{c} = [0, \frac{1}{3}, \frac{2}{3}, 1]^T$ and $\beta_1 = \beta_2 = \beta_3 = E$, the coefficients $(\mathbf{A}, \mathbf{U}, \mathbf{B}, \mathbf{V})$

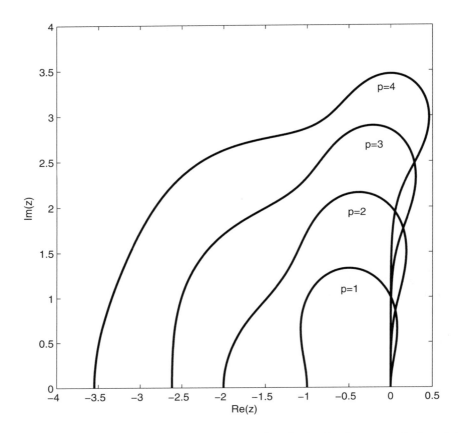

Figure 7.9.9 Regions of absolute stability of methods of order $p = 1, 2, 3$, and 4 with maximal intervals of absolute stability on the imaginary axis

of this method are

$$
\begin{bmatrix}
0 & 0 & 0 & 0 & 1 & 0 & 0 & 0 \\
-\frac{221}{907} & 0 & 0 & 0 & 1 & \frac{311}{539} & \frac{1}{18} & \frac{1}{162} \\
-\frac{13993}{4190} & \frac{2323}{779} & 0 & 0 & 1 & \frac{929}{907} & -\frac{1045}{1354} & -\frac{417}{3586} \\
-\frac{3125}{428} & \frac{967}{157} & \frac{2099}{2852} & 0 & 1 & \frac{727}{517} & -\frac{701}{343} & -\frac{572}{1687} \\
-\frac{5452}{573} & -\frac{571}{2270} & \frac{377}{2455} & -\frac{239}{2751} & 1 & \frac{3456}{323} & \frac{1239}{2180} & \frac{711}{3743} \\
-\frac{5470}{141} & \frac{9773}{246} & \frac{377}{2455} & -\frac{239}{2751} & 0 & 0 & -\frac{24749}{2019} & -\frac{382}{225} \\
\frac{4622}{445} & -\frac{6132}{257} & \frac{1912}{141} & -\frac{239}{2751} & 0 & 0 & 0 & -\frac{357}{554} \\
-\frac{4163}{456} & \frac{5531}{152} & -\frac{6899}{152} & \frac{8267}{456} & 0 & 0 & 0 & 0
\end{bmatrix}.
$$

The region of absolute stability of this method is plotted in Fig. 7.9.9.

Finally, consider the method of order 4. It follows from (7.8.9) and (7.8.10) with $r = 2$ $(p = 4)$ that the maximal $\overline{y}$ at which the boundary of the stability region of $R_4(z; \eta)$ intersects the positive part of the imaginary axis is at $\overline{y} = 2\sqrt{3}$, which corresponds to $\overline{\eta} = \frac{5}{6}$. We can also verify that $|R_4(iy; \overline{\eta})| \leq 1$ for all $y \in [-\overline{y}, \overline{y}]$. The error constant of this method is $E = \frac{1}{720}$, and choosing $\mathbf{c} = [0, \frac{1}{4}, \frac{1}{2}, \frac{3}{4}, 1]^T$ and $\beta_1 = \beta_2 = \beta_3 = \beta_4 = E$, the coefficients of this method are

$$
\mathbf{A} = \begin{bmatrix}
0 & 0 & 0 & 0 & 0 \\
\frac{385}{5568} & 0 & 0 & 0 & 0 \\
-\frac{1037}{1151} & \frac{1245}{1186} & 0 & 0 & 0 \\
-\frac{1187}{634} & \frac{1387}{1126} & \frac{293}{334} & 0 & 0 \\
-\frac{1702}{651} & \frac{840}{1249} & \frac{1242}{697} & \frac{998}{2029} & 0
\end{bmatrix},
$$

$$
\mathbf{U} = \begin{bmatrix}
1 & 0 & 0 & 0 & 0 \\
1 & \frac{495}{2737} & \frac{1}{32} & \frac{1}{384} & \frac{1}{6144} \\
1 & \frac{596}{1697} & -\frac{163}{1186} & -\frac{15}{1253} & -\frac{1}{7719} \\
1 & \frac{661}{1288} & -\frac{369}{793} & -\frac{249}{3199} & -\frac{98}{11807} \\
1 & \frac{463}{693} & -\frac{2423}{2611} & -\frac{606}{2813} & -\frac{130}{4089}
\end{bmatrix},
$$

$$
\mathbf{B} = \begin{bmatrix}
\frac{5027}{998} & -\frac{773}{1828} & \frac{358}{419} & -\frac{206}{299} & \frac{449}{2025} \\
-\frac{9324}{131} & \frac{17301}{241} & \frac{358}{419} & -\frac{206}{299} & \frac{449}{2025} \\
\frac{12207}{188} & -\frac{3755}{28} & \frac{7034}{101} & -\frac{206}{299} & \frac{449}{2025} \\
-\frac{16908}{271} & \frac{12583}{62} & -\frac{13329}{61} & \frac{5674}{73} & \frac{449}{2025} \\
\frac{4304}{45} & -\frac{20096}{45} & \frac{11488}{15} & -\frac{25856}{45} & \frac{7184}{45}
\end{bmatrix},
$$

$$
\mathbf{V} = \begin{bmatrix}
1 & -\frac{2881}{720} & \frac{1099}{2321} & \frac{314}{2013} & \frac{137}{3758} \\
0 & 0 & -\frac{4953}{290} & -\frac{767}{434} & -\frac{123}{4625} \\
0 & 0 & 0 & -\frac{628}{183} & -\frac{517}{876} \\
0 & 0 & 0 & 0 & -\frac{320}{669} \\
0 & 0 & 0 & 0 & 0
\end{bmatrix}.
$$

The region of absolute stability of this method is plotted in Fig. 7.9.9.

7.10   CONSTRUCTION OF A- AND L-STABLE METHODS WITH IRKS

In this section we describe the construction of A- and L-stable GLMs (7.1.1) with IRKS. We recall from Section 7.5 that stability function $R(z)$ of GLM of order p with IRKS satisfies the relation

$$R(z) = \frac{P(z)}{(1 - \lambda z)^{p+1}} = \exp(z) - Ez^{p+1} + O(z^{p+2})$$

(compare (7.5.6)), where

$$P(z) = \sum_{n=0}^{p} M_{n,p+1}(\lambda)z^n + \epsilon z^{p+1}$$

(compare (7.5.13)), and the polynomials $M_{n,m}(\lambda)$ are defined by (7.5.8). Assuming that $\epsilon = 0$, it follows that

$$\lim_{z \to -\infty} R(z) = 0,$$

so that if the corresponding method is A-stable, it is automatically also L-stable. It also follows from (7.5.14) that the error constant E of the resulting method is given by

$$E = M_{p+1,p+1}(\lambda).$$

To investigate A-stability, we reformulate $P(z)$ in terms of Laguerre polynomials. Using (7.5.10), we obtain

$$P(z) = \sum_{n=0}^{p} M_{n,p+1}(\lambda)z^n = \sum_{n=0}^{p} (-\lambda)^n L_{n,p+1-n}\left(\frac{1}{\lambda}\right) z^n,$$

and substituting $s = p + 1$, it follows from property (7.5.9) of generalized Laguerre polynomials that

$$P(z) = \sum_{n=0}^{s-1} (-\lambda)^n (-1)^{s-n} L_s^{(s-n)}\left(\frac{1}{\lambda}\right) z^n = (-1)^s \sum_{n=0}^{s-1} L_s^{(s-n)}\left(\frac{1}{\lambda}\right)(\lambda z)^n.$$

Observe that this expression is equivalent to the polynomial $P(z)$ defined by formula (2.7.5). Hence, we can conclude that stability properties of the resulting GLMs with IRKS are the same as those of stiffly accurate SDIRK methods of order $p = s - 1$. In particular, the resulting methods are A-stable for values of the parameter λ listed in the second column of Table 2.7.2.

To construct specific examples of A- and L-stable methods with IRKS, we again use the algorithm described at the end of Section 7.5. In this algorithm the matrix Γ can be computed more conveniently without the need to utilize matrices $\mathbf{I}_r$ and $\mathbf{I}_c$. It follows from formulas (7.4.8) and (7.4.12), with $\mathbf{S}$ replaced by $\mathbf{K}$, that the matrix $\mathbf{F} = \exp(\mathbf{K}(\mathbf{I} + \lambda\mathbf{K})^{-1})$ takes the form

$$
\mathbf{F} = \begin{bmatrix}
1 & N_1(\lambda) & \cdots & & N_{p-2}(\lambda) & N_{p-1}(\lambda) & N_p(\lambda) \\
0 & 1 & N_1(\lambda) & \cdots & & N_{p-2}(\lambda) & N_{p-1}(\lambda) \\
\vdots & 0 & 1 & N_1(\lambda) & \cdots & & N_{p-2}(\lambda) \\
\vdots & \vdots & \ddots & \ddots & \ddots & & \vdots \\
0 & \vdots & & \ddots & \ddots & \ddots & \vdots \\
0 & 0 & \cdots & \cdots & 0 & 1 & N_1(\lambda) \\
0 & 0 & 0 & \cdots & \cdots & 0 & 1
\end{bmatrix}.
$$

Hence, the matrix $\Gamma = \mathbf{I}_r\mathbf{I}_c\mathbf{F}\mathbf{I}_r\mathbf{I}_c + \delta\mathbf{e}_1^T$ takes the form

$$
\Gamma = \begin{bmatrix}
\epsilon + \lambda M_{p,p}(\lambda) & 0 & \cdots & \cdots & 0 & 0 & 0 \\
N_p(\lambda) & 1 & N_1(\lambda) & \cdots & \cdots & N_{p-2}(\lambda) & N_{p-1}(\lambda) \\
\vdots & 0 & 1 & N_1(\lambda) & \cdots & \cdots & N_{p-2}(\lambda) \\
\vdots & \vdots & \ddots & \ddots & \ddots & & \vdots \\
N_3(\lambda) & \vdots & & \ddots & \ddots & \ddots & \vdots \\
N_2(\lambda) & 0 & \cdots & \cdots & 0 & 1 & N_1(\lambda) \\
N_1(\lambda) & 0 & 0 & \cdots & \cdots & 0 & 1
\end{bmatrix}.
$$

It can be verified using (7.4.16) that

$$
(\mathbf{JF} - \mathbf{FJ})(\mathbf{K} + \mathbf{e}_{p+1}\mathbf{e}_1^T) = \begin{bmatrix}
0 & -N_1(\lambda) & \cdots & -N_p(\lambda) \\
N_p(\lambda) & 0 & \cdots & 0 \\
\vdots & \vdots & \ddots & \vdots \\
N_1(\lambda) & 0 & \cdots & 0
\end{bmatrix},
$$

and it follows that $\Gamma^* := \mathbf{F} + (\mathbf{JF} - \mathbf{FJ})(\mathbf{K} + \mathbf{e}_{p+1}\mathbf{e}_1^T)$ is equal to Γ except for the $(1,1)$ element. As a result, we can compute Γ from the formula

$$
\gamma_{ij} = \begin{cases} \epsilon + \lambda M_{p,p}(\lambda), & i = j = 1, \\ \gamma_{ij}^*, & \text{otherwise.} \end{cases}
$$

Here γ_{ij} and γ_{ij}^* are elements of Γ and Γ^*, respectively. This alternative way to compute Γ is also applicable in the case of explicit methods (i.e., if $\lambda = 0$).

7.11   EXAMPLES OF A- AND L-STABLE METHODS WITH IRKS

In this section we present coefficient matrices $\mathbf{A}$, $\mathbf{U}$, $\mathbf{B}$, and $\mathbf{V}$ of GLMs with IRKS of order $p = 1$, 2, 3, and 4. The abscissa vector $\mathbf{c}$ is chosen according to (7.9.1) and the parameter λ so that the resulting method is A-stable. The ranges of parameters λ for which this is the case for particular orders are given in Table 2.7.2. We always choose $\epsilon = 0$ so that the resulting methods are also L-stable. As in Section 7.9, the coefficients $\beta_1, \beta_2, \ldots, \beta_p$ are chosen according to (7.9.2), which can again be justified by (7.9.3) and (7.9.4). The unit lower triangular matrix $\mathbf{L}$ is chosen as $\mathbf{L} = \mathbf{I}$, so that the formula for $\widetilde{\mathbf{B}}$ again simplifies to (7.9.5).

Method with $p = q = 1$, $s = 2$, $\mathbf{c} = [0, 1]^T$, $\lambda = \frac{1}{2}$, $\epsilon = 0$, and

$$\beta_1 = E = -\tfrac{1}{4}.$$

$$
\left[
\begin{array}{c|c}
\mathbf{A} & \mathbf{U} \\
\hline
\mathbf{B} & \mathbf{V}
\end{array}
\right]
=
\left[
\begin{array}{cc|cc}
\frac{1}{2} & 0 & 1 & -\frac{1}{12} \\
-\frac{4}{5} & \frac{1}{2} & 1 & \frac{13}{10} \\
\hline
-\frac{17}{16} & \frac{5}{6} & 1 & \frac{7}{4} \\
-\frac{1}{4} & \frac{5}{4} & 0 & 0
\end{array}
\right].
$$

Method with $p = q = 2$, $s = 3$, $\mathbf{c} = [0, \frac{1}{2}, 1]^T$, $\lambda = \frac{1}{4}$, $\epsilon = 0$, and

$$\beta_1 = \beta_2 = E = -\tfrac{7}{192}.$$

$$
\left[
\begin{array}{c|c}
\mathbf{A} & \mathbf{U} \\
\hline
\mathbf{B} & \mathbf{V}
\end{array}
\right]
=
\left[
\begin{array}{ccc|ccc}
\frac{1}{4} & 0 & 0 & 1 & -\frac{1}{4} & 0 \\
-\frac{123}{770} & \frac{1}{4} & 0 & 1 & \frac{631}{1540} & 0 \\
-\frac{1496}{1987} & \frac{1391}{1631} & \frac{1}{4} & 1 & \frac{756}{1163} & -\frac{690}{3911} \\
\hline
-\frac{1376}{1157} & \frac{315}{269} & \frac{68}{1277} & 1 & \frac{1297}{1344} & -\frac{140}{1009} \\
-\frac{1572}{1093} & \frac{938}{531} & \frac{483}{719} & 0 & 0 & -\frac{560}{1009} \\
\frac{55}{48} & -\frac{103}{24} & \frac{151}{48} & 0 & 0 & 0
\end{array}
\right].
$$

Method with $p = q = 3$, $s = 4$, $\mathbf{c} = [0, \frac{1}{3}, \frac{2}{3}, 1]^T$, $\lambda = \frac{1}{4}$, $\epsilon = 0$, and

$$\beta_1 = \beta_2 = \beta_3 = E = \tfrac{1}{256}.$$

$$\left[\begin{array}{c|c} \mathbf{A} & \mathbf{U} \\ \hline \mathbf{B} & \mathbf{V} \end{array}\right] = \left[\begin{array}{cccc|cccc} \frac{1}{4} & 0 & 0 & 0 & 1 & -\frac{1}{4} & 0 & 0 \\ \frac{732}{1009} & \frac{1}{4} & 0 & 0 & 1 & -\frac{829}{1291} & -\frac{1}{36} & -\frac{5}{648} \\ \frac{556}{311} & -\frac{427}{804} & \frac{1}{4} & 0 & 1 & -\frac{1759}{2094} & \frac{187}{804} & \frac{85}{3643} \\ \frac{1253}{393} & -\frac{1048}{571} & \frac{598}{1767} & \frac{1}{4} & 1 & -\frac{658}{699} & \frac{306}{481} & \frac{573}{8374} \\ \hline \frac{2412}{1927} & -\frac{3329}{966} & \frac{339}{1391} & \frac{775}{2087} & 1 & \frac{1981}{768} & \frac{980}{879} & \frac{379}{3204} \\ \frac{3517}{682} & -\frac{3161}{585} & \frac{101}{2479} & \frac{1084}{899} & 0 & 0 & \frac{2909}{1855} & \frac{253}{1344} \\ \frac{800}{1003} & -\frac{587}{11358} & -\frac{9685}{1831} & \frac{1722}{379} & 0 & 0 & 0 & -\frac{403}{4312} \\ -\frac{2277}{256} & \frac{9135}{256} & -\frac{11439}{256} & \frac{4581}{256} & 0 & 0 & 0 & 0 \end{array}\right].$$

Method with $p = q = 4$, $s = 5$, $\mathbf{c} = [0, \frac{1}{4}, \frac{1}{2}, \frac{3}{4}, 1]^T$, $\lambda = \frac{1}{2}$, $\epsilon = 0$, and

$$\beta_1 = \beta_2 = \beta_3 = \beta_4 = E = -\frac{11}{48}.$$

$$\mathbf{A} = \begin{bmatrix} \frac{1}{2} & 0 & 0 & 0 & 0 \\ \frac{365}{8177} & \frac{1}{2} & 0 & 0 & 0 \\ -\frac{49}{14306} & \frac{409}{4768} & \frac{1}{2} & 0 & 0 \\ \frac{10823}{127} & -\frac{1519}{9} & \frac{17638}{211} & \frac{1}{2} & 0 \\ \frac{36065}{117} & -\frac{22593}{37} & \frac{24179}{80} & -\frac{89}{10395} & \frac{1}{2} \end{bmatrix},$$

$$\mathbf{U} = \begin{bmatrix} 1 & -\frac{1}{2} & 0 & 0 & 0 \\ 1 & -\frac{467}{1585} & -\frac{3}{32} & -\frac{5}{384} & -\frac{7}{6144} \\ 1 & -\frac{456}{5537} & -\frac{414}{2827} & -\frac{71}{1601} & -\frac{43}{5351} \\ 1 & \frac{293}{1364} & \frac{443}{1455} & -\frac{1327}{253} & -\frac{1553}{1173} \\ 1 & \frac{49}{76} & \frac{645}{418} & -\frac{2291}{122} & \frac{3347}{705} \end{bmatrix},$$

$$\mathbf{B} = \begin{bmatrix} -\frac{11225}{77} & \frac{36677}{121} & -\frac{10124}{69} & -\frac{2774}{225} & \frac{311}{96} \\ -\frac{7607}{51} & \frac{23873}{85} & -\frac{22201}{239} & \frac{8666}{167} & \frac{1211}{86} \\ -\frac{10968}{131} & \frac{69590}{667} & \frac{2971}{40} & -\frac{15939}{122} & \frac{7153}{200} \\ \frac{10431}{158} & -\frac{9664}{37} & \frac{43573}{108} & -\frac{25580}{89} & \frac{9575}{121} \\ \frac{1528}{15} & -\frac{7072}{15} & \frac{4016}{5} & -\frac{8992}{15} & \frac{2488}{15} \end{bmatrix},$$

$$
\mathbf{V} = \begin{bmatrix}
1 & -\frac{313}{599} & \frac{3481}{851} & \frac{2688}{247} & \frac{2723}{1033} \\
0 & 0 & \frac{753}{364} & \frac{3615}{332} & \frac{318}{119} \\
0 & 0 & 0 & \frac{1222}{167} & \frac{2907}{1525} \\
0 & 0 & 0 & 0 & \frac{404}{1369} \\
0 & 0 & 0 & 0 & 0
\end{bmatrix}.
$$

The methods above are just examples of A- and L-stable GLMs, and we are not claiming that they are "optimal" in any other sense. Additional examples of such methods can be found in the literature [50, 77, 80, 81, 167, 293]. The construction of optimal GLMs with IRKS is the main challenge ahead.

7.12   STIFFLY ACCURATE METHODS WITH IRKS

This section follows the presentation by Wright [293]. The GLM (7.1.1) written componentwise takes the form

$$
Y_i^{[n]} = h \sum_{j=1}^{s} a_{ij} f(Y_j^{[n]}) + \sum_{j=1}^{s} u_{ij} y_j^{[n-1]},
$$
$$
y_i^{[n]} = h \sum_{j=1}^{s} b_{ij} f(Y_j^{[n]}) + \sum_{j=1}^{s} v_{ij} y_j^{[n-1]},
$$

(7.12.1)

$i = 1, 2, \ldots, s$, $s = p + 1$, $n = 1, 2, \ldots, N$. Assuming that

$$
a_{p+1,j} = b_{1j}, \quad u_{p+1,j} = v_{1j}, \quad j = 1, 2, \ldots, p+1,
$$

(7.12.2)

it follows that the last internal stage $Y_{p+1}^{[n]}$ is equal to the first external stage $y_1^{[n]}$:

$$
Y_{p+1}^{[n]} = y_1^{[n]}.
$$

(7.12.3)

Since for method (7.12.1) of order p and stage order $q = p$ such that $y^{[0]} = z(t_0, h) + O(h^p)$, we have

$$
Y_{p+1}^{[n]} = y(t_{n-1} + c_{p+1}h) + O(h^p)
$$

and

$$
y_1^{[n]} = y(t_n) + O(h^p)
$$

(compare Theorem 2.4.3), condition (7.12.3) necessarily implies that the last component of the vector $\mathbf{c}$ is $c_{p+1} = 1$. Again assuming (7.12.3) we also have

$$
y_2^{[n]} = hy'(t_n) + O(h^p) = hf(y(t_n)) + O(h^p) = hf(Y_{p+1}^{[n]}) + O(h^p),
$$

and to enforce that this condition is satisfied exactly, that is, that

$$y_2^{[n]} = hf(Y_{p+1}^{[n]}), \qquad (7.12.4)$$

we assume that

$$b_{2j} = \delta_{p+1,j}, \quad v_{2j} = 0, \quad j = 1, 2, \ldots, p+1, \qquad (7.12.5)$$

where $\delta_{p+1,j}$ is the Kronecker delta. To analyze conditions (7.12.2) and (7.12.5) further it is convenient to partition the matrices $\mathbf{A}$, $\mathbf{U}$, $\mathbf{B}$, and $\mathbf{V}$ into rows as follows

$$\mathbf{A} = \begin{bmatrix} \mathbf{a}_1^T \\ \vdots \\ \mathbf{a}_{p+1}^T \end{bmatrix}, \quad \mathbf{U} = \begin{bmatrix} \mathbf{u}_1^T \\ \vdots \\ \mathbf{u}_{p+1}^T \end{bmatrix}, \quad \mathbf{B} = \begin{bmatrix} \mathbf{b}_1^T \\ \vdots \\ \mathbf{b}_{p+1}^T \end{bmatrix}, \quad \mathbf{V} = \begin{bmatrix} \mathbf{v}_1^T \\ \vdots \\ \mathbf{v}_{p+1}^T \end{bmatrix}.$$

Then these conditions take the vector form

$$\mathbf{a}_{p+1}^T = \mathbf{b}_1^T, \quad \mathbf{u}_{p+1}^T = \mathbf{v}_1^T \qquad (7.12.6)$$

and

$$\mathbf{b}_2^T = \mathbf{e}_{p+1}^T, \quad \mathbf{v}_2^T = \mathbf{0}. \qquad (7.12.7)$$

Assuming that GLM (7.12.1) has IRKS and given that $c_{p+1} = 1$, it follows from (7.1.6) and (7.1.7) that

$$\mathbf{u}_{p+1}^T = \mathbf{t}_{p+1}^T - \mathbf{a}_{p+1}^T \mathbf{C} \mathbf{K}, \quad \mathbf{v}_1^T = \mathbf{t}_{p+1}^T - \mathbf{b}_1^T \mathbf{C} \mathbf{K},$$

where

$$\mathbf{t}_{p+1} = \begin{bmatrix} 1 & \dfrac{1}{1!} & \dfrac{1}{2!} & \cdots & \dfrac{1}{p!} \end{bmatrix}^T$$

is the last row of the matrix $\mathbf{C}$ (recall that $c_{p+1} = 1$) and the first row of the matrix $\mathbf{E} = \exp(\mathbf{K})$. These matrices are defined in Section 7.1. These relations imply that it is sufficient to impose only the first condition of (7.12.6) since the second condition is then satisfied automatically. Comparing the second rows of (7.1.7), we also have

$$\mathbf{v}_2^T = \mathbf{t}_p^T - \mathbf{b}_2^T \mathbf{C} \mathbf{K},$$

where $\mathbf{t}_p^T$ is the second row of $\mathbf{E}$, which is given by

$$\mathbf{t}_p = \begin{bmatrix} 0 & 1 & \dfrac{1}{1!} & \cdots & \dfrac{1}{(p-1)!} \end{bmatrix}^T.$$

Assuming that $\mathbf{b}_2^T = \mathbf{e}_{p+1}^T$, it follows that

$$\mathbf{v}_2^T = \mathbf{t}_p^T - \mathbf{e}_{p+1}^T \mathbf{C} \mathbf{K} = \mathbf{t}_p^T - \mathbf{t}_{p+1}^T \mathbf{K} = \mathbf{t}_p^T - \mathbf{t}_p^T = \mathbf{0}.$$

Hence, similarly as before, the first condition of (7.12.7) implies that the second condition is satisfied automatically. We show next that for GLMs with IRKS, the condition $\mathbf{b}_2^T = \mathbf{e}_{p+1}^T$ and the additional condition $\beta_p = 0$ implies the first condition of (7.12.6). Indeed, comparing the second rows of the IRKS relation $\mathbf{BA} = \mathbf{XB}$, we obtain

$$\mathbf{b}_2^T \mathbf{A} = \mathbf{b}_1^T - \beta_p \mathbf{b}_{p+1}^T = \mathbf{b}_1^T.$$

Hence, if $\mathbf{b}_2^T = \mathbf{e}_{p+1}^T$, then $\mathbf{a}_{p+1}^T = \mathbf{e}_{p+1}^T \mathbf{A} = \mathbf{b}_2^T \mathbf{A}$, and it follows that $\mathbf{a}_{p+1}^T = \mathbf{b}_1^T$. The considerations above motivate the following definition.

Definition 7.12.1 *GLM (7.12.1) with IRKS has strict stiff accuracy if the following conditions are satisfied:*

$$c_{p+1} = 1, \quad \mathbf{b}_2^T = \mathbf{e}_{p+1}^T, \quad and \quad \beta_p = 0.$$

The following lemma will aid the description of GLMs with IRKS and strict stiff accuracy.

Lemma 7.12.2 (Wright [293]) *Given the shifting matrices $\mathbf{J}$ and $\mathbf{K}$ defined in Section 7.1 and the matrix $\mathbf{K}^-$ defined in Section 7.2, we have the relation*

$$\mathbf{K} \exp(\lambda \mathbf{K}^-) \mathbf{J} \exp(-\lambda \mathbf{K}^-) = \mathbf{I} - \lambda \mathbf{K} - \mathbf{e}_{p+1} \mathbf{e}_{p+1}^T. \tag{7.12.8}$$

Proof: It follows from Lemma 7.3.3 that

$$\mathbf{X}(\alpha, \beta) = \Psi(\mathbf{J} + \lambda \mathbf{I})\Psi^{-1},$$

and using formula (7.3.9) for Ψ and (7.3.10) for Ψ^{-1}, we obtain

$$\mathbf{X}(\alpha, \beta) = \beta(\mathbf{K}) \exp(\lambda \mathbf{K}^-)(\mathbf{J} + \lambda \mathbf{I}) \exp(-\lambda \mathbf{K}^-)\beta^{-1}(\mathbf{K}).$$

Multiplying this relation by $\beta^{-1}(\mathbf{K})$ from the left and by $\beta(\mathbf{K})$ from the right yields

$$\beta^{-1}(\mathbf{K})\mathbf{X}(\alpha, \beta)\beta(\mathbf{K}) = \exp(\lambda \mathbf{K}^-)\mathbf{J} \exp(-\lambda \mathbf{K}^-) + \lambda \mathbf{I}.$$

Using Lemma 7.3.7, we have

$$\mathbf{X}(\alpha, \beta) = \beta(\mathbf{K})\mathbf{X}(\alpha\beta, 1)\beta^{-1}(\mathbf{K}),$$

where $\mathbf{X}(\alpha\beta, 1)$ is the companion matrix corresponding to $\alpha\beta$. Hence,

$$\mathbf{X}(\alpha\beta, 1) - \lambda \mathbf{I} = \exp(\lambda \mathbf{K}^-)\mathbf{J} \exp(-\lambda \mathbf{K}^-).$$

Multiplying this equation from the left by the matrix $\mathbf{K}$ and taking into account that

$$\mathbf{KX}(\alpha\beta, 1) = \mathbf{I} - \mathbf{e}_{p+1} \mathbf{e}_{p+1}^T,$$

formula (7.12.8) follows. This completes the proof. ∎

Let the unit lower triangular matrix $\mathbf{L}^{-1}$ appearing in formula (7.7.6) for the matrix $\widetilde{\mathbf{B}}$ be represented by

$$
\mathbf{L}^{-1} =
\left[
\begin{array}{c|ccccc}
1 & 0 & 0 & \cdots & 0 & 0 \\
\hline
0 & 1 & 0 & \cdots & 0 & 0 \\
0 & \tau_{2,1} & 1 & \cdots & 0 & 0 \\
\vdots & \vdots & \vdots & \ddots & \vdots & \vdots \\
0 & \tau_{p-1,1} & \tau_{p-1,2} & \cdots & 1 & 0 \\
0 & \tau_{p,1} & \tau_{p,2} & \cdots & \tau_{p,p-1} & 1
\end{array}
\right].
$$

The characterization of GLMs with IRKS and strict stiff accuracy is given in the following result.

Theorem 7.12.3 (Wright [293]) *Assume that $c_{p+1} = 1$ and $\beta_p = 0$. Then GLM (7.1.1) with IRKS has strict stiff accuracy for any parameters τ_{ij}, $i = 2, 3, \ldots, p-1$, $j = 1, 2, \ldots, p-2$, if the last row $\mathbf{L}_{p+1}^{-1}$ of the matrix $\mathbf{L}^{-1}$ is given by*

$$
\mathbf{L}_{p+1}^{-1} = \frac{1}{\mathbf{e}_2^T \beta(\mathbf{K})\Lambda} \, \mathbf{e}_2^T \beta(\mathbf{K}) \exp(\lambda \mathbf{K}^-). \tag{7.12.9}
$$

Proof: It follows from $\mathbf{B} = \Psi \widetilde{\mathbf{B}}$ and (7.7.6) that the coefficient matrix $\mathbf{B}$ takes the form

$$
\mathbf{B} = \Psi \mathbf{L} \Delta \big(\mathbf{L}^{-1} \Gamma \mathbf{L} \mathcal{U} (\Omega \mathbf{L})^{-1} \big) \mathcal{L} (\Omega \mathbf{L})^{-1},
$$

and we have to show that

$$
\mathbf{b}_2^T = \mathbf{e}_2^T \mathbf{B} = \mathbf{e}_2^T \Psi \mathbf{L} \Delta \big(\mathbf{L}^{-1} \Gamma \mathbf{L} \mathcal{U} (\Omega \mathbf{L})^{-1} \big) \mathcal{L} (\Omega \mathbf{L})^{-1} = \mathbf{e}_{p+1}^T.
$$

We first compute $\mathbf{e}_2^T \Psi \mathbf{L}$. Using (7.3.9) and (7.12.9), we have

$$
\mathbf{e}_2^T \Psi \mathbf{L} = \mathbf{e}_2^T \beta(\mathbf{K}) \exp(\lambda \mathbf{K}^-) = (\mathbf{e}_2^T \beta(\mathbf{K})\Lambda) \mathbf{L}_{p+1}^{-1} \mathbf{L},
$$

and since $\mathbf{L}_{p+1}^{-1} = \mathbf{e}_{p+1}^T \mathbf{L}$, it follows that

$$
\mathbf{e}_2^T \beta(\mathbf{K}) \exp(\lambda \mathbf{K}^-) \mathbf{L} = (\mathbf{e}_2^T \beta(\mathbf{K})\Lambda) \mathbf{e}_{p+1}^T. \tag{7.12.10}
$$

This formula implies that only the last row of

$$
\Delta \big(\mathbf{L}^{-1} \Gamma \mathbf{L} \mathcal{U} (\Omega \mathbf{L})^{-1} \big) \mathcal{L} (\Omega \mathbf{L})^{-1}
$$

needs to be considered. This row is equal to

$$
\mathbf{e}_{p+1}^T \Delta \big(\mathbf{L}^{-1} \Gamma \mathbf{L} \mathcal{U} (\Omega \mathbf{L})^{-1} \big) \mathcal{L} (\Omega \mathbf{L})^{-1} = \mathbf{e}_{p+1}^T \mathbf{L}^{-1} \Gamma \mathbf{L} (\Omega \mathbf{L})^{-1} = \mathbf{e}_{p+1}^T \mathbf{L}^{-1} \Gamma \Omega^{-1},
$$

and the condition $\mathbf{B}_2^T = \mathbf{e}_{p+1}^T$ takes the form

$$
(\mathbf{e}_2^T \beta(\mathbf{K})\Lambda) \mathbf{e}_{p+1}^T \mathbf{L}^{-1} \Gamma \Omega^{-1} = \mathbf{e}_{p+1}^T
$$

or

$$(\mathbf{e}_2^T \beta(\mathbf{K})\Lambda)\mathbf{e}_{p+1}^T \mathbf{L}^{-1}\Gamma = \mathbf{e}_{p+1}^T \Omega.$$

Using equation (7.12.10), the preceding relation can be rewritten in the form

$$\mathbf{e}_2^T \beta(\mathbf{K}) \exp(\lambda \mathbf{K}^-)\Gamma = \mathbf{e}_{p+1}^T \Omega. \tag{7.12.11}$$

To prove (7.12.11), we consider the right and left sides of this equation separately. Substituting the formula for Ω and then the formula for Ψ, the right-hand side of (7.12.11) takes the form

$$\mathbf{e}_{p+1}^T \Omega = \mathbf{e}_{p+1}^T \mathbf{C}\big(\beta(\mathbf{K})\mathbf{e}_{p+1}\mathbf{e}_1^T + \mathbf{K}\beta(\mathbf{K}) \exp(\lambda \mathbf{K}^-)\big).$$

Since $c_{p+1} = 1$, we have $\mathbf{e}_{p+1}^T \mathbf{C} = \mathbf{e}_1^T \mathbf{E}$ and exchanging the order of $\mathbf{K}$ and $\beta(\mathbf{K})$, we obtain

$$\mathbf{e}_{p+1}^T \Omega = \mathbf{e}_1^T \mathbf{E}\beta(\mathbf{K})\mathbf{e}_{p+1}\mathbf{e}_1^T + \mathbf{e}_1^T \mathbf{E}\beta(\mathbf{K})\mathbf{K} \exp(\lambda \mathbf{K}^-). \tag{7.12.12}$$

We now consider the left-hand side of (7.12.11). Since $\beta(\mathbf{K})$ and $\exp(\lambda \mathbf{K}^-)$ are upper triangular, the first element of the vector $\mathbf{e}_2^T \beta(\mathbf{K}) \exp(\lambda \mathbf{K}^-)$ is equal to zero. This implies that the left-hand side of (7.12.11) is not affected by the first row of the matrix Γ. We have

$$\Gamma = \mathbf{I}_c \mathbf{I}_r \mathbf{FI}_r \mathbf{I}_c + \delta \mathbf{e}_1^T \equiv \mathbf{F} + \delta \mathbf{e}_1^T.$$

It can be verified that $\mathbf{JFe}_{p+1} \equiv \delta$; compare the explicit expression for $\mathbf{F}$ in Section 7.10, which implies that $\mathbf{JFe}_{p+1}\mathbf{e}_1^T \equiv \delta \mathbf{e}_1^T$. Hence, also substituting the formula for $\mathbf{F}$, we obtain the following equivalence relation for the matrix Γ:

$$\Gamma \equiv \exp(-\lambda \mathbf{K}^-)\mathbf{E} \exp(\lambda \mathbf{K}^-) + \mathbf{J} \exp(-\lambda \mathbf{K}^-)\mathbf{E} \exp(\lambda \mathbf{K}^-)\mathbf{e}_{p+1}\mathbf{e}_1^T.$$

Taking into account that $\exp(\lambda \mathbf{K}^-)\mathbf{e}_{p+1} = \Lambda$, where the vector Λ is defined in Section 7.5, we get

$$\Gamma = \exp(-\lambda \mathbf{K}^-)\mathbf{E} \exp(\lambda \mathbf{K}^-) + \mathbf{J} \exp(-\lambda \mathbf{K}^-)\mathbf{E}\Lambda \mathbf{e}_1^T. \tag{7.12.13}$$

Since the left-hand side of (7.12.11) does not depend on the first row of Γ, we can substitute the equivalence relation (7.12.13) for Γ which leads to

$$\begin{aligned}
\mathbf{e}_2^T \beta(\mathbf{K}) \exp(\lambda \mathbf{K}^-)\Gamma = {} & \mathbf{e}_2^T \beta(\mathbf{K})\mathbf{E} \exp(-\lambda \mathbf{K}^-) \\
& + \mathbf{e}_2^T \beta(\mathbf{K}) \exp(\lambda \mathbf{K}^-)\mathbf{J} \exp(-\lambda \mathbf{K}^-)\mathbf{E}\Lambda \mathbf{e}_1^T.
\end{aligned}$$

We have $\mathbf{e}_2^T = \mathbf{e}_1^T \mathbf{K}$, and exchanging the order of matrices $\mathbf{K}$, $\mathbf{E}$, and $\beta(\mathbf{K})$ the preceding relation can be written in the form

$$\begin{aligned}
\mathbf{e}_2^T \beta(\mathbf{K}) \exp(\lambda \mathbf{K}^-)\Gamma = {} & \mathbf{e}_1^T \mathbf{E}\beta(\mathbf{K})\mathbf{K} \exp(-\lambda \mathbf{K}^-) \\
& + \mathbf{e}_1^T \beta(\mathbf{K})\mathbf{K} \exp(\lambda \mathbf{K}^-)\mathbf{J} \exp(-\lambda \mathbf{K}^-)\mathbf{E}\Lambda \mathbf{e}_1^T.
\end{aligned}$$

Using Lemma 7.12.2, it follows that

$$\mathbf{e}_2^T \beta(\mathbf{K}) \exp(\lambda \mathbf{K}^-) \Gamma = \mathbf{e}_1^T \mathbf{E} \beta(\mathbf{K}) \mathbf{K} \exp(-\lambda \mathbf{K}^-)$$
$$+ \mathbf{e}_1^T \beta(\mathbf{K})(\mathbf{I} - \lambda \mathbf{K}) \mathbf{E} \Lambda \mathbf{e}_1^T - \mathbf{e}_1^T \beta(\mathbf{K}) \mathbf{e}_{p+1} \mathbf{e}_{p+1}^T \mathbf{E} \Lambda \mathbf{e}_1^T.$$

It can be verified using the relation $\Lambda - \lambda \mathbf{K} \Lambda = \mathbf{e}_{p+1}$ that

$$\mathbf{e}_1^T \beta(\mathbf{K})(\mathbf{I} - \lambda \mathbf{K}) \mathbf{E} \Lambda \mathbf{e}_1^T = \mathbf{e}_1^T \beta(\mathbf{K}) \mathbf{E} (\Lambda - \lambda \mathbf{K} \Lambda) \mathbf{e}_1^T = \mathbf{e}_1^T \beta(\mathbf{K}) \mathbf{E} \mathbf{e}_{p+1} \mathbf{e}_1^T.$$

We also have

$$\mathbf{e}_1^T \beta(\mathbf{K}) \mathbf{e}_{p+1} = \beta_p = 0.$$

Hence, the left-hand side of (7.12.11) can be written as

$$\mathbf{e}_2^T \beta(\mathbf{K}) \exp(\lambda \mathbf{K}^-) = \mathbf{e}_1^T \mathbf{E} \beta(\mathbf{K}) \mathbf{K} \exp(\lambda \mathbf{K}^-) + \mathbf{e}_1^T \mathbf{E} \beta(\mathbf{K}) \mathbf{e}_{p+1} \mathbf{e}_1^T. \quad (7.12.14)$$

Comparing (7.12.12) and (7.12.14), equation (7.12.11) follows. This completes the proof.   ∎

These methods can be generated using the algorithm described at the end of Section 7.7, where the last row $\mathbf{L}_{p+1}^{-1}$ of a unit lower triangular matrix $\mathbf{L}^{-1}$ is computed according to (7.12.9). The examples of such methods of order $p = 1, 2, 3,$ and 4 are listed below.

Method with $p = q = 1$, $s = 2$, $\mathbf{c} = [\frac{1}{2}, 1]^T$, $\lambda = \frac{2}{5}$, $\epsilon = 0$, and

$$\beta_1 = 0, \quad E = -\frac{7}{50}.$$

$$\left[\begin{array}{c|c} \mathbf{A} & \mathbf{U} \\ \hline \mathbf{B} & \mathbf{V} \end{array} \right] = \left[\begin{array}{cc|cc} \frac{2}{5} & 0 & 1 & \frac{1}{10} \\ \frac{12}{25} & \frac{2}{5} & 1 & \frac{3}{25} \\ \hline \frac{12}{25} & \frac{2}{5} & 1 & \frac{3}{25} \\ 0 & 1 & 0 & 0 \end{array} \right].$$

Method with $p = q = 2$, $s = 3$, $\mathbf{c} = [\frac{1}{2}, \frac{3}{4}, 1]^T$, $\lambda = \frac{1}{4}$, $\epsilon = 0$, and

$$\beta_1 = E = -\frac{7}{192}, \quad \beta_2 = 0.$$

$$\left[\begin{array}{c|c} \mathbf{A} & \mathbf{U} \\ \hline \mathbf{B} & \mathbf{V} \end{array} \right] = \left[\begin{array}{ccc|ccc} \frac{1}{4} & 0 & 0 & 1 & \frac{1}{4} & 0 \\ \frac{1591}{9516} & \frac{1}{4} & 0 & 1 & \frac{211}{634} & \frac{31}{3053} \\ \frac{391}{1937} & \frac{337}{1803} & \frac{1}{4} & 1 & \frac{587}{1625} & \frac{95}{10689} \\ \hline \frac{391}{1937} & \frac{337}{1803} & \frac{1}{4} & 1 & \frac{587}{1625} & \frac{95}{10689} \\ 0 & 0 & 1 & 0 & 0 & 0 \\ -\frac{599}{492} & -\frac{3052}{738} & \frac{192}{41} & 0 & \frac{935}{1476} & 0 \end{array} \right].$$

Method with $p = q = 3$, $s = 4$, $\mathbf{c} = [\frac{1}{4}, \frac{1}{2}, \frac{3}{4}, 1]^T$, $\lambda = \frac{1}{4}$, $\epsilon = 0$, and

$$\beta_1 = \beta_2 = E = \tfrac{1}{256}, \quad \beta_3 = 0, \quad \tau_{21} = 0.$$

$$
\left[\begin{array}{c|c} \mathbf{A} & \mathbf{U} \\ \hline \mathbf{B} & \mathbf{V} \end{array}\right] =
\left[\begin{array}{cccc|cccc}
\frac{1}{4} & 0 & 0 & 0 & 1 & 0 & -\frac{1}{32} & -\frac{1}{192} \\
\frac{345}{292} & \frac{1}{4} & 0 & 0 & 1 & -\frac{68}{73} & -\frac{345}{1168} & -\frac{193}{4077} \\
\frac{1795}{669} & -\frac{347}{826} & \frac{1}{4} & 0 & 1 & -\frac{1592}{903} & -\frac{509}{1387} & -\frac{495}{15797} \\
\frac{951}{770} & -\frac{961}{695} & \frac{664}{1987} & \frac{1}{4} & 1 & \frac{1025}{1819} & \frac{589}{1542} & \frac{199}{2429} \\
\hline
\frac{951}{770} & -\frac{961}{695} & \frac{664}{1987} & \frac{1}{4} & 1 & \frac{1025}{1819} & \frac{589}{1542} & \frac{199}{2429} \\
0 & 0 & 0 & 1 & 0 & 0 & 0 & 0 \\
-\frac{5477}{201} & \frac{4142}{337} & -\frac{731}{164} & \frac{260}{69} & 0 & \frac{9623}{615} & \frac{1228}{989} & -\frac{506}{1605} \\
-\frac{2378}{23} & \frac{6515}{96} & -\frac{28557}{770} & \frac{1024}{69} & 0 & \frac{5604}{97} & \frac{3252}{665} & -\frac{1228}{989}
\end{array}\right].
$$

Method with $p = q = 4$, $s = 5$, $\mathbf{c} = [\frac{1}{5}, \frac{2}{5}, \frac{3}{5}, \frac{4}{5}, 1]^T$, $\lambda = \frac{1}{2}$, $\epsilon = 0$, and

$$\beta_1 = \beta_2 = \beta_3 = E = -\tfrac{11}{480}, \quad \beta_4 = 0, \quad \tau_{21} = \tau_{31} = \tau_{32} = 0.$$

$$
\mathbf{A} = \left[\begin{array}{ccccc}
\frac{1}{2} & 0 & 0 & 0 & 0 \\
-\frac{359}{1919} & \frac{1}{2} & 0 & 0 & 0 \\
\frac{549}{881} & -\frac{1427}{3395} & \frac{1}{2} & 0 & 0 \\
\frac{3403}{593} & -\frac{3831}{1111} & \frac{683}{1376} & \frac{1}{2} & 0 \\
-73 & \frac{2734}{77} & \frac{2581}{891} & -\frac{1883}{416} & \frac{1}{2}
\end{array}\right],
$$

$$
\mathbf{U} = \left[\begin{array}{ccccc}
1 & -\frac{3}{10} & -\frac{2}{25} & -\frac{13}{1500} & -\frac{3}{5000} \\
1 & \frac{126}{1447} & -\frac{239}{2894} & -\frac{40}{1563} & -\frac{83}{20661} \\
1 & -\frac{69}{671} & -\frac{349}{4562} & -\frac{214}{6517} & -\frac{89}{9947} \\
1 & -\frac{1875}{754} & -\frac{3028}{20705} & -\frac{85}{29057} & -\frac{229}{15970} \\
1 & \frac{5785}{146} & \frac{691}{303} & -\frac{423}{788} & -\frac{1028}{25015}
\end{array}\right],
$$

$$\mathbf{B} = \begin{bmatrix} -73 & \frac{2734}{77} & \frac{2581}{891} & -\frac{1883}{416} & \frac{1}{2} \\ 0 & 0 & 0 & 0 & 1 \\ \frac{15391}{378} & \frac{32605}{337} & -\frac{7181}{54} & \frac{4283}{99} & \frac{414}{163} \\ -\frac{4261}{50} & \frac{36328}{69} & -\frac{19093}{36} & \frac{28065}{178} & \frac{916}{163} \\ -\frac{35635}{46} & \frac{33584}{17} & -\frac{47656}{29} & \frac{12825}{29} & \frac{1920}{163} \end{bmatrix},$$

$$\mathbf{V} = \begin{bmatrix} 1 & \frac{5785}{146} & \frac{691}{303} & -\frac{423}{788} & -\frac{1028}{25015} \\ 0 & 0 & 0 & 0 & 0 \\ 0 & -\frac{13930}{277} & -\frac{1455}{454} & \frac{1267}{999} & \frac{198}{2303} \\ 0 & -\frac{91340}{1231} & -\frac{78}{11} & \frac{1942}{697} & \frac{421}{2108} \\ 0 & -\frac{3007}{260} & -\frac{6837}{460} & \frac{1863}{319} & \frac{751}{1794} \end{bmatrix}.$$

In all the examples above we have chosen the parameter λ so that the resulting methods are A-stable. Since we have also chosen $\epsilon = 0$, all these methods are also L-stable. Other examples of A- and L-stable GLMs with IRKS and strict stiff accuracy are available in the literature [77, 81, 293]. Preliminary tests of methods with strict stiff accuracy indicate that they perform extremely well compared with RK or linear multistep methods. The design of efficient software based on such methods is a subject of current work.

CHAPTER 8

IMPLEMENTATION OF GLMS WITH IRKS

8.1 VARIABLE STEP SIZE FORMULATION OF GLMS

To formulate GLMs (7.1.1) on the nonuniform grid

$$t_0 < t_1 < \cdots < t_N, \quad t_N \geq T,$$

and to investigate error propagation of these methods, it is convenient to partition the coefficient matrices $\mathbf{A}$, $\mathbf{U}$, $\mathbf{B}$, and $\mathbf{V}$ as follows:

$$\left[\begin{array}{c|c} \mathbf{A} & \mathbf{U} \\ \hline \mathbf{B} & \mathbf{V} \end{array} \right] = \left[\begin{array}{c|cc} A & e & U \\ \hline b^T & 1 & v^T \\ B & 0 & V \end{array} \right], \tag{8.1.1}$$

where $e = [1, \ldots, 1]^T \in \mathbb{R}^{p+1}$, $b \in \mathbb{R}^{p+1}$, $v \in \mathbb{R}^p$, and

$$A \in \mathbb{R}^{(p+1)\times(p+1)}, \quad U \in \mathbb{R}^{(p+1)\times p}, \quad B \in \mathbb{R}^{p\times(p+1)}, \quad V \in \mathbb{R}^{p\times p}.$$

We also denote by $c = [c_1, \ldots, c_{p+1}]^T$ the abscissa vector of the method. It is also convenient to partition correspondingly the vector $\mathbf{y}^{[n-1]}$ of external

General Linear Methods for Ordinary Differential Equations. By Zdzisław Jackiewicz **417**
Copyright © 2009 John Wiley & Sons, Inc.

approximations

$$\mathbf{y}^{[n-1]} = \begin{bmatrix} y_{n-1} \\ z^{[n-1]} \end{bmatrix},$$

where $z^{[n-1]}$ is now an approximation to the Nordsieck vector $z(t_{n-1}, h_n)$, $h_n = t_n - t_{n-1}$, defined by

$$z(t, h) := \begin{bmatrix} hy'(t) \\ h^2 y''(t) \\ \vdots \\ h^p y^{(p)}(t) \end{bmatrix}. \tag{8.1.2}$$

Observe that this definition of the Nordsieck vector differs from that given by (7.1.2), since (8.1.2) does not include the solution component $y(t)$. The resulting GLM now takes the form

$$Y^{[n]} = (e \otimes I)y_{n-1} + h_n(A \otimes I)F(Y^{[n]}) + (U \otimes I)z^{[n-1]},$$

$$y_n = y_{n-1} + h_n(b^T \otimes I)F(Y^{[n]}) + (v^T \otimes I)z^{[n-1]}, \tag{8.1.3}$$

$$\overline{z}^{[n]} = h_n(B \otimes I)F(Y^{[n]}) + (V \otimes I)z^{[n-1]},$$

$n = 1, 2, \ldots, N$, where $\overline{z}^{[n]}$ is an approximation to the Nordsieck vector $z(t_n, h_n)$ and I is the identity matrix of dimension m. We recall that as in Section 2.1, the vectors $Y^{[n]}$, $F(Y^{[n]})$, and $z^{[n-1]}$ are defined by

$$Y^{[n]} = \begin{bmatrix} Y_1^{[n]} \\ \vdots \\ Y_s^{[n]} \end{bmatrix}, \quad F(Y^{[n]}) = \begin{bmatrix} f(Y_1^{[n]}) \\ \vdots \\ f(Y_s^{[n]}) \end{bmatrix}, \quad z^{[n-1]} = \begin{bmatrix} z_1^{[n-1]} \\ \vdots \\ z_p^{[n-1]} \end{bmatrix},$$

$s = p + 1$. The definition of GLM given by (8.1.3) is not yet complete and we have to prescribe how to compute an approximation $z^{[n]}$ to the Nordsieck vector $z(t_n, h_{n+1})$, corresponding to a new step size h_{n+1} from t_n to t_{n+1}. This will be done using the rescale and modify strategy introduced by Butcher and Jackiewicz [71] and later refined by Butcher et al. [75]. This strategy is described in Section 8.5.

The solution component y_n of method (8.1.3) acts quite differently from the remaining components $\overline{z}^{[n]}$ or $z^{[n]}$, which approximate Nordsieck vectors $z(t_n, h_n)$ or $z(t_n, h_{n+1})$, respectively. As a result, partitioning the method as in (8.1.3) makes it considerably more convenient to understand error propagation and to discuss various implementation issues related to GLMs with IRKS. Error propagation of GLMs (8.1.3) is discussed in Section 8.3.

8.2 STARTING PROCEDURES

General-purpose codes based on GLMs will usually change the step size automatically as well as the order of the method as the integration progresses from step to step, adjusting the step size and increasing or decreasing the order according to the smoothness of the right-hand side of the differential system. The integration usually starts with the formula of order 1, for which the required approximations to the initial value $y(t_0)$ and initial derivative $y'(t_0)$ are readily available from the initial condition y_0 and the relation

$$y'(t_0) = f\big(y(t_0)\big) = f(y_0).$$

However, one may also wish to design codes based on GLMs of fixed order, in which case starting procedures are required to compute an approximation $z^{[0]}$ to the Nordsieck vector $z(t_0, h_1)$, to start the integration with an initial step size h_1. Starting procedures are also required if one wants to compare a particular GLM of fixed order with currently available code based, for example, on an RK or DIMSIM of the same order.

Starting procedures for DIMSIMs of fixed order, discussed in Chapters 3 and 4, were developed by VanWieren [273, 274, 275]. In this section we describe the construction of starting procedures for GLMs (8.1.3).

For GLMs (8.1.3) of order p with initial step size $h_1 = h$, we have to determine the starting vector $z^{[0]}$ such that

$$z^{[0]} = z(t_0, h) + O(h^{p+1}) = \begin{bmatrix} hy'(t_0) \\ h^2 y''(t_0) \\ \vdots \\ h^p y^{(p)}(t_0) \end{bmatrix} + O(h^{p+1}).$$

This vector will be computed from the formulas

$$\overline{Y}_i = y_0 + h \sum_{j=1}^{p} \overline{a}_{ij} f(\overline{Y}_j),$$

$$z_i^{[0]} = h \sum_{j=1}^{p} \overline{b}_{ij} f(\overline{Y}_j), \tag{8.2.1}$$

$i = 1, 2, \ldots, p$, where $\overline{A} = [\overline{a}_{ij}] \in \mathbb{R}^{p \times p}$ and $\overline{B} = [\overline{b}_{ij}] \in \mathbb{R}^{p \times p}$ are determined by requiring that

$$\overline{Y}_i = y(t_0 + \overline{c}_i h) + O(h^{p+1}),$$

$$z_i^{[0]} = h^i y^{(i)}(t_0) + O(h^{p+1}), \tag{8.2.2}$$

$i = 1, 2, \ldots, p$, where $\overline{c} = [\overline{c}_1, \ldots, \overline{c}_p]^T \in \mathbb{R}^p$ is a given abscissa vector of the starting procedure (8.2.1). This starting procedure can be specified by a partitioned matrix

$$\left[\begin{array}{c|c} e & \overline{A} \\ \hline \overline{c} & B \end{array}\right].$$

Substituting $y_0 = y(t_0)$ and (8.2.2) into (8.2.1), we obtain

$$y(t_0 + \overline{c}_i h) = y(t_0) + h \sum_{j=1}^{p} \overline{a}_{ij} y'(t_0 + \overline{c}_j h) + O(h^{p+1}),$$

$$h^i y^{(i)}(t_0) = h \sum_{j=1}^{p} \overline{b}_{ij} y'(t_0 + \overline{c}_j h) + O(h^{p+1}),$$

$i = 1, 2, \ldots, p$. Expanding $y(t_0 + \overline{c}_i h)$ and $y'(t_0 + \overline{c}_j h)$ into a Taylor series around t_0 and collecting terms with the same powers of h, it follows that

$$\sum_{k=1}^{p} \left(\frac{\overline{c}_i^k}{k!} - \sum_{j=1}^{p} \overline{a}_{ij} \frac{\overline{c}_j^{k-1}}{(k-1)!} \right) h^k y^{(k)}(t_0) = O(h^{p+1}),$$

$$\sum_{k=1}^{p} \left(\delta_{ik} - \sum_{j=1}^{p} \overline{b}_{ij} \frac{\overline{c}_j^{k-1}}{(k-1)!} \right) h^k y^{(k)}(t_0) = O(h^{p+1}),$$

$i, k = 1, 2, \ldots, p$. These relations lead to the following systems of equations for the coefficients $\overline{a}_{ij}$ and $\overline{b}_{ij}$:

$$\sum_{j=1}^{p} \overline{a}_{ij} \frac{\overline{c}_j^{k-1}}{(k-1)!} = \frac{\overline{c}_i^k}{k!},$$

$$\sum_{j=1}^{p} \overline{b}_{ij} \frac{\overline{c}_j^{k-1}}{(k-1)!} = \delta_{ik}, \tag{8.2.3}$$

$i, k = 1, 2, \ldots, p$. Setting

$$\overline{C} = \left[\begin{array}{ccccc} e & \overline{c} & \dfrac{\overline{c}^2}{2!} & \cdots & \dfrac{\overline{c}^{p-1}}{(p-1)!} \end{array} \right] \in \mathbb{R}^{p \times p},$$

(8.2.3) can be rewritten in vector form as

$$\overline{A}\,\overline{C} = \overline{C}J,$$

$$B\,\overline{C} = I. \tag{8.2.4}$$

Here J is a shifting matrix of dimension p defined as in Section 7.1, and I is the identity matrix of dimension p. Assuming that the abscissa vector $\overline{c}$ has distinct components, the unique solution to (8.2.4) is given by

$$\overline{A} = \overline{C}J\,\overline{C}^{-1}, \quad B = \overline{C}^{-1}.$$

Starting procedures of order $p = 2$, 3, and 4 obtained in this way are given below.

Starting procedure of order $p = 2$:

$$
\left[\begin{array}{c|c} e & \overline{A} \\ \hline \overline{c} & \overline{B} \end{array}\right] = \left[\begin{array}{c|cc} 1 & 0 & 0 \\ 1 & 1 & 0 \\ \hline 0 & 1 & 0 \\ 1 & -1 & 1 \end{array}\right].
$$

Starting procedure of order $p = 3$:

$$
\left[\begin{array}{c|c} e & \overline{A} \\ \hline \overline{c} & \overline{B} \end{array}\right] = \left[\begin{array}{c|ccc} 1 & 0 & 0 & 0 \\ 1 & \frac{1}{8} & \frac{1}{2} & -\frac{1}{8} \\ 1 & -\frac{1}{2} & 2 & -\frac{1}{2} \\ \hline 0 & 1 & 0 & 0 \\ \frac{1}{2} & -3 & 4 & -1 \\ 1 & 4 & -8 & 4 \end{array}\right].
$$

Starting procedure of order $p = 4$:

$$
\left[\begin{array}{c|c} e & \overline{A} \\ \hline \overline{c} & \overline{B} \end{array}\right] = \left[\begin{array}{c|cccc} 1 & 0 & 0 & 0 & 0 \\ 1 & \frac{5}{36} & \frac{2}{9} & -\frac{1}{36} & 0 \\ 1 & \frac{1}{3} & -\frac{2}{9} & \frac{7}{9} & -\frac{2}{9} \\ 1 & \frac{5}{4} & -3 & \frac{15}{4} & -1 \\ \hline 0 & 1 & 0 & 0 & 0 \\ \frac{1}{3} & -\frac{11}{2} & 9 & -\frac{9}{2} & 1 \\ \frac{2}{3} & 18 & -45 & 36 & -9 \\ 1 & -27 & 81 & -81 & 27 \end{array}\right].
$$

Starting procedures for GLMs of order p were also considered by Wright [293] and Huang [167]. Wright's approach [293] requests, in effect, that the starting vector $z^{[0]}$ satisfies a stronger condition than that required for the method of order p; that is,

$$z^{[0]} = z(t_0, h) + O(h^{p+2}).$$

The starting procedure takes the form

$$\widehat{Y}_i = y_0 + h \sum_{j=1}^{p+1} \widehat{a}_{ij} f(\widehat{Y}_j), \quad i = 1, 2, \ldots, p+1,$$

$$z_i^{[0]} = h \sum_{j=1}^{p+1} \widehat{b}_{ij} f(\widehat{Y}_j), \qquad i = 1, 2, \ldots, p,$$

(8.2.5)

where now $\widehat{\mathbf{A}} = [\widehat{a}_{ij}] \in \mathbb{R}^{(p+1) \times (p+1)}$, $\widehat{\mathbf{B}} = [\widehat{b}_{ij}] \in \mathbb{R}^{p \times (p+1)}$, and $\widehat{Y}_i$ approximates $y(t_0 + \widehat{c}_i h)$ for some vector $\widehat{\mathbf{c}} = [\widehat{c}_1, \ldots, \widehat{c}_{p+1}]^T \in \mathbb{R}^{p+1}$. Substituting

$$\widehat{Y}_i = y(t_0 + \widehat{c}_i h) + O(h^{p+2}),$$

$$z_i^{[0]} = h^i y^{(i)}(t_0) + O(h^{p+2}),$$

into (8.2.5), expanding $y(t_0 + \widehat{c}_i h)$ and $y'(t_0 + \widehat{c}_j h)$ into a Taylor series around t_0, and collecting terms of order h^i for $i = 1, 2, \ldots, p+1$, leads to systems of linear equations for $\widehat{a}_{ij}$ and $\widehat{b}_{ij}$ of the form

$$\sum_{j=1}^{p+1} \widehat{a}_{ij} \frac{\widehat{c}_j^{k-1}}{(k-1)!} = \frac{\widehat{c}_i^k}{k!}, \quad i, k = 1, 2, \ldots, p+1,$$

and

$$\sum_{j=1}^{p+1} \widehat{b}_{ij} \frac{\widehat{c}_j^{k-1}}{(k-1)!} = \delta_{ik}, \quad i = 1, 2, \ldots, p, \quad k = 1, 2, \ldots, p+1.$$

These systems can be written in vector form,

$$\widehat{\mathbf{A}} \widehat{\mathbf{C}} = \widehat{\mathbf{C}} \mathbf{J}$$

and

$$\widehat{\mathbf{B}} \widehat{\mathbf{C}} = \mathbf{I}_0,$$

where $\mathbf{I}_0 = [\, I \,|\, 0 \,] \in \mathbb{R}^{p \times (p+1)}$, the matrix $\widehat{\mathbf{C}}$ is defined by

$$\widehat{\mathbf{C}} = \left[\, \mathbf{e} \quad \widehat{\mathbf{c}} \quad \frac{\widehat{\mathbf{c}}^2}{2!} \quad \cdots \quad \frac{\widehat{\mathbf{c}}^p}{p!} \,\right] \in \mathbb{R}^{(p+1) \times (p+1)},$$

and the matrix $\mathbf{J}$ is defined as in Section 7.1. Assuming that the abscissa vector $\widehat{\mathbf{c}}$ has distinct components, the unique solution to these systems is given by

$$\widehat{\mathbf{A}} = \widehat{\mathbf{C}} \mathbf{J} \widehat{\mathbf{C}}^{-1}, \quad \widehat{\mathbf{B}} = \mathbf{I}_0 \widehat{\mathbf{C}}^{-1}.$$

The examples of starting procedures obtained in this way are given by Wright [293]. Observe that, in general, starting methods obtained using the approach presented in this section or the approach by Wright [293] are implicit. Huang [167] proposed an approach to the construction of starting procedures, in which the coefficient matrix corresponding to $\overline{A}$ or $\widehat{\mathbf{A}}$ is lower triangular, with the parameter λ on the diagonal, which is equal to the corresponding parameter of the main method.

8.3   ERROR PROPAGATION FOR GLMS

In this section we investigate error propagation of GLMs (8.1.3) up to terms of order $p+2$. Error propagation of GLMs including only terms up to the order $p+1$ was discussed before by Butcher and Jackiewicz [71, 72] and Wright [293].

Following Butcher et al. [75], assume that the input quantities y_{n-1} and $z^{[n-1]}$ to the current step from t_{n-1} to $t_n = t_{n-1} + h_n$ satisfy the relations

$$
\begin{aligned}
y_{n-1} &= y(t_{n-1}), \\[2mm]
z^{[n-1]} &= z(t_{n-1}, h_n) - (\beta \otimes I)h_n^{p+1}y^{(p+1)}(t_{n-1}) \\[2mm]
&\quad - (\gamma \otimes I)h_n^{p+2}y^{(p+2)}(t_{n-1}) \\[2mm]
&\quad - (\delta \otimes I)h_n^{p+2}\frac{\partial f}{\partial y}\big(y(t_{n-1})\big)y^{(p+1)}(t_{n-1}) + O(h_n^{p+3}),
\end{aligned}
$$

(8.3.1)

where $y(t)$ is the solution to the differential system; β, γ, and δ are some vectors; and I is the identity matrix of dimension m, the dimension of the differential system (2.1.1). We then try to determine an error constant E and constants F and G such that the output quantities y_n and $\overline{z}^{[n]}$ computed in the step satisfy

$$
\begin{aligned}
y_n &= y(t_n) - E\,h_n^{p+1}y^{(p+1)}(t_n) - F\,h_n^{p+2}y^{(p+2)}(t_n) \\[2mm]
&\quad - G\,h_n^{p+2}\frac{\partial f}{\partial y}\big(y(t_n)\big)y^{(p+1)}(t_n) + O(h_n^{p+3}), \\[2mm]
\overline{z}^{[n]} &= \overline{z}(t_n, h_n) - (\beta \otimes I)h_n^{p+1}y^{(p+1)}(t_n) \\[2mm]
&\quad - (\gamma \otimes I)h_n^{p+2}y^{(p+2)}(t_n) \\[2mm]
&\quad - (\delta \otimes I)h_n^{p+2}\frac{\partial f}{\partial y}\big(y(t_n)\big)y^{(p+1)}(t_n) + O(h_n^{p+3}),
\end{aligned}
$$

(8.3.2)

for the same vectors β, γ, and δ, and the Nordsieck vector $\bar{z}(t_n, h_n)$ corresponding to the solution $\bar{y}(t)$ of the initial value problem

$$\bar{y}'(t) = f(\bar{y}(t)), \quad t \in [t_n, t_{n+1}],$$

$$\bar{y}(t_n) = y_n,$$

$t_{n+1} = t_n + h_{n+1}$. Here h_{n+1} is a new step size.

The error constant E was introduced in Section 7.5 as a term of order $p+1$ in the approximation of the exponential function $\exp(z)$ by the stability function $R(z)$ of GLM (7.1.1) (compare equation (7.5.6)). Later in this section we demonstrate that the two concepts of the error constant E – the one introduced in (7.5.6) and the one introduced in (8.3.2) – are equivalent. We also demonstrate that the vector β appearing in (8.3.1) and (8.3.2) is composed of the coefficients $\beta_p, \beta_{p-1}, \ldots, \beta_1$ of the doubly companion matrix $\mathbf{X} = \mathbf{X}(\alpha, \beta)$.

Recall that the abscissa vector of the method is $c = [c_1, \ldots, c_{p+1}]^T$, and let $c^\nu = [c_1^\nu, \ldots, c_{p+1}^\nu]^T$. We have the following theorem.

Theorem 8.3.1 (Butcher et al. [75]) *Assume that method (8.1.3) has order p and stage order $q = p$. Then the error constant E and the constants F and G appearing in (8.3.2) are given by*

$$E = \frac{1}{(p+1)!} - \frac{b^T c^p}{p!} + v^T \beta, \tag{8.3.3}$$

$$F = \frac{1}{(p+2)!} - E - \frac{b^T c^{p+1}}{(p+1)!} + v^T \gamma, \tag{8.3.4}$$

and

$$G = b^T \xi + v^T \delta, \tag{8.3.5}$$

where ξ is the vector of stage errors defined by

$$\xi = \frac{c^{p+1}}{(p+1)!} - A \frac{c^p}{p!} + U\beta. \tag{8.3.6}$$

Moreover, the vectors β, γ, and δ appearing in (8.3.1) and (8.3.2) are given by

$$\beta = (I - V)^{-1} \left(t_p - B \frac{c^p}{p!} \right), \tag{8.3.7}$$

$$\gamma = (I - V)^{-1} \left(\hat{t}_p - \beta - B \frac{c^{p+1}}{(p+1)!} \right), \tag{8.3.8}$$

and

$$\delta = (I - V)^{-1} (B\xi - E e_1), \tag{8.3.9}$$

with $e_1 = [1, 0, \ldots, 0]^T \in \mathbb{R}^p$, E and ξ given by (8.3.3) and (8.3.6), and the vectors t_p and $\hat{t}_p$ defined by

$$t_p = \left[\begin{array}{cccc} \dfrac{1}{p!} & \dfrac{1}{(p-1)!} & \cdots & \dfrac{1}{1!} \end{array} \right]^T, \quad \hat{t}_p = \left[\begin{array}{cccc} \dfrac{1}{(p+1)!} & \dfrac{1}{p!} & \cdots & \dfrac{1}{2!} \end{array} \right]^T.$$

Proof: Since

$$h_n \, \overline{y}'(t_n) \;=\; h_n f\Big(y(t_n) - E \, h_n^{p+1} y^{(p+1)}(t_n) + O(h_n^{p+2})\Big)$$

$$= \; h_n \, y'(t_n) - E \, h_n^{p+2} \frac{\partial f}{\partial y}\big(y(t_n)\big) y^{(p+1)}(t_n) + O(h_n^{p+3})$$

and

$$h_n^k \, \overline{y}^{(k)}(t_n) = h_n^k \, y^{(k)}(t_n) + O(h_n^{p+3}), \quad k = 2, 3, \ldots, p+1,$$

it follows that

$$\overline{z}(t_n, h_n) = z(t_n, h_n) - E(e_1 \otimes I) h_n^{p+2} \frac{\partial f}{\partial y}\big(y(t_n)\big) y^{(p+1)}(t_n) + O(h_n^{p+3}). \quad (8.3.10)$$

It follows from the assumption that the stage order q is equal to the order p that there exists a vector $\xi \in \mathbb{R}^{p+1}$ of stage errors such that

$$Y^{[n]} = y(t_{n-1} + ch_n) - (\xi \otimes I) \, h_n^{p+1} y^{(p+1)}(t_{n-1}) + O(h_n^{p+2}). \quad (8.3.11)$$

This relation implies that

$$h_n F(Y^{[n]}) \;=\; h_n y'(t_{n-1} + ch_n)$$

$$- (\xi \otimes I) \, h_n^{p+2} \frac{\partial f}{\partial y}\big(y(t_{n-1})\big) y^{(p+1)}(t_{n-1}) + O(h_n^{p+3}). \quad (8.3.12)$$

Substituting (8.3.1), (8.3.2), (8.3.10), (8.3.11), and (8.3.12) into (8.1.3), we obtain

$$y(t_{n-1} + ch_n) - (\xi \otimes I) h_n^{p+1} y^{(p+1)}(t_{n-1})$$

$$= \; (e \otimes I) y(t_{n-1}) + h_n (A \otimes I) y'(t_{n-1} + ch_n) \quad (8.3.13)$$

$$+ (U \otimes I) \Big(z(t_{n-1}, h_n) - \beta \, h_n^{p+1} \Big) + O(h_n^{p+2}),$$

$$y(t_n) - E \, h_n^{p+1} y^{(p+1)}(t_n) - F \, h_n^{p+2} y^{(p+2)}(t_n) - G \, h_n^{p+2} \frac{\partial f}{\partial y}\big(y(t_n)\big) y^{(p+1)}(t_n)$$

$$= \; y(t_{n-1}) + (b^T \otimes I) \Big(h_n y'(t_{n-1} + ch_n) - (\xi \otimes I) \frac{\partial f}{\partial y}\big(y(t_n)\big) y^{(p+1)}(t_n) \Big)$$

$$+ (v^T \otimes I) \Big(z(t_{n-1}, h_n) - (\beta \otimes I) h_n^{p+1} y^{(p+1)}(t_{n-1})$$

$$- (\gamma \otimes I) h_n^{p+2} y^{(p+2)}(t_{n-1}) - (\delta \otimes I) h_n^{p+2} \frac{\partial f}{\partial y}\big(y(t_{n-1})\big) y^{(p+1)}(t_{n-1}) \Big)$$

$$+ O(h_n^{p+2}),$$

$$(8.3.14)$$

and

$$
z(t_n, h_n) - E\, h_n^{p+2}(e_1 \otimes I)\frac{\partial f}{\partial y}\big(y(t_n)\big)y^{(p+1)}(t_n) - (\beta \otimes I)h_n^{p+1}y^{(p+1)}(t_n)
$$

$$
- (\gamma \otimes I)h_n^{p+2}y^{(p+2)}(t_n) - (\delta \otimes I)h_n^{p+2}\frac{\partial f}{\partial y}\big(y(t_n)\big)y^{(p+1)}(t_n)
$$

$$
= (B \otimes I)\bigg(h_n y'(t_{n-1} + ch_n) - (\xi \otimes I)h_n^{p+2}\frac{\partial f}{\partial y}\big(y(t_{n-1})\big)y^{(p+1)}(t_{n-1}) \bigg)
$$

$$
+ (V \otimes I)\bigg(z(t_{n-1}, h_n) - (\beta \otimes I)h_n^{p+1}y^{(p+1)}(t_{n-1})
$$

$$
- (\gamma \otimes I)\, h_n^{p+2}y^{(p+2)}(t_{n-1}) - (\delta \otimes I)\, h_n^{p+2}\frac{\partial f}{\partial y}\big(y(t_{n-1})\big)y^{(p+1)}(t_{n-1}) \bigg)
$$

$$
+ O(h_n^{p+3}).
$$

$$(8.3.15)$$

Expanding $y(t_{n-1}+ch_n)$ and $y(t_n)$ in (8.3.13) and (8.3.14) into a Taylor series around t_{n-1} leads to

$$
\sum_{j=0}^{p}\frac{c^j \otimes I}{j!}h_n^j y^{(j)}(t_{n-1}) + \bigg(\Big(\frac{c^{p+1}}{(p+1)!} - \xi\Big) \otimes I \bigg)h_n^{p+1}y^{(p+1)}(t_{n-1})
$$

$$
= (e \otimes I)y(t_{n-1}) + (A \otimes I)\sum_{j=1}^{p}\frac{c^{j-1} \otimes I}{(j-1)!}h_n^j y^{(j)}(t_{n-1})
$$

$$
+ \bigg(\Big(A\frac{c^p}{p!} - U\beta\Big) \otimes I \bigg)h_{n-1}^{p+1}y^{(p+1)}(t_{n-1}) + (U \otimes I)z(t_{n-1}, h_n) + O(h_n^{p+2})
$$

and

$$
\sum_{j=0}^{p}\frac{1}{j!}h_n^j y^{(j)}(t_{n-1}) + \Big(\frac{1}{(p+1)!} - E \Big)h_n^{p+1}y^{(p+1)}(t_{n-1})
$$

$$
= y(t_{n-1}) + (b^T \otimes I)\sum_{j=1}^{p}\frac{c^{j-1} \otimes I}{(j-1)!}h_n^j y^{(j)}(t_{n-1})
$$

$$
+ \Big(\frac{b^T c^p}{p!} - v^T\alpha \Big)h_n^{p+1}y^{(p+1)}(t_{n-1}) + (v^T \otimes I)z(t_{n-1}, h_n) + O(h_n^{p+2}).
$$

Since method (8.1.3) has order p and stage order $q = p$, all terms up to order p in the expressions above cancel out and comparing terms of order $p + 1$ corresponding to $h_n^{p+1}y^{(p+1)}(t_{n-1})$ and of order $p + 2$ corresponding to $h_n^{p+2}y^{(p+2)}(t_{n-1})$ and $h_n^{p+2}\frac{\partial f}{\partial y}(y(t_{n-1}))y^{(p+1)}(t_{n-1})$, we obtain (8.3.3), (8.3.4), (8.3.5), and (8.3.6). To obtain expressions for the vectors β, γ, and δ we need the following relationship between $z(t_n, h_n)$ and $z(t_{n-1}, h_n)$:

$$
\begin{aligned}
z(t_n, h_n) = {}& T_p\, z(t_{n-1}, h_n) + t_p\, h_n^{p+1} y^{(p+1)}(t_{n-1}) \\
& + \widehat{t_p}\, h_n^{p+2} y^{(p+2)}(t_{n-1}) + O(h_n^{p+3}),
\end{aligned}
\tag{8.3.16}
$$

where

$$
T_p = \begin{bmatrix}
1 & 1 & \frac{1}{2!} & \cdots & \frac{1}{(p-1)!} \\
0 & 1 & 1 & \cdots & \frac{1}{(p-2)!} \\
0 & 0 & 1 & \cdots & \frac{1}{(p-3)!} \\
\vdots & \vdots & \vdots & \ddots & \vdots \\
0 & 0 & 0 & \cdots & 1
\end{bmatrix} \in \mathbb{R}^{p \times p}.
$$

(This matrix was denoted by **E** in Section 7.1). This relationship can be verified by expanding $h_n^j y^{(j)}(t_n)$ appearing in $z(t_n, h_n)$ into a Taylor series around the point t_{n-1}. Substituting (8.3.16) into (8.3.15), we obtain

$$
(T_p \otimes I) z(t_{n-1}, h_n) + (t_p \otimes I) h_n^{p+1} y^{(p+1)}(t_{n-1}) + (\widehat{t_p} \otimes I) h_n^{p+2} y^{(p+2)}(t_{n-1})
$$

$$
- E\, h_n^{p+2} (e_1 \otimes I) \frac{\partial f}{\partial y}\big(y(t_{n-1})\big) y^{(p+1)}(t_{n-1})
$$

$$
- (\beta \otimes I) h_n^{p+1}\Big(y^{(p+1)}(t_{n-1}) + h_n y^{(p+2)}(t_{n-1}) \Big)
$$

$$
- (\gamma \otimes I) h_{n-1}^{p+2} y^{(p+2)}(t_{n-1}) - (\delta \otimes I) h_n^{p+2} \frac{\partial f}{\partial y}\big(y(t_{n-1})\big) y^{(p+1)}(t_{n-1})
$$

$$
= (B \otimes I)\Bigg(\sum_{j=1}^{p+2} \frac{c^{j-1} \otimes I}{(j-1)!}\, h_n^j y^{(j)}(t_{n-1})
$$

$$
- (\xi \otimes I) h_n^{p+2} \frac{\partial f}{\partial y}\big(y(t_{n-1})\big) y^{(p+1)}(t_{n-1}) \Bigg)
$$

$$
+ (V \otimes I)\Bigg(z(t_{n-1}, h_n) - (\beta \otimes I) h_n^{p+1} y^{(p+1)}(t_{n-1})
$$

$$
- (\gamma \otimes I) h_n^{p+2} y^{(p+2)}(t_{n-1}) - (\delta \otimes I) h_n^{p+2} \frac{\partial f}{\partial y}\big(y(t_{n-1})\big) y^{(p+1)}(t_{n-1}) \Bigg)
$$

$$
+ O(h_n^{p+3}).
$$

As before, terms up to the order p cancel out, and comparing terms of order $p+1$ corresponding to $h_n^{p+1} y^{(p+1)}(t_{n-1})$ and of order $p+2$ corresponding to $h_n^{p+2} y^{(p+2)}(t_{n-1})$ and $h_n^{p+2} \frac{\partial f}{\partial y}(y(t_{n-1})) y^{(p+1)}(t_{n-1})$, we obtain (8.3.7), (8.3.8), and (8.3.9) with E and ξ given by (8.3.3) and (8.3.6). This completes the proof. ∎

Observe that in formula (8.3.3) for the error constant E of the method, the term $1/(p+1)! - b^T c^p/p!$ corresponds to the error generated by the method itself in the current step from t_{n-1} to t_n, and $v^T\beta$ corresponds to the error inherited from previous steps.

The next result investigates the relationship between the error constant E defined by (8.3.3) and the error constant E defined by the relation (7.5.6) in Section 7.5. We have the following theorem.

Theorem 8.3.2 (compare [72]) *Assume that GLM (8.1.3) has order p and stage order $q = p$. Then the error constant E defined by (8.3.3) is equal to the error constant E defined by (7.5.6).*

Proof: Denote the error constant defined by (8.3.3) by $\widetilde{E}$. We will show that $\widetilde{E} = E$, where E is defined by (7.5.6). Consider method (8.1.3) with constant step size $h_n = h$. Applying method (8.1.3) with $h_n = h$ to the linear test equation

$$y' = \xi y, \quad t \geq 0,$$

we obtain

$$\mathbf{y}^{[n]} = \mathbf{M}(z)\mathbf{y}^{[n-1]},$$

$n = 1, 2, \ldots$, where $\mathbf{M}(z) = \mathbf{V} + z\mathbf{B}(\mathbf{I} - z\mathbf{A})^{-1}\mathbf{U}$ is the stability matrix and $z = h\xi$. Assuming that y_{n-1} and $z^{[n-1]}$ satisfy (8.3.1) and that y_n and $y^{[n]}$ satisfy (8.3.2) with $E = \widetilde{E}$, we obtain

$$\begin{bmatrix} y(t_n) - \widetilde{E}h^{p+1}y^{(p+1)}(t_n) \\ z(t_n,h) - \beta h^{p+1}y^{(p+1)}(t_n) \end{bmatrix}$$

$$= \mathbf{M}(z)\begin{bmatrix} y(t_{n-1}) \\ z(t_{n-1},h) - \beta h^{p+1}y^{(p+1)}(t_{n-1}) \end{bmatrix} + O(h^{p+2}).$$

Since the solution of the test equation is given by $y(t) = e^{\xi t}y(0)$, it can be verified that this relation can be written in the form

$$\begin{bmatrix} \widetilde{E} \\ \beta \end{bmatrix} z^{p+1}e^z + \left(\mathbf{M}(z) - e^z\mathbf{I}\right)\begin{bmatrix} 1 \\ Z \end{bmatrix} - \mathbf{M}(z)\begin{bmatrix} 0 \\ \beta \end{bmatrix} z^{p+1} = O(z^{p+2}), \quad (8.3.17)$$

where the vector Z is defined by

$$Z = \begin{bmatrix} z & z^2 & \cdots & z^p \end{bmatrix}^T$$

and $\mathbf{I}$ is the identity matrix of dimension $p+1$. The equation (8.3.17) simplifies to

$$\mathbf{M}(z)\begin{bmatrix} 1 \\ Z - \beta z^{p+1} \end{bmatrix} - (1 - \widetilde{E}z^{p+1})e^z\begin{bmatrix} 1 \\ \dfrac{Z - \beta z^{p+1}}{1 - \widetilde{E}z^{p+1}} \end{bmatrix} = O(z^{p+2}),$$

and taking into account that

$$\frac{Z - \beta z^{p+1}}{1 - \widetilde{E}z^{p+1}} = Z - \beta z^{p+1} + O(z^{p+2})$$

and $z^{p+1}e^z = z^{p+1} + O(z^{p+2})$, we obtain

$$\left(\mathbf{M}(z) - (e^z - \widetilde{E}z^{p+1})\mathbf{I}\right)\begin{bmatrix} 1 \\ \widetilde{W}(z) \end{bmatrix} = O(z^{p+2}), \qquad (8.3.18)$$

where $\widetilde{W}(z) = Z - \beta z^{p+1}$. It follows from Theorem 7.2.4 that the stability matrix $\mathbf{M}(z)$ of the GLM (8.1.3) has only one nonzero eigenvalue $R(z)$, given by formula (7.2.11). As in the proof of Lemma 7.2.1, we can assume without loss of generality that the eigenvector of $\mathbf{M}(z)$ corresponding to this eigenvalue $R(z)$ can be scaled so that its first component is equal to 1. This leads to the relation

$$\left(\mathbf{M}(z) - R(z)\mathbf{I}\right)\begin{bmatrix} 1 \\ W(z) \end{bmatrix} = 0.$$

It follows from the discussion in Section 7.5 that this stability function $R(z)$ of GLM (8.1.3) also satisfies (7.5.6), and substituting this equation into the relation above, we obtain

$$\left(\mathbf{M}(z) - (e^z - Ez^{p+1})\mathbf{I}\right)\begin{bmatrix} 1 \\ W(z) \end{bmatrix} = O(z^{p+2}). \qquad (8.3.19)$$

Set

$$P_{\widetilde{E}}(z) = e^z - \widetilde{E}z^{p+1}, \quad P_E(z) = e^z - Ez^{p+1}.$$

Clearly,

$$P_{\widetilde{E}}(z) - P_E(z) = O(z^{p+1})$$

and we will show that this implies that

$$P_{\widetilde{E}}(z) - P_E(z) = O(z^{p+2}).$$

The idea of the proof of this fact is based on the application of the power method for computing eigenvalues and eigenvectors, where in each iteration, one more power of z is gained. We have

$$\mathbf{M}(z) = \begin{bmatrix} M_{11}(z) & M_{12}(z) \\ M_{21}(z) & M_{22}(z) \end{bmatrix}$$

$$= \begin{bmatrix} 1 + zb^T(I - zA)^{-1}e & v^T + zb^T(I - zA)^{-1}U \\ zB(I - zA)^{-1}e & V + zB(I - zA)^{-1}U \end{bmatrix}$$

and it follows from (8.3.18) and (8.3.19) that

$$M_{11}(z) - P_{\widetilde{E}}(z) - M_{12}(z)\widetilde{W}(z) = O(z^{p+2}),$$
$$M_{21}(z) + \big(M_{22}(z) - P_{\widetilde{E}}(z)I\big)\widetilde{W}(z) = O(z^{p+2}) \tag{8.3.20}$$

and

$$M_{11}(z) - P_E(z) - M_{12}(z)W(z) = O(z^{p+2}),$$
$$M_{21}(z) + \big(M_{22}(z) - P_E(z)I\big)W(z) = O(z^{p+2}). \tag{8.3.21}$$

Here I is an identity matrix of dimension p. Subtracting the first equations of (8.3.20) and (8.3.21), we obtain

$$P_{\widetilde{E}}(z) - P_E(z) = M_{12}(z)\big(\widetilde{W}(z) - W(z)\big) + O(z^{p+2}). \tag{8.3.22}$$

Similarly, subtracting the second equations of (8.3.20) and (8.3.21), after some computations we obtain

$$\big(P_E(z)I - M_{22}(z)\big)\big(\widetilde{W}(z) - W(z)\big)$$
$$= \big(P_{\widetilde{E}}(z) - P_E(z)\big)\widetilde{W}(z) + O(z^{p+2}). \tag{8.3.23}$$

It follows from $P_{\widetilde{E}}(z) - P_E(z) = O(z^{p+1})$, $\widetilde{W}(z) = O(z)$, and equation (8.3.23) that $\widetilde{W}(z) - W(z) = O(z^{p+2})$. Substituting this into equation (8.3.22), we obtain $P_{\widetilde{E}}(z) - P_E(z) = O(z^{p+2})$. This implies that $\widetilde{E} = E$, which is our claim. ∎

The next result describes the structure of the vector β appearing in (8.3.1) and (8.3.2).

Theorem 8.3.3 (Wright [293]) *For GLM (8.1.3) of order p and stage order $q = p$ and with IRKS, the vector β appearing in (8.3.1) and (8.3.2) is given by*

$$\beta = \begin{bmatrix} \beta_p & \beta_{p-1} & \cdots & \beta_1 \end{bmatrix}^T,$$

where $\beta_p, \beta_{p-1}, \ldots, \beta_1$ are free parameters of the doubly companion matrix $\mathbf{X} = \mathbf{X}(\alpha, \beta)$.

Proof: Substituting formulas (7.1.6) and (7.1.7) for the coefficient matrices $\mathbf{U}$ and $\mathbf{V}$ into the relation $\mathbf{BA} \equiv \mathbf{XV} - \mathbf{VX}$, we obtain

$$\mathbf{B}(\mathbf{C} - \mathbf{ACK}) \equiv \mathbf{X}(\mathbf{E} - \mathbf{BCK}) - \mathbf{VX}.$$

It follows from Lemma 7.4.2 that $\mathbf{BA} = \mathbf{XB}$ and the relation above simplifies to

$$\mathbf{BC} \equiv \mathbf{XE} - \mathbf{VX}.$$

Comparing the last columns of this relation, we obtain

$$\mathbf{BC}e_{p+1} \equiv \mathbf{XE}e_{p+1} - \mathbf{VX}e_{p+1}. \tag{8.3.24}$$

It can be verified that

$$\mathbf{C}e_{p+1} = \frac{c^p}{p!}, \quad \mathbf{E}e_{p+1} = \begin{bmatrix} t_p \\ 1 \end{bmatrix}, \quad \mathbf{X}e_{p+1} = \begin{bmatrix} -\alpha_{p+1} - \beta_{p+1} \\ -\beta \end{bmatrix},$$

and partitioning the matrices $\mathbf{B}$, $\mathbf{X}$, and $\mathbf{V}$ appropriately, relation (8.3.24) can be written in the form

$$\begin{bmatrix} b^T \\ B \end{bmatrix} \frac{c^p}{p!} \equiv \begin{bmatrix} -\alpha^T & -\alpha_{p+1} - \beta_{p+1} \\ I & -\beta \end{bmatrix} \begin{bmatrix} t_p \\ 1 \end{bmatrix}$$
$$+ \begin{bmatrix} 1 & v^T \\ 0 & V \end{bmatrix} \begin{bmatrix} \alpha_{p+1} + \beta_{p+1} \\ \beta \end{bmatrix},$$

where

$$\alpha = \begin{bmatrix} \alpha_1 & \alpha_2 & \cdots & \alpha_p \end{bmatrix}^T.$$

Comparing the last p components in this equivalence relation, we obtain the equation

$$B\frac{c^p}{p!} = t_p - \beta + V\beta,$$

which is equivalent to (8.3.7). This completes the proof. ∎

It follows from Theorems 8.3.2 and 8.3.3 that the free parameters of GLMs (8.1.3) with IRKS are directly related to the accuracy of the resulting formulas. In the algorithm for construction of these methods presented at the end of Section 7.5, the free parameter ϵ is related to the error constant $\widetilde{E} = E$ through equation (7.5.14), that is,

$$\epsilon = M_{p+1,p+1}(\lambda) - E,$$

and the parameters $\beta_p, \beta_{p-1}, \ldots, \beta_1$ are related to the errors of the external approximations $\overline{z}^{n]}$ through the relation

$$\overline{z}^{[n]} = \overline{z}(t_n, h_n) - (\beta \otimes I)h_n^{p+1} y^{(p+1)}(t_n) + O(h_n^{p+2})$$

(compare (8.3.2) and Theorem 8.3.3). As a result, these parameters control directly the accuracy of the resulting methods. The corresponding effects on the stability of these formulas are discussed in Sections 7.8 and 7.9 for the explicit methods and in Sections 7.10, 7.11, and 7.12 for the implicit methods.

8.4   ESTIMATION OF LOCAL DISCRETIZATION ERROR AND ESTIMATION OF HIGHER ORDER TERMS

It follows from (8.3.2) that the local discretization error of GLM (8.1.3) at the point t_n is given by

$$\mathrm{le}(t_n) = E h_n^{p+1} y^{(p+1)}(t_n) + O(h_n^{p+2}), \tag{8.4.1}$$

where the error constant E is defined by (8.3.3). We look for the estimate of the quantity $h_n^{p+1} y^{(p+1)}(t_n)$, which corresponds to the principal part of $\mathrm{le}(t_n)$, in the form

$$h_n^{p+1} y^{(p+1)}(t_n) = (\varphi^T \otimes I) h_n F(Y^{[n]}) + (\psi^T \otimes I) z^{[n-1]} + O(h_n^{p+2}),$$

where $\varphi \in \mathbb{R}^{p+1}$ and $\psi \in \mathbb{R}^p$ are some vectors. To compute accurately the input vector $z^{[n]}$ for the next step corresponding to the new step size, we will also have to investigate terms of order $p+2$ in the expression above. Therefore, we consider a more precise estimate of $h_n^{p+1} y^{(p+1)}(t_n)$, which is of the form

$$
\begin{aligned}
h_n^{p+1} y^{(p+1)}(t_n) &= (\varphi^T \otimes I) h_n F(Y^{[n]}) + (\psi^T \otimes I) z^{[n-1]} \\
&\quad + C_1 h_n^{p+2} y^{(p+2)}(t_n) + C_2 h_n^{p+2} \frac{\partial f}{\partial y}\big(y(t_n)\big) y^{(p+1)}(t_n) \\
&\quad + O(h_n^{p+3}),
\end{aligned}
\tag{8.4.2}
$$

where C_1 and C_2 are constants. We have the following result.

Theorem 8.4.1 (compare [75]) *For GLM (8.1.3) of order p and stage order $q = p$, the vectors φ and ψ satisfy the system of equations*

$$
\begin{aligned}
\frac{\varphi^T c^{j-1}}{(j-1)!} + \psi_j &= 0, \quad j = 1, 2, \ldots, p, \\
\frac{\varphi^T c^p}{p!} - \psi^T \beta &= 1.
\end{aligned}
\tag{8.4.3}
$$

Here ψ_j stands for the jth component of ψ. Furthermore, the constants C_1 and C_2 are given by

$$
\begin{aligned}
C_1 &= 1 - \frac{\varphi^T c^{p+1}}{(p+1)!} + \psi^T \gamma, \\
C_2 &= \varphi^T \xi + \psi^T \delta.
\end{aligned}
\tag{8.4.4}
$$

Proof: Substituting (8.3.12) and (8.3.1) into (8.4.2) and expanding $y^{(p+1)}(t_n)$, $y'(t_{n-1} + ch_n)$ into a Taylor series around t_{n-1}, we obtain

$$h_n^{p+1} y^{(p+1)}(t_{n-1}) + h_n^{p+2} y^{(p+2)}(t_{n-1})$$

$$= (\varphi^T \otimes I)\left(\sum_{j=1}^{p} \frac{c^{j-1} \otimes I}{(j-1)!} h_n^j y^{(j)}(t_{n-1}) + \frac{c^p \otimes I}{p!} h_n^{p+1} y^{(p+1)}(t_{n-1}) \right.$$

$$+ \frac{c^{p+1} \otimes I}{(p+1)!} h_n^{p+2} y^{(p+2)}(t_{n-1}) - (\xi \otimes I) h_n^{p+2} \frac{\partial f}{\partial y}\big(y(t_{n-1})\big) y^{(p+1)}(t_{n-1}) \Big)$$

$$+ (\psi^T \otimes I)\Big(z(t_{n-1}, h_n) - (\beta \otimes I) h_n^{p+1} y^{(p+1)}(t_{n-1})$$

$$- (\gamma \otimes I) h_n^{p+2} y^{(p+2)}(t_{n-1}) - (\delta \otimes I) h_{n-1}^{p+2} \frac{\partial f}{\partial y}\big(y(t_{n-1})\big) y^{(p+1)}(t_{n-1}) \Big)$$

$$+ C_1 h_n^{p+2} y^{(p+2)}(t_{n-1}) + C_2 h_n^{p+2} \frac{\partial f}{\partial y}\big(y(t_{n-1})\big) y^{(p+1)}(t_{n-1}) + O(h_n^{p+3}).$$

Comparing terms up to order $p+1$ leads to the system of equations (8.4.3). Comparing terms of order $p+2$ corresponding to

$$h_n^{p+1} y^{(p+2)}(t_{n-1}) \quad \text{and} \quad h_n^{p+1} \frac{\partial f}{\partial y}\big(y(t_{n-1})\big) y^{(p+1)}(t_{n-1}),$$

we obtain the expressions (8.4.4) for the constants C_1 and C_2. ∎

We can also compute an estimate of $h_n^{p+1} y^{(p+1)}(t_n)$ in terms of $h_n F(Y^{[n]})$ and $\overline{z}^{[n]}$, that is, an estimate of the form

$$h_n^{p+1} y^{(p+1)}(t_n) = (\overline{\varphi}^T \otimes I) h_n F(Y^{[n]}) + (\overline{\psi}^T \otimes I) \overline{z}^{[n]}$$

$$+ \overline{C}_1 h_n^{p+2} y^{(p+2)}(t_n) + \overline{C}_2 h_n^{p+2} \frac{\partial f}{\partial y}\big(y(t_n)\big) y^{(p+1)}(t_n) \tag{8.4.5}$$

$$+ O(h_n^{p+3}),$$

where $\overline{\varphi} \in \mathbb{R}^{p+1}$, $\overline{\psi} \in \mathbb{R}^p$, and $\overline{C}_1$ and $\overline{C}_2$ are constants. We have the following result.

Theorem 8.4.2 *For GLM (8.1.3) of order p and stage order $q = p$, the vectors $\overline{\varphi}$ and $\overline{\psi}$ satisfy the system of equations*

$$\frac{\overline{\varphi}^T (c - e)^{j-1}}{(j-1)!} + \overline{\psi}_j = 0, \quad j = 1, 2, \ldots, p,$$

$$\frac{\overline{\varphi}^T (c - e)^p}{p!} - \overline{\psi}^T \beta = 1. \tag{8.4.6}$$

Here $\overline{\psi}_j$ stands for the jth component of $\overline{\psi}$. Furthermore, the constants $\overline{C}_1$ and $\overline{C}_2$ are given by

$$\overline{C}_1 = -\frac{\overline{\varphi}^T(c-e)^{p+1}}{(p+1)!} + \overline{\psi}^T\gamma,$$
$$\overline{C}_2 = \overline{\varphi}^T\xi + \overline{\psi}_1 + \overline{\psi}^T\delta. \tag{8.4.7}$$

Proof: Substituting

$$h_n F(Y^{[n]}) = h_n y'\big(t_n + (c-e)h_n\big) - (\xi \otimes I)h_n^{p+2}\frac{\partial f}{\partial y}\big(y(t_n)\big)y^{(p+1)}(t_n) + O(h_n^{p+3})$$

and formula (8.3.2) for $\overline{z}^{[n]}$, then taking (8.3.10) into account, and expanding $y'(t_n + (c-e)h_n)$ into a Taylor series around t_n, we obtain

$$h_n^{p+1}y^{(p+1)}(t_n) = \sum_{j=1}^{p+2}\frac{\overline{\varphi}^T(c-e)^{j-1}}{(j-1)!}h_n^j y^{(j)}(t_n)$$

$$- \overline{\varphi}^T\xi h_n^{p+2}\frac{\partial f}{\partial y}\big(y(t_n)\big)y^{(p+1)}(t_n) + \sum_{j=1}^{p}\overline{\psi}_j h_n^j y^{(j)}(t_n)$$

$$- \overline{\psi}_1 h_n^{p+2}\frac{\partial f}{\partial y}\big(y(t_n)\big)y^{(p+1)}(t_n) - \overline{\psi}^T\beta h_n^{p+1}y^{(p+1)}(t_n)$$

$$- \overline{\psi}^T\gamma h_n^{p+2}y^{(p+2)}(t_n) - \overline{\psi}^T\delta h_n^{p+2}\frac{\partial f}{\partial y}\big(y(t_n)\big)y^{(p+1)}(t_n)$$

$$+ \overline{C}_1 h_n^{p+2}y^{(p+2)}(t_n) + \overline{C}_2 h_n^{p+2}\frac{\partial f}{\partial y}\big(y(t_n)\big)y^{(p+1)}(t_n) + O(h_n^{p+3}).$$

Comparing terms up to order $p+1$, we obtain system (8.4.6). Comparing terms of order $p+2$, we obtain (8.4.7). This completes the proof. ∎

Observe that systems (8.4.3) and (8.4.6) of $p+1$ equations with $2p+1$ unknowns φ_j, ψ_j or $\overline{\varphi}_j, \overline{\psi}_j$ are always solvable with respect to φ or $\overline{\varphi}$ for any choice of the vectors ψ or $\overline{\psi}$ if the abscissa vector c has distinct components.

Next we will look for the estimates of the quantities of order $p+2$:

$$h_n^{p+2}y^{(p+2)}(t_n) \quad \text{and} \quad h_n^{p+2}\frac{\partial f}{\partial y}\big(y(t_n)\big)y^{(p+1)}(t_n),$$

in the form

$$h_n^{p+2}y^{(p+2)}(t_n) = (\widetilde{\varphi}^T \otimes I)h_n F(Y^{[n]}) + (\widetilde{\psi}^T \otimes I)z^{[n-1]} + O(h_n^{p+3}) \tag{8.4.8}$$

and

$$h_n^{p+2}\frac{\partial f}{\partial y}\big(y(t_n)\big)y^{(p+1)}(t_n) = (\widehat{\varphi}^T \otimes I)h_n F(Y^{[n]}) + (\widehat{\psi}^T \otimes I)z^{[n-1]}$$
$$+ O(h_n^{p+3}), \tag{8.4.9}$$

where $\widetilde{\varphi}, \widehat{\varphi} \in \mathbb{R}^{p+1}$ and $\widetilde{\psi}, \widehat{\psi} \in \mathbb{R}^{p}$. We have the following theorem.

Theorem 8.4.3 (compare [75]) *For GLM (8.1.3) of order p and stage order $q = p$, the vectors $\widetilde{\varphi}$, $\widehat{\varphi}$ and $\widetilde{\psi}$, $\widehat{\psi}$ satisfy the systems of equations*

$$\frac{\widetilde{\varphi}^T c^{j-1}}{(j-1)!} + \widetilde{\psi}_j = 0, \quad j = 1, 2, \ldots, p,$$

$$\frac{\widetilde{\varphi}^T c^p}{p!} - \widetilde{\psi}^T \beta = 0, \quad \frac{\widetilde{\varphi}^T c^{p+1}}{(p+1)!} - \widetilde{\psi}^T \gamma = 1, \qquad (8.4.10)$$

$$\widetilde{\varphi}^T \xi + \widetilde{\psi}^T \delta = 0$$

and

$$\frac{\widehat{\varphi}^T c^{j-1}}{(j-1)!} + \widehat{\psi}_j = 0, \quad j = 1, 2, \ldots, p,$$

$$\frac{\widehat{\varphi}^T c^p}{p!} - \widehat{\psi}^T \beta = 0, \quad \frac{\widehat{\varphi}^T c^{p+1}}{(p+1)!} - \widehat{\psi}^T \gamma = 0, \qquad (8.4.11)$$

$$\widehat{\varphi}^T \xi + \widehat{\psi}^T \delta = -1.$$

Proof: The proof of this theorem is similar to the proof of Theorem 8.4.1 and is therefore omitted. ∎

Observe that (8.4.10) and (8.4.11) are systems of $p+3$ equations in $2p+1$ unknowns, which usually have solutions for $p \geq 2$. The solution corresponding to the method of order $p = q = 2$ constructed in Section 7.9 is given in Section 8.6.

We could also derive, if desired, estimates of the quantities

$$h_n^{p+2} y^{(p+2)}(t_n) \quad \text{and} \quad h_n^{p+2} \frac{\partial f}{\partial y}\big(y(t_n)\big) y^{(p+1)}(t_n)$$

in terms of $F(Y^{[n]})$ and $\overline{z}^{[n]}$ as was done in the case of the estimate of the quantity $h_n^{p+1} y^{(p+1)}(t_n)$ in formula (8.4.5). However, we do not pursue this topic further.

8.5 COMPUTING THE INPUT VECTOR OF EXTERNAL APPROXIMATIONS FOR THE NEXT STEP

After the step from t_{n-1} to t_n is accepted, we have computed y_n and the vector $\overline{z}^{[n]}$ such that

$$y_n = y(t_n) - E\, h_n^{p+1} y^{(p+1)}(t_n) + O(h_n^{p+3}) \qquad (8.5.1)$$

and

$$\overline{z}^{[n]} = \overline{z}(t_n, h_n) - (\beta \otimes I)h_n^{p+1}y^{(p+1)}(t_n)$$

$$- (\gamma \otimes I)h_n^{p+2}y^{(p+2)}(t_n) \tag{8.5.2}$$

$$- (\delta \otimes I)h_n^{p+2}\frac{\partial f}{\partial y}\big(y(t_n)\big)y^{(p+1)}(t_n) + O(h_n^{p+3}),$$

where the error constant E and vectors β, γ, and δ are given by (8.3.3), (8.3.7), (8.3.8), and (8.3.9). We also have available estimates of

$$h_n^{p+1}y^{(p+1)}(t_n), \quad h_n^{p+2}y^{(p+2)}(t_n), \quad \text{and} \quad h_n^{p+2}\frac{\partial f}{\partial y}\big(y(t_n)\big)y^{(p+1)}(t_n).$$

As explained in Section 8.4, these estimates, which are denoted by $\delta_1(t_n)$, $\delta_2(t_n)$, and $\delta_3(t_n)$, take the form

$$\delta_1(t_n) = (\varphi^T \otimes I)h_n F(Y^{[n]}) + (\psi^T \otimes I)z^{[n-1]}, \tag{8.5.3}$$

$$\delta_2(t_n) = (\widetilde{\varphi}^T \otimes I)h_n F(Y^{[n]}) + (\widetilde{\psi}^T \otimes I)z^{[n-1]}, \tag{8.5.4}$$

and

$$\delta_3(t_n) = (\widehat{\varphi}^T \otimes I)h_n F(Y^{[n]}) + (\widehat{\psi}^T \otimes I)z^{[n-1]}, \tag{8.5.5}$$

where φ, ψ satisfy (8.4.3), $\widetilde{\varphi}, \widetilde{\psi}$ satisfy (8.4.10), and $\widehat{\varphi}, \widehat{\psi}$ satisfy (8.4.11). To proceed with integration from t_n to $t_{n+1} = t_n + h_{n+1}$, where h_{n+1} is a new step size chosen according to some step size changing strategy, we have to compute a new input vector $z^{[n]}$ such that

$$z^{[n]} = z(t_n, h_{n+1}) - (\beta \otimes I)h_{n+1}^{p+1}y^{(p+1)}(t_n)$$

$$- (\gamma \otimes I)h_{n+1}^{p+2}y^{(p+2)}(t_n) \tag{8.5.6}$$

$$- ((\delta + E\,e_1) \otimes I)h_{n+1}^{p+2}\frac{\partial f}{\partial y}\big(y(t_n)\big)y^{(p+1)}(t_n) + O(h_{n+1}^{p+3}).$$

The presence of the term involving Ee_1 can be justified by formula (8.3.10). Such a vector will be computed by rescaling and modifying the vector $\overline{z}^{[n]}$ according to the formula

$$z^{[n]} = \big(D(r_n) \otimes I\big)\overline{z}^{[n]} + \big(\theta_1(r_n) \otimes I\big)\delta_1(t_n)$$

$$+ \big(\theta_2(r_n) \otimes I\big)\delta_2(t_n) + \big(\theta_3(r_n) \otimes I\big)\delta_3(t_n), \tag{8.5.7}$$

where $r_n = h_{n+1}/h_n$, $D(r)$ is the rescaling matrix defined by

$$D(r) = \text{diag}\big(r, r^2, \ldots, r^p\big),$$

and $\theta_1(r)$, $\theta_2(r)$, and $\theta_3(r)$ are appropriate vectors that are determined by comparing (8.5.6) with (8.5.7). Observe that the ratio of step sizes is now denoted by r_n, which differs from the notation in Chapters 4 and 5, where this ratio was denoted by δ_n. Using formula (8.5.2) and the relations

$$\delta_1(t_n) \;=\; h_n^{p+1} y^{(p+1)}(t_n) - C_1 h_n^{p+2} y^{(p+2)}(t_n)$$

$$- \; C_2 h_n^{p+2} \frac{\partial f}{\partial y}\big(y(t_n)\big) y^{(p+1)}(t_n) + O(h_n^{p+3}),$$

$$\delta_2(t_n) = h_n^{p+2} y^{(p+2)}(t_n) + O(h_n^{p+3}),$$

and

$$\delta_3(t_n) = h_n^{p+2} \frac{\partial f}{\partial y}\big(y(t_n)\big) y^{(p+1)}(t_n) + O(h_n^{p+3}),$$

where the constants C_1 and C_2 are defined by (8.4.4), this leads to

$$\big(D(r_n) \otimes I\big)\bigg(z(t_n, h_n) - (\beta \otimes I) h_n^{p+1} y^{(p+1)}(t_n) - (\gamma \otimes I) h_n^{p+2} y^{(p+2)}(t_n)$$

$$- \big((\delta + E\, e_1) \otimes I\big) h_n^{p+2} \frac{\partial f}{\partial y}\big(y(t_n)\big) y^{(p+1)}(t_n) + O(h_n^{p+3})\bigg)$$

$$+ \big(\theta_1(r_n) \otimes I\big)\bigg(h_n^{p+1} y^{(p+1)}(t_n) - C_1 h_n^{p+2} y^{(p+2)}(t_n)$$

$$- C_2 h_n^{p+2} \frac{\partial f}{\partial y}\big(y(t_n)\big) y^{(p+1)}(t_n)\bigg)$$

$$+ \big(\theta_2(r_n) \otimes I\big) h_n^{p+2} y^{(p+2)}(t_n) + \big(\theta_3(r_n) \otimes I\big) h_n^{p+2} \frac{\partial f}{\partial y}\big(y(t_n)\big) y^{(p+1)}(t_n)$$

$$= \; z(t_n, h_{n+1}) - (\beta \otimes I) r_n^{p+1} h_{n-1}^{p+1} y^{(p+1)}(t_n)$$

$$- (\gamma \otimes I) r_n^{p+2} h_n^{p+2} y^{(p+2)}(t_n)$$

$$- \big((\delta + E\, e_1) \otimes I\big) r_n^{p+2} h_n^{p+2} \frac{\partial f}{\partial y}\big(y(t_n)\big) y^{(p+1)}(t_n) + O(h_n^{p+3}).$$

We have

$$\big(D(r_n) \otimes I\big) z(t_n, h_n) = z(t_n, h_{n+1}),$$

and comparing terms of order $p+1$ corresponding to $h_n^{p+1} y^{(p+1)}(t_n)$ and of order $p+2$ corresponding to

$$h_n^{p+2} y^{(p+2)}(t_n) \quad \text{and} \quad h_n^{p+2} \frac{\partial f}{\partial y}\big(y(t_n)\big) y^{(p+1)}(t_n),$$

we obtain

$$\theta_1(r) = \big(D(r) - r^{p+1} I\big)\beta, \qquad (8.5.8)$$

$$\theta_2(r) = \big(D(r) - r^{p+2}I\big)\gamma + C_1\,\theta_1(r), \tag{8.5.9}$$

and

$$\theta_3(r) = \big(D(r) - r^{p+2}I\big)(\delta + E\,e_1) + C_2\,\theta_1(r). \tag{8.5.10}$$

As mentioned in Section 8.1, the formula (8.5.7) with $\theta_1(r)$, $\theta_2(r)$, and $\theta_3(r)$ defined by (8.5.8), (8.5.9), and (8.5.10), respectively, is the integral part of GLM (8.1.3).

8.6 ZERO-STABILITY ANALYSIS

In this section we analyze the zero-stability properties of the method (8.1.3) with $z^{[n]}$ defined by (8.5.7), $\delta_1(t_n)$, $\delta_2(t_n)$, $\delta_3(t_n)$ defined by (8.5.3), (8.5.4), (8.5.5), and $\theta_1(r_n)$, $\theta_2(r_n)$, $\theta_3(r_n)$ defined by (8.5.8), (8.5.9) and (8.5.10), respectively. Applying this method to the test equation

$$y' = 0, \qquad t \geq 0,$$
$$y(0) = 1,$$

on the nonuniform grid $\{t_n\}$, we obtain

$$y_n = y_{n-1} + v^T z^{[n-1]}$$

and

$$z^{[n]} = D(r_n)\,V\,z^{[n-1]} + \theta_1(r_n)\delta_1(t_n) + \theta_2(r_n)\delta_2(t_n) + \theta_3(r_n)\delta_3(t_n).$$

Substituting the formulas for $\delta_1(t_n)$, $\delta_2(t_n)$, and $\delta_3(t_n)$, we obtain

$$z^{[n]} = M(r_n)\,z^{[n-1]}, \tag{8.6.1}$$

$n = 1, 2, \ldots$, where the amplification matrix $M(r)$ is defined by

$$M(r) = D(r)V + \theta_1(r)\psi^T + \theta_2(r)\widetilde{\psi}^T + \theta_3(r)\widehat{\psi}^T. \tag{8.6.2}$$

Hence,

$$z^{[n]} = M(r_n)M(r_{n-1})\cdots M(r_1)z^{[0]}$$

and it follows that the zero-stability of method (8.1.3), (8.5.7) is equivalent to the uniform boundedness of the product of matrices:

$$M(r_n)M(r_{n-1})\cdots M(r_1).$$

To find the conditions under which this is the case, we follow the approach proposed by Guglielmi and Zennaro [138, 139]. This approach is based on the theory of the joint spectral radius and the notion of a polytope norm of a family of matrices. According to this theory the zero-stability of (8.1.3),

(8.5.7) would follow if we can construct a polytope norm $\|\cdot\|_*$ in $\mathbb{R}^p$ such that for the induced matrix norm denoted by the same symbol, we have

$$\left\|M(r)\right\|_* \leq 1 \tag{8.6.3}$$

for $r \in [0, r^*]$ for some r^*. These polytope norms are defined by their unit balls in $\mathbb{R}^p$. Set

$$r^* = \max\left\{r : \ \rho\big(M(r)\big) \leq 1\right\},$$

where $\rho(M)$ is the spectral radius of the matrix M. As explained by Guglielmi and Zennaro [138], in many cases these polytope norms $\|\cdot\|_*$ can be found by successively applying the matrix $M(r^*)$ to the set of vectors

$$\mathcal{S} = \left\{e_1, e_2, \ldots, e_p\right\},$$

where e_i are canonical basis vectors in $\mathbb{R}^p$. If

$$M^j(r)P, \quad P \in \mathcal{S}, \quad j = 1, 2, \ldots,$$

are contained in a common convex hull, symmetric with respect to the origin, of some points in $\mathbb{R}^p$ for $r \in [0, r^*]$, this convex hull defines the unit ball of the polytope norm $\|\cdot\|_*$ satisfying (8.6.3). This process was illustrated by Guglielmi and Zennaro [139] for the variable step size three-step backward differentiation method, by Butcher and Jackiewicz [71] for some GLMs of order $p = 2$, and by Jackiewicz et al. [180] for some two-step W-methods of order $p = 2$. In a recent paper by Butcher et al. [75] this process was illustrated for the composition of Adams-Bashforth methods of order 2 and 3 over two steps of size $h/2$, and then using PECE scheme with the composite Adams-Bashforth method as a predictor and the Adams-Moulton method of appropriate order as a corrector, and for GLMs with IRKS of order 2 and 3. In what follows we illustrate this process for explicit and implicit methods with $p = q = 2$ and $s = 3$. We consider first the explicit method derived by Butcher and Jackiewicz [72], whose coefficients are presented in Section 7.9. For this method the abscissa vector $c = [0, \frac{1}{2}, 1]^T$, and for easy reference, the coefficients are given again below.

$$\left[\begin{array}{c|cc} A & e & U \\ \hline b^T & 1 & v^T \\ \hline B & 0 & V \end{array}\right] = \left[\begin{array}{ccc|ccc} 0 & 0 & 0 & 1 & 0 & 0 \\ \frac{279}{574} & 0 & 0 & 1 & \frac{4}{287} & \frac{1}{8} \\ \frac{81}{14105} & \frac{1968}{2015} & 0 & 1 & \frac{8}{455} & \frac{47}{4030} \\ \hline \frac{608663}{499968} & -\frac{2009}{35712} & \frac{455}{2304} & 1 & -\frac{241}{672} & \frac{41}{124} \\ \hline -\frac{113815}{71424} & \frac{85567}{35712} & \frac{455}{2304} & 0 & 0 & -\frac{1177}{2976} \\ \frac{17}{24} & -\frac{41}{12} & \frac{65}{24} & 0 & 0 & 0 \end{array}\right].$$

$$\tag{8.6.4}$$

It can be verified using (8.3.3), (8.3.7), (8.3.8), and (8.3.9) that the error constant is $E = \frac{7}{96}$ and the vectors β, γ, and δ are

$$\beta = \begin{bmatrix} \frac{7}{96} \\ \frac{7}{96} \end{bmatrix}, \quad \gamma = \begin{bmatrix} -\frac{2177}{285696} \\ \frac{3}{64} \end{bmatrix}, \quad \delta = \begin{bmatrix} \frac{329}{142848} \\ \frac{191}{9216} \end{bmatrix}.$$

We choose the vector ψ as $\psi = [0, -v_{11}/\beta_1]^T$. This choice will be motivated in Section 8.8, where the construction of unconditionally stable methods is discussed. Then solving (8.4.3), (8.4.10), and (8.4.11), we obtain

$$\varphi = \begin{bmatrix} \frac{113815}{5208} \\ -\frac{85567}{2604} \\ \frac{1849}{168} \end{bmatrix}, \quad \widetilde{\varphi} = \begin{bmatrix} -\frac{916034185}{46171524} \\ -\frac{33566249}{3297966} \\ \frac{1262495}{212772} \end{bmatrix}, \quad \widehat{\varphi} = \begin{bmatrix} \frac{513580145}{6595932} \\ -\frac{29738735}{471138} \\ \frac{429065}{30396} \end{bmatrix}$$

and

$$\psi = \begin{bmatrix} 0 \\ \frac{1177}{217} \end{bmatrix}, \quad \widetilde{\psi} = \begin{bmatrix} \frac{2989248}{124117} \\ -\frac{464258}{549661} \end{bmatrix}, \quad \widehat{\psi} = \begin{bmatrix} -\frac{511680}{17731} \\ \frac{1369810}{78523} \end{bmatrix}.$$

The amplification matrix $M(r)$ given by formula (8.6.2) takes the form

$$\begin{bmatrix} \frac{11r(r-1)(104304+104304r+182149r^2)}{851088} & \frac{r(294867408+39150823r^2-477122599r^3)}{361833984} \\ \frac{r^2(r-1)(9156768+3162703r)}{5957616} & \frac{r^2(r-1)(541382496+815438257r)}{2532837888} \end{bmatrix}.$$

Applying the procedure described above, it can be verified that condition (8.6.3) is satisfied for $r \in [0, r^*]$, $r^* \approx 1.11366$, for the polytope norm $\| \cdot \|_*$ whose unit ball with vertices P_1, P_2, P_3, P_4, P_5, P_6, P_7, and P_8 is plotted in Fig. 8.6.1. The coordinates of the points P_i are

$$P_1 = -P_5 = [1, 0]^T, \quad P_2 = -P_6 = [0.9713, 0.0807]^T,$$

$$P_3 = -P_7 = [0, 1]^T, \quad P_4 = -P_8 = [-0.7303, 0.3000]^T.$$

For comparison we have also analyzed zero-stability properties of the method (8.6.4) with the vector $z^{[n]}$ of external approximations given by

$$z^{[n]} = D(r_n)\overline{z}^{[n]} + \theta_1(r_n)\delta_1(t_n), \tag{8.6.5}$$

where $\theta_1(r)$ is defined by (8.5.8) and $\delta_1(t_n)$ is defined by (8.5.3). This corresponds to the rescale and modify approach introduced in [71]. It can be verified that for the method (8.1.3), (8.6.5) the amplification matrix $\widetilde{M}(r)$ is given by

$$\widetilde{M}(r) = D(r)V + \theta_1(r)\psi^T,$$

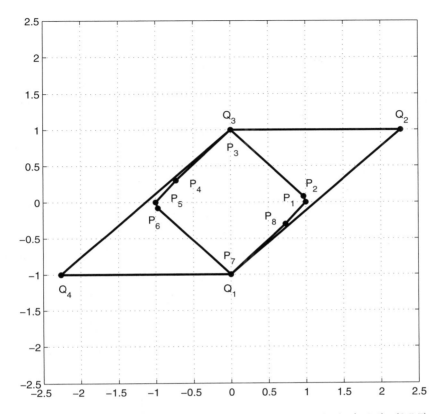

Figure 8.6.1   Unit balls in the polytope norms for methods (8.6.4), (8.5.7) and (8.6.4), (8.6.5)

which for the method (8.6.4) with $z^{[n]}$ defined by (8.6.5) takes the form

$$
\widetilde{M}(r) = \begin{bmatrix} 0 & -\dfrac{1177 r^3}{2976} \\[2ex] 0 & \dfrac{1177 r^2 (r-1)}{2976} \end{bmatrix}.
$$

It follows that the condition (8.6.3) is satisfied for $r \in [0, r^*]$, $r^* \approx 1.7895$ for the polytope norm whose unit ball with vertices Q_1, Q_2, Q_3, and Q_4 is also plotted in Fig. 8.6.1. The coordinates of the points Q_i are

$$
Q_1 = -Q_3 = [0, -1]^T, \quad Q_2 = -Q_4 = [2.2664, 1]^T.
$$

We can observe that for the specific method (8.6.4), the rescale and modify approach of Butcher and Jackiewicz [71] leads to better zero-stability properties than does the approach described in this section. However, this happens at the expense of losing the ability of estimating the higher order terms of the form

$$h_n^{p+2} y^{(p+2)}(t_n) \quad \text{and} \quad h_n^{p+2} \frac{\partial f}{\partial y}(y(t_n)) y^{(p+1)}(t_n).$$

The construction of highly stable methods (possibly unconditionally stable) which also allow for the estimation of terms of order $p + 2$ is the subject of current work.

We consider next the implicit A- and L-stable method with simple coefficients derived by Wright [293]. For this method $c = [0, \frac{1}{2}, 1]^T$, $\lambda = \frac{1}{4}$, and $\epsilon = 0$. The coefficients of this method are given by

$$
\left[
\begin{array}{c|cc}
A & e & U \\
\hline
b^T & 1 & v^T \\
\hline
B & 0 & V
\end{array}
\right]
=
\left[
\begin{array}{ccc|ccc}
\frac{1}{4} & 0 & 0 & 1 & -\frac{1}{4} & 0 \\
\frac{1}{4} & \frac{1}{4} & 0 & 1 & 0 & 0 \\
\frac{1}{2} & \frac{1}{4} & \frac{1}{4} & 1 & 0 & \frac{1}{8} \\
\frac{1}{2} & -\frac{1}{8} & \frac{1}{2} & 1 & \frac{1}{8} & \frac{1}{16} \\
\frac{1}{2} & -\frac{1}{2} & 1 & 0 & 0 & \frac{1}{4} \\
0 & -2 & 2 & 0 & 0 & 0
\end{array}
\right].
\tag{8.6.6}
$$

It can be verified using (8.3.3), (8.3.7), (8.3.8), and (8.3.9) that the error constant is $E = -\frac{7}{192}$ and the vectors β, γ, and δ are

$$
\beta = \left[\begin{array}{c} \frac{1}{4} \\ \frac{1}{8} \end{array} \right], \quad
\gamma = \left[\begin{array}{c} -\frac{1}{8} \\ -\frac{1}{24} \end{array} \right], \quad
\delta = \left[\begin{array}{c} \frac{3}{32} \\ \frac{5}{48} \end{array} \right].
$$

We again choose $\psi = [0, -v_{11}/\beta_1]^T$. Solving (8.4.3), (8.4.10), and (8.4.11), we also obtain

$$
\varphi = \left[\begin{array}{c} -4 \\ 4 \\ 0 \end{array} \right], \quad
\widetilde{\varphi} = \left[\begin{array}{c} \frac{160}{3} \\ -\frac{184}{3} \\ \frac{56}{3} \end{array} \right], \quad
\widehat{\varphi} = \left[\begin{array}{c} \frac{224}{3} \\ -\frac{224}{3} \\ \frac{64}{3} \end{array} \right]
$$

and

$$
\psi = \left[\begin{array}{c} 0 \\ -2 \end{array} \right], \quad
\widetilde{\psi} = \left[\begin{array}{c} -\frac{32}{3} \\ 12 \end{array} \right], \quad
\widehat{\psi} = \left[\begin{array}{c} -\frac{64}{3} \\ 16 \end{array} \right].
$$

The amplification matrix $M(r)$ given by formula (8.6.2) takes the form

$$
M(r) = \left[
\begin{array}{cc}
-\frac{r(r-1)(r^2-8r-8)}{9} & \frac{r(7r^3-12r^2+8)}{12} \\
\frac{2r^2(r-1)(8r+7)}{9} & -\frac{r^2(r-1)(7r+19)}{6}
\end{array}
\right].
$$

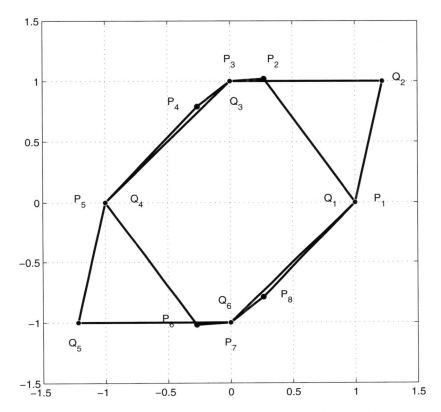

Figure 8.6.2   Unit balls in the polytope norms for methods (8.6.6), (8.5.7) and (8.6.6), (8.6.5)

Applying the procedure described above, it can be verified that the condition (8.6.3) is satisfied for $r \in [0, r^*]$, $r^* \approx 1.13614$, for the polytope norm $\| \cdot \|_*$ whose unit ball with vertices P_1, P_2, P_3, P_4, P_5, P_6, P_7, and P_8 is plotted in Fig. 8.6.2. The coordinates of the points P_i are

$$P_1 = -P_5 = [1, 0]^T, \quad P_2 = -P_6 = [0.2715, 1.0188]^T,$$

$$P_3 = -P_7 = [0, 1]^T, \quad P_4 = -P_8 = [-0.2628, 0.7894]^T.$$

For comparison we have also analyzed zero-stability properties of method (8.6.6) with the vector $z^{[n]}$ of external approximations given by (8.6.5), where again $\theta_1(r)$ is defined by (8.5.8) and $\delta_1(t_n)$ is defined by (8.5.3). It can be verified that for the method (8.1.3), (8.6.5) the amplification matrix given by $\widetilde{M}(r) = D(r)V + \theta_1(r)\psi^T$ for the method (8.6.6) with $z^{[n]}$ defined by (8.6.5) takes the form

$$\widetilde{M}(r) = \begin{bmatrix} 0 & \frac{r^3}{4} \\ 0 & \frac{r^2(r-1)}{2} \end{bmatrix}.$$

It follows that condition $(8.6.3)$ is satisfied for $r \in [0, r^*]$, $r^* \approx 1.69562$ for the polytope norm whose unit ball with vertices Q_1, Q_2, Q_3, Q_4, Q_5, and Q_6 is also plotted in Fig. 8.6.2. The coordinates of the points Q_i are

$$Q_1 = -Q_4 = [0, -1]^T, \quad Q_2 = -Q_5 = [1.2188, 1]^T, \quad Q_3 = -Q_6 = [0, 1]^T.$$

As before, we can observe that for the specific method $(8.6.6)$ the rescale and modify approach of Butcher and Jackiewicz [71] leads to better zero-stability properties than does the approach described in this section.

The refinement of the technique of Guglielmi and Zennaro [138, 139] was proposed by Butcher and Heard [62] in the context of backward differentiation methods for ODEs.

8.7   TESTING THE RELIABILITY OF ERROR ESTIMATION AND ESTIMATION OF HIGHER ORDER TERMS

In this section we test experimentally the reliability of the estimates of

$$h_n^{p+1} y^{(p+1)}(t_n), \quad h_n^{p+2} y^{(p+2)}(t_n), \quad \text{and} \quad h_n^{p+2}\frac{\partial f}{\partial y}(y(t_n)) y^{(p+1)}(t_n)$$

derived in Section 8.4 in a variable step size environment. These experiments were performed for explicit GLM $(8.6.4)$ derived by Butcher and Jackiewicz [72] and given in Section 7.9. This method was applied to the van der Pohl equation

$$y_1' = y_2, \qquad\qquad y_1(0) = 2,$$

$$y_2' = (1 - y_1^2)y_2 - y_1, \quad y_2(0) = 0,$$

$t \in [0, 8]$, with the standard step size changing strategy based on the formula

$$h_{n+1} = \left(\frac{0.5\,\text{Tol}}{\|\text{est}(p, t_n)\|} \right)^{1/(p+1)} h_n,$$

$p = 2$, without any limiters or exceptions (compare, e.g., [143, 257, 261]). Here Tol is a given accuracy tolerance and $\text{est}(p, t_n)$ is the estimate of the local discretization error given by

$$\text{est}(p, t_n) = E\,\delta_1(t_n),$$

where E is the error constant of the method and $\delta_1(t_n)$ is defined by $(8.5.3)$. We have plotted in the top graph of Fig. 8.7.1 $\|\text{le}(p, t_n)\|$, and $\|\text{est}(p, t_n)\|$,

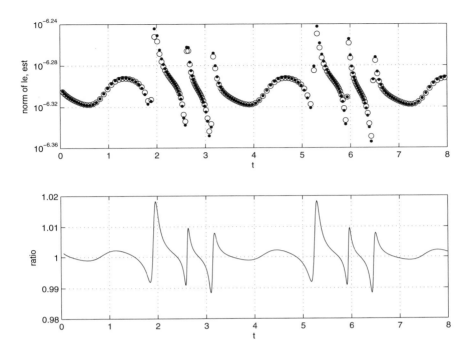

Figure 8.7.1   Top graph: the quantities $\|\text{le}(p, t_n)\|$ (symbol "o") and $\|\text{est}(p, t_n)\|$ (symbol "·") versus t for Tol $= 10^{-6}$; every fifth point is plotted. Bottom graph: ratio(p, t_n) versus t

where $\text{le}(p, t_n)$ is the local error of the method. This local error was computed using the code `ode45` from the Matlab ODE suite [263] on the interval $[t_{n-1}, t_n]$ with RelTol $=$ AbsTol $= 0.01 \cdot$ Tol following a suggestion by Hull et al. [168]. We have also plotted in the bottom graph of Fig. 8.7.1 the ratio(p, t_n) defined by

$$\text{ratio}(p, t_n) = \frac{\|\text{est}(p, t_n)\|}{\|\text{le}(p, t_n)\|}.$$

We have plotted in the top graph of Fig. 8.7.2 $h_n^{p+2}\|y^{(p+2)}(t_n)\|$ and $\|\delta_2(t_n)\|$, where $\delta_2(t_n)$ is given by (8.5.4). We have plotted in the bottom graph of Fig. 8.7.2 the ratio of these quantities. Similarly, we have plotted in the top graph of Fig. 8.7.3 $h_n^{p+2}\|\frac{\partial f}{\partial y}(y(t_n))y^{(p+1)}(t_n)\|$ and $\|\delta_3(t_n)\|$, where $\delta_3(t_n)$ is defined by (8.5.5). We have plotted in the bottom graph of Fig. 8.7.3 the ratio of these quantities. The quantities

$$h_n^{p+2}y^{(p+2)}(t_n) \quad \text{and} \quad h_n^{p+2}\frac{\partial f}{\partial y}(y(t_n))y^{(p+1)}(t_n)$$

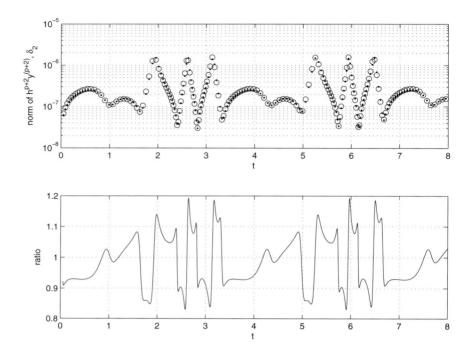

Figure 8.7.2   Top graph: the quantities $h_n^{p+2}\|y^{(p+2)}(t_n)\|$ (symbol "○") and $\|\delta_2(t_n)\|$ (symbol "·") versus t for Tol $= 10^{-6}$; every fifth point is plotted. Bottom graph: the ratio of these quantities versus t

were computed using the systems

$$y_1'' = y_2', \quad y_2'' = -2y_1 y_1' y_2 + (1 - y_1^2)y_2' - y_1',$$

$$y_1''' = y_2'', \quad y_2''' = -2y_1'^2 y_2 - 2y_1 y_1'' y_2 - 4y_1 y_1' y_2' + (1 - y_1^2)y_2'' - y_1'',$$

and

$$y_1^{(4)} = y_2''',$$

$$y_2^{(4)} = -6y_1' y_1'' y_2 - 6y_1'^2 y_2' - 2y_1 y_1''' y_2 - 6y_1 y_1'' y_2'$$

$$- 6y_1 y_1' y_2'' + (1 - y_1^2)y_2''' - y_1'''$$

obtained by successive differentiation of the van der Pohl equation (1.8.1) and the relation

$$\frac{\partial f}{\partial y}(y) = \begin{bmatrix} \frac{\partial f_1}{\partial y_1} & \frac{\partial f_1}{\partial y_2} \\ \frac{\partial f_2}{\partial y_1} & \frac{\partial f_2}{\partial y_2} \end{bmatrix} = \begin{bmatrix} 0 & 1 \\ -2y_1 y_2 - 1 & 1 - y_1^2 \end{bmatrix}.$$

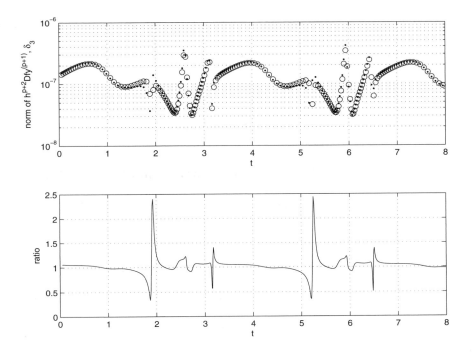

Figure 8.7.3   Top graph: the quantities $h_n^{p+2}\|\frac{\partial f}{\partial y}(y(t_n))y^{(p+1)}(t_n)\|$ (symbol "$\circ$") and $\|\delta_3(t_n)\|$ (symbol "$\cdot$") versus t for $\text{Tol} = 10^{-6}$; every fifth point is plotted. Bottom graph: the ratio of these quantities versus t

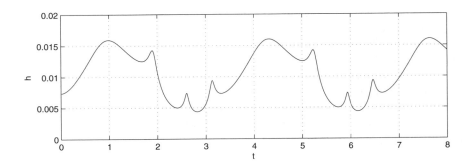

Figure 8.7.4   Step size pattern for the method of order $p = 2$ applied to the van der Pohl equation with adaptive step size control for $\text{Tol} = 10^{-6}$

All these figures correspond to the step size pattern displayed in Fig. 8.7.4.

Examining these figures we can observe that the quality of the estimation of the local discretization error for the method of order $p = 2$ is exceptionally good. The quality of the estimators for the higher order terms of the form

$$h_n^{p+2} y^{(p+2)}(t_n) \quad \text{and} \quad h_n^{p+2} \frac{\partial f}{\partial y}\big(y(t_n)\big) y^{(p+1)}(t_n)$$

is not as high as for the term $h_n^{p+1} y^{(p+1)}(t_n)$, but is still very good.

8.8   UNCONDITIONAL STABILITY ON NONUNIFORM MESHES

This section follows the presentation by Butcher and Jackiewicz [73]. As observed in Section 8.6, for GLMs (8.1.3) with the input vector $z^{[n]}$ for the next step from t_n to t_{n+1} defined by

$$z^{[n]} = \big(D(r_n) \otimes I\big)\overline{z}^{[n]} + \big(\theta(r_n) \otimes I\big)\delta(t_n),$$

where $\theta(r) = \theta_1(r)$ is defined by (8.5.8) and $\delta(t) = \delta_1(t)$ is defined by (8.5.3), that is,

$$\delta(t_n) = (\varphi^T \otimes I) h_n F(Y^{[n]}) + (\psi^T \otimes I) z^{[n-1]},$$

the requirement of zero-stability usually leads to some restrictions on the ratio of step sizes $r_n = h_{n+1}/h_n$. Similarly, some restrictions on r_n usually also hold if $\delta(t) = \overline{\delta}(t)$, where $\overline{\delta}(t)$ is defined in terms of $F(Y^{[n]})$ and $\overline{z}^{[n]}$ by the formula

$$\overline{\delta}(t_n) = (\overline{\varphi}^T \otimes I) h_n F(Y^{[n]}) + (\overline{\psi}^T \otimes I) \overline{z}^{[n]},$$

with $\overline{\varphi}$ and $\overline{\psi}$ satisfying system (8.4.6). In this section we adopt a different approach and try to construct GLMs with reliable error estimates and with an appropriately defined input vector $z^{[n]}$ for the next step from t_n to t_{n+1}, so that the overall numerical algorithm consisting of (8.1.3) and the formula for $z^{[n]}$ is unconditionally stable for any step size pattern. This approach is based on constructing the vector of external approximations $z^{[n]}$ by making different corrections to the components of the vector $\overline{z}^{[n]}$ in such a way that the errors in these components are asymptotically constant. To be more specific, the components $z_i^{[n]}$ of $z^{[n]}$ are defined by

$$z_i^{[n]} = r_n^i \overline{z}_i^{[n]} + (r_n^i - r_n^{p+1})\beta_i \delta_i(t_n),$$

$i = 1, 2, \ldots, p$, where $\delta_i(t_n)$ is the family of the estimators to $h_n^{p+1} y^{(p+1)}(t_n)$, and β_i are the components of the vector β. This relation can be written in the vector form

$$z^{[n]} = (D(r_n) \otimes I)\overline{z}^{[n]} + \Big((D(r_n) - r_n^{p+1} I) D_\beta \otimes I\Big)\Delta(t_n), \tag{8.8.1}$$

where

$$D_\beta = \text{diag}(\beta_1, \ldots, \beta_p) \quad \text{and} \quad \Delta(t_n) = \begin{bmatrix} \delta_1(t_n) \\ \vdots \\ \delta_p(t_n) \end{bmatrix}.$$

Similarly to (8.4.2) and (8.4.5), we will look for $\delta_i(t_n)$ in the form

$$\delta_i(t_n) = (\varphi_i^T \otimes I)h_n F(Y^{[n]}) + (\psi_i^T \otimes I)z^{[n-1]} \qquad (8.8.2)$$

or

$$\delta_i(t_n) = (\overline{\varphi}_i^T \otimes I)h_n F(Y^{[n]}) + (\overline{\psi}_i^T \otimes I)\overline{z}^{[n]}, \qquad (8.8.3)$$

where $\varphi_i, \overline{\varphi}_i \in \mathbb{R}^{p+1}$ and $\psi_i, \overline{\psi}_i \in \mathbb{R}^p$. However, in contrast to the algorithm for the computation of $z^{[n]}$ described in Section 8.5 (compare also [71]), it will be possible to choose ψ_i or $\overline{\psi}_i$ to accomplish unconditional stability. This is explained in the remainder of this section.

It can be verified that the application of (8.1.3) and (8.8.1) to the test equation $y' = 0$, $t \geq 0$, $y(0) = 1$, leads to the relation

$$z^{[n]} = \left(D(r_n)V + (D(r_n) - r_n^{p+1}I)D_\beta\Psi\right)z^{[n-1]}$$

if the $\delta_i(t_n)$ are defined by (8.8.2) or

$$z^{[n]} = \left(D(r_n) + (D(r_n) - r_n^{p+1}I)D_\beta\overline{\Psi}\right)Vz^{[n-1]}$$

if the $\delta_i(t_n)$ are defined by (8.8.3). Here

$$\Psi = \begin{bmatrix} \psi_1^T \\ \vdots \\ \psi_p^T \end{bmatrix} \quad \text{and} \quad \overline{\Psi} = \begin{bmatrix} \overline{\psi}_1^T \\ \vdots \\ \overline{\psi}_p^T \end{bmatrix}.$$

The unconditional stability would then follow if we could choose a scalar function $\xi(r)$ and the matrices Ψ or $\overline{\Psi}$ such that

$$D(r_n)V + \left(D(r_n) - r_n^{p+1}I\right)D_\beta\Psi = \xi(r_n)V$$

or

$$\left(D(r_n) + (D(r_n) - r_n^{p+1}I)D_\beta\overline{\Psi}\right)V = \xi(r_n)V.$$

This can be accomplished clearly if we define $\xi(r) = r^{p+1}$ and if Ψ or $\overline{\Psi}$ satisfies the equation

$$D_\beta \Psi = -V \qquad (8.8.4)$$

or

$$D_\beta \overline{\Psi}V = -V. \qquad (8.8.5)$$

Assuming that D_β is nonsingular, this leads to

$$\Psi = -D_\beta^{-1}V \tag{8.8.6}$$

or

$$\overline{\Psi} = -D_\beta^{-1}. \tag{8.8.7}$$

Observe that (8.8.4) and (8.8.5) may still have solutions even if the matrix D_β is singular. Once the matrices Ψ or $\overline{\Psi}$ are chosen, the vector φ_i or $\overline{\varphi}_i$ can be determined so that

$$\delta_i(t_n) = h_n^{p+1} y^{(p+1)}(t_n) + O(h_n^{p+2}),$$

$i = 1, 2, \ldots, p$.

We now describe the construction of the family of error estimators $\delta_i(t_n)$, $i = 1, 2, \ldots, p$. We can argue as in Section 8.4 that $h_n^{p+1} y^{(p+1)}(t_n)$ can be estimated by (8.8.2) or (8.8.3) if the vectors φ_i, ψ_i, or $\overline{\varphi}_i$, $\overline{\psi}_i$, $i = 1, 2, \ldots, p$, satisfy systems of linear equations of the form (8.4.3) or (8.4.6). Introducing the matrices $\mathbf{C}_p$ and $\overline{\mathbf{C}}_p$ of dimension $(p + 1) \times p$ defined by

$$\mathbf{C}_p = \left[\begin{array}{ccccc} e & c & \dfrac{c^2}{2!} & \cdots & \dfrac{c^{p-1}}{(p-1)!} \end{array} \right]^T,$$

$$\overline{\mathbf{C}}_p = \left[\begin{array}{ccccc} e & c-e & \dfrac{(c-e)^2}{2!} & \cdots & \dfrac{(c-e)^{p-1}}{(p-1)!} \end{array} \right]^T,$$

the linear systems corresponding to (8.4.3) and (8.4.6) can be written in a more compact form as

$$\varphi_i^T \mathbf{C}_p + \psi_i^T = 0, \qquad\qquad \overline{\varphi}_i^T \overline{\mathbf{C}}_p + \overline{\psi}_i^T = 0,$$

$$\varphi_i^T \dfrac{c^p}{p!} - \psi_i^T \beta = 1, \qquad \text{or} \qquad \overline{\varphi}_i^T \dfrac{(c-e)^p}{p!} - \overline{\psi}_i^T \beta = 1,$$

$i = 1, 2, \ldots, p$. Setting

$$\Phi = \left[\begin{array}{c} \varphi_1^T \\ \vdots \\ \varphi_p^T \end{array} \right], \qquad \overline{\Phi} = \left[\begin{array}{c} \overline{\varphi}_1^T \\ \vdots \\ \overline{\varphi}_p^T \end{array} \right],$$

the systems above can be written in the vector form

$$\Phi\, \mathbf{C}_{p+1} + \left[\begin{array}{c|c} \Psi & -\Psi\beta \end{array} \right] = \left[\begin{array}{c|c} 0 & e \end{array} \right] \tag{8.8.8}$$

or

$$\overline{\Phi}\, \overline{\mathbf{C}}_{p+1} + \left[\begin{array}{c|c} \overline{\Psi} & -\overline{\Psi}\beta \end{array} \right] = \left[\begin{array}{c|c} 0 & e \end{array} \right], \tag{8.8.9}$$

where $e \in \mathbb{R}^p$. Solving (8.8.8) or (8.8.9) for Φ or $\overline{\Phi}$, we obtain

$$\Phi = \left[\begin{array}{c|c} -\Psi & \Psi\alpha + e \end{array} \right] \mathbf{C}_{p+1}^{-1} \tag{8.8.10}$$

or

$$\overline{\Phi} = \left[\begin{array}{c|c} -\overline{\Psi} & \overline{\Psi}\alpha + e \end{array} \right] \overline{\mathbf{C}}_{p+1}^{-1}. \tag{8.8.11}$$

It was already demonstrated in this section that if Ψ or $\overline{\Psi}$ satisfies (8.8.4) or (8.8.5), then GLM (8.1.3) combined with (8.8.1) is unconditionally stable for any step size pattern. Moreover, assuming that the matrix D_β is nonsingular, this leads to formulas (8.8.6) and (8.8.7). Substituting these into (8.8.10) or (8.8.11), respectively, and taking into account that $D_\beta^{-1}\beta = e$, we obtain closed-form expressions for matrix Φ or $\overline{\Phi}$:

$$\Phi = \left[\begin{array}{c|c} D_\beta^{-1} V & D_\beta^{-1} V\beta + e \end{array} \right] \mathbf{C}_{p+1}^{-1}$$

or

$$\overline{\Phi} = \left[\begin{array}{c|c} D_\beta^{-1} & 0 \end{array} \right] \overline{\mathbf{C}}_{p+1}^{-1},$$

whose rows represent the vectors φ_i^T or $\overline{\varphi}_i^T$, $i = 1, 2, \ldots, p$, appearing in (8.8.2) or (8.8.3).

Summing up the discussion above, GLM (8.1.3) combined with (8.8.1) is unconditionally stable if $\delta_i(t_n)$, $i = 1, 2, \ldots, p$, are defined by (8.8.2) or (8.8.3), that is, if $\Delta(t_n)$ is given by

$$\Delta(t_n) = (\Phi \otimes I)h_n F(Y^{[n]}) + (\Psi \otimes I)z^{[n-1]}$$

or

$$\Delta(t_n) = (\overline{\Phi} \otimes I)h_n F(Y^{[n]}) + (\overline{\Psi} \otimes I)\overline{z}^{[n]}.$$

The local discretization error

$$\mathrm{le}(t_n) = E\, h_n^{p+1} y^{(p+1)}(t_n) + O(h_n^{p+2})$$

of method (8.1.3) can be estimated by the quantities

$$\mathrm{est}(t_n) = E\, \delta_i(t_n), \quad i = 1, 2, \ldots, p,$$

where E is the error constant of the method. These estimates are asymptotically correct as $h \to 0$. However, this is not sufficient for implicit methods for stiff systems, whose efficient implementation requires the construction of error estimates that are also accurate and reliable for "large" step sizes compared with certain characteristics of the problem. This is a difficult problem which is not discussed here since its solution requires techniques that differ from those employed in this section. This problem was discussed in the context of DIMSIMs in Section 4.3 (compare also [177]), where we used the approach by Shampine and Baca [258]. This approach is based on the design of the "filtered" error estimates of the form

$$\text{est}^*(t_n) = (I - h_n \lambda J)^{-\mu} \text{est}(t_n),$$

where λ is the diagonal element of the coefficient matrix A, J is the Jacobian of problem (2.1.1), and μ is an appropriate integer. Another approach to error estimation for implicit RK methods for stiff systems has been proposed by Swart and Söderlind [268].

It can be verified that for method (8.6.4), the matrices Φ, Ψ, $\overline{\Phi}$, and $\overline{\Psi}$ are given by

$$\Phi = \begin{bmatrix} \varphi_1^T \\ \varphi_2^T \end{bmatrix} = \begin{bmatrix} \frac{113815}{5208} & -\frac{85567}{2604} & \frac{1849}{168} \\ 4 & -8 & 4 \end{bmatrix}, \quad \Psi = \begin{bmatrix} \psi_1^T \\ \psi_2^T \end{bmatrix} = \begin{bmatrix} 0 & \frac{1177}{217} \\ 0 & 0 \end{bmatrix}$$

and

$$\overline{\Phi} = \begin{bmatrix} \overline{\varphi}_1^T \\ \overline{\varphi}_2^T \end{bmatrix} = \begin{bmatrix} 0 & 0 & \frac{96}{7} \\ \frac{96}{7} & -\frac{384}{7} & \frac{288}{7} \end{bmatrix}, \quad \overline{\Psi} = \begin{bmatrix} \overline{\psi}_1^T \\ \overline{\psi}_2^T \end{bmatrix} = \begin{bmatrix} -\frac{96}{7} & 0 \\ 0 & -\frac{96}{7} \end{bmatrix}.$$

Similarly, for method (8.6.6) these matrices take the form

$$\Phi = \begin{bmatrix} \varphi_1^T \\ \varphi_2^T \end{bmatrix} = \begin{bmatrix} -4 & 4 & 0 \\ 4 & -8 & 4 \end{bmatrix}, \quad \Psi = \begin{bmatrix} \psi_1^T \\ \psi_2^T \end{bmatrix} = \begin{bmatrix} 0 & -2 \\ 0 & 0 \end{bmatrix},$$

and

$$\overline{\Phi} = \begin{bmatrix} \overline{\varphi}_1^T \\ \overline{\varphi}_2^T \end{bmatrix} = \begin{bmatrix} 0 & 0 & 8 \\ 4 & -16 & 12 \end{bmatrix}, \quad \overline{\Psi} = \begin{bmatrix} \overline{\psi}_1^T \\ \overline{\psi}_2^T \end{bmatrix} = \begin{bmatrix} -8 & 0 \\ 0 & -4 \end{bmatrix}.$$

Observe that for both methods

$$\Phi = \overline{\Phi} + \overline{\Psi} B \quad \text{and} \quad \Psi = \overline{\Psi} V.$$

We would like to stress again that the overall numerical algorithms consisting of (8.6.4) or (8.6.6) combined with the formula for $z^{[n]}$ defined by (8.8.1) with $\delta_i(t_n)$, $i = 1, 2$, given by (8.8.2) or (8.8.3) are unconditionally zero-stable for any step size pattern.

8.9   NUMERICAL EXPERIMENTS

Methods (8.6.4) and (8.6.6) implemented as described in Section 8.8 are unconditionally stable for arbitrary nonuniform meshes. This property also has a desirable effect on the quality of local error estimation. To illustrate this point, these methods with $z^{[n]}$ defined by (8.8.1) were applied first to the well-known variant of the Prothero-Robinson test problem

$$y'(t) = \lambda(y - e^{\mu t}) + \mu e^{\mu t}, \quad t \in [t_0, T],$$

$$y(t_0) = y_0 = 1,$$

(8.9.1)

with $\lambda = -\mu = -0.1$, $t_0 = 0$, $T = 20$, which was used by Butcher and Jackiewicz [70, 71] in the context of a more general class of Nordsieck methods.

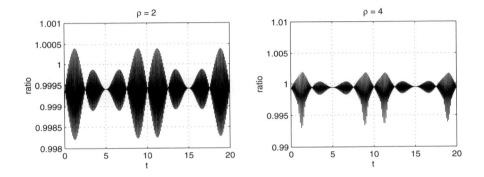

Figure 8.9.1   The ratio$(t_n) := |\mathrm{le}(t_n)|/|E\delta_2(t_n)|$ versus t for the explicit method (8.6.4) with vector of external approximations $z^{[n]}$ defined by (8.8.1) applied to (8.9.1) for $\delta_2(t_n) = \varphi_2^T h_n F(Y^{[n]}) + \psi_2^T \overline{y}^{[n]}$

To estimate the local discretization error $\mathrm{le}(t_n)$, we can choose $\delta_1(t_n)$ defined in terms of φ_1, ψ_1 or $\overline{\varphi}_1$, $\overline{\psi}_1$, or $\delta_2(t_n)$ defined in terms of φ_2, ψ_2 or $\overline{\varphi}_2$, $\overline{\psi}_2$. These vectors are given in Section 8.8. We have verified numerically that the estimates $\delta_2(t_n)$ are somewhat more accurate than those based on $\delta_1(t_n)$. These methods were implemented for a quite demanding, periodic step size pattern chosen according to the formula

$$h_{n+1} = \rho^{(-1)^k \sin(8\pi t_n/(X - x_0)) \cos(2\pi t_n/(T - t_0))} h_n,$$

$$h_0 = (T - t_0)/N, \quad N = 800,$$

(8.9.2)

$n = 0, 1, \ldots, N - 1$, where

$$k = \begin{cases} 1 & \text{for} \quad n = 0 \text{ or } 1 \mod(4), \\ 2 & \text{for} \quad n = 2 \text{ or } 3 \mod(4), \end{cases}$$

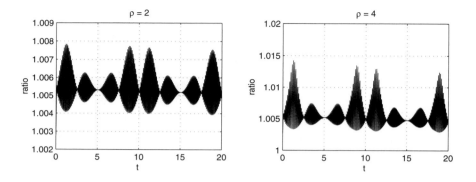

Figure 8.9.2 The ratio$(t_n) := |\mathrm{le}(t_n)|/|E\delta_2(t_n)|$ versus t for the implicit method (8.6.6) with vector of external approximations $z^{[n]}$ defined by (8.8.1) applied to (8.9.1) for $\delta_2(t_n) = \varphi_2^T h_n F(Y^{[n]}) + \psi_2^T \overline{y}^{[n]}$

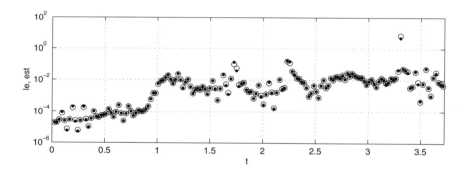

Figure 8.9.3 The norm of the local discretization error $\|\mathrm{le}(t_n)\|$ (symbol "o") and the norm of the local error estimate $\|E\,\delta_2(t_n)\|$ (symbol "·") versus x for method (8.6.6) with $z^{[n]}$ defined by (8.8.1) applied to the problem ROPE with a step size pattern defined by (8.9.2) for $\rho = 2$

and $\rho = 2$ or $\rho = 4$. This choice of k ensures that the step size is increased twice successively, then decreased twice over the interval of integration, and that the step size oscillates around an approximately constant value. In Fig. 8.9.1 we have plotted the ratio between the norm of the exact local error $|\mathrm{le}(t_n)|$ and the norm of the local error estimates $|E\delta_2(t_n)|$ with the φ_2 and ψ_2 given in Section 8.8 for method (8.6.4). For this figure the ranges of $|\mathrm{le}(t_n)|$ are

$$1.894 \cdot 10^{-10} \le |\mathrm{le}(t_n)| \le 4.683 \cdot 10^{-8}, \quad 2.747 \cdot 10^{-11} \le |\mathrm{le}(t_n)| \le 2.371 \cdot 10^{-7}$$

for $\rho = 2$ and $\rho = 4$, respectively. In Fig. 8.9.2 we present the corresponding results for the method (8.6.6). For this figure the ranges of $|\text{le}(t_n)|$ are

$$9.504 \cdot 10^{-11} \leq |\text{le}(t_n)| \leq 2.360 \cdot 10^{-8}, \quad 1.376 \cdot 10^{-11} \leq |\text{le}(t_n)| \leq 1.199 \cdot 10^{-7}$$

for $\rho = 2$ and $\rho = 4$, respectively. For both figures the ranges of h_n are

$$1.272 \cdot 10^{-2} \leq h_n \leq 4.778 \cdot 10^{-2}, \quad 6.056 \cdot 10^{-3} \leq h_n \leq 9.068 \cdot 10^{-2}$$

for $\rho = 2$ and $\rho = 4$, respectively.

These figures confirm a very high quality of the error estimators $\delta_2(t_n)$. For explicit methods with $\rho = 2$, the estimator differs by less than 0.2% from the local discretization error. For more rapidly changing step sizes, corresponding to $\rho = 4$, the difference can be as high as 1%, but this can be judged to be satisfactory from the point of view of a possible step size control algorithm. For a similar experiment performed for implicit methods with $\rho = 2$, the estimator understates the local truncation error by less than 1%. Even for $\rho = 4$, the understatement is less than 2%. It can also be verified that the quality of local error estimation is much higher than that previously reported [70, 71] for the class of Nordsieck methods with $p = q = r - 1 = s$.

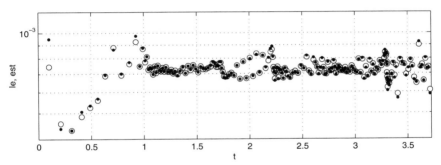

Figure 8.9.4   The norm of the local discretization error $\|\text{le}(t_n)\|$ (symbol "o") and the norm of the local error estimate $\|E\,\delta_2(t_n)\|$ (symbol ".") versus t for method (8.8.4) applied to the problem ROPE with adaptive step size control for Tol $= 10^{-3}$

In Fig. 8.9.3 we present the results of numerical experiments for the same step size pattern with $\rho = 2$ for method (8.6.6) applied to the problem ROPE defined in Section 1.2. On this figure we have plotted at every fifth step the norm of the local discretization error $\|\text{le}(t_n)\|$ using the symbol "o" and the norm of the local error estimate $\|E\,\delta_2(x_n)\|$ using the symbol "." versus x. In Fig. 8.9.4 we present $\|\text{le}(t_n)\|$ and $\|E\,\delta_2(t_n)\|$ versus x, and in Fig. 8.9.5 we present the step size pattern for the ROPE problem, where the step size is chosen adaptively according to the formula

$$h_{n+1} = 0.9 \left(\frac{\text{Tol}}{\|\text{est}(t_n)\|} \right)^{1/3} h_n,$$

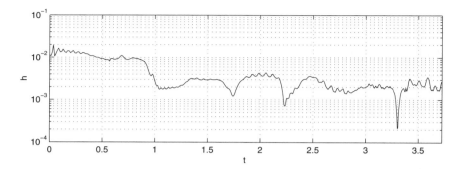

Figure 8.9.5   Step size pattern for method (8.6.4) applied to the problem ROPE with adaptive step size control for Tol $= 10^{-3}$

without limiters or exceptions. Here Tol is a given error tolerance. The results on these figures correspond to Tol $= 10^{-3}$, where in Fig. 8.9.4 we have plotted every eighth step. We again observe very good agreement between local errors and local error estimates.

As discussed in Section 8.8, error estimation and control for stiff differential systems require the construction of error estimates which are not only asymptotically correct as $h \to 0$, but which are also accurate and reliable for "large" step sizes. The construction of such error estimators for DIMSIMs is discussed in Section 4.3. The construction of such estimators for GLMs with IRKS is the topic of current work.

8.10   LOCAL ERROR ESTIMATION FOR STIFFLY ACCURATE METHODS

In this section we describe the approach proposed by Butcher and Podhaisky [77] to the estimation of $h_n^{p+1} y^{(p+1)}(t_n)$ and $h_n^{p+2} y^{(p+2)}(t_n)$ for strictly stiffly accurate GLMs with IRKS discussed in Section 7.12.

It follows from the localizing assumptions (8.3.1) discussed in Section 8.3 that $Y^{[n]}$ and $h_n F(Y^{[n]})$ satisfy (8.3.11) and (8.3.12), which, for convenience, are reproduced here:

$$Y^{[n]} = y(t_{n-1} + ch_n) - (\xi \otimes I)\, h_n^{p+1} y^{(p+1)}(t_{n-1}) + O(h_n^{p+2}), \qquad (8.10.1)$$

$$h_n F(Y^{[n]}) \;=\; h_n y'(t_{n-1} + ch_n)$$

$$- \;(\xi \otimes I)\, h_n^{p+2} \frac{\partial f}{\partial y}\big(y(t_{n-1})\big) y^{(p+1)}(t_{n-1}) \qquad (8.10.2)$$

$$+ \; O(h_n^{p+3}).$$

Here the vector ξ depends on the method and does not change from step to step. This vector is given by formula (8.3.6) with the vector β given by (8.3.7). We also recall that strictly stiffly accurate GLMs with IRKS satisfy conditions (7.12.6) and (7.12.7) and that $c_{p+1} = 1$. It can be verified that these conditions imply that the first components β_1, γ_1, and δ_1 of the vectors β, γ, and δ defined by (8.3.7), (8.3.8), and (8.3.9) are equal to zero. Hence, it follows from (8.3.1) that the second component of the vector $\mathbf{y}^{[n-1]}$ or the first component of the vector $z^{[n-1]}$ satisfies the relation

$$z_1^{[n-1]} = h_n y'(t_{n-1}) + O(h_n^{p+3}). \tag{8.10.3}$$

Following Butcher and Podhaisky [77], we consider the estimation of scaled derivatives $h_n^{p+1} y^{(p+1)}(t_n)$ and $h_n^{p+2} y^{(p+2)}(t_n)$, or the quantities related to these, using linear combinations of the form

$$\eta(t_n) = \varphi_0 z_1^{[n-1]} + (\varphi^T \otimes I) h_n F(Y^{[n]}), \tag{8.10.4}$$

where $\varphi_0 \in \mathbb{R}$ and $\varphi \in \mathbb{R}^{p+1}$. Substituting (8.10.2) into (8.10.4) and expanding $y'(t_{n-1} + c h_n)$ into a Taylor series around t_{n-1}, we obtain

$$\begin{aligned}
\eta(t_n) &= \big(\varphi_0 + (\varphi^T \otimes I)(e \otimes I)\big) h_n y'(t_{n-1}) \\
&\quad + (\varphi^T \otimes I) \sum_{j=1}^{p+1} \frac{c^j \otimes I}{j!} h_n^{j+1} y^{(j+1)}(t_{n-1}) \\
&\quad - (\varphi^T \otimes I)(\xi \otimes I) h_n^{p+2} \frac{\partial f}{\partial y}(y(t_{n-1})) y^{(p+1)}(t_{n-1}) + O(h_n^{p+3})
\end{aligned}$$

or

$$\begin{aligned}
\eta(t_n) &= (\varphi_0 + \varphi^T e) h_n y'(t_{n-1}) + \sum_{j=1}^{p+1} \frac{\varphi^T c^j}{j!} h_n^{j+1} y^{(j+1)}(t_{n-1}) \\
&\quad - \varphi^T \xi h_n^{p+2} \frac{\partial f}{\partial y}(y(t_{n-1})) y^{(p+1)}(t_{n-1}) + O(h_n^{p+3}).
\end{aligned} \tag{8.10.5}$$

Assume first that

$$\begin{aligned}
\eta(t_n) = \widehat{\eta}(t_n) &= h_n^{p+1} y^{(p+1)}(t_n) \\
&= h_n^{p+1} y^{(p+1)}(t_{n-1}) + h_n^{p+2} y^{(p+2)}(t_{n-1}) + O(h_n^{p+3})
\end{aligned} \tag{8.10.6}$$

and define $\widehat{\varphi}_0$ and $\widehat{\varphi}$ as the solution to the linear system

$$\widehat{\varphi}_0 + \widehat{\varphi} e = 0,$$

$$\frac{\widehat{\varphi}^T c^j}{j!} = 0, \quad j = 1, 2, \ldots, p-1,$$

$$\frac{\widehat{\varphi}^T c^p}{p!} = 1, \qquad \frac{\widehat{\varphi}^T c^{p+1}}{(p+1)!} = 1. \tag{8.10.7}$$

This system was obtained by comparing the terms $h_n^j y^{(j)}(t_{n-1})$ on both sides of (8.10.5) with $\eta(t_n) = \widehat{\eta}(t_n)$ defined by (8.10.6). Assume next that

$$\eta(t_n) = \widetilde{\eta}(t_n) = h_n^{p+2} y^{(p+2)}(t_n) = h_n^{p+2} y^{(p+2)}(t_{n-1}) + O(h_n^{p+3}) \qquad (8.10.8)$$

and define $\widetilde{\varphi}_0$ and $\widetilde{\varphi}$ as the solution to the linear system

$$\widetilde{\varphi}_0 + \widetilde{\varphi}e = 0,$$

$$\frac{\widetilde{\varphi}^T c^j}{j!} = 0, \quad j = 1, 2, \ldots, p, \qquad (8.10.9)$$

$$\frac{\widetilde{\varphi}^T c^{p+1}}{(p+1)!} = 1.$$

As before, this system was obtained by comparing the terms of order h_n^j on both sides of (8.10.5) with $\eta(t_n) = \widetilde{\eta}(t_n)$ defined by (8.10.8).

Systems (8.10.7) and (8.10.9) for $\widehat{\varphi}$ and $\widetilde{\varphi}$ can be written more compactly as

$$\begin{bmatrix} \widehat{\varphi}^T \\ \widetilde{\varphi}^T \end{bmatrix} \begin{bmatrix} c & \dfrac{c^2}{2!} & \cdots & \dfrac{c^{p+1}}{(p+1)!} \end{bmatrix} = \begin{bmatrix} 0 & \cdots & 0 & 1 & 1 \\ 0 & \cdots & 0 & 0 & 1 \end{bmatrix}.$$

Assuming that the abscissa vector c has distinct components, the matrix on the left-hand side of the relation above is nonsingular, and this system has a unique solution $\widehat{\varphi}$ and $\widetilde{\varphi}$. The scalars $\widehat{\varphi}_0$ and $\widetilde{\varphi}_0$ are then defined by

$$\widehat{\varphi}_0 = -\widehat{\varphi}^T e, \quad \widetilde{\varphi}_0 = -\widetilde{\varphi}^T e.$$

We compute next the expressions $\widehat{\varphi}^T \xi$ and $\widetilde{\varphi}^T \xi$ appearing on the right-hand side of (8.10.5) and distinguish two cases: $\widetilde{\varphi}^T e = 0$ and $\widetilde{\varphi}^T e \neq 0$. If $\widetilde{\varphi}^T e = 0$ we can estimate $h_n^{p+1} y^{(p+1)}(t_n)$ using (8.10.4) with $\eta(t_n) = \widehat{\eta}(t_n) - \theta \widetilde{\eta}(t_n)$ for any $\theta \in \mathbb{R}$; that is,

$$h_n^{p+1} y^{(p+1)}(t_{n-1}) + (1-\theta) h_n^{p+2} y^{(p+2)}(t_{n-1})$$

$$= (\widehat{\varphi}_0 - \theta \widetilde{\varphi}_0) z_1^{[n-1]} + \big((\widehat{\varphi}^T - \theta \widetilde{\varphi}^T) \otimes I\big) h_n F(Y^{[n]}) + O(h_n^{p+3}),$$

which can also be written in the form

$$h_n^{p+1} y^{(p+1)}(t_n - \theta h_n) = (\widehat{\varphi}_0 - \theta \widetilde{\varphi}_0) z_1^{[n-1]} + \big((\widehat{\varphi}^T - \theta \widetilde{\varphi}^T) \otimes I\big) h_n F(Y^{[n]}) + O(h_n^{p+3}).$$

We can estimate $h_n^{p+2} y^{(p+2)}(t_n)$ just by using (8.10.4) with $\eta(t_n) = \widetilde{\eta}(t_n)$; that is,

$$h_n^{p+2} y^{(p+2)}(t_n) = \widetilde{\varphi}_0 z_1^{[n-1]} + (\widetilde{\varphi}^T \otimes I) h_n F(Y^{[n]}) + O(h_n^{p+3}).$$

Consider next the case $\widetilde{\varphi}^T e \neq 0$. To estimate $h_n^{p+1} y^{(p+1)}(t_n)$, we use (8.10.4) with

$$\varphi_0 = \widehat{\varphi}_0 - \theta \widetilde{\varphi}_0, \quad \varphi = \widehat{\varphi} - \theta \widetilde{\varphi},$$

and choose θ to eliminate in (8.10.5) with $\eta(t_n) = \widehat{\eta}(t_n) - \theta\,\widetilde{\eta}(t_n)$ the contribution from the term

$$h_n^{p+2}\frac{\partial f}{\partial y}\big(y(t_{n-1})\big)y^{(p+1)}(t_{n-1}).$$

This leads to $\widehat{\varphi}^T\xi - \theta\,\widetilde{\varphi}^T\xi = 0$, or

$$\theta = \widehat{\varphi}^T\xi/\widetilde{\varphi}^T\xi.$$

Similarly to the previous case, the estimated quantity $\eta(t_n)$ takes the form

$$
\begin{aligned}
\eta(t_n) \;=\;& h_n^{p+1}y^{(p+1)}(t_{n-1}) + (1-\theta)h_n^{p+2}y^{(p+2)}(t_{n-1}) \\[2mm]
&+\; O(h_n^{p+3}) \hspace{4cm} (8.10.10)\\[2mm]
=\;& h_n^{p+1}y^{(p+1)}(t_n - \theta\,h_n) + O(h_n^{p+3}).
\end{aligned}
$$

To estimate $h_n^{p+2}y^{(p+2)}(t_n)$ we use the difference between $\eta(t_n)$ and a suitably scaled value of $\eta(t_{n-1})$ computed in the preceding step from t_{n-2} to t_{n-1}. It follows from (8.10.10) that

$$
\begin{aligned}
\eta(t_{n-1}) \;=\;& h_{n-1}^{p+1}y^{(p+1)}(t_{n-1} - \theta\,h_{n-1}) + O(h_{n-1}^{p+3}) \\[2mm]
=\;& \frac{h_n^{p+1}}{r_{n-1}^{p+1}}y^{(p+1)}\left(t_n - \left(\frac{\theta}{r_{n-1}} + 1\right)h_n\right) + O(h_n^{p+3}),
\end{aligned}
$$

where $r_{n-1} = h_n/h_{n-1}$, and we have

$$
\begin{aligned}
\eta(t_n) \;-\;& r_{n-1}^{p+1}\eta(t_{n-1}) \\[2mm]
=\;& h_n^{p+1}\left(y^{(p+1)}(t_n - \theta\,h_n) - y^{(p+1)}\left(t_n - \left(\frac{\theta}{r_{n-1}} + 1\right)h_n\right)\right) \\[2mm]
&+\; O(h_n^{p+3}) \\[2mm]
=\;& \frac{r_{n-1} + \theta - r_{n-1}\theta}{r_{n-1}}h_n^{p+2}y^{(p+2)}(t_n) + O(h_n^{p+3}).
\end{aligned}
$$

Hence, the estimate of $h_n^{p+2}y^{(p+2)}(t_n)$ is given by

$$h_n^{p+2}y^{(p+2)}(t_n) = \frac{r_{n-1}}{r_{n-1} + \theta - r_{n-1}\theta}\left(\eta(t_n) - r_{n-1}^{p+1}\eta(t_{n-1})\right) + O(h_n^{p+3}).$$

Butcher and Podhaisky [77] implemented in Matlab a family of stiffly accurate GLMs with IRKS of order $1 \le p \le 4$, which were derived in [77] in a variable step size variable order environment, where the step size and order were chosen using the estimates of $h_n^{p+1} y^{(p+1)}(t_n)$ and $h_n^{p+2} y^{(p+2)}(t_n)$ discussed in this section. The order changing strategy for this experimental code is discussed by Butcher and Podhaisky [77, Section 4] and additional implementation details are described by Butcher et al. [64]. Butcher and Podhaisky report in [77] that the observed relative errors

$$\left| \frac{y_{estimated}^{(p+1)} - y^{(p+1)}}{y^{(p+1)}} \right|, \quad \left| \frac{y_{estimated}^{(p+2)} - y^{(p+2)}}{y^{(p+2)}} \right|,$$

on the Prothero-Robinson problem (1.7.1) are almost always less than 10^{-4} for the $p+1$ derivative and less than 10^{-2} for the $p+2$ derivative, and rarely, and only after order increases, relative errors as large as 10^{-1} were observed.

8.11 SOME REMARKS ON RECENT WORK ON GLMS

GLMs with IRKS examined in Chapters 7 and 8 have many desirable properties that are not shared by other classes of methods. They have the same stability properties as RK methods of the same order, and this allows the construction of explicit methods with large regions of absolute stability and implicit methods that are A- and L-stable. They allow for accurate, efficient, and reliable estimation of local discretization errors and can achieve unconditional stability for any step size pattern. Moreover, they can be constructed using only linear operations, which makes it possible to examine vast collections of methods trying to identify formulas that are optimal in some sense.

However, the definition of optimal formulas and their construction within the class of GLMs with IRKS or within other classes of GLMs are the main challenges ahead and these are topics of recent work. Recent work is also related to efficient implementation of various methods in variable step size, variable order environments using sophisticated controllers for step size and order selection. This should lead to high quality codes for both nonstiff and stiff differential systems. The design and testing of such software based on various classes of GLMs is the subject of current work.

Recent work in this area is also related to the construction of various classes of GLMs with stability properties stronger than A- or L-stability: namely, methods that are algebraically stable. Various stability concepts for GLMs are reviewed in Section 2.9. So far this problem has been solved for MRK methods (see [28, 145] and Section 2.9). A promising new approach to investigating the algebraic stability of GLMs was proposed recently by Hewitt and Hill [154, 155]. This approach was also reviewed in Section 2.9, and its applicability to special cases of GLMs, such as DIMSIMs, TSRK methods, peer methods, and GLMs with IRKS, is currently under investigation.

References

1. M. Abramowitz and I.A. Stegun, *Handbook of Mathematical Functions with Formulas, Graphs, and Mathematical Tables*, National Bureau of Standards Applied Mathematics Series 55, U.S. Government Printing Office, Washington, DC, 1972.

2. R.C. Aiken, Editor, *Stiff Computation*, Oxford University Press, New York, 1985.

3. A. Albert, Conditions for positive and non-negative definiteness in terms of pseudoinverses, *SIAM J. Appl. Math.* 17(1969), 434–440.

4. P. Albrecht, Numerical treatment of O.D.E.s: the theory of *A*-methods, *Numer. Math.* 47(1985), 59–87.

5. P. Albrecht, A new theoretical approach to Runge–Kutta methods, *SIAM J. Numer. Anal.* 24(1987), 391–406.

6. P. Albrecht, Elements of a general theory of composite integration methods, *Appl. Math. Comput.* 31(1989), 1–17.

7. P. Albrecht, The Runge-Kutta theory in a nutshell, *SIAM J. Numer. Anal.* 33 (1996), 1712–1735.

8. P. Albrecht, The common basis of the theories of linear cyclic methods and Runge-Kutta methods, *Appl. Numer. Math.* 22(1996), 3–21.

9. R. Alexander, Diagonally implicit Runge-Kutta methods for stiff o.d.e.'s, *SIAM J. Numer. Anal.* 14(1977), 1006–1021.

10. E.L. Allgower and K. Georg, *Numerical Continuation Methods: An Introduction*, Series in Computational Mathematics 13, Springer-Verlag, Berlin, 1990.

11. E.L. Allgower and K. Georg, Continuation and path following, *Acta Numerica* (1993), 1–64.

12. R.F. Arenstorf, Periodic solutions of the restricted three-body problem representing analytic continuations of Keplerian elliptic motions, *Amer. J. Math.* 85(1963), 27–35.

13. R.F. Arenstorf, New periodic solutions of the plane three-body problem corresponding to elliptic motion in the lunar theory, *J. Differential Equations* 4(1968), 202–256.

14. U.M. Ascher and L.R. Petzold, *Computer Methods for Ordinary Differential Equations and Differential-Algebraic Equations*, SIAM, Philadelphia, 1998.

15. Z. Bartoszewski and Z. Jackiewicz, Construction of two-step Runge-Kutta methods of high order for ordinary differential equations, *Numer. Algorithms* 18(1998), 51–70.

16. Z. Bartoszewski and Z. Jackiewicz, Toward a two-step Runge-Kutta code for nonstiff differential systems, *Appl. Math. (Warsaw)* 28(2001), 353–365.

17. Z. Bartoszewski and Z. Jackiewicz, Nordsieck representation of two-step Runge-Kutta methods for ordinary differential equations, *Appl. Numer. Math.* 53(2005), 149–163.

18. Z. Bartoszewski and Z. Jackiewicz, Derivation of continuous explicit two-step Runge-Kutta methods of order three, *J. Comput. Appl. Math.* 205(2007), 764–776.

19. A. Bellen, Z. Jackiewicz, and M. Zennaro, Local error estimation for singly-implicit formulas by two-step Runge-Kutta methods, *BIT* 32(1992), 104–117.

20. W.J. Beyn, On invariant closed curves for one-step methods, *Numer. Math.* 51(1987), 103–122.

21. G. Birkhoff and G.C. Rota, *Ordinary Differential Equations*, Wiley, New York, 1978.

22. P. Bogacki and L.F. Shampine, A 3(2) pair of Runge-Kutta formulas, *Appl. Math. Lett.* 2(1989), 1–9.

23. P.N. Brown, G.D. Byrne, and A.C. Hindmarsh, VODE, a variable-coefficient ODE solver, *SIAM J. Sci. Statist. Comput.* 10(1989), 1038–1051.

24. M. Brown and J. Hershenov, Periodic solutions of finite difference equations, *Quart. Appl. Math.* 35(1977), 139–147.

25. P.N. Brown and A.C. Hindmarsh, Reduced storage matrix methods in stiff ODE systems, *Appl. Math. Comput.* 31(1989), 40–91.

26. D.S. Bunch, D.M. Gay, and R.E. Welsch, Algorithm 717, subroutines for maximum likelihood and quasi-likelihood estimation of parameters in nonlinear regression models, *ACM Trans. Math. Software* 19(1993), 109–130.

27. K. Burrage, A special family of Runge-Kutta methods for solving stiff differential equations, *BIT* 18(1978), 22–41.

28. K. Burrage, High order algebraically stable Runge-Kutta methods, *SIAM J. Numer. Anal.* 24(1987), 106–115.

29. K. Burrage, Order properties of implicit multivalue methods for ordinary differential equations, *IMA J. Numer. Anal.* 8(1988), 43–69.

30. K. Burrage, The dichotomy of stiffness: pragmatism versus theory, *Appl. Math. Comput.* 31(1989), 92–111.

31. K. Burrage, *Parallel and Sequential Methods for Ordinary Differential Equations*, Oxford Science Publications, Clarendon Press, Oxford, UK, 1995.

32. K. Burrage and J.C. Butcher, Stability criteria for implicit Runge-Kutta methods, *SIAM J. Numer. Anal.* 16(1979), 46–57.

33. K. Burrage and J.C. Butcher, Non-linear stability of a general class of differential equation methods, *BIT* 20(1980), 185–203.

34. K. Burrage, J.C. Butcher, and F.H. Chipman, An implementation of singly-implicit Runge-Kutta methods, *BIT* 20(1980), 326–340.

35. K. Burrage and P. Moss, Simplifying assumptions for the order of partitioned multivalue methods, *BIT* 20(1980), 452–465.

36. K. Burrage and P.W. Sharp, A class of variable-step explicit Nordsieck multivalue methods, *SIAM J. Numer. Anal.* 31(1994), 1434–1451.

37. J.C. Butcher, A modified multistep method for the numerical integration of ordinary differential equations, *J. Assoc. Comput. Mach.* 12(1965), 124–135.

38. J.C. Butcher, A stability property of implicit Runge-Kutta methods, *BIT* 15(1975), 358–361.

39. J.C. Butcher, A transformed implicit Runge-Kutta method, *J. Assoc. Comput. Mach.* 26(1979), 731–738.

40. J.C. Butcher, General linear method: A survey, *Appl. Numer. Math.* 1(1985), 273–284.

41. J.C. Butcher, *The Numerical Analysis of Ordinary Differential Equations. Runge-Kutta and General Linear Methods*, Wiley, New York, 1987.

42. J.C. Butcher, Linear and non-linear stability of general linear methods, *BIT* 27(1987), 182–189.

43. J.C. Butcher, The equivalence of algebraic stability and *AN*-stability, *BIT* 27(1987), 510–533.

44. J.C. Butcher, Diagonally-implicit multi-stage integration methods, *Appl. Numer. Math.* 11(1993), 347–363.

45. J.C. Butcher, General linear methods for the parallel solution of ordinary differential equations, *World Sci. Ser. Appl. Anal.* 2(1993), 99–111.

46. J.C. Butcher, A transformation for the analysis of DIMSIMs, *BIT* 34(1994), 25–32.

47. J.C. Butcher, An introduction to "almost Runge-Kutta" methods, *Appl. Numer. Math.* 24(1997), 331–342.

48. J.C. Butcher, Order and stability of parallel methods for stiff problems, *Adv. Comput. Math.* 7(1997), 79–96.

49. J.C. Butcher, ARK methods up to order five, *Numer. Algorithms* 17(1998), 193–221.

50. J.C. Butcher, General linear methods for stiff differential equations, *BIT* 41(2001), 240–264.

51. J.C. Butcher, Software issues for ordinary differential equations, *Numer. Algorithms* 31(2002), 401–418.

52. J.C. Butcher, *Numerical Methods for Ordinary Differential Equations*, Wiley, Chichester, 2003.

53. J.C. Butcher, General linear methods, *Acta Numerica* 15(2006), 157–256.

54. J.C. Butcher, Thirty years of G-stability, *BIT* 46(2006), 479–489.

55. J.C. Butcher and P. Chartier, A generalization of singly-implicit Runge-Kutta methods, *Appl. Numer. Math.* 24(1997), 343–350.

56. J.C. Butcher and P. Chartier, The effective order of singly-implicit Runge-Kutta methods, *Numer. Algorithms* 20(1999), 269–284.

57. J.C. Butcher, P. Chartier, and Z. Jackiewicz, Nordsieck representation of DIMSIMs, *Numer. Algorithms* 16(1997), 209–230.

58. J.C. Butcher, P. Chartier, and Z. Jackiewicz, Experiments with a variable-order type 1 DIMSIM code, *Numer. Algorithms* 22(1999), 237–261.

59. J.C. Butcher and D.J.L. Chen, ESIRK methods and variable step size, *Appl. Numer. Math.* 28(1998), 193–207.

60. J.C. Butcher and F.H. Chipman, Generalized Padé approximations to the exponential functions, *BIT* 32(1992), 118–130.

61. J.C. Butcher and M.T. Diamantakis, DESIRE: diagonally extended singly implicit Runge-Kutta effective order methods, *Numer. Algorithms* 17(1998), 121–145.

62. J.C. Butcher and A.D. Heard, Stability of numerical methods for ordinary differential equations, *Numer. Algorithms* 31(2002), 75–85.

63. J.C. Butcher and A.T. Hill, Linear multistep methods as irreducible general linear methods, *BIT* 46(2006), 5–19.

64. J.C. Butcher, S.J.Y. Huang, and H. Podhaisky, Some implementation questions for general linear methods, in preparation.

65. J.C. Butcher and Z. Jackiewicz, Diagonally implicit general linear methods for ordinary differential equations, *BIT* 33(1993), 452–472.

66. J.C. Butcher and Z. Jackiewicz, Construction of diagonally implicit general linear methods of type 1 and 2 for ordinary differential equations, *Appl. Numer. Math.* 21(1996), 385–415.

67. J.C. Butcher and Z. Jackiewicz, Implementation of diagonally implicit multistage integration methods for ordinary differential equations, *SIAM J. Numer. Anal.* 34(1997), 2119–2141.

68. J.C. Butcher and Z. Jackiewicz, Construction of high order diagonally implicit multistage integration methods for ordinary differential equations, *Appl. Numer. Math.* 27(1998), 1–12.

69. J.C. Butcher and Z. Jackiewicz, A reliable error estimation for diagonally implicit multistage integration methods, *BIT* 41(2001), 656–665.

70. J.C. Butcher and Z. Jackiewicz, Error estimation for Nordsieck methods, *Numer. Algorithms* 31(2002), 75–85.

71. J.C. Butcher and Z. Jackiewicz, A new approach to error estimation for general linear methods, *Numer. Math.* 95(2003), 487–502.

72. J.C. Butcher and Z. Jackiewicz, Construction of general linear methods with Runge-Kutta stability properties, *Numer. Algorithms* 36(2004), 53–72.

73. J.C. Butcher and Z. Jackiewicz, Unconditionally stable general linear methods for ordinary differential equations, *BIT* 44(2004), 557–570.

74. J.C. Butcher, Z. Jackiewicz, and H.D. Mittelmann, A nonlinear optimization approach to the construction of general linear methods of high order, *J. Comput. Appl. Math.* 81(1997), 181–196.

75. J.C. Butcher, Z. Jackiewicz, and W.M. Wright, Error propagation for general linear methods for ordinary differential equations, *J. Complexity* 23(2007), 560–580.

76. J.C. Butcher and N. Moir, Experiments with a new fifth order method, *Numer. Algorithms* 33(2003), 137–151.

77. J.C. Butcher and H. Podhaisky, On error estimation in general linear methods for stiff ODEs, *Appl. Numer. Math.* 56(2006), 345–357.

78. J.C. Butcher and S. Tracogna, Order conditions for two-step Runge-Kutta methods, *Appl. Numer. Math.* 24(1997), 351–364.

79. J.C. Butcher and W.M. Wright, A transformation relating explicit and diagonally-implicit general linear methods, *Appl. Numer. Math.* 44(2003), 313–327.

80. J.C. Butcher and W.M. Wright, The construction of practical general linear methods, *BIT* 43(2003), 695–721.

81. J.C. Butcher and W.M. Wright, Applications of doubly companion matrices, *Appl. Numer. Math.* 56(2006), 358–373.

82. G.D. Byrne, Parameters for pseudo Runge-Kutta methods, *Comm. ACM* 10(1967), 102–104.

83. G.D. Byrne, Pragmatic experiments with Krylov methods in the stiff ODE setting, in: *Computational Ordinary Differential Equations* (J.R. Cash and I. Gladwell, eds.), Clarendon Press, Oxford, UK, 1992, pp. 323–356.

84. G.D. Byrne and R.J. Lambert, Pseudo Runge-Kutta methods involving two points, *J. Assoc. Comput. Mach.* 13(1966), 114–123.

85. J.R. Cash, On the integration of stiff systems of O.D.E.s using extended backward differentiation formulae, *Numer. Math.* 34(1980), 235–246.

86. J.R. Cash, The integration of stiff initial value problems in ODEs using modified extended backward differentiation formulae, *Comput. Math. Appl.* 9(1983), 645–657.

87. J.R. Cash, Split linear multistep methods for the numerical integration of stiff differential systems, *Numer. Math.* 42(1983), 299–310.

88. J.R. Cash and S. Considine, An MEBDF code for stiff initial value problems, *ACM Trans. Math. Software* 18(1992), 142–155.

89. J.R. Cash and S. Considine, Algorithm 703. MEBDF: A FORTRAN subroutine for solving first-order systems of stiff initial value problems for ordinary differential equations, *ACM Trans. Math. Software* 18(1992), 156–158.

90. R. Caira, C. Costabile, and F. Costabile, A class of pseudo Runge-Kutta methods, *BIT* 30(1990), 642–649.

91. J. Chollom and Z. Jackiewicz, Construction of two-step Runge-Kutta methods with large regions of absolute stability, *J. Comput. Appl. Math.* 157(2003), 125–137.

92. D. Conte, R. D'Ambrosio, and Z. Jackiewicz, Two-step Runge-Kutta methods with quadratic stability functions, submitted.

93. D. Conte, R. D'Ambrosio, and Z. Jackiewicz, Construction of two-step Runge-Kutta methods of high order with quadratic stability functions, in preparation.

94. W.A. Coppel, *Stability and Asymptotic Behavior of Differential Equations*, D.C. Heath, Boston, 1965.

95. W.A. Coppel, *Dichotomies in Stability Theory*, Lecture Notes in Mathematics 629, Springer-Verlag, New York, 1978.

96. M. Crouzeix, Sur la B-stabilité des méthodes de Runge-Kutta, *Numer. Math.* 32(1979), 75–82.

97. G. Dahlquist, Convergence and stability in the numerical integration of ordinary differential equations, *Math. Scand.* 4(1956), 33–53.

98. G. Dahlquist, Stability and error bounds in the numerical integration of ordinary differential equations, *Kungl. Tekn. Högsk. Handl. Stockholm* 130(1959).

99. G. Dahlquist, A special stability problem for linear multistep methods, *BIT* 3(1963), 27–43.

100. G. Dahlquist, Error analysis for a class of methods for stiff non-linear initial value problems, in: *Numerical Analysis*, Proc. Dundee Conf., 1975 (G.A. Watson, ed.), Lecture Notes in Mathematics 506, Springer-Verlag, Berlin, 1976, pp. 60–72.

101. G. Dahlquist, *G*-stability is equivalent to *A*-stability, *BIT* 18(1978), 384–401.

102. G. Dahlquist, On one-leg multistep methods, *SIAM J. Numer. Anal.* 20(1983), 1130–1138.

103. R. D'Ambrosio, Two-step collocation methods for ordinary differential equations, Ph.D. thesis, University of Salerno, in preparation.

104. R. D'Ambrosio, M. Ferro, Z. Jackiewicz, and B. Paternoster, Two-step almost collocation methods for ordinary differential equations, to appear in *Numer. Algorithms*.

105. R. D'Ambrosio and Z. Jackiewicz, Continuous two-step Runge-Kutta methods for ordinary differential equations, in preparation.

106. R. D'Ambrosio and Z. Jackiewicz, Construction and implementation of highly stable two-step collocation methods, in preparation.

107. P.J. Davis and P. Rabinowitz, *Methods of Numerical Integration*, Academic Press, New York, 1984.

108. K. Dekker, Algebraic stability of general linear methods, Tech. Rep. 25, Computer Science Department, University of Auckland, New Zealand, 1981.

109. K. Dekker and J.G. Verwer, *Stability of Runge-Kutta methods for stiff nonlinear differential equations*, CWI Monographs, North-Holland, Amsterdam, 1984.

110. J.W. Demmel, *Applied Numerical Linear Algebra*, SIAM, Philadelphia, 1997.

111. J.E. Dennis, D.M. Gay, and R.E. Welsch, An adaptive nonlinear least-squares algorithm, *ACM Trans. Math. Software* 7(1981), 348–368.

112. J.E. Dennis, D.M. Gay, and R.E. Welsch, Algorithm 573, NL2SOL - an adaptive nonlinear least-squares algorithm, *ACM Trans. Math. Software* 7(1981), 369–383.

113. J.E. Dennis and R.B. Schnabel, *Numerical Methods for Unconstrained Optimization and Nonlinear Equations*, Prentice-Hall, Englewood Cliffs, NJ, 1983.

114. P. Deuflhard, B. Fiedler, and P. Kunkel, Efficient numerical path following beyond critical points, *SIAM J. Numer. Anal.* 24(1987), 912–927.

115. J. Donelson and E. Hansen, Cyclic composite multistep predictor-corrector methods, *SIAM J. Numer. Anal.* 8(1971), 137–157.

116. J.R. Dormand, *Numerical Methods for Differential Equations. A Computational Approach*, CRC Press, New York, 1996.

117. J.R. Dormand and P.J. Prince, A family of embedded Runge-Kutta formulae, *J. Comput. Appl. Math.* 6(1980), 19–26.

118. R.J. Duffin, Algorithms for classical stability problems, *SIAM Rev.* 11(1969), 196–213.

119. H. Dym and H.P. McKean, *Fourier Series and Integrals*, Academic Press, New York, 1972.

120. G. Edgar, *Measure, Topology, and Fractal Geometry*, Springer-Verlag, New York, 1995.

121. B.L. Ehle, High order A-stable methods for the numerical solution of systems of D.E.'s, *BIT* 8(1968), 276–278.

122. B.L. Ehle, A-stable methods and Páde approximations to the exponential, *SIAM J. Math. Anal.* 4(1973), 671–680.

123. T. Eirola, Invariant curves of one-step methods, *BIT* 28(1988), 113-122.

124. T. Eirola and O. Nevanlinna, What do multistep methods approximate? *Numer. Math.* 53(1988), 559–569.

125. W.H. Enright and T.E. Hull, Comparing numerical methods for the solution of stiff systems of ODEs arising in chemistry, in: *Numerical Methods for Differential Systems, Recent Developments in Algorithms, Software and Applications* (L. Lapidus and W.E. Schiesser, eds.), Academic Press, New York, 1976, pp. 10–48.

126. R.F. Enenkel, DIMSEM-diagonally implicit single-eigenvalue methods for the solution of stiff ordinary differential equations on parallel computers, Ph.D. thesis, University of Toronto, 1996.

127. R.F. Enenkel and K.R. Jackson, DIMSEMs-diagonally implicit single-eigenvalue methods for the numerical solution of stiff ODEs on parallel computers, *Adv. Comput. Math.* 7(1997), 97–133.

128. J. Field and R.M. Noyes, Oscillations in chemical systems. 4. Limit cycle behavior in a model of a real chemical-reaction, *J. Chem. Phys.* 60(1974), 1877–1884.

129. D.M. Gay, A trust-region approach to linearly constrained optimization, in: *Numerical Analysis*, Proc. Dundee Conf., 1983, (D.F. Griffiths, ed.), Lecture Notes in Mathematics 1066, Springer-Verlag, Berlin, 1984, pp. 72–105.

130. C.W. Gear, Hybrid methods for initial value problems in ordinary differential equations, *SIAM J. Numer. Anal.* 2(1965), 69–86.

131. C.W. Gear, The automatic integration of ordinary differential equations, *Comm. ACM* 14(1971), 176–179.

132. C.W. Gear, Algorithm 407 DIFSUB for solution of ordinary differential equations [D2], *Comm. ACM* 14(1971), 185–190.

133. C.W. Gear, *Numerical Initial Value Problems in Ordinary Differential Equations*, Prentice-Hall, Englewood Cliffs, NJ, 1971.

134. I. Gladwell, L.F. Shampine, and R.W. Brankin, Automatic selection of the initial step size for an ODE solver, *J. Comput. Appl. Math.* 18(1987), 175–192.

135. W.B. Gragg and H.J. Stetter, Generalized multistep predictor-corrector methods, *J. Assoc. Comput. Mach.* 11(1964), 188–209.

136. G.H. Golub and C.F. Van Loan, *Matrix Computations*, Johns Hopkins University Press, Baltimore, MD, 1996.

137. T.H. Gronwall, Note on the derivatives with respect to a parameter of the solutions of a system of differential equations, *Ann. Math.* 2(1919), 292–296.

138. N. Guglielmi and M. Zennaro, On the asymptotic properties of a family of matrices, *Linear Algebra Appl.* 322(2001), 169–192.

139. N. Guglielmi and M. Zennaro, On the zero-stability of variable step size multistep methods: the spectral radius approach, *Numer. Math.* 88(2001), 445–458.

140. K. Gustafson, M. Lundh, and G. Söderlind, A PI step size control for the numerical solution of ordinary differential equations, *BIT* 28(1988), 270-287.

141. E. Hairer, A. Iserles, and J.M. Sanz-Serna, Equilibria of Runge-Kutta methods, *Numer. Math.* 58(1990), 243–254.

142. E. Hairer, C. Lubich, and G. Wanner, *Geometric Numerical Integration. Structure-Preserving Algorithms for Ordinary Differential Equations*, Springer Verlag, New York, 2002.

143. E. Hairer, S.P. Nørsett, and G. Wanner, *Solving Ordinary Differential Equations I: Nonstiff Problems*, Springer-Verlag, New York 1993.

144. E. Hairer and G. Wanner, Multistep-multistage-multiderivative methods for ordinary differential equations, *Computing* 11(1973), 114–123.

145. E. Hairer and G. Wanner, *Solving Ordinary Differential Equations II: Stiff and Differential-Algebraic Problems*, Springer-Verlag, New York 1991.

146. E. Hairer and G. Wanner, *Solving Ordinary Differential Equations II: Stiff and Differential-Algebraic Problems*, 2nd rev. ed., Springer-Verlag, New York, 1996.

147. E. Hairer and G. Wanner, Order conditions for general two-step Runge-Kutta methods, *SIAM J. Numer. Anal.* 34(1997), 2087–2089.

148. E. Hairer and G. Wanner, Stiff differential equations solved by Radau methods, *J. Comput. Appl. Math.* 111(1999), 93–111.

149. A. Halanay, *Differential Equations. Stability, Oscillations, Time Lags*, Academic Press, New York, 1966.

150. G. Hall, Equilibrium states of Runge-Kutta schemes, *ACM Trans. Math. Software* 11(1985), 289–301.

151. G. Hall, Equilibrium states of Runge-Kutta schemes: Part II, *ACM Trans. Math. Software* 12(1986), 183–192.

152. G. Hall and D.J. Higham, Analysis of step size selection schemes for Runge-Kutta codes, *IMA J. Numer. Anal.* 8(1988), 305–310.

153. P. Henrici, *Discrete Variable Methods in Ordinary Differential Equations*, Wiley, New York, 1962.

154. L.L. Hewitt and A.T. Hill, Algebraically stable general linear methods and the G-matrix, *BIT* 49(2009), 93–111.

155. L.L. Hewitt and A.T. Hill, Algebraically stable diagonally implicit general linear methods, submitted.

156. D.J. Higham, The tolerance proportionality of adaptive ODE solvers, *J. Comput. Appl. Math.* 45(1993), 227–236.

157. D.J. Higham and G. Hall, Embedded Runge-Kutta formulas with stable equilibrium states, *J. Comput. Appl. Math.* 29(1990), 25–33.

158. A.T. Hill, Nonlinear stability of general linear methods, *Numer. Math.* 103(2006), 611–629.

159. A.C. Hindmarsh, ODEPACK, a systematized collection of ODE solvers, *IMACS Trans. Sci. Comput.* 1(1983), 55–64.

160. R.A. Horn and C.R. Johnson, *Matrix Analysis*, Cambridge University Press, New York, 1985.

161. M.E. Hosea, A new recurrence for computing Runge-Kutta truncation error coefficients, *SIAM J. Numer. Anal.* 32(1995), 1989–2001.

162. M.E. Hosea and L.F. Shampine, Efficiency comparisons of methods for integrating ODEs, *Comput. Math. Appl.* 28(1994), 45–55.

163. A.S. Householder, *The Theory of Matrices in Numerical Analysis*, Dover, New York, 1964.

164. P.J. van der Houwen, *Construction of Integration Formulas for Initial Value Problems*, North-Holland, Amsterdam, 1977.

165. P.J. van der Houwen and B.P. Sommeijer, On the internal stability of explicit, m-stage Runge-Kutta methods for large m-values, *Z. Angew. Math. Mech.* 60(1980), 479–485.

166. P.J. van der Houwen and B.P. Sommeijer, A special class of multistep Runge-Kutta methods with extended real stability interval, *IMA J. Numer. Anal.* 2(1982), 183–209.

167. S.J.Y. Huang, Implementation of general linear methods for stiff ordinary differential equations, Ph.D. thesis, The University of Auckland, New Zealand, 2004.

168. T.E. Hull, W.H. Enright, B.M. Fellen, and A.E. Sedgwick, Comparing numerical methods for ordinary differential equations, *SIAM J. Numer. Anal.* 9(1972), 603–637.

169. W.H. Hundsdorfer and B.I. Steininger, Convergence of linear multistep and one-leg methods for stiff nonlinear initial value problems, *BIT* 31(1991), 124–143.

170. W. Hundsdorfer and J.G. Verwer, *Numerical Solution of Time-Dependent Advection-Diffusion-Reaction Equations*, Springer Series in Computational Mathematics 33, Springer-Verlag, Berlin, 2003.

171. A. Iserles, Stability and dynamics of numerical methods for nonlinear ordinary differential equations, *IMA J. Numer. Anal.* 10(1990), 1–30.

172. A. Iserles, *A First Course in the Numerical Analysis of Differential Equations*, Cambridge University Press, Cambridge, UK, 1996.

173. A. Iserles, On the method of Newmann series for highly oscillatory equations, *BIT* 44(2004), 473–488.

174. A. Iserles, On the numerical analysis of rapid oscillation, *CRM Proc. Lecture Notes* 39(2005), 149–163.

175. G. Izzo and Z. Jackiewicz, Continuous two-step peer methods, in preparation.

176. Z. Jackiewicz, Step-control stability of diagonally implicit multistage integration methods, *New Zealand J. Math.* 29(2000), 193–201.

177. Z. Jackiewicz, Implementation of DIMSIMs for stiff differential systems, *Appl. Numer. Math.* 42(2002), 251–267.

178. Z. Jackiewicz, Construction and implementation of general linear methods for ordinary differential equations. A review, *J. Sci. Comput.* 25(2005), 29–49.

179. Z. Jackiewicz and H.D. Mittelmann, Exploiting structure in the construction of DIMSIMs, *J. Comput. Appl. Math.* 107(1999), 233-239.

180. Z. Jackiewicz, H. Podhaisky, and R. Weiner, Construction of highly stable two-step W-methods for ordinary differential equations, *J. Comput. Appl. Math.* 167(2004), 389–403.

181. Z. Jackiewicz, R. Renaut, and A. Feldstein, Two-step Runge-Kutta methods, *SIAM J. Numer. Anal.* 28(1991), 1165–1182.

182. Z. Jackiewicz, R. Renaut, and M. Zennaro, Explicit two-step Runge-Kutta methods, *Appl. Math.* 40(1995), 433–456.

183. Z. Jackiewicz and S. Tracogna, A general class of two-step Runge-Kutta methods for ordinary differential equations, *SIAM J. Numer. Anal.* 32(1995), 1390–1427.

184. Z. Jackiewicz and S. Tracogna, Variable step size continuous two-step Runge-Kutta methods for ordinary differential equations, *Numer. Algorithms* 12(1996), 347–368.

185. Z. Jackiewicz and R. Vermiglio, General linear methods with external stages of different orders, *BIT* 36(1996), 688–712.

186. Z. Jackiewicz, R. Vermiglio, and M. Zennaro, Variable step size diagonally implicit multistage integration methods for ordinary differential equations, *Appl. Numer. Math.* 16(1995), 343–367.

187. Z. Jackiewicz, R. Vermiglio, and M. Zennaro, Regularity properties of Runge-Kutta methods for ordinary differential equations, *J. Comput. Appl. Math.* 22(1996), 251–262.

188. Z. Jackiewicz, R. Vermiglio, and M. Zennaro, Regularity properties of multistage integration methods, *J. Comput. Appl. Math.* 87(1997), 285–302.

189. Z. Jackiewicz and J.H. Verner, Derivation and implementation of two-step Runge-Kutta pairs, *Japan J. Indust. Appl. Math.* 19(2002), 227–248.

190. Z. Jackiewicz and M. Zennaro, Variable step size explicit two-step Runge-Kutta methods, *Math. Comput.* 59(1992), 421–438.

191. U. Kirchgraber, Multi-step methods are essentially one-step methods, *Numer. Math.* 48(1986), 85–90.

192. P.E. Kloeden and J. Lorenz, Stable attracting sets in dynamical systems and in their one-step discretizations, *SIAM J. Numer. Anal.* 23(1986), 986–995.

193. P.E. Kloeden and J. Lorenz, A note on multistep methods and attracting sets of dynamical systems, *Numer. Math.* 56(1990), 667–673.

194. J.D. Lambert, *Computational Methods in Ordinary Differential Equations*, Wiley, New York, 1973.

195. J.D. Lambert, *Numerical Methods for Ordinary Differential Systems*, Wiley, New York, 1991.

196. P. Lancaster and L. Rodman, *Algebraic Riccati Equations*, Oxford University Press, New York, 1995.

197. P. Lancaster and M. Tismenetsky, *The Theory of Matrices*, 2nd ed., Academic Press, Orlando, FL, 1985.

198. J. Lang and J.G. Verwer, ROS3P—An accurate third-order Rosenbrock solver designed for parabolic problems, *BIT* 41(2001), 731–738.

199. L. Lapidus and J.H. Seinfeld, *Numerical Solution of Ordinary Differential Equations*, Academic Press, New York, 1971.

200. G.J. Lastman, R.A. Wentzell, and A.C. Hindmarsh, Numerical solution of a bubble cavitation problem, *J. Comput. Phys.* 28(1978), 56–64.

201. S. Lefschetz, *Differential Equations: Geometric Theory*, Interscience, New York, 1957.

202. B. Leimkuhler, Estimating waveform relaxation convergence, *SIAM J. Sci. Comput.* 14(1993), 872–889.

203. B. Leimkuhler and S. Reich, *Simulating Hamiltonian Dynamics*, Cambridge University Press, Cambridge, UK, 2004.

204. B. Leimkuhler and A. Ruehli, Rapid convergence of waveform relaxation, *Appl. Numer. Math.* 11(1993), 211-224.

205. E. Lelarasmee, The waveform relaxation methods for the time domain analysis of large scale nonlinear dynamical systems, Ph.D. thesis, University of California, Berkeley, 1982.

206. E. Lelarasmee, A. Ruehli, and A. Sangiovanni-Vincentelli, The waveform relaxation methods for time domain analysis of large scale integrated circuits, *IEEE Trans. CAD IC Syst.* 1(1982), 131–145.

207. K. Levenberg, A method for the solution of certain problems in least squares, *Quart. Appl. Math.* 2(1944), 164–168.

208. R.J. LeVeque, *Finite Difference Methods for Ordinary and Partial Differential Equations: Steady-State and Time-Dependent Problems*, SIAM, Philadelphia, 2007.

209. E.N. Lorenz, On the prevalence of aperiodicity in simple systems: Global analysis, *Proc. Biennial Sem. Canad. Math. Congr., Univ. Calgary, Calgary, Alta., 1978*, Lecture Notes in Mathematics 755, Springer-Verlag, Berlin, 1979, pp. 53–75.

210. A.I. Markushevich, *Theory Of Functions of a Complex Variable*, Chelsea, New York, 1985.

211. D. Marquardt, An algorithm for least-squares estimation of nonlinear parameters, *SIAM J. Appl. Math.* 11(1963), 431–441.

212. J.E. Marsden, *Elementary Classical Analysis*, W.H. Freeman, New York, 1974.

213. U. Miekkala and O. Nevanlinna, Convergence of dynamic iteration method for initial value problems, *SIAM J. Sci. Statist. Comput.* 8(1987), 459–482.

214. J.J.H. Miller, On the location of zeros of certain classes of polynomials with applications to numerical analysis, *J. Inst. Math. Appl.* 8(1971), 397–406

215. W.L. Miranker, *Numerical Methods for Stiff Equations and Singular Perturbation Problems*, D. Reidel, Boston, 1981.

216. C.B. Moler, *Numerical Computing with MATLAB*, SIAM, Philadelphia, 2004.

217. J.J. Moré, The Levenberg-Marquardt algorithm: implementation and theory in: *Numerical Analysis*, Proc. Dundee Conf., 1977, Lecture Notes in Mathematics 630, Springer-Verlag, Berlin 1978, pp. 105–116.

218. O. Nevanlinna, Remarks on Picard-Lindelöf iteration. I, *BIT* 29(1989), 328–346.

219. O. Nevanlinna, Remarks on Picard-Lindelöf iteration. II, *BIT* 29(1989), 535–562.

220. O. Nevanlinna, Linear acceleration of Picard-Lindelöf iteration, *Numer. Math.* 57(1990), 147–156.

221. O. Nevanlinna and F. Odeh, Multiplier techniques for linear multistep methods, *Numer. Funct. Anal. Optim.* 3(1981), 377–423.

222. A. Nordsieck, On numerical integration of ordinary differential equations, *Math. Comput.* 16(1962), 22–49.

223. A. Nordsieck, Automatic numerical integration of ordinary differential equations, *AMS Proc. Symp. Appl. Math.* 15(1963), 241–250.

224. J. Oliver, A curiosity of low-order explicit Runge-Kutta methods, *Math. Comput.* 29(1975), 1032–1036.

225. B. Orel, Runge-Kutta methods with real eigenvalues, Ph.D. thesis, Department of Mathematics, University of Ljubljana, Slovenia, 1991.

226. S.S. Oren, Self-scaling variable metric algorithms without line search for unconstrained minimization, *Math. Comput.* 27(1973), 873–885.

227. J.M. Ortega, *Matrix Theory: A Second Course*, Plenum Press, New York, 1987.

228. J.M. Ortega, *Numerical Analysis: A Second Course*, SIAM, Philadelphia, 1990.

229. J.M. Ortega and W.C. Rheinboldt, *Iterative Solution of Nonlinear Equations in Several Variables*, Academic Press, New York, 1970.

230. B. Owren, Continuous explicit Runge-Kutta methods with applications to ordinary and delay differential equations, Ph.D. thesis, Norges Tekniske Høgskole, Trondheim, 1989.

231. B. Owren and M. Zennaro, Order barriers for continuous explicit Runge-Kutta methods, *Math. Comput.* 56(1991), 645–661.

232. B. Owren and M. Zennaro, Derivation of efficient, continuous, explicit Runge-Kutta methods, *SIAM J. Sci. Statist. Comput.* 13(1992), 1488–1501.

233. B. Owren and M. Zennaro, Continuous explicit Runge-Kutta methods, in: *Computational Ordinary Differential Equations*, (London, 1989), Inst. Math. Appl. Conf. Ser. New Ser. 39, Oxford University Press, New York, 1992, pp. 97–105.

234. J. Pan, Local error estimation of multi-implicit Runge-Kutta methods by using two-step Runge-Kutta formulae, M.Sc. thesis, Department of Mathematics, Arizona State University, Tempe, 1992.

235. G. Peano, Démonstration de l'intégrabilité des équations différentielles ordinaires, *Math. Ann.* 43(1893), 553–568.

236. O. Perron, Ein neuer Existenzbeweis für die Integrale eines Systems gewöhnlicher Differentialgleichungen, *Math. Ann.* 78(1918), 378–384.

237. H. Podhaisky, R. Weiner, and B.A. Schmitt, Two-step W-methods for stiff ODE systems, *Vietnam J. Math.* 30(2002), 591–603.

238. H. Podhaisky, R. Weiner, and B.A. Schmitt, Rosenbrock-type 'Peer' two-step methods, *Appl. Numer. Math.* 53(2005), 409–420.

239. H. Podhaisky, R. Weiner, and B.A. Schmitt, Linearly-implicit two-step methods and their implementation in Nordsieck form, *Appl. Numer. Math.* 56(2006), 374–387.

240. A. Prothero and A. Robinson, On the stability and accuracy of one-step methods for solving stiff systems of ordinary differential equations, *Math. Comput.* 28(1974), 145–162.

241. A. Quarteroni, R. Sacco, and F. Saleri, *Numerical Mathematics*, Springer Verlag, New York, 2000.

242. K. Radhakrishnan and A.C. Hindmarsh, Description and use of LSODE, the Livermore solver for ordinary differential equations, *NASA Ref. Publ. 1327, (Lawrence Livermore National Laboratory Rep. UCRL-ID-113855)*, 1993.

243. R.A. Renaut-Williamson, Numerical solution of hyperbolic partial differential equations, Ph.D. thesis, Cambridge University, Cambridge, UK, 1985.

244. R.A. Renaut, Two-step Runge-Kutta methods and hyperbolic partial differential equations, *Math. Comput.* 55(1990), 563–579.

245. W.C. Rheinboldt and J.V. Burkardt, A locally-parametrized continuation process, *ACM Trans. Math. Software* 9(1983), 215–235.

246. W.C. Rheinboldt and J.V. Burkardt, Algorithm 596: a program for a locally-parametrized continuation process, *ACM Trans. Math. Software* 9(1983), 236–241.

247. H.H. Robertson, The solution of a set of reaction rate equations, in: *Numerical Analysis: An Introduction* (J. Walsh, Ed.), Academic Press, New York, 1966, pp. 178–182.

248. E. Schäfer, A new approach to explain the "high irradiance responses" of photomorphogenesis on the basis of phytochrome, *J. Math. Biol.* 2(1975), 41–56.

249. R. Scherer, A necessary condition for B-stability, *BIT* 19(1979), 111-115.

250. B.A. Schmitt and R. Weiner, Parallel two-step W-methods with peer variables, *SIAM J. Numer. Anal.* 42(2004), 265–282.

251. B.A. Schmitt, R. Weiner, and K. Erdmann, Implicit parallel peer methods for stiff initial value problems, *Appl. Numer. Math.* 53(2005), 457–470.

252. B.A. Schmitt, R. Weiner, and H. Podhaisky, Multi-implicit peer two-step W-methods for parallel time integration, *BIT* 45(2005), 197–217.

253. J. Schur, Über Potenzreihen die im Innern des Einheitskreises beschränkt sind, *J. Reine Angew. Math.* 147(1916), 205–232.

254. L.F. Shampine, Implementation of implicit formulas for the solution of ODEs, *SIAM J. Sci. Statist. Comput.* 1(1980), 103–118.

255. L.F. Shampine, The step sizes used by one-step codes for ODEs, *Appl. Numer. Math.* 1(1985), 95–106.

256. L.F. Shampine, Tolerance proportionality in ODE codes, in: *Numerical Methods for Ordinary Differential Equations*, (A. Bellen, C.W. Gear, and E. Russo, eds.), Lecture Notes in Mathematics 1386, Springer-Verlag, New York, 1989, pp. 118–136.

257. L.F. Shampine, *Numerical Solution of Ordinary Differential Equations*, Chapman & Hall, New York, 1994.

258. L.F. Shampine and S. Baca, Error estimators for stiff differential equations, *J. Comput. Appl. Math.* 11(1984), 197–207.

259. L.F. Shampine and C.W. Gear, A user's view of solving stiff ordinary differential equations, *SIAM Rev.* 21(1979), 1–17.

260. L.F. Shampine and I. Gladwell, Software based on explicit RK formulas, *Appl. Numer. Math.* 22(1996), 293–308.

261. L.F. Shampine, I. Gladwell, and S. Thompson, *Solving ODEs with MATLAB*, Cambridge University Press, Cambridge, UK, 2003.

262. L.F. Shampine and M.K. Gordon, *Computer Solution of Ordinary Differential Equations: The Initial Value Problem*, W.H. Freeman, San Francisco, 1975.

263. L.F. Shampine and M.W. Reichelt, The Matlab ODE suite, *SIAM J. Sci. Comput.* 18(1997), 1–22.

264. R.D. Skeel, Equivalent forms of multistep formulas, *Math. Comput.* 33(1979), 1229–1250.

265. B.P. Sommeijer, L.F. Shampine, and J.G. Verwer, RKC: An explicit solver for parabolic PDEs, *J. Comput. Appl. Math.* 88(1997), 315–326.

266. G. Steinebach and P. Rentrop, An adaptive method of lines approach for modelling flow and transport in rivers, in: *Adaptive Methods of Lines* (A. Vande Wouver, P. Sauzez, and W.E. Schiesser, eds.), Chapman & Hall, London, 2001, pp. 181–205.

267. D. Stoffer, General linear methods: connection to one step methods and invariant curves, *Numer. Math.* 64(1993), 395–407.

268. J.J.B. de Swart and G. Söderlind, On the construction of error estimators for implicit Runge-Kutta methods, *J. Comput. Appl. Math.* 86(1997), 347–358.

269. *The PORT Mathematical Software Library*, 3rd ed., AT&T Laboratories, Murray Hill, NJ, 1984.

270. S. Tracogna, A general class of two-step Runge-Kutta methods for ordinary differential equations, Ph.D. thesis, Arizona State University, Tempe, 1996.

271. S. Tracogna, Implementation of two-step Runge-Kutta methods for ordinary differential equations, *J. Comput. Appl. Math.* 76(1997), 113–136.

272. S. Tracogna and B. Welfert, Two-step Runge-Kutta: Theory and practice, *BIT* 40(2000), 775–799.

273. J. VanWieren, Using diagonally implicit multistage integration methods for solving ordinary differential equations, Ph.D. thesis, Arizona State University, Tempe, 1997.

274. J. VanWieren, Using diagonally implicit multistage integration methods for solving ordinary differential equations. Part 1: Introduction and explicit methods, *Rep. NAWCWPNS TP 8340*, Naval Air Warfare Center Weapons Division, China Lake, CA, 1997.

275. J. VanWieren, Using diagonally implicit multistage integration methods for solving ordinary differential equations. Part 2: Implicit methods, *Rep. NAWCWPNS TP 8356*, Naval Air Warfare Center Weapons Division, China Lake, CA, 1997.

276. S. Vandewalle, *Parallel Multigrid Waveform Relaxation for Parabolic Problems*, B.G. Teubner, Stuttgart, Germany, 1993.

277. J.H. Verner, Starting methods for two-step Runge-Kutta methods of stage-order 3 and order 6, *J. Comput. Appl. Math.* 185(2006), 292–307.

278. J.H. Verner, Improved starting methods for two-step Runge-Kutta methods of stage-order $p - 3$, *Appl. Numer. Math.* 56(2006), 388–396.

279. J.G. Verwer, Multipoint multistep Runge-Kutta methods: I. On a class of two-step methods for parabolic equations, *Rep. NW 30/76*, Mathematisch Centrum, Department of Numerical Mathematics, Amsterdam, the Netherlands, 1976.

280. J.G. Verwer, Multipoint multistep Runge-Kutta methods: II. The construction of a class of stabilized three-step methods for parabolic equations, *Rep. NW31/76*, Mathematisch Centrum, Department of Numerical Mathematics, Amsterdam, the Netherlands, 1976.

281. J.G. Verwer, An implementation of a class of stabilized explicit methods for the time integration of parabolic equations, *ACM Trans. Math. Software* 6(1980), 188–205.

282. R.E. Vinograd, On a criterion of instability in the sense of Lyapunov of the solutions of a linear system of ordinary differential equations (in Russian), *Dokl. Akad. Nauk. SSSR* 84(1952), 201–204.

283. D.A. Voss and M.J. Casper, Efficient split linear multistep methods for stiff ordinary differential equations, *SIAM J. Sci. Statist. Comput.* 10(1989), 990–999.

284. L.T. Watson, S.C. Billups, and A.P. Morgan, HOMPACK: a suite of codes for globally convergent homotopy algorithms, *ACM Trans. Math. Software* 13(1987), 281–310.

285. J.M. Watt, The asymptotic discretization error of a class of methods for solving ordinary differential equations, *Proc. Cambridge Philos. Soc.* 61(1967), 461–472.

286. R. Weiner, K. Biermann, B.A. Schmitt, and H. Podhaisky, Explicit two-step peer methods, *Comput. Math. Appl.* 55(2008), 609–619.

287. R. Weiner, B.A. Schmitt, and H. Podhaisky, ROWMAP — a ROW-code with Krylov techniques for large stiff ODEs, *Appl. Numer. Math.* 25(1997), 303–319.

288. R. Weiner, B.A. Schmitt, and H. Podhaisky, Parallel 'Peer' two-step W-methods and their application to MOL-systems, *Appl. Numer. Math.* 48(2004), 425–439.

289. R. Weiner, B.A. Schmitt, H. Podhaisky, and S. Jebens, Superconvergent explicit two-step peer methods, *J. Comput. Appl. Math.* 223(2009), 753–764.

290. J.K. White and A. Sangiovanni-Vincentelli, *Relaxation Techniques for the Simulation of VLSI Circuts*, Kluwer, Boston, 1987.

291. W. Wright, General linear methods for ordinary differential equations, M.Sc. thesis, University of Auckland, New Zealand, 1999.

292. W. Wright, The construction of order 4 DIMSIMs for ordinary differential equations, *Numer. Algorithms* 26 (2001), 123–130.

293. W. Wright, General linear methods with inherent Runge-Kutta stability, Ph.D. thesis, University of Auckland, New Zealand, 2002.

294. W.M. Wright, Explicit general linear methods with inherent Runge-Kutta stability, *Numer. Algorithms* 31(2002), 381–399.

295. M. Zennaro, Natural continuous extensions of Runge-Kutta methods, *Math. Comput.* 46(1986), 119–133.

296. M. Zennaro, Natural Runge-Kutta and projection methods, *Numer. Math.* 53(1988), 423–438.

297. Z. Zlatev, Modified diagonally implicit Runge-Kutta methods, *SIAM J. Sci. Statist. Comput.* 2(1981), 321–334.

Index